ASTRONOMY AND
ASTROPHYSICS LIBRARY

Series Editors: M. O. Harwit, R. Kippenhahn, J.-P. Zahn

ASTRONOMY AND ASTROPHYSICS LIBRARY

Series Editors: M. O. Harwit, R. Kippenhahn, J.-P. Zahn

H. Scheffler   H. Elsässer

Physics of the Galaxy and Interstellar Matter

Translated by A. H. Armstrong

With 207 Figures

Springer-Verlag Berlin Heidelberg New York
London Paris Tokyo

Professor Dr. Helmut Scheffler

Landessternwarte, Königstuhl, D-6900 Heidelberg, FRG

Professor Dr. Hans Elsässer

Max-Planck-Institut für Astronomie, Königstuhl, D-6900 Heidelberg, FRG

Translator
Arthur H. Armstrong

"Everglades", Brimpton Common, Reading, RG7 4RY, Berks., UK

Series Editors

Martin O. Harwit

Cornell University
Department of Astronomy
Space Sciences Building
Ithaca, NY 14853-6801, USA

Rudolf Kippenhahn

Max-Planck-Institut für
Physik und Astrophysik
Institut für Astrophysik
Karl-Schwarzschild-Straße 1
D-8046 Garching, FRG

Jean-Paul Zahn

Université Paul Sabatier
Observatoires du Pic-du-Midi
et de Toulouse
14, Avenue Edouard-Belin
F-31400 Toulouse, France

Cover picture: Sagittarius A. Filamentary structure near the Galactic Center. Radio continuum observations at 20 cm with the Very Large Array (VLA) radio telescope in New Mexico. Observers: F. Yusef-Zadeh, M. R. Morris and D. R. Chance – Courtesy of NRAO/AUI.

Title of the original German edition: *Bau und Physik der Galaxis*
© Bibliographisches Institut AG, Zürich 1982

ISBN 3-540-17314-5 Springer-Verlag Berlin Heidelberg New York
ISBN 0-387-17314-5 Springer-Verlag New York Berlin Heidelberg

Library of Congress Cataloging-in-Publication Data. Scheffler, Helmut, 1928 –. Physics of the galaxy and interstellar matter. (Astronomy and astrophysics library). Translation of: Bau und Physik der Galaxis. Includes index. 1. Galaxies. 2. Interstellar matter. 3. Astrometry. I. Elsässer, Hans, 1929 –. II. Title. III. Series. QB857.S3414 1987   523.1'12   86-31616

© Springer-Verlag Berlin Heidelberg 1987
Printed in Germany

Printing and binding: Konrad Triltsch, Graphischer Betrieb, Würzburg
2153/3150-543210

Preface to the English Edition

The present book is a translation of the original German edition (published in 1982) with some minor corrections and improvements. The guide to supplementary and advanced literature given in the Appendix, however, has been brought up to date.

This book is addressed primarily to students taking astronomy as a principal or subsidiary subject, and to scientists of related fields, but amateur astronomers should also be able to profit from it. For most chapters an elementary knowledge of mathematics and physics will be sufficient, however, Chaps. 5 and 6 impose somewhat greater requirements. In addition the reader should already be acquainted with the basic concepts of stellar physics as treated in introductory books, including the spectral types, the system of stellar magnitudes and colours, absolute magnitudes and luminosities, the Hertzsprung-Russell diagram and its interpretation.

A modern textbook should use SI units. On the other hand, the use of the cgs system is still the prevailing custom in astrophysics — together with the special units of astronomy: length is quoted in parsecs [pc], mass in solar masses [$\mathcal{M}_\odot$] and time in years [a]. We have therefore compromised and employed both cgs and SI units in this book, whichever was the appropriate choice in each instance. A table for conversion of cgs units into SI units and vice versa is given in the Appendix.

Our thanks are due to Springer-Verlag for undertaking the English edition and to A.H. Armstrong, who not only translated the book but in the course of doing so offered helpful suggestions which led to the improvement of several passages.

Heidelberg, June 1987 *H. Scheffler · H. Elsässer*

Preface to the German Edition

In recent astronomical literature there has up to now been a lack — not only in the German language — of a systematic textbook which sets out the methods and results of studies of our galaxy over the whole field at an intermediate level. To remedy this situation is the chief aim of this book, which therefore has a similar objective to that of our *Physik der Sterne und der Sonne*, published by Bibliographisches Institut, Mannheim, Wien, Zürich[1]. Its contents span a significant part of astronomy, from spherical astronomy and stellar statistics to the observation and theory of interstellar matter, to the dynamics of the Milky Way system as a whole.

Like the *Physik der Sterne und der Sonne*, this book is also addressed primarily to students of astronomy, whether as principal subject or subsidiary subject, and to interested applied scientists. We hope that amateur astronomers will also be able to profit from it. A similar division of the subject matter appeared appropriate here as is adopted in the *Physik der Sterne und der Sonne*. In the first three chapters the empirical research is placed in the foreground as a foundation. Where possible without special development of the theory, a preview is often given here of the most recent quantitative interpretation, by way of clarification. An elementary knowledge of mathematics and physics will usually suffice for comprehension. Chapters 4 and 5, which are primarily devoted to theory, impose somewhat greater requirements, generally corresponding to about the level of a first degree in physics.

Since research on the galaxy has sometimes followed tortuous paths and is even today in a fluid state, the whole is preceded by an introduction, which provides both a review of the historical development of our science and a preview of the content of the five chapters of the book.

The necessity of arranging the material in a one-dimensional sequence inevitably requires a few compromises. Thus the separation of observation and theory has the result that a particular topic is referred to at two or three places in the book. For example, interstellar polarisation is discussed in Chaps. 3 (observations), 4 (interpretation by the optics of small particles) and 5 (theory of the orientation of elongated dust particles). We hope that this drawback is mitigated by the frequent cross-references provided. Another difficulty arises in the treatment of the interstellar gas. Whilst it is convenient to discuss the distribution and motion of the stars *after* the physics of the stars, a similar sep-

---

[1] A completely revised English edition of this book is now (1987) in preparation.

aration for the interstellar gas would scarcely be practical. The questions of the "physical state" on the one hand and the "spatial distribution and motion" on the other cannot here as a rule be discussed separately. It is necessary, for example, in deriving the density distribution of the interstellar neutral hydrogen to make assumptions concerning its kinematics, and for detailed conclusions on its temperature one needs at the same time a discussion of its spatial structure. For this reason we have treated the physical state as well as the spatial distribution and kinematics of the interstellar gas in the closest possible association within *one* chapter (4).

In the selection of material we again tried in the first place to take account of clearly established scientific principles. The abundance of recent empirical and theoretical results of intensive research in all fields close to those treated here made it appear inevitable, however, in some cases to pick out those results which are today under discussion. References (authors' names, dates), which are otherwise given in the text only for outstanding results, are in these cases the rule and can then usually be followed up with the help of the literature references given in the Appendix. This collection of supplementary and more detailed literature should be an initial guide for those readers who wish to probe more deeply into a particular part of the field.

The relatively large number of formulae arising in the various regions of astronomy and astrophysics has the result that in a few cases different quantities must be denoted by the same symbol. Thus it was unacceptable, for example, to denote the interstellar extinction and the first of Oort's constants of galactic rotation by symbols other than the familiar A. Similarly we use ϕ as the polar angle and as the function symbol, say, for the luminosity function of the stars. In other cases the limited number of available letters compels the multiple allocation of these symbols − but in different sections. The reader who is able to follow the presentation, however, should experience few difficulties arising from the notation. In order to maintain continuity with our *Physik der Sterne und der Sonne*, in this book we generally use the Gaussian cgs system − as is still the prevailing custom in astrophysics − together with the special units of astronomy: length is quoted in parsecs [pc], mass in solar masses [$\mathcal{M}_\odot$] and time in years [a].

We are very much indebted to T. Lederle and R. Wielen for critical reviews of Chaps. 1 and 5, respectively. We also thank T. Schmidt, who read the text of the entire manuscript and suggested various improvements.

For permission to reproduce illustrations we thank W.J. Altenhoff, K. Birkle, D. Downes, H.H. Hippelein, H. Jahreiss, G. Lingenfelder, C. Madsen, K. Mattila, P.G. Mezger, G. Münch, J. Solf and J. Staude.

Mrs. B. Schwander typed all the versions of the entire manuscript. Complete drawings were prepared by Mrs. K. Dorn. Mrs. D. Gayer and Mr. W. Neumann produced the photographs of the completed drawings and most of the half-tone illustrations for reproduction. For this we are grateful to all of them. We thank the publisher again for constant pleasant cooperation.

Heidelberg, April 1982 *H. Scheffler · H. Elsässer*

Contents

1. Introductory Survey

Astronomy has for a long time past been engaged in the exploration of the galaxy. With the passage of time, new aspects have repeatedly come to the foreground of interest and nowadays, as before, objects and processes in our Milky Way system are central themes of research. The following introduction contains a brief survey of the diverse questions and their history. This should facilitate the approach to the more detailed presentations in later chapters.

1.1 The Stellar System

Since the invention of the telescope it has been known for certain that the phenomenon of the Milky Way is produced by very many faint stars. Interpretations of it as a vast, superior system of stars, however, were not put forward until about the middle of the 18th century, at first in a tentative and speculative manner. In 1750 Thomas Wright proposed two alternative possibilities as explanations of the Milky Way: an annular arrangement of stars around the sun, and also, in his opinion, the more likely model of a very large thin spherical shell filled with stars including the sun. Immanuel Kant (*Allgemeine Naturgeschichte und Theorie des Himmels,* 1755) and then Johann Heinrich Lambert (*Cosmologische Briefe,* 1761) introduced the hypothesis of a flat disc-shaped stellar system for the first time. According to this, the sun and its planets were at the centre of the main plane of the system, so that for an observer on Earth the stars appear concentrated on a broad band that spanned the whole sphere of the heavens along a great circle.

The problem of the Milky Way was first investigated using the methods of empirical research by William Herschel (1738–1822). In the field of view of his reflecting telescope he carried out star counts. Under the assumptions of: (1) equal real brightness of all stars, (2) uniform spatial density and (3) brightness falling off with the square of distance, he deduced that the sun is located near the centre of a roughly lens-shaped stellar system. The inference concerning the position of the sun, as also the estimates of the system's dimensions, indeed failed later to be confirmed. Nevertheless, William Herschel with his method of star counts was the founder of stellar statistics.

The problem of *"classical" stellar statistics,* to deduce the spatial distribution of stars in a given direction from the number of stars as a function of the apparent brightness, received its mathematical formulation first from F.G.W. Struve (1847) and then, at the beginning of the 20th century, chiefly from H. von Seeliger, K. Schwarzschild and J.C. Kapteyn. This treatment now also took into account that the absolute brightnesses of stars scatter over a broad band.

Between about 1890 and 1920 it was possible to treat the following observations with the new methods: (1) cataloguing of the entire sky for the positions and apparent brightnesses of all stars with magnitude up to 10, and in selected fields up to even higher magnitudes; and (2) distance determinations, mainly by statistical methods, from which, in conjunction with the apparent magnitudes, the distribution function of absolute magnitudes was derived. The results obtained for the spatial distribution of stars were qualitatively similar to those of William Herschel. With regard to the extent of the stellar system in its plane of symmetry and the position of the sun, however, the resulting picture was demonstrated by the end of the twenties to be false. In reality Man had comprehended only a small part of the Milky Way system.

The main reasons for failure were the small amount of observational data and the neglect of the strong general attenuation of light in the immediate vicinity of the galactic plane through interstellar dust – this was first established in 1930. Today we know that the interstellar extinction of light from the stars on the way from the true galactic centre to the sun amounts on average to about 20 magnitudes, or a factor of about 10^{-8}! The general star counts down to successive brightness levels known at that time could therefore provide evidence only on the "local star distribution" in a certain neighbourhood of the sun.

An additional shortcoming of the classical methods arose from the fact that all stars of various absolute magnitudes were included, whereas there is only an extraordinarily weak statistical relationship between apparent magnitude and distance. The further development of the method, therefore, apart from taking into account the interstellar extinction, consisted in restricting it to the consideration, at any one time, of the narrowest possible spectral groups with a relatively small scatter of absolute magnitudes. It is of course for this reason that a spectral classification or equivalent photometric classification of all stars in the chosen field down to the lowest possible brightness level is necessary. The attainable range naturally depends strongly on the absolute magnitudes of the star types included: with a limiting visual magnitude $m_V \approx 13^m$ one can obtain the star distribution in the galactic plane, for example for types F8V-G5V up to a distance of about 500 pc, and for types B5V-A2V up to about 2 kpc, whereas the distance to the centre of the system is currently believed to be about 10 kpc!

Decisive progress, in particular the determination of the true dimensions of the galaxy and the position of the sun in relation to the centre, came with the idea of obtaining, first of all, the *broad structure of the system,* by focussing attention on stellar objects with especially great absolute brightnesses, which can therefore be observed at very great distances. The impetus moreover came from the strong growth of astronomy in the USA after the turn of the century. Of special significance were the studies of the distribution of globular star clusters.

2

H. Shapley pointed out in 1917 that, just from the apparent distribution of globular clusters on the sphere, it must be concluded that (1) these objects belong to the system and (2) the centre of the system lies at a relatively great distance from the sun in the direction of the constellation Sagittarius. The globular clusters are symmetrically arranged about the galactic plane, hence (1); most of the objects are found in *one* half of the sky, and strongly concentrated in the direction of the constellation Sagittarius, hence (2).

In addition to the advantage of high total luminosity, the globular clusters offered the possibility of comparatively good photometric distance determination with the help of their RR Lyrae stars. The absolute magnitudes of these "cluster variables" were known approximately. In the more distant globular clusters these stars were indeed no longer to be seen, but here one could apply the knowledge gained in the nearer globular clusters concerning the luminosity of the brightest group stars or even of the whole group. The interstellar extinction, which was at first neglected, has a less serious effect than for the star counts in the Milky Way because the globular clusters are mostly at a greater distance from the galactic plane and their light therefore follows a shorter path through the comparatively shallow dust stratum.

Shapley found a nearly spherical distribution of globular clusters, with significantly greater extent than that shown by the above-mentioned stellar system of the first classical studies based on star counts. The sun now assumes a very eccentric location (Fig. 1.1). Later work has confirmed this result and given a distance for the sun from the centre of about 10 kpc. The conclusions drawn on the basis of the older general star counts therefore applied to a small portion of an outer and nearly plane parallel stratum of the "galactic disc" round about the region of the sun.

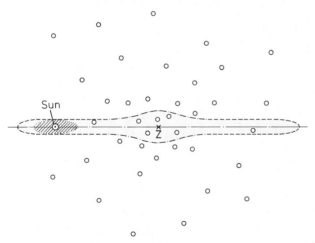

Fig. 1.1. Cross-section of the stellar system perpendicular to the galactic plane of symmetry and passing through the galactic centre (Z) and the region of the sun (schematic). The boundary of the "galactic disc" is shown by a *broken line* (thickness of the disc exaggerated); in the central region there is a "bulge" (*broken line*). The globular star clusters are represented by *open circles*. The classical studies comprehended only the hatched region around the sun ("local stellar system")

3

Fig. 1.2. The spiral galaxy M 51 in the constellation Canes Venatici by red light. (Photograph obtained with the 1.2 m telescope in the Calar Alto Observatory in Spain of the Max-Planck-Institut für Astronomie, Heidelberg)

Other stellar objects with relatively great absolute brightness, whose distribution in space can be studied by the method employed by Shapley, are, besides the RR Lyrae stars occurring in isolation and a few other variable types, the open star clusters, the O- and B-stars and the supergiants of later spectral type. For the young open star clusters and the OB stars, clear indications were established for the first time of a spiral structure, as had already been known for a long time for extra-galactic systems (Figs. 1.2, 3).

A further important key to the solution of the classical galactic problem emerged from studies of *stellar motions*. Herschel had already ascertained the apex of the sun's motion relative to neighbouring stars, from systematic po-

Fig. 1.3. The spiral galaxy M 101 in the Great Bear. (Photograph obtained with the 2.2 m telescope in the Calar Alto Observatory in Spain of the Max-Planck-Institut für Astronomie, Heidelberg)

sition changes (proper motions) of the stars. The application of the Doppler principle to the determination of the "radial" components of stellar motions along the line of sight from the displacement of the Fraunhofer lines led later to determination of the velocity of the sun.

"Radial velocities" were first measured to a satisfactory degree of accuracy by H.C. Vogel in 1890 using photographic spectra. The accumulation of data in the twenties engendered a fruitful discussion of the frequency distribution of the spatial velocities of stars in the neighbourhood of the sun. This showed that: (1) most stars have small velocities below about $30\,\mathrm{km\,s^{-1}}$ with an approximately random distribution of directions and (2) the motions of the remaining stars with high velocities, above about $60\,\mathrm{km\,s^{-1}}$, are however predominantly directed towards one side of the sky (asymmetry of high stellar velocities).

An explanation was offered in 1926 by the Swedish astronomer B. Lindblad. The observed effect is easily explained if (a) the sun and the other neighbouring stars of low velocity are circulating about a far distant centre and (b) the stars of higher velocity are not, or only weakly, taking part in the circulation. The direction of the centre of circulation of the system must lie perpendicular to the direction of the asymmetry, but within the galactic plane. The direction of the centre is therefore in the direction of Sagittarius, in which Shapley had located the centre of the system of globular clusters.

Lindblad distinguished several rotationally symmetric sub-systems with different high rotation velocities and flattening. The high-velocity stars and globular clusters are observed in profusion in the high galactic latitudes and there form slightly flattened sub-systems, which scarcely rotate: the individual objects move in arbitrarily inclined paths about the galactic centre.

At about the same time as Lindblad, the Dutchman J.H. Oort was investigating stellar motions. Oort studied the small systematic effects which must be observable as a result of rotation. Even if the flat sub-system rotates as a rigid body, this must still be detectable: the components of the proper motions parallel to the galactic equator should contain a systematic part. For example, to an observer in an astronomical coordinate system, participating in the galactic rotation but not itself rotating, the direction towards the centre of the galaxy would appear to turn right round during one galactic rotation. Actually such an effect of about $0''.002$ per year was already known at that time. Against the very much greater precessional motion of the Earth's axis, however, such a small quantity is detected only with great difficulty.

Oort had still another argument: if the circulating velocity of the stars decreases with distance from the centre of mass, as in the Kepler motion of the planets around the sun, thus producing shear in the velocity field, then this should also be observed by a circulating observer as a systematic variation of the proper motions and radial velocities of the stars along the galactic equator. This effect of the differential galactic rotation should take the form of a double wave; in the direction towards the centre and the anti-centre, for example, it should vanish in the radial velocities and attain a maximum in the proper motions (Sect. 3.4.1). Oort could show in 1927 that this effect is in fact superposed on the individual stellar motions. The stellar motions had accordingly given yet another argument for the direction of the galactic centre. Taking the distance from the centre to the sun as between 8 and 10 kpc, the interpretation of the observations leads to a circulating velocity in the region of the sun in the range of 200 to 250 km s^{-1}. From this orbital velocity and the distance from the centre of the galaxy one immediately obtains a first estimate of the gravitational mass contained inside the solar orbit of an order of magnitude of 10^{11} solar masses.

The dependence of the circulating velocity of the stars of the galactic disc upon the distance from the centre has so far been determined only for a region around the sun of a few kpc diameter. Evidence for the whole disc, however, is obtained, under certain assumptions, from the interstellar gas (see below). Discussion of these results together with theoretical arguments led to the conclusion that the spatial velocity field of the stars and of the interstellar material, in addition to the pure rotation, also shows the effects of a spiral structure of the system: the spiral arms of the galaxy are probably density waves, which move relative to the stars and the interstellar material.

The understanding that the galaxy consists of sub-systems of different degrees of flattening and different kinematics brought a new approach, as the theory of stellar evolution made possible statements on the age of each member. The empirical basis was the colour-magnitude diagrams (CMD) of star clusters. The differences discovered by comparing the CMDs of open clusters and normal stars in the neighbourhood of the sun on the one hand, and of globular clusters on the other hand, could be chiefly explained as a difference in age: open star clusters with O- and B-stars are the youngest objects ($\lesssim 10^7$ years), globular clusters are the oldest ($\approx 10^{10}$ years).

Next, W. Baade succeeded in 1944 in resolving into individual stars the inner region of the Andromeda galaxy, which is similar to our Milky Way system, and showed, on the basis of their position in the CMD, that these stars are very similar to the brightest red giants of globular clusters. Moreover, he showed that the star mixture in the spiral arms of this galaxy is somewhat similar to that found in open star clusters with O- and B-stars. Baade deduced from this that there are two fundamentally different stellar populations: Population I contains the stars of the flat sub-systems fully partaking in the galactic rotation, including in particular the typical spiral arm objects, which can still be very young; Population II consists entirely of very old globular clusters and other objects distributed approximately spherically and concentrated near the galactic centre, and in particular the stars of the central region of the galaxy.

A few years later, the chemical constitution was recognised as a further distinguishing feature: comparison of the observed CMD of globular clusters with calculations of stellar evolution showed that these objects must contain about two orders of magnitude lower proportions of the heavy elements (from carbon) than the open clusters and normal stars of the neighbourhood of the sun. This yielded a relationship between chemical constitution and age.

Baade's model soon proved an over-simplification. With improvements in observations and theory it became clear that refinement of the classification was necessary. At a conference on stellar populations in the Vatican in 1957 an extended scheme was developed. It is described at the conclusion of the third chapter and discussed in connection with ideas on the evolution of the stars and the galaxy as a whole.

1.2 Interstellar Matter

In the outer regions of the galactic disc, in the neighbourhood of our solar system, the ratio of the average distance between stars to star diameters is of the order of magnitude $1\,\mathrm{pc}{:}10^6\,\mathrm{km} = 3 \times 10^{13}{:}10^6 = 3 \times 10^7{:}1$. Only a tiny part of space, therefore, of the order of the cube of the reciprocal of the value of this ratio, viz. 10^{-22}, is occupied by matter in the form of stars. The realisation of the true distance of the stars and of their dimensions at once raised the question whether the overwhelmingly large remaining part of space also contained matter in other than stellar form.

We have known the answer for only a few decades: between the stars there are great masses of gas and small solid particles – interstellar dust – with a share of the total mass of the galaxy amounting to 5–10%; in the neighbourhood of the sun the mass contribution of the interstellar matter is even around 30%. The density of this medium however amounts on average to only about one atom per cubic centimetre and it reached its greatest concentration, just before the beginning of the formation of the protostars, at only 10^{-18} times the mean density of the sun. It is therefore not surprising that this matter was almost completely overlooked in the classic visual observations of stellar astronomy. The study of interstellar matter is consequently a very young branch of galactic astronomy. In spite of the great progress that this attractive field has made today, the results obtained are often of a tentative character. The quality of the picture obtained therefore betrays somewhat the short history of its development.

The manifestations of interstellar matter are very varied. The basic reason for this is the inhomogeneous structure of the medium: density, temperature, constitution of atoms, ions, molecules and solid particles (dust), and its state of motion vary considerably from region to region. Only two phenomena are directly available to visual observation: the diffuse bright nebulae and the partial extinction of starlight by "dark clouds" in the visible rugged structure of the Milky Way. Examples are shown in Figs. 1.4–6.

In the year 1811, on the basis of his observations of "nebulae" with his great reflecting telescope, William Herschel had already arrived at the conclusion that the abundance of nebulous matter must exceed all expectations. Herschel was also perplexed by the "holes in the starry sky" caused by the dark clouds of interstellar dust. Afterwards the explanation was put forward that the bright nebulae were distant stellar systems, but most astronomers – even up to the twenties of this century – were inclined to assume a completely empty space between the stars, in which the few objects, later known as gas- and dust-clouds, were only isolated phenomena. The discovery of a narrow absorption line of interstellar origin, the famous "stationary" calcium line in the spectrum of the spectroscopic binary star δ Orionis, by J. Hartmann in 1904, was recognised as important by only a very few researchers and was at first explained as local absorption in the immediate neighbourhood of the star.

The first convincing observational evidence on the widespread nature of interstellar phenomena was produced by the introduction of photography into systematic observation of the sky. Long exposure photographs revealed a far clearer picture of these phenomena than could be obtained from visual observation. Many weak diffuse nebulae and dark clouds, hitherto unknown, were discovered (E.E. Barnard, F. Ross, M. Wolf). Moreover the photographs of spectra of diffuse nebulae showed that there were two different types: "emission nebulae", the optical radiation of which consisted of individual emission lines from atoms or ions (for example: Fig. 1.4), and "reflection nebulae" which simply reflected the light of neighbouring stars and therefore gave continuous spectra (for example: Fig. 1.7). In the dark clouds, on the other hand, a

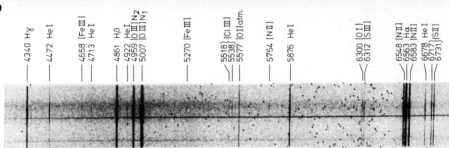

Fig. 1.4. (a) The Orion nebula by the light of the hydrogen line Hα. (Palomar Observatory Photograph). (b) Spectrum of the Orion nebula (part) taken with an electromagnetically focussed image intensifier. Slit width 2″, slit length 190″, dispersion on the original 60 Å/mm. In about the middle of the slit there was a star whose continuous spectrum shows as a thin thread. However, a weak nebular continuum is also perceptible. The scattered black spots are caused by the impingement of residual ions in the vacuum of the image intensifier tube on the cathode ("ion-flashes"). (From Solf)

Fig. 1.5. The "Coalsack", a large dark cloud in the southern Milky Way, and the constellation of the Southern Cross. (From Mattila)

Fig. 1.6. Caption see opposite page

Fig. 1.7. Reflection nebulae around the Pleiades. (From Barnard, 1927)

weakening of the continuous starlight was demonstrated (M. Wolf, 1923). The growing evidence in about 1930 for the existence of a general widespread interstellar medium led to the definitive change of approach: for gas components, the observation of a systematic increase in the strength of sharp interstellar absorption lines in stellar spectra with distance (O. Struve, 1928) and for dust components, a comparison of photometric and geometric determinations of distance of open star clusters, from which resulted a general interstellar extinction of almost one magnitude per 1 kpc (R. Trümpler, 1930). The year 1930 accordingly marked the approximate starting point of the study of interstellar matter as a new branch of astronomy.

Of basic importance for the understanding of the light of diffuse nebulae was the empirical law, discovered by E. Hubble in 1922, that in the case of an *emission nebula* the star which is the actual energy source of the nebula under consideration is of a spectral type earlier than B 1. Only these very hot stars

---

Fig. 1.6. Wide-angle picture of the southern Milky Way between Aquila and Carina in the red spectral region (6100 to 6900 Å). The bright star cloud in the middle of the Milky Way arch indicates the direction of the galactic centre. Half-right the "Coalsack" can be seen as a dark rectangle. The exposure was made on the site of the European Southern Observatory in Chile with a spherical mirror camera. The camera itself and its three supporting struts are in the light inlet and are therefore also visible. (From Schlosser, Schmidt-Kaler, Hunecke, 1979)

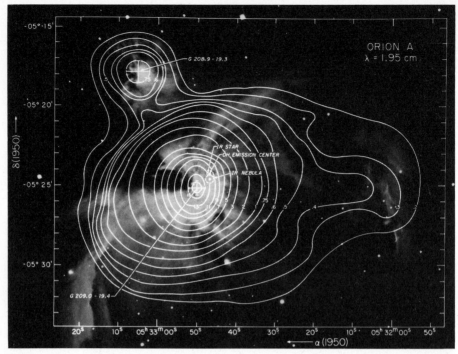

Fig. 1.8. Contour map of the observed radio continuum radiation at 1.95 cm wavelength (radio source Orion A) on a Hα exposure of the Orion nebula. The centre of the radio contours almost coincides with the Trapezium stars. (After Mezger and Altenhoff)

radiate sufficiently strongly in the extreme ultraviolet region to ionise hydrogen, the most abundant element in interstellar gas, thereby producing a thin thermal plasma with a temperature of about 10^4 K, the lines from which we observe in the nebula light. After A.S. Eddington's early studies (1926) of the questions of the ionisation and temperature of interstellar matter, B. Strömgren showed in 1939 that the region of ionised hydrogen round the hot star should have a relatively sharp outer boundary, as had indeed been observed many times.

Appearances in the visible region are of course usually modified considerably by intervening interstellar dust, and the true shape of these regions of ionised gas, now named H II regions[1], cannot be recognised without other information. An undistorted picture of the structure of H II regions was first obtained – in the sixties – from measurements of their continuous radio emission, which passes unhindered through dust. Figure 1.8 shows as an example of this the Orion nebula. This emission can be explained as thermal Bremsstrahlung which arises from encounters between free electrons and ions (free-free transitions). Besides the radio continuum one also finds radio emission lines, arising from recombination processes of the ions, mainly hydrogen, through transitions

---

[1] Strictly one should say the H^+ regions, since H II is the name of the spectrum of H^+, which does not exist. Rather, one often observes here the H I spectrum, arising after formation of neutral hydrogen by recombination.

between the high and very densely distributed energy levels, for example from n=110 to n=109. They are known as radio-recombination lines.

The radiation from cool stars with spectral types from about B 1 does not suffice for the generation of an extended H II region. If there is enough interstellar matter in the neighbourhood, however, these stars can nevertheless give rise to the appearance of *reflection nebulae* through the scattering of the starlight by the dust components. In emission nebulae likewise, of course, the dust component causes a weak continuous scattered light in addition to the line radation. In red exposures of the Milky Way emission nebulae stand out especially strongly through the predominance of the Hα line of hydrogen at λ 6563 Å, whereas reflection nebulae are more clearly shown on blue exposures: they reflect mainly the blue continuum of the neighbouring bright stars with spectral types B 1 to A.

Quantitative statements on the general *interstellar extinction of starlight* were indeed already being made in the thirties, the decisive material for this being first produced by the modern photoelectric measurement techniques. The most significant results were the comparisons of the observed energy distribution in the spectra of various far distant stars of the same spectral type, and hence with the same true energy distributions. There was a reddening of the starlight which increased systematically with distance. The extinction increased with decreasing wavelength, about proportionally to λ^{-1} in the visible region. Measurements using artificial satellites later confirmed this in the ultraviolet region also. Interpretation of the first results on the reddening in the visible region had already led to the conclusion that the cause of the phenomenon involved only small solid particles with diameters of the order of magnitude of the wavelength of light and possibly still smaller particles. The true nature of these particles ("dirty ice", silicates, graphite, etc.) is of course to this day still not fully elucidated.

The interstellar reddening is mostly concentrated in the region near the galactic equator. Through extension of the measurements to the far infrared, where the extinction finally becomes vanishingly small, one can also determine the actual total amount of light attenuation, usually expressed in magnitudes for the visual region ($\lambda \approx 5400$ Å) by the extinction A_V. In the galactic plane one finds an average value for A_V of 1 to 2 magnitudes per 1 kpc path length, with large variations, however, from region to region. The cloudy distribution of dust particles corresponds broadly with the density distribution, discussed further below, of the far more massive gas component of the interstellar medium; the thickness of the dust layer extends effectively for only about 200 pc.

The nearest especially dense concentrations of gas and dust appear as dark clouds in the picture. In the visible, and even more so in the ultraviolet, the view of distant galactic or even extragalactic objects in low galactic latitudes is denied us: the range of vision does not extend as far as the galactic centre, and in a belt between about +15° and −15° galactic latitude practically no extragalactic stellar systems are found (E. Hubble, 1936; see Fig. 1.9). Our Milky Way system, viewed from outside in a direction parallel to the galactic

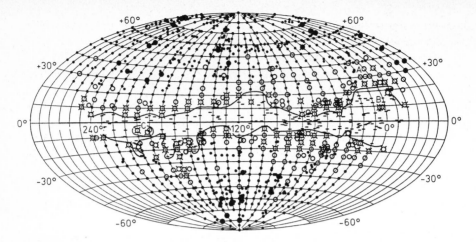

Fig. 1.9. Distribution of extragalactic stellar systems on the sphere in galactic coordinates with the line of symmetry of the Milky Way as the base circle (galactic equator 0°–0°). *Small points*: normal density; *large points*: excess; *circles*: deficit; *circles with radial lines*: gross deficit; *horizontal lines*: no galaxies. (After Hubble, 1934)

plane, must offer an appearance similar to the picture of a distant galaxy shown in Fig. 1.10.

Through reflection of a part of the incident stellar radiation, the interstellar dust particles together produce a general galactic scattered light, which it has been possible to measure in more recent times with increasing reliability; the observations nowadays extend into the far ultraviolet. In particular, even in the dark clouds the background between the weak stars is not completely dark.

Fig. 1.10. Spiral galaxy seen "edge-on" (Palomar Observatory Photograph)

Such measurements provide excellent information on the scattering properties of the particles.

An additional phenomenon produced by the interstellar dust is a weak, but nowadays very accurately measurable, *interstellar polarisation of the starlight*. This effect due to the solid particles was discovered in 1948 by W.A. Hiltner and J.S. Hall in the search for a polarisation of stellar origin theoretically predicted by S. Chandrasekhar, which must cancel out over the whole symmetric star disc and which it was hoped it would be possible to observe at eclipsing variables. Measurements on a rather large number of stars did indeed produce positive results, but left no doubt as to the interstellar origin of the linear polarisation detected: in several regions of the Milky Way the light from all the observed stars showed nearly the same direction of polarisation (see Fig. 1.11).

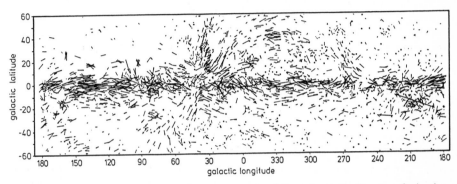

Fig. 1.11. Summary presentation of measurements of the interstellar linear polarisation of the starlight in galactic coordinates: b=galactic latitude, l=galactic longitude. In the position of every observed star its polarisation is represented by a short line showing by its direction and length the polarisation angle and strength. Stars with polarisation strength below 0.08 per cent are indicated by a small circle. (After Mathewson and Ford, 1970)

This evidence forces one to the conclusion that at least some of the solid particles are non-spherical, so their shape is somewhat elongated, and their major axes must be to a certain extent similarly oriented in space. It was conjectured that this is caused by interstellar magnetic fields, the structure of which was then studied for the first time.

This hypothesis soon received a strong boost by the observation of Faraday rotation of the linearly polarised radiation found from distant discrete radio sources, which must arise from these spatial magnetic fields during its passage.

Let us now return to the phenomena of the interstellar gas. The presence of diffuse gas outside the emission nebulae was at first indicated only by the observation of its interstellar absorption lines in the visible region of the stellar spectra. Here we find lines representing a few atoms and ions, such as Na and Ca^+, and also the diatomic molecules CN and CH. The most abundant element, hydrogen, however, was not observed: far away from H II regions the H atoms are in the ground state and therefore absorb first in the ultraviolet, beginning with the Lyman α line at 1216 Å. The proof of interstellar absorption lines of neutral hydrogen and also of hydrogen molecules therefore first came from observations from rockets and particularly successfully from satellites of the OAO series (Orbiting Astronomical Observatory). In the seventies interstellar lines were also discovered from unexpectedly highly ionised elements, such as two O VI-lines, which must arise in a widely diffused very hot component (10^5–10^6 K) of the interstellar medium.

The spectral observations in the visible region had at an early stage already provided important evidence on the motions and structure of the absorbing gas. J.S. Plaskett and J.A. Pearce had shown in 1933 that the Doppler shifts of the interstellar absorption lines follow on the average the double wave produced by the differential galatic rotation; thus the gas takes part in the galactic rotation. High-resolution spectrograms then showed that the lines usually consisted of several sharp components, the number of which increased with the path length of the light in the galactic disc (C.S. Beals, 1936; W.S. Adams, 1943). These observations gave strong support to the idea that the gas − as well as the dust − also has a cloud structure.

Interstellar matter has come, thanks to the developments in observational technology which occurred after the second World War, to be one of the flourishing research fields of astronomy. Above all, unexpected advances resulted from measurements of *radiofrequency radiation of the interstellar gas,* which made possible for the first time penetration of the whole galactic disc, and right up to the present surprising discoveries are constantly being made.

The investigation of the galactic structure and kinematics was profoundly influenced in the fifties by the measurements of line emission of the interstellar neutral hydrogen at 21 cm wavelength. This radiospectral line is due to a hyperfine structure transition in the ground state of neutral H atoms. After a prediction by H.C. van Hulst in 1944, the emission was established almost simultaneously in 1951 by three different groups of workers in the Netherlands, the USA and Australia. The significance of this discovery arises from the fact that a radio line, in contrast to the continuous radio emission which was established at about the same time, provides evidence via the Doppler shift of the velocity field of the gas, and hence also, under suitable assumptions, of the spatial distribution of the gas.

The first systematic series of observations showed that the 21 cm emission comes from a comparatively shallow cloudy layer around the galactic plane, the same region as that where the interstellar dust was concentrated. In the direction of the galactic centre the layer becomes optically thick for the line centre. The radiation temperature of about 100 K measured there accordingly represented an approximate value for the kinetic temperature of the neutral hydrogen. Moreover, the rotation law could now be derived for a much greater region than the optical observations had previously allowed. On the basis of these results from the line profiles, the first model for the density distribution of gas in the whole galactic disc could be constructed with clear indications of a spiral structure (Fig. 1.12). The resulting picture was so convincing that its failings only became apparent one and a half decades later.

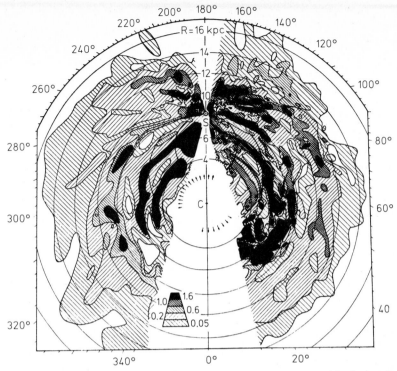

Fig. 1.12. The pattern for the distribution of interstellar neutral hydrogen in the galactic plane, deduced from the first systematic series of observations of the 21 cm line radiation, and published by Oort, Kerr, and Westerhout (1958). The assumption was made that the interstellar gas moves in circular paths around the galactic centre (distance of the sun from the centre $R = 8.2$ kpc). For both white sectors in the directions 0° (centre) and 180° (anticentre) no conclusions are possible. The density is indicated by different amounts of hatching in levels from 0.05 to 1.6 H atoms per cm^3

New evidence on the fine structure of the gas distribution came from observations of the 21 cm line profile in absorption which is produced by galactic hydrogen in the continuous spectra of discrete galactic and extragalactic radio sources. Like the optical interstellar absorption lines in stellar spectra, the 21 cm absorption line can often be resolved into several components produced by clouds intersected by the line of sight.

The continuous radiofrequency radiation which was discovered first is caused by sources of very different structure (point sources, extended sources, general background) and spectral composition. Karl Jansky, an American radio-engineer, who discovered cosmic radio waves accidentally in 1932, was already able to show that the radiation came from a source spread over the whole Milky Way, with greatest strength in the direction of Sagittarius. Another American radio-engineer, Grote Reber, constructed in 1944 the first radio map of the Milky Way, whose publication gave a substantial impulse for the start of galactic radioastronomy. With the increase in sensitivity, and above all in directional

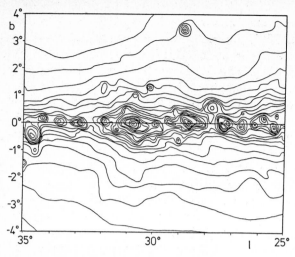

Fig. 1.13. Part of a contour map of galactic radio continuum radiation at 1414 MHz. Angular resolution capability 11 minutes of arc. The middle horizontal line $b = 0°$ indicates the position of the galactic equator (b = galactic latitude, l = galactic longitude). The strongly concentrated H II regions here stand out clearly as discrete radio sources. (After Altenhoff et al., 1970)

resolution, of the radiotelescopes, the interstellar galactic sources could later be clearly distinguished from the other radio sources (Fig. 1.13) and the following picture emerged. The H II regions of the thin galactic gas-dust-disc produce continuous Bremsstrahlung in a very narrow belt, the strength of which falls off rapidly for wavelengths above 1 m. This emission is superposed on a nonthermal radiation dominated by long waves, with significantly little concentration in the galactic plane. This is produced by the motions of relativistic electrons in magnetic fields; as discrete interstellar sources supernova remnants stand out.

The search for other radio lines introduced in the sixties a new batch of discoveries. Line radiation from interstellar OH molecular radicals was measured at 18 cm wavelength in 1963, and in 1968 came the first discovery of lines from polyatomic molecules, of ammonia. Primarily from observations in the area of the H II regions, up to now more than 40 sorts of interstellar molecules have been revealed by their line emissions in the microwave region, from simple molecules such as CO up to complex organic compounds such as, for example, formamide, $HCONH_2$.

These molecules are not found in the H II regions themselves, but in larger and more massive concentrations of matter, which are impenetrable to optical radiation. The molecules owe the possibility of their long term existence to the high dust densities in these very cold clouds. The interstellar radiation field, which would soon dissociate all these molecules, is here almost completely extinguished; in less dense clouds one often finds only the simple and stable molecules, such as CO and OH. The H II regions are, moreover, seen to be the localities of the great molecular clouds, in which the condensation to the onset of star formation is at an advanced stage.

Radio continuum observations with high angular resolution have also revealed small so-called compact H II regions with particularly high densities, many of which do not show up in the visible region as emission nebulae. We are dealing here with the early stages of H II regions. An interesting peculiarity appears with strong line emissions of OH and H_2O molecules, which emanate from extremely small angular regions, with linear dimensions of the order of one Astronomical Unit. This radiation was explained with the assumption of a natural maser mechanism, possibly located in the shells of newly born massive stars.

Thicks clouds cannot be observed using the 21 cm line. Hydrogen must be present here mainly in molecular form. Actually, one generally finds a deficit in 21 cm radiation in the directions of dark clouds. The H_2 molecule unfortunately has no radio spectrum and its lines lying in the far ultraviolet are not observable in thick clouds on account of the strong extinction.

At about the same time as the microwave spectroscopy of interstellar molecules the technique of astronomical infrared observations was being developed and applied, amongst other things, to the study of H II regions. The compact H II regions often showed up as strong *infrared sources* with a continuous spectrum resembling that of a black body at a few hundred Kelvin. One recognised here the thermal emission of circum- or inter-stellar dust heated up by one or more massive young stars which are not visible on account of high extinction. Observations in the far infrared also provide evidence of emission from protostars at temperatures of less than a hundred Kelvin. Recently thermal emission of the wide spread cold interstellar dust has been measured by the "Infra-Red Astronomical Satellite" (IRAS) at wavelengths up to $100\,\mu$m. Maps of its distribution on the sphere clearly show a strong concentration to the galactic plane (Fig. 1.14).

Space technology offers the possibility of measuring directly the primary components of the long recognised *cosmic rays* (V.F. Hess, 1912) before their modification by the geomagnetic field. Cosmic rays consist of charged particles

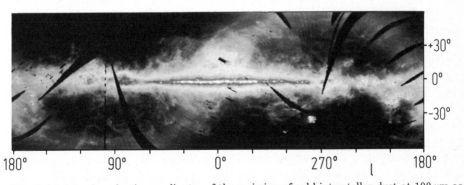

Fig. 1.14. Map in galactic coordinates of the emission of cold interstellar dust at $100\,\mu$m as detected by the Infra-Red Astronomical Satellite (IRAS). Also shown is emission from solar system dust confined to the zodiac and therefore following approximately a sine curve. Black areas were not surveyed. (By courtesy of E.R. Deul and W.B. Burton)

with relativistic velocities, chiefly protons, α-particles and electrons, which flow through interstellar space and in so doing interact reciprocally with the ambient magnetic field and also the interstellar matter. The light electrons are deflected and retarded by the interstellar magnetic field which generates the already noticed nonthermal radiofrequency radiation (synchrotron radiation) from the galactic disc, in so far as it does not come from discrete sources (supernova remnants). For sufficiently energetic particles this Bremsstrahlung can also be very shortwave, in the region of x- or even gamma-rays. High energy gamma-quanta (energy $> 10\,\mathrm{MeV}$) can also arise in connection with collisions between nucleons of the cosmic rays and nuclei of interstellar atoms, namely in the decay of the neutral π-mesons generated therein.

In 1975 the first reliable measurements were made on this diffuse *gamma radiation from interstellar space*, by a few research teams using satellites of NASA and the European Space Agency (ESA). The distribution on the sphere clearly showed the galactic disc and also — besides discrete sources — enhanced radiation from directions tangential to the spiral arms, in which the material density is higher than in the regions between the arms. The non-thermal component of cosmic radiofrequency radiation (without discrete sources) produces a very similar distribution, because this radiation arises mostly from the retardation of relativistic electrons of cosmic radiation in the magnetic fields of the galactic spiral arms. The spectrum of the gamma radiation so produced is consistent with its formation by π°-meson decay.

1.3 Dynamics and Evolution

The following Chaps. 2 to 4 will describe the stellar and interstellar phenomena of our galaxy, whilst Chaps. 5 and 6 will be concerned with their explanation and physics. In particular we shall be enquiring into the dynamics, and therefore into the forces involved and the nature of their effects.

To obtain a deeper understanding of the empirically established relationships between the spatial distribution of stars and their motions, the mechanics of systems of many gravitating point masses has been studied (*stellar dynamics*). For the instantaneous configuration and the timewise evolution of such a system it is crucially important whether impacts and close encounters of the individual stars occur. In the star field of the galaxy this is generally not the case. The motion of the individual star is therefore governed by the gravitational field arising from the total of all the stars of the Milky Way system. It may be otherwise in the densely packed star clusters and the kernel of the galaxy. There, from time to time, so much energy will be transferred to the "collision partner" in a close encounter that the escape velocity will be exceeded and the star with its complement of energy will escape from the system.

The theory of stellar dynamics, even if studied statistically, quickly runs into considerable mathematical difficulties. Analytic solutions can be found only under considerable limitations of generality. It is possible, however, to reproduce the velocity distribution of the stars in general terms. From the stellar motions perpendicular to the galactic plane one can deduce the mass density in the neighbourhood of the sun and then, with the known rotation curve, a model of the mass distribution in the whole galactic disc.

Important progress has been made since the seventies through the use of large and very fast computers. It has thus become possible to simulate whole stellar systems numerically. In this way new answers could be obtained to numerous questions which are difficult to handle analytically, such as the stability of star clusters or of the whole galactic disc, or the origin of the spiral structure.

The quantitative theory of the interstellar medium is today still far behind the stage reached by the physics of stars, with its theory of stellar structure and stellar evolution. The difficulty lies in the fact that the interstellar medium involves a complicated structure, with the mutual interaction of several components, and it is often governed by nonlinear dynamics: in the thin neutral gas the sonic velocity is about $1\,\mathrm{km\,s}^{-1}$, so that nearly all the observed motions have supersonic velocities and shock waves therefore arise; moreover there are interactions with the interstellar magnetic field.

Before one can deal with the dynamics of interstellar matter, its physical state must be understood from the observations with the help of theoretical considerations. How high are the densities and temperatures, how is it distributed in space and how does it behave kinematically? An important topic is the heat balance of the interstellar gas: how is it heated and cooled? The kinetic temperature of the neutral hydrogen (H I) in interstellar space amounts to about 10^2 K; in the ionised H II regions it reaches about 10^4 K. Insight into the essential thermal processes makes it possible to clear up these matters of fact and allows one to understand, for example, why the temperature of the H II regions is practically constant and independent of the surface temperature of the exciting star. The heat balance of the different gas components outside the H II regions, in the presence of the dust and the interstellar radiation field, including the high energy particle stream of cosmic rays, is also determined by dynamical processes.

Other questions which are addressed in Chaps. 5 and 6, concern the chemical composition of the interstellar matter − by and large one finds the same results as for the Population I stars if one overlooks differences in detail − the state of molecular clouds and the role of the interstellar dust particles.

In the dynamics of the interstellar medium notable advances have been made since the fifties, again largely thanks to the use of fast electronic computers. As already mentioned, the occurrence of shock fronts is an important phenomenon, both for the timewise evolution of an H II region in the neighbourhood of a newly formed hot star, and for the expansion of a supernova shell into the surrounding interstellar medium. Detailed numerical computations in recent times have provided new insight here.

The theory of the interstellar matter can naturally not leave out of account the fact that the diffuse medium is embedded in a stellar system of overwhelming mass. The effects of the gravitational field of the stars can therefore be important, as for example in the galactic spiral structure. Thus, the density wave theory of spiral structure formulated by Lin and Shu in the sixties stated that in the gravitational field of the galaxy spiral-shaped waves would occur, which would have only about half the velocity of the galactic rotation. The observed concentration of the interstellar clouds in the spiral arms can then be explained as a consequence of the moving of a thin intercloud medium into these waves of higher gravitational potential followed by condensation. The empirical proof of a far extended hot component of the interstellar medium ($\approx 10^6$ K, density about 10^{-3} cm^{-3}), which was probably produced by the shock fronts of supernovae, shows that dynamical processes play a far greater role than has previously been assumed.

Another problem region of the theory, which concerns the relation between the interstellar matter and that concentrated in the stars, is the process of star formation, in which the matter passes by condensation from one state into the other. Computations current today are attempting to follow through the evolution of a gravitationally unstable cloud right up to star formation, and thereby to obtain insight into the essential physical processes accompanying it. The formation and evolution of the galaxy as a whole also belongs here, since it is assumed that the protogalaxy consisted of a vast, almost spherical, gas cloud filled with diffuse matter at very low density, in which the first star formation occurred about 10^{10} years ago. The empirical findings on the stellar populations give an important clue here as to how an initially spherical cloud has evolved into a strongly flattened rotating disc with the characteristic features to be observed in our galaxy today.

2. Positions, Motions and Distances of the Stars – Concepts and Methods

The bodies of the planetary system, including the interplanetary matter, can be investigated nowadays by direct measurements at the locality of the object by using space probes. Our knowledge of the structure of the universe beyond the limits of the solar system, on the other hand, depends exclusively on the radiation fields of the cosmic objects (including particle streams and possibly gravitational waves) observable at the location of a terrestrial or "extra-terrestrial" observer.

In order to deduce the spatial arrangement of the emitting objects one needs the directions of their radiation and their distances. Even when one knows nothing of the true radiation flux at the locality of the object, deductions can be made concerning its distance from measurements of the change of direction of the radiation as a result of a known change of position of the observer or of the object. All fundamental determinations of distance are therefore based on measurements of direction.

This is true in particular of stars, which in practice appear as points. The desired changes in position of the observer are provided by the motion of the earth around the sun and by the motion of the whole solar system through space (parallax). The method is therefore based on knowledge of the dimensions of the earth's orbit and of the velocity of the sun through space relative to a selection of stars considered as representative, respectively. The "solar motion" can be obtained from the Doppler shifts of the spectra of these stars – they supply the relative velocity components along the line of sight – with appropriate averaging. The case of a known change of position of the object arises for the members of one of the groups of stars extending sufficiently across the sphere, with equal and parallel velocity vectors: here the motion of the group can be obtained in magnitude and direction from measurements of the Doppler effect in combination with a geometric determination of the destination direction, which again is based on measurements of direction (moving-cluster parallax).

The distances of stars of various physical types measured by these methods form the basis for the derivation of the magnitudes of the intrinsic absolute radiation output of these objects, upon which to build far reaching secondary methods of distance determination – through comparison of "apparent" and "true" brightness. These make possible quantitative statements on the spatial distribution of stellar objects and the limits of the stellar system, they provide also a measuring rod for deriving the distribution of the interstellar gas from its line radiations in the radiofrequency region. It can therefore be said that our present knowledge of the dimensions of the Milky Way system, of the scale of its structure, and partially of its internal motions, is finally based on determinations of the directions of stars in combination with measurements of the Doppler effect.

The starting point for deducing the direction changes of a cosmic object is a reproducible coordinate system to which the direction measurements carried out at different times can be referred. Since the observations are made directly on the directions only, one chooses spatial polar coordinates, with origin coinciding with the locality of the observer. The distance coordinate (radius vector) is for the time being still undecided, so that one can speak of positions on the heavenly sphere, a sphere of indeterminate radius centred on the locality of the observer. Since the angle between two directions is defined in principle as the length of a circular arc one comes to adopt the term "spherical coordinates".

The notional concept of an appropriate spherical coordinate system and the realisation of a rotation-free system fixed in space are problems of "Spherical Astronomy" and of "Astrometry". In this chapter we shall therefore first of all treat the relevant parts of these two oldest branches of astronomy including the essential concepts for describing stellar motions in space. The second section is then concerned with the various methods of determining the distances of stars.

2.1 Positions and Motions

2.1.1 Astronomical Coordinate Systems

For the definition of a spherical coordinate system one needs a base plane through the origin, which describes a great circle on the heavenly sphere, and a fixed direction point or null point on this great circle. Then the position of a celestial object is uniquely defined as a point on the surface of the sphere by two arcs measured on great circles. According to the problem, different systems will be used, with different choices of base plane and of the null point for measurement on the corresponding base circle.

Horizontal System: In the most obvious reference system for the practical observer, the comparatively easily determined local direction of gravity is used as the vertical axis; the upper and lower points on the heavenly sphere lying in this direction are called the zenith and the nadir. The plane through the position of the observer and perpendicular to the vertical is called the horizon plane and forms the base plane of the system. One of the two angular coordinates used to define the position of a point-like cosmic object is the *zenith distance* z, the angle between the zenith as reference direction and the object. Sometimes its complement, the altitude $h = 90° - z$ is also used. Great circles through the zenith and the nadir are called vertical circles; zenith distance and altitude can therefore be denoted as arcs on a vertical circle (Fig. 2.1). Circles

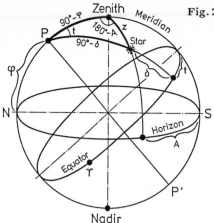

Fig. 2.1. Horizontal system and equatorial system

of equal zenith distance, or altitude, and hence circles parallel to the horizon, which are not great circles, are also called almucantars or parallels of altitude.

For the second angular coordinate one uses the angle between the plane of the vertical circle through the object and the plane of a fixed vertical circle which passes through a null direction in the horizontal plane, namely the south direction (in the northern hemisphere) or, expressed otherwise, the arc on the horizon circle between the vertical circle through the object and the south point. This arc is called the *azimuth A*, and is measured clockwise (through west, north, east) from $0°$ to $360°$ (left system)[1].

The south point is fixed by the fact that all the celestial objects attain their greatest altitude, or smallest zenith distance, at the vertical circle through this point (S) in the course of their apparent daily motion through the earth's rotation. One calls this the upper culmination or transit; it follows that the lower culmination occurs on the opposite (northern) half of the same vertical circle. This great circle also contains the two end points of the axis of the daily rotation of the sky on the sphere, the north and the south *celestial poles* (P and P' in Fig. 2.1), and is known as the observer's *meridian*. The vertical circle on which lie the horizon points with $A = 90°$ and $270°$, the west and east points, is called the *prime vertical*.

Dependent Equatorial System: The fundamental plane of a reference system for the determination of the spatial position and motion of a celestial body should be free of the earth's rotation and at least in the first approximation attached to the fixed stars. This requirement is fulfilled by the plane lying perpendicular to the rotation axis (terrestrial equator). The great circle of the sphere corresponding to this is known as the celestial equator. The angle between the axis through both celestial poles and the horizon plane is the altitude of the pole φ (Fig. 2.1); it is equal to the geographical latitude of the location

---

[1] In geodesy the numbering begins at north.

of the observer. Great circles of the celestial sphere through both celestial poles are called hour-circles. Analogously to the altitude in the horizontal system, the *declination* δ of an object is defined as the arc on an hour-circle between the equator and the object direction, measured in degrees, positive towards the north pole and negative towards the south pole. The small circles of constant declination are again known as parallels of declination. Stars on parallels with polar distance $90 - \delta < \varphi$, and hence declinations $\delta > 90 - \varphi$ remain always above the horizon and are called circumpolar stars.

As the second coordinate one uses the arc on the celestial equator between the hour-circle through the south point, the meridian, and the hour-circle of the object; it is called the *hour angle t*, since it is a measure of the time passed since its upper culmination. The hour angle is reckoned clockwise when seen from the celestial north pole, like the azimuth (left system), and is usually measured in time; it increases from 0 to 24 hours, corresponding to $0°$ to $360°$, and at the lower culmination it amounts to 12 hours or $180°$.

For conversion from horizontal to equatorial coordinates and conversely, the following transformation formulae are used:

$$\cos \delta \sin t = \sin z \sin A$$
$$\sin \delta \quad = \cos z \sin \varphi - \sin z \cos \varphi \cos A \qquad (2.1)$$
$$\cos \delta \cos t = \cos z \cos \varphi + \sin z \sin \varphi \cos A$$

and

$$\sin z \sin A = \cos \delta \sin t$$
$$\cos z \quad = \sin \varphi \sin \delta + \cos \varphi \cos \delta \cos t \qquad (2.2)$$
$$\sin z \cos A = \cos \varphi \sin \delta - \sin \varphi \cos \delta \cos t \; .$$

These relations follow directly from the spherical trigonometric formulae derived in Appendix A, on replacing the angles a, b, c, α and γ, respectively, by z, $90° - \varphi$, $90° - \delta$, t and $180° - A$ (cf. Fig. 2.1 and Fig. A.1): equations (2.2) come from (A.3), (A.6) and (A.4), the first one of (2.1) is identical with the first one of (2.2), and both the other ones of (2.1) come from (A.7) and (A.5).

Independent Equatorial System, Sidereal Time: The declination is independent of the time. In order to have a time-independent quantity as the second coordinate instead of t, and hence to obtain a reference system completely free of the earth's rotation, one takes as the nullpoint on the celestial equator, the *First Point of Aries* or *vernal equinox*. That is the point at which the centre of the solar disc crosses the celestial equator in spring, coming from the south, or in other words, the "ascending node" of the *ecliptic,* the course of the annual journey of the sun across the celestial sphere of fixed stars from west, through south, to east. The arc on the equator from the vernal equinox (Υ) to the hour-circle of the object is called the *Right Ascension α*, often also abbreviated to AR (for ascensio recta). The right ascension is reckoned counter-clockwise, and so

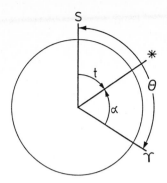

Fig. 2.2. Relationship between the hour angle t, right ascension α and sidereal time Θ. View of the equatorial plane from the celestial north pole

from west, through south, to east from 0 to 24 hours or from 0° to 360° (right system).

The hour angle of a distant object, practically fixed in space, is also a measure of time, with the period of the earth's rotation as natural unit, whereas the right ascension is the fixed difference between the hour angle of the vernal equinox and that of the object. The transformation between the two equatorially systems, $\alpha \leftrightarrow t$, therefore requires the knowledge of the hour angle of the vernal equinox. This is defined as the *sidereal time* Θ. We have (Fig. 2.2)

$$\Theta = \alpha + t \ . \tag{2.3}$$

The sidereal time is a local time, being an hour angle from the observer's meridian, that is to say, it depends upon the geographical longitude of the observer's position. To train an equatorially mounted telescope (dependent equatorial system) on an object of known right ascension (and declination) one needs a clock set to the local sidereal time, from whose reading Θ the hour angle to be adjusted $t = \Theta - \alpha$ can be obtained. For an object in transit ($t = 0$) we have $\Theta = \alpha$, or in other words the sidereal time is equal to the "culminating right ascension".

For a complete astrometric observation one needs statements of position and time. We therefore go briefly into the relationship between sidereal time and solar time. During the year the sun moves along the great circle of the ecliptic against the direction of the apparent daily motion of the stars. The ecliptic is inclined to the equator at about $23\frac{1}{2}$ degrees; this angle is called the "obliquity of the ecliptic". The intersection points of the two great circles are called the vernal and autumnal equinoxes (where day and night are equal). The ecliptic serves also as the base plane of a further spherical coordinate system, the Ecliptical System (ecliptical latitude β and longitude λ analogous to δ and α, defined with the vernal equinox as the null point for λ), which however is used only for objects of the solar system. The hour angle of the midpoint of the sun $t_\odot$ fails to represent a proportionate measure of time for two reasons: (1) because the motion of the sun does not follow the equator and (2) because the sun's motion on the ecliptic – as the result of the earth's motion on an elliptical orbit round the sun – does not occur at a uniform angular velocity. To obtain a uniform measure of solar time, a fictitious mean sun has been introduced which moves with constant angular velocity along the celestial equator and synchronises with the real sun at the vernal equinox. If $\bar{t}_\odot$ denotes the hour angle of this mean sun, then the definition of *mean solar time* (local time) is $\bar{t}_\odot + 12^h$. The difference from the apparent solar time $t_\odot + 12^h$, in the sense of apparent time minus mean time, is called the *equation of time* $= t_\odot - \bar{t}_\odot = \bar{\alpha}_\odot - \alpha_\odot$ [from (2.1)], where $\bar{\alpha}_\odot$ and $\alpha_\odot$ are the right ascensions of the mean sun and the true sun, respectively. The magnitude of the equation of time does not exceed 17^m at any time of year.

To convert from mean solar time to sidereal time and conversely, one has to take account of the ratio of the two time units and of the difference between the two starting points. The mean sun moves relative to the vernal equinox through $360°$ counterclockwise in 365.2422 mean solar days (tropical year), so that a mean day is increased by an angle of $24^h/365.2422 = 3^m56\overset{s}{.}555$. A mean solar day therefore corresponds to $24^h3^m56\overset{s}{.}555$ in sidereal time. The sidereal time at mean midday, $\bar{t}_\odot = 0$, is given by (2.3) as $\Theta = \bar{\alpha}_\odot$. The right ascension of the mean sun $\bar{\alpha}_\odot$ is known from the mean time, so that for each day the mutual position of the starting point in the other time scale is known. The astronomical almanacs give Θ for the (zero) meridian of Greenwich daily for 0 hours Greenwich Mean Local Time (GMT) = Universal Time (UT). If $\Theta_{\lambda,\mathrm{UT}}$ is the sidereal time at a place with east longitude λ (time measure) at Universal Time UT, and in particular $\Theta_{0,0}$ is the sidereal time on the zero meridian at 0^h UT, then

$$\Theta_{\lambda,\mathrm{UT}} = \Theta_{0,0} + \mathrm{UT} + 3^m56\overset{s}{.}555\frac{\mathrm{UT}}{24^h} + \lambda \ .$$

To obtain a first idea as to which stars are observable at different times of the year, it is useful to know the sidereal time at midnight (= culminating right ascension): at the start of spring (21 March) and autumn (23 September) the local sidereal times at 0^h mean local time are 12^h and 0^h, respectively; for the summer and winter solstices (22 June and 22 December) they are 18^h and 6^h.

Galactic Coordinates: Because of its exact reproducibility, all measurements of star positions are referred primarily to the equatorial system. In investigations into the structure and relative motion in the Milky Way system, however, it is often convenient to refer the discussion of the data to a coordinate system that has the plane of symmetry of the galaxy as its fundamental plane. It turns out that the sun lies very close to the middle plane defined by means of neutral hydrogen (see Chap. 1), within error limits of about $\pm 20\,\mathrm{pc}$. The section of the sphere by this plane is therefore practically a great circle; it is called the galactic equator and runs along the visible band of the Milky Way. As the analog of the declination, the angular distance from the galactic equator − counted positive on the north side and negative on the south side − is introduced as the *galactic latitude b*. On the galactic equator the *galactic longitude l* is reckoned counter clockwise from a null direction from $0°$ to $360°$ (Fig. 2.3).

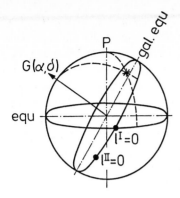

Fig. 2.3. Definition of galactic coordinates

The system of galactic coordinates used nowadays is specified by international convention with the equatorial coordinates of the galactic north pole at

$$\alpha = 12^{\mathrm{h}}49^{\mathrm{m}} , \quad \delta = +27\overset{\circ}{.}4 \ (1950.0)$$

and, as the null direction for longitude, the point on the galactic equator which lies at a distance of 33° (reckoned clockwise) from the ascending node of the galactic equator over the celestial equator. This point has the equatorial coordinates

$$\alpha = 17^{\mathrm{h}}42^{\mathrm{m}}\!.4 , \quad \delta = 28\overset{\circ}{.}92 \ (1950.0) \ .$$

Here the expression enclosed in brackets indicates that these values refer to the position of the equatorial system at the beginning of the year 1950 (see below in Sects. 2.1.2, 3). The specified null point for longitude has been determined by optical and radio observations as the direction of the centre of the galaxy; it practically coincides with the position of the intense radio source Sagittarius A (Sects. 3.1.3, 4.2.2). The inclination of the galactic equator to the celestial equator amounts to $i = 62\overset{\circ}{.}6$; the right ascension of the ascending node is $\alpha_\Omega = 18^{\mathrm{h}}49^{\mathrm{m}} = 282\overset{\circ}{.}25$.

The transformation formulae for the conversion from (α, δ) into (l, b) are:

$$
\begin{aligned}
\cos b \cos(l - 33^\circ) &= \cos \delta \cos(\alpha - 282\overset{\circ}{.}25) \\
\sin b &= \sin \delta \cos 62\overset{\circ}{.}6 - \cos \delta \sin 62\overset{\circ}{.}6 \sin(\alpha - 282\overset{\circ}{.}25) \\
\cos b \sin(l - 33^\circ) &= \sin \delta \sin 62\overset{\circ}{.}6 \\
&\quad + \cos \delta \cos 62\overset{\circ}{.}6 \sin(\alpha - 282\overset{\circ}{.}25) \ .
\end{aligned}
\tag{2.4}
$$

To derive these formulae from equations (A.3), (A.7) and (A.5) we replace a, b, c, α and γ by $90° - \delta$, $i = 62\overset{\circ}{.}6$, $90° - b$, $90° - (l - 33°)$ and $(\alpha - 282\overset{\circ}{.}25) + 90°$ (cf. Fig. 2.3 with Fig. A.1).

For the transformation of (α, δ) into (l, b) and conversely, detailed tables are given by Torgård (1961). Approximate values are shown in the charts given in Appendix C.

In the older literature yet other galactic coordinate systems have been used. The fundamental plane there rests on the apparent stellar distribution in the region of the Milky Way which is strongly modified by interstellar extinction. Between about 1932 and 1960 a system was most often used which is today known as the "old galactic coordinate system", and whose longitudes and latitudes are indicated by l^I and b^I. The null point was the ascending node of the "old" galactic equator over the celestial equator. The supplanting galactic coordinates of the new system were at first (from ~ 1960) denoted by l^{II}, b^{II}, but the superscripts have since been dropped. The point $l = 0°$, $b = 0°$, the direction of the centre, is given in the old system by $l^I = 327°.7$ and $b = -1°.40$. Tables for the transformation of old into new galactic coordinates and conversely are given in the tables of Torgård (1961) mentioned above.

2.1.2 Temporal Changes of the Star Coordinates

Precession: The gravitational effects of the sun and moon on the slightly flattened body of the earth lead to variations of the rotation axis and hence of the celestial equator relative to the ecliptic. The secular portion of this spin variation, i.e. the portion which is proportional to time, is called the *lunisolar precession,* the remaining periodic portion is called nutation. The influence of the planets on the orbital motion of the earth round the sun causes in addition a small variation of the ecliptic compared with an inertial system, in which the observed motions of all the planets can be described by Newtonian mechanics with high accuracy. This variation is called *planetary precession.*

The effect of lunisolar precession is illustrated in Fig. 2.4a. The celestial pole P moves on a parallel circle of the ecliptic system around the pole of the ecliptic Π, causing the equinoxes to "advance". With the movement of the vernal equinox Υ the ecliptical longitudes and also the right ascensions of all objects fixed in space are increased. The annual lunisolar precession in ecliptical longitude (Fig. 2.4b), the movement $\Upsilon_1 \Upsilon_2$, is denoted by ψ and amounts to

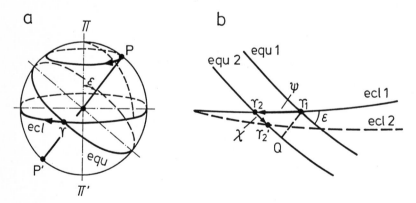

Fig. 2.4. (a) The motion of the celestial north pole P and the vernal equinox Υ on account of the lunisolar precession. (b) Definition of the annual precession in longitude ψ and of the annual precession due to the planets χ. (Explanation in text)

about 50″, a complete circulation of the pole lasts about 26 000 years ("Platonic Year"). This rotation of the coordinate system can be resolved into two components: a rotation of the equatorial system about the axis of the celestial sphere PP' as z-axis (cf. Appendix A, Fig. A.3) $\omega_3 = \Upsilon_2 Q = -\psi \cos \varepsilon$ (positive spin angle denotes right-handed spin) and a rotation about the perpendicular axis through the equatorial point $\alpha = 90°$ as y-axis $\omega_2 = \Upsilon_1 Q = +\psi \sin \varepsilon$. There is no rotation about the equinoxial axis (x-axis): $\omega_1 = 0$. The weak variable annual precession due to the planets $\Upsilon_2 \Upsilon_2' = \chi \approx 0''1$ per year works against the lunisolar precession and can be allowed for by writing $\omega_3 = -\psi \cos \varepsilon + \chi$. The obliquity of the ecliptic ε changes only very insignificantly, at present it is decreasing by about $\approx 0''5$ per year; in 1900 it amounted to $23°27'8''26$, at the beginning of the year 2000 (usually written 2000.0) it will be $\varepsilon = 23°26'21''45$.

The annual advance of the vernal equinox; $p = \psi - \chi \cos \varepsilon$ is called the general precession in longitude.

The changes in the right ascension and declination are given directly if one sets the values of the rotational components $\omega_1, \omega_2, \omega_3$ in (A.16) of Appendix A which are valid for $\omega_i \ll 1$, and L, B are set equal to α, δ. The results are written in the form

$$\dot{\alpha} = p_\alpha = m + n \sin \alpha \tan \delta$$
$$\dot{\delta} = p_\delta = n \cos \alpha , \tag{2.5}$$

where m and n, the components of the so-called general precession in α and δ, per year are given by

$$m = \psi \cos \varepsilon - \chi \approx 3^{\rm s}07 \approx 46''$$
$$n = \psi \sin \varepsilon \approx 1^{\rm s}34 \approx 20'' . \tag{2.6}$$

These quantities vary slowly with time. For the time interval Δt one has approximately for small rotations

$$\Delta\alpha = p_\alpha \Delta t , \quad \Delta\delta = p_\delta \Delta t . \tag{2.7}$$

These formulae break down in the neighbourhood of the celestial pole $\delta \approx \pm 90°$. Tables for conveniently finding the components p_α and p_δ of the annual precession as functions of α and δ are to be found, for example, in Sect. 8.1.1.6 in Schaifers and Voigt (1982).

For higher accuracy and longer time intervals one uses the rigorous transformation formulae between two arbitrary positions of the vernal equinox or applies the series expansion

$$\Delta\alpha = \alpha - \alpha_0 = {\rm I}_\alpha(t - t_0) + {\rm II}_\alpha(t - t_0)^2 + {\rm III}_\alpha(t - t_0)^3 + \ldots \tag{2.8}$$

with ${\rm I}_\alpha = (d\alpha/dt)_0 = $ annual precession in α, ${\rm II}_\alpha = (d^2\alpha/dt^2)_0/2$ and ${\rm III}_\alpha = (d^3\alpha/dt^3)_0/6$; $\Delta\delta = \delta - \delta_0$ is expressed similarly. The third term becomes appreciable only for long time intervals $\Delta t = t - t_0$. We cannot go into the calculation of II_α and III_α here; tables for them are also available.

The practical basis for all the foregoing applications of the precession formulae, and hence also of the relevant tables existing in the literature, were up to recent times the values for ψ and χ deduced by S. Newcomb from nineteenth century observations, which correspond to a general precession in longitude $p = 5025\rlap{.}''641$ (per century, for 1900). From present knowledge this value is about $1''$ too small. The general basic concepts for finding ψ and χ are described in Sect. 2.1.3. The material of the new star catalogues gave a correction $\Delta\psi = +1\rlap{.}''1$ (see Sect. 3.4.2), which together with improvements on χ on the basis of more recent results on planetary masses gives a value $p = 5029\rlap{.}''0966$ for 2000.0. This value was recommended in 1976 by the International Astronomical Union (IAU) and is used since 1984. For the components of the general precession in α and δ the following expressions have been derived (in each case per century):

$$m = 4612\rlap{.}''4362 + 2\rlap{.}''79312\,T - 0\rlap{.}''000278\,T^2$$
$$n = 2004\rlap{.}''3109 - 0\rlap{.}''85330\,T - 0\rlap{.}''000217\,T^2$$

where T denotes the time difference in Julian centuries ($= 36525$ days) from 2000.0.

Nutation: Upon the precession motion of the celestial pole on a circle around the pole of the ecliptic with constant velocity, there is superposed the nutation as a periodic oscillation in which not only the ecliptical longitude but also the ecliptical latitude of the celestial pole varies: the actual celestial pole describes an approximately elliptic motion about the so-called mean celestial pole moving under the influence of the precession. The instantaneous position of the pole is described by the nutation in longitude $\Delta\lambda(t)$ and the nutation in obliquity $\Delta\varepsilon(t)$. Referring to Figs. 2.4a,b one can easily see that this deviation corresponds to a rotation of the equatorial system with components about the x-, y- and z-axes: $\omega_1 = -\Delta\varepsilon$, $\omega_2 = +\Delta\lambda \sin\varepsilon$, $\omega_3 = -\Delta\lambda \cos\varepsilon$, where the ω_i here denote small angles. Equations (A.16) therefore lead to the following nutation formulae:

$$\Delta\alpha = \Delta\lambda(\cos\varepsilon + \sin\varepsilon \tan\delta \sin\alpha) - \Delta\varepsilon \tan\delta \cos\alpha$$
$$\Delta\delta = \Delta\lambda \sin\varepsilon \cos\alpha + \Delta\varepsilon \sin\alpha \tag{2.9}$$

$\Delta\varepsilon(t)$ and $\Delta\lambda(t)$ consist of several terms with differing periodicities. The term with the largest amplitude reads $+9\rlap{.}''21 \cos\Omega$ and $-17\rlap{.}''24 \sin\Omega$, respectively, where Ω is the ecliptical longitude of the ascending node of the path of the moon on the ecliptic. In consequence of the influence of the sun on the lunar orbit, inclined at about $5°$ to the ecliptic, the node goes backwards right round the ecliptic once in 18.6 years, so that Ω goes from $0°$ to $360°$. The celestial pole therefore in 18.6 years describes an ellipse on the sphere, with its centre at the mean pole and major semi-axis (parallel to a circle of constant longitude λ) $\Delta\varepsilon = 9''\!.21$ and minor semi-axis (on a circle parallel to the ecliptic) $\Delta\lambda \sin \varepsilon = 17''\!.24 \sin \varepsilon = 6''\!.86$. The numerical factor of the main term of $\Delta\varepsilon(t)$ of magnitude $N = 9''\!.21$ is known as the *Nutation Constant*. The exact values of $\Delta\varepsilon(t)$ and $\Delta\lambda(t)$ are given for the beginning of each day in the astronomical almanacs.

Annual Parallax, Astronomical Unit of Distance: Associated with the motion of the earth around the sun there are, in a geocentric reference system, small variations in the positions of the stars over the annual period. The angle subtended by the major semi-axis of the earth's orbit a at a star on a perpendicular line of sight is called the parallax p of the star. It is related to the distance r of the star by the equation

$$\frac{a}{r} = p \; ,$$

because even for the nearest stars $p < 1''$ (see Sect. 2.2.1).

The mean earth-sun distance $= a$ is defined as the Astronomical Unit (AU) for the bodies of the solar system. It has a value $1\,\mathrm{AU} = 149\,597\,870\,\mathrm{km}$. For the stars one uses the larger unit, the parsec (pc), the distance at which $1\,\mathrm{AU}$ subtends an angle of $1''$. Accordingly, since $p \approx \sin p \approx p'' \sin 1'' = p''/206\,265$,

$$1\,\mathrm{pc} = 206\,265\,\mathrm{AU} = 3.086 \times 10^{18}\,\mathrm{cm} \; .$$

The larger units $1\,\mathrm{kpc} = 10^3\,\mathrm{pc}$ and $1\,\mathrm{Mpc} = 10^6\,\mathrm{pc}$ are also used. The relationship with the light-year (l.y.) is $1\,\mathrm{pc} = 3.26\,\mathrm{l.y.}$ If r is expressed in pc, then we have simply

$$r = \frac{1}{p''} \; . \tag{2.10}$$

The variation of the star position due to parallax can be calculated in the following way. Transforming to a heliocentric reference system and using the equations for earth motion in a circular orbit in rectilinear ecliptical coordinates (Fig. 2.5a) leads to the displacement components

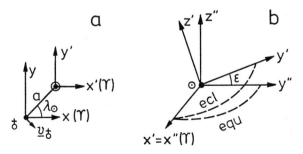

Fig. 2.5. (a) Derivation of (2.11); **(b)** derivation of (2.12). Symbol for the earth ♁

$$\Delta x = x' - x = -a \cos \lambda_\odot , \quad \Delta y = y' - y = -a \sin \lambda_\odot . \tag{2.11}$$

$\lambda_\odot$ denotes the ecliptical longitude of the sun. On passing to rectangular equatorial coordinates (Fig. 2.5b), Δx remains unchanged, whilst Δy acquires a factor $\cos \varepsilon$ and a finite z-component arises: $z'' - z' = (y' - y) \sin \varepsilon$. In this case therefore

$$\Delta x = -a \cos \lambda_\odot , \quad \Delta y = -a \sin \lambda_\odot \cos \varepsilon ,$$
$$\Delta z = -a \sin \lambda_\odot \sin \varepsilon . \tag{2.12}$$

To obtain the observable changes of the ecliptical and equatorial coordinates in the geocentric system, for the tangential components (A.11) and (A.12) in Appendix A can be used, with $\dot{x}, \dot{y}, \dot{z}$ replaced there by $-\Delta x, -\Delta y, -\Delta z$ from (2.11) and (2.12) respectively and L, B identified with λ, β and α, δ respectively ($v_r = 0$).

In the ecliptical system, with $t_\lambda = r \cos \beta \, \Delta\lambda$, $t_\beta = r \, \Delta\beta$ and $a/r = p$, one finds for the tangential components on the sphere:

$$\xi = \frac{t_\lambda}{r} = \Delta\lambda \cos \beta = p \sin(\lambda_\odot - \lambda) ,$$
$$\eta = \frac{t_\beta}{r} = \Delta\beta = p \cos(\lambda_\odot - \lambda) . \tag{2.13}$$

It follows in particular that

$$\frac{\xi^2}{p^2} + \frac{\eta^2}{p^2 \sin^2 \beta} = 1 , \tag{2.14}$$

i.e. the star describes (in the plane tangential to the sphere) a little ellipse about the star position for $r = \infty$, whose major axis lies parallel to the ecliptic and is equal to $2p$. For a star at the pole of the ecliptic ($\beta = 90°$) we get (for a circular orbit of the earth) a circle; a star in the ecliptic ($\beta = 0°$) moves to and fro along an arc of length $2p$ once a year.

In equatorial coordinates one obtains for the variations $\Delta\alpha$ and $\Delta\delta$, in the sense of geocentric minus heliocentric,

$$t_\alpha/r = \Delta\alpha \cos \delta = p(\cos \alpha \cos \varepsilon \cos \lambda_\odot - \sin \alpha \cos \lambda_\odot)$$
$$t_\delta/r = \Delta\delta = p(\cos \delta \sin \varepsilon \sin \lambda_\odot - \cos \alpha \sin \delta \cos \lambda_\odot$$
$$- \sin \alpha \sin \delta \cos \varepsilon \sin \lambda_\odot) . \tag{2.15}$$

With p in seconds of arc, $\Delta\alpha$ and $\Delta\delta$ are also given in these units.

Aberration: Because of the finite velocity of light, c, the direction of electromagnetic radiation perceived by an observer moving with velocity v will in general vary. The angle γ between the direction of motion and the direction of the light source (Fig. 2.6a) is reduced, when $v \ll c$, by the quantity

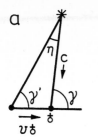

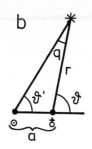

Fig. 2.6. Comparison of the effects of the annual aberration (**a**) and the annual parallax (**b**)

$$\eta = \gamma - \gamma' = \frac{v}{c} \sin \gamma' \ . \tag{2.16}$$

As a result of the motion of the earth round the sun — with approximately constant velocity $v = v_\oplus$ on a circle — there is an effect similar to the annual parallax on the geocentric star coordinates. This is evident from a comparison of Figs. 2.6a and b. By application of the sine rule, $\sin q / \sin(180° - \vartheta) = a/r = p$ or, since $\sin q \approx q$,

$$q = \vartheta - \vartheta' = p \sin \vartheta' \ . \tag{2.17}$$

The base lines of the triangles shown in Figs. 2.6a and b are at right-angles to one another (circular orbit !) whilst both the base-lines lie in the plane of the ecliptic. Accordingly the expressions for the coordinate changes due to annual aberration analogous to (2.13) to (2.15) must be obtainable simply from these formulae, by replacing the parallax p by the so-called *Constant of Aberration* $A = \bar{v}_\oplus/c$, with $\bar{v}_\oplus = $ the mean orbital velocity of the earth, and $\lambda_\odot$ by $\lambda_\odot - 90°$. Specifically one obtains

$$
\begin{aligned}
\Delta\alpha \cos \delta &= A(\cos \alpha \cos \varepsilon \cos \lambda_\odot + \sin \alpha \sin \lambda_\odot) \\
\Delta\delta &= -A(\cos \delta \sin \varepsilon \cos \lambda_\odot + \cos \alpha \sin \delta \sin \lambda_\odot \\
&\quad - \sin \alpha \sin \delta \cos \varepsilon \cos \lambda_\odot) \ .
\end{aligned}
\tag{2.18}
$$

The motion of the star in the tangential plane turns out to be an "aberration ellipse" with the major semi-axis A parallel to the ecliptic, but it runs 90° out of phase with the parallactic ellipse (Fig. 2.7). The Constant of Aberration is found to be $A = 20''5$, so that the annual aberration is a far greater effect than the annual parallax.

The motion of the observer as a result of the earth's rotation produces a diurnal aberration, the constant of which amounts to $0''32 \cos \varphi$ ($\varphi =$ geographical latitude of the observer). This must in particular be taken into consideration for precise absolute determination of α, since on this account every star passes through the meridian about $0^s021 \cos\varphi/\cos \delta$ late (eastwards motion of the observer).

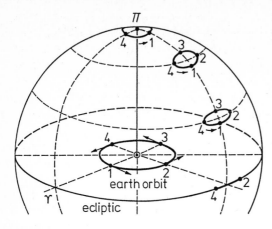

Fig. 2.7. Aberration paths on the sphere for stars with various ecliptical latitudes. The digits 1 to 4 show positions of the earth at times a quarter of a year apart, and the corresponding positions of the stars on the aberration ellipses. The parallactic ellipses occupy the same positions, but the star positions are at *2,3,4,1* instead of *1,2,3,4*, respectively

Apparent, True and Mean Place: Under the influence of precession and nutation the equatorial system introduced in Sect. 2.1 changes its position with time. The specification of the (α, δ) spherical coordinate system used accordingly always requires a time statement. α and δ are therefore said to refer to the true *equinox*, i.e. to the positions of the vernal equinox and equatorial plane at a definite time. If α and δ are freed from the nutation (2.9) at this time under consideration, then these refer to a fictitious, mean equatorial system influenced only by precession or, as we say, to the mean equinox at this point in time.

Direct determinations of α and δ for a star through absolute observations – of which more will be said under (Sect. 2.1.4) – always refer to the instantaneous true equinox (at the time of observation). Of course, the position statements are at first distorted by refraction in the earth's atmosphere, as well as by the annual and diurnal aberration and annual parallax. To distinguish the different stages of reduction one uses the following definitions:

Apparent Place: Observed position, corrected for refraction and diurnal aberration

(geocentric star position);

Mean Place: Heliocentric position referred to a specified mean equinox, generally that of the beginning of a year.

The Apparent Place is therefore the Mean Place distorted by (annual) aberration and parallax and the influence of Nutation – the allowance for the diurnal aberration already in the Apparent Position has a practical purpose: it is conveniently done at the same time as the correction for instrumental errors.

Series of absolute observations carried out at different times and at different observatories produce true positions which in general refer to different true equinoxes (at different observation times). For the compiling of a catalogue of star positions it is necessary that all positions be referred to the mean equinox at the beginning of the relevant year of observation by taking account of the corresponding precession and nutation, and subsequently to refer them to an unique Normal Equinox, for example, at the beginning of the year 1950; they are then denoted 1950.0. That the coordinates themselves also refer to a definite observation time – called in astrometry the "epoch" of the observations – is naturally a separate matter. If, between the coordinates of a star belonging to different observation times, but referred to the same Normal Equinox, there exist significant differences, then they are interpreted as *"proper motion"*.

Proper Motions: The individual positions of a star (α, δ) do actually vary because of the "tangential components" perpendicular to the line of sight of the individual velocities of the stars through space. These proper motions, referred to a year (or a century), are expressed by the arc μ along a great circle between the two positions at the times t and $t+1$ year and the position angle P of the associated direction (Fig. 2.8). Apart from terms of higher order, which are appreciable only for large proper motions (near objects), the equatorial components are given by (seconds of arc per year)

$$\mu''_\alpha = \mu'' \sin P \ , \quad \mu''_\delta = \mu'' \cos P \ . \tag{2.19}$$

Here $\mu''_\alpha = 15 \, \mu^s_\alpha \cos \delta$, if μ^s_α is the variation – measured on the equator – of α measured in time seconds (cf. Fig. A.2, where $\varrho = \cos B$).

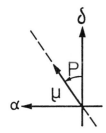

Fig. 2.8. Concept of proper motion. The position angle P goes from $0°–360°$ through north $\rightarrow$ east $\rightarrow$ south $\rightarrow$ west

 The amounts of the measured proper motions are in general less than $0''.1$ per year. Reliable values can therefore as a rule be deduced only from position determinations with epochs separated by about a decade. The largest proper motion found up to now amounts to $10''.3$ per year (Barnard's star, a nearby red dwarf).
 From the absolute positions (α, δ) of a star the so-called *absolute proper motion* is deduced. In order to find as many stars as possible with large proper motions one makes a direct comparison of photographs of the same field which were taken at different times. In this, the faintest and hence generally far distant stars serve as reference points. The results are then only *relative proper motions,* in which for example the participation of the observer in a pure rotation of the entire stellar system would give a zero result. Absolute proper motions therefore demand the specification of an astronomical coordinate system which does not take part in the galactic rotation and which is entirely free from spin (Sect. 2.1.4).

2.1.3 Space Velocities and Solar Motion

The derivation of the spatial velocity vector of a star is based on the determination of the velocity component in the line of sight, the so-called radial velocity, and of the two perpendicular components of the tangential velocity, which are obtained from the proper motion and the distance.

Radial Velocity: This component of the motion of a cosmic object is obtained by measurement of the Doppler shift $\Delta\lambda = \lambda - \lambda_0$ of lines in its spectrum (rest wavelength λ_0) and application of the Doppler formula for velocities which are small compared with the velocity of light c:

$$\frac{v'_r}{c} = \frac{\Delta\lambda}{\lambda_0} \ . \tag{2.20}$$

Here v'_r is the radial velocity relative to an observer on the earth's surface; v'_r is positive if $\Delta\lambda = \lambda - \lambda_0$ is positive, and hence with a red shift. Transformation to v_r, referred to the sun, involves making allowance for the components of the orbital velocity and the rotation of the earth. If v_a denote the component of the orbital velocity of the earth in the direction of the star and v_d the component of the velocity of the observer as a result of the earth's rotation in the direction of the star, then it follows that

$$v_r = v'_r + v_a + v_d \ . \tag{2.21}$$

For the calculation of v_a the expression (A.10) given in Appendix A can be applied. If one replaces $\dot{x}$, $\dot{y}$, $\dot{z}$ by the components of the orbital velocity of the earth, then this expression gives v_a directly. Since the geocentric rectangular coordinates of the sun in the equatorial system X, Y, Z (in AU) are given for each day in the astronomical almanacs, and the first differences ΔX, ΔY, ΔZ, can be used for the practically equal $\dot{X}$, $\dot{Y}$, $\dot{Z}$, apart from the conversion factor from AU/day to km s^{-1}: $1\,\mathrm{AU}/86\,400 = 1731.5$, so $\dot{x}$, $\dot{y}$, $\dot{z}$ can be directly replaced by $-\Delta X$, $-\Delta Y$, $-\Delta Z$. Equation (A.10), with $(L, B) = (\alpha, \delta)$, then gives

$$\begin{aligned} v_a = &- 1731.5(\Delta X \cos\alpha \cos\delta + \Delta Y \sin\alpha \cos\delta \\ &+ \Delta Z \sin\delta) \ [\mathrm{km\,s}^{-1}] \ . \end{aligned} \tag{2.22}$$

Assuming the earth's surface to be a sphere with the velocity of a point on the equator of $0.465\,\mathrm{km\,s}^{-1}$, one obtains for v_d the expression

$$v_d = -0.465 \cos\varphi \cos\delta \sin t \ [\mathrm{km\,s}^{-1}] \ . \tag{2.23}$$

Here φ is the geographical latitude of the position of the observer, and t is the hour angle of the object.

The mean error of the best radial velocity determinations for stars nowadays is about $\pm 0.5 \, \mathrm{km \, s^{-1}}$. Radial velocity measurements for about 25 000 stars are available, published in various catalogues. The most extensive source is the General Catalogue of Stellar Radial Velocities (GCSRV) of R.E. Wilson. Compilations of further catalogues are given in Voigt (1965) and Schaifers and Voigt (1982); see list for further reading.

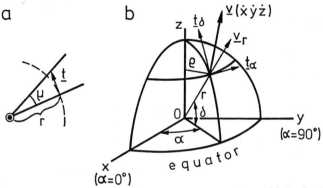

Fig. 2.9. (a) Definition of the tangential velocity t. (b) Derivation of the rectangular components $\dot{x}$, $\dot{y}$, $\dot{z}$ of the spatial velocity

Tangential Velocity and Proper Motion: We denote the components of the tangential velocity of a star (Fig. 2.9a) in the direction of increasing right ascension and declination by t_α and t_δ respectively. One obtains these quantities by multiplying the corresponding components of the proper motion by the distance r. Using the units seconds of arc per year and parsecs, we obtain

$$
\begin{aligned}
t_\alpha &= K r \mu''_\alpha \; [\mathrm{km \, s^{-1}}] \\
t_\delta &= K r \mu''_\delta \; [\mathrm{km \, s^{-1}}] \qquad \text{with}
\end{aligned}
\tag{2.24}
$$

$$
K = \frac{(1 \, \mathrm{pc} \rightarrow \mathrm{km}) \cdot (1'' \rightarrow \mathrm{arc})}{(1 \, \mathrm{year} \rightarrow \mathrm{s})} = 4.738
\tag{2.25}
$$

and $\mu''_\alpha = 15 \, \mu^s_\alpha \cos \delta$. As an example the numerical values for Sirius $= \alpha$ Canis Major (α CMa) are tabulated below (t_α, t_δ, v_r in $\mathrm{km \, s^{-1}}$):

μ^s_α	μ''_α	μ''_δ	$r[\mathrm{pc}]$	t_α	t_δ	v_r
$-0\overset{s}{.}0379$	$-0\overset{''}{.}545$	$-1\overset{''}{.}211$	2.65	-6.84	-15.21	-8

Components of the Spatial Velocity $\dot{x}$, $\dot{y}$, $\dot{z}$ in the Equatorial System: The spatial velocity is defined to be the vector $\boldsymbol{v}(\dot{x}, \dot{y}, \dot{z})$, referred to the coordinate system with origin at the centre of gravity of the solar system (Fig. 2.9b).

The absolute value of v is given in the case of Sirius by $v = \sqrt{t_\alpha^2 + t_\delta^2 + v_r^2}$ $= 21.0\,\mathrm{km\,s^{-1}}$. For the derivation of the relationship between $(\dot{x}, \dot{y}, \dot{z})$ and $(t_\alpha, t_\delta, v_r)$ let us refer to Appendix A. The angles L, B used there correspond to α, δ. Equations (A.13), (A.10) to (A.12) and (A.14), with the indicated substitutions, then lead directly to

$$\begin{pmatrix} \dot{x} \\ \dot{y} \\ \dot{z} \end{pmatrix} = (\gamma_{ik}) \begin{pmatrix} t_\alpha \\ t_\delta \\ v_r \end{pmatrix} \quad \text{and} \quad \begin{pmatrix} t_\alpha \\ t_\delta \\ v_r \end{pmatrix} = (\gamma_{ki}) \begin{pmatrix} \dot{x} \\ \dot{y} \\ \dot{z} \end{pmatrix} \quad \text{with} \quad (2.26)$$

$$(\gamma_{ik}) = \begin{pmatrix} -\sin\alpha & -\cos\alpha\sin\delta & \cos\alpha\cos\delta \\ \cos\alpha & -\sin\alpha\sin\delta & \sin\alpha\cos\delta \\ 0 & \cos\delta & \sin\delta \end{pmatrix}. \quad (2.27)$$

Solar Motion and Local Standard of Rest: The choice of the sun as origin of the coordinate system concedes arbitrarily to this star a preferred position. In order to avoid this, one defines a local standard of rest in which the mean values of the three velocity components of a representative selection of stars in a certain neighbourhood of the sun shall vanish. If we denote the velocity components of the stars in the local standard of rest by u, v, w, then (with a bar denoting the mean over all the selected stars)

$$\begin{aligned}
u &= \dot{x} - \bar{\dot{x}} & \bar{u} &= 0 & u_\odot &= -\bar{\dot{x}} \\
v &= \dot{y} - \bar{\dot{y}} \quad \text{and hence} \quad & \bar{v} &= 0 \quad \text{and} \quad & v_\odot &= -\bar{\dot{y}} \\
w &= \dot{z} - \bar{\dot{z}} & \bar{w} &= 0 & w_\odot &= -\bar{\dot{z}}
\end{aligned} \qquad (2.28)$$

since $(\dot{x}_\odot, \dot{y}_\odot, \dot{z}_\odot) = (0,0,0)$. The mean velocity point $(\bar{\dot{x}}, \bar{\dot{y}}, \bar{\dot{z}})$ is taken as the centroid of the motions of the chosen collection of stars. The vector (u, v, w) is called the *Peculiar Velocity* of the star. The peculiar velocity of the sun $(u_\odot, v_\odot, w_\odot)$ is also called the local solar motion. The destination direction of the sun's motion with the equatorial coordinates $A_\odot$ and $D_\odot$ is called the *Apex*, the opposite point the *Antapex*. The relation between $u_\odot$, $v_\odot$, $w_\odot$ and the apex coordinates and the size of the solar motion $S_\odot = \sqrt{u_\odot^2 + v_\odot^2 + w_\odot^2}$ follows (Fig. 2.10)

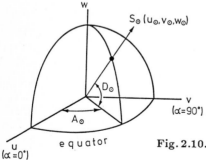

Fig. 2.10. Derivation of the apex of the sun's motion

$$u_\odot = S_\odot \cos D_\odot \cos A_\odot$$
$$v_\odot = S_\odot \cos D_\odot \sin A_\odot$$
$$w_\odot = S_\odot \sin D_\odot \; . \tag{2.29}$$

We shall go into the determination of the apex $(A_\odot, D_\odot)$ and of the speed $S_\odot$ in Sect. 3.2.3.

In more recent investigations one usually passes to a *galactically oriented coordinate system*. In place of $\dot{x}$, $\dot{y}$, $\dot{z}$ one then deals with $\dot{\xi}$, $\dot{\eta}$, $\dot{\zeta}$, where

$\dot{\xi}=$ component in direction $l = 0°$, $b = 0°$ (galactic centre)
$\dot{\eta}=$ component in direction $l = 90°$, $b = 0°$ (direction of galactic rotation)
$\dot{\zeta}=$ component in direction $b = 90°$ (galactic north pole).

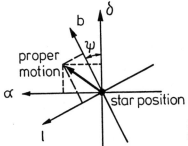

Fig. 2.11. Calculation of the proper motion components μ_l, μ_b in galactic coordinates

To derive $\dot{\xi}$, $\dot{\eta}$, $\dot{\zeta}$ one first calculates the proper motion components in the galactic coordinates (l, b) : $\mu_l'' = \dot{l} \cos b$ and μ_b''. From Fig. 2.11 one finds that

$$\mu_l'' = \mu_\alpha'' \sin \psi + \mu_\delta'' \cos \psi$$
$$\mu_b'' = -\mu_\alpha'' \cos \psi + \mu_\delta'' \sin \psi \; . \tag{2.30}$$

The "parallactic angle" ψ introduced here follows from the equations

$$-\cos b \cos \psi = \cos \delta_0 \sin \delta \cos(\alpha_0 - \alpha) - \sin \delta_0 \sin \delta$$
$$-\cos b \sin \psi = \cos \delta_0 \sin(\alpha_0 - \alpha) \; . \tag{2.31}$$

(Comparison of Fig. 2.1 with Fig. A.1, where this angle is denoted by β, and application of the formulae of spherical trigonometry.) Here α_0, δ_0, are the equatorial coordinates of the galactic north pole; ψ lies always in the same quadrant as $\alpha - \alpha_0$.

The transformation from $t_l = Kr\mu_l''$, $t_b = Kr\mu_b''$ and v_r to $\dot{\xi}$, $\dot{\eta}$, $\dot{\zeta}$ follows analogously to (2.26) with (2.27). One has to replace t_α, t_δ there by t_l, t_b and also α and δ by l and b, respectively. The components of the peculiar velocity in the galactic coordinate system will be denoted by U, V, W; the equations analogous to (2.28) are valid.

2.1.4 The Fundamental Astronomical Coordinate System

For the study of the motions of celestial bodies one needs a reference system which can be treated as being at rest. The problem of absolute motion, and with it the question of an absolute standard of rest, has been the object of various thought experiments and practical investigations. The currently used Fundamental Astronomical Coordinate System was defined on the basis of observations extending over more than 200 years, so that motion events could be specified with greater accuracy according to the laws of Newtonian mechanics[2]. It was therefore a matter of approximating to an inertial system, in which the law of inertia holds. The time parameter, treated as an independent variable, the inertial time, is called in the astronomical almanacs the *dynamical time* (prior to 1984: ephemeris time).

An essential basis is formed by the positions of selected stars measured with the greatest possible accuracy. Since even the stars move relative to an inertial system, a catalogue of star positions at a certain point in time, without statements concerning their proper motions, can always only form a reference system for this point of time (epoch). A coordinate system that is valid for a longer period of time is possible only after gaining knowledge of the star motions.

Realisation of an Astronomical Coordinate System at an Epoch: The practical specification of a reference system, which is defined through the equatorial plane and the null direction lying in it and through the vernal equinox at a certain point of time, consists in direct measurements of the spherical coordinates α and δ of a number of celestial bodies relative to this system, which we can, without serious limitation, think of as carried out at the same time. For these so-called absolute determinations, it suffices to have the meridian circle, a telescope with eye-piece micrometer which is rotatable only about an east-west axis, and whose optical axis therefore always lies in the meridian plane. Graduated circles attached to the telescope enable one, after fixing the zenith direction with a mercury horizon, to read the (apparent) zenith distance of a culminating star. By using a clock adjusted to sidereal time one can also determine the time of meridian passage of the star.

Figure 2.12a illustrates the principle of the *measurement of absolute declination*. The measured ("apparent") zenith distance must first of all be freed from the influence of refraction in the earth's atmosphere, of which no more will be said here. From the reduced ("true") zenith distance of the star z in the meridian, the declination is given by

$$\delta = \begin{cases} \varphi - z & \text{for} \quad \delta < \varphi \quad \text{(star south of the zenith)} \\ \varphi + z & \text{for} \quad \delta > \varphi \quad \text{(star north of the zenith)} \end{cases} .$$

---

[2] The deviations from Newtonian mechanics to be expected from Relativity Theory, in practice always very small, are noticeable only in the motions of quite a few objects and are there taken into consideration.

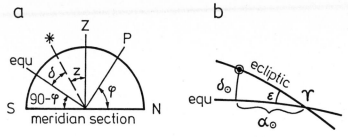

Fig. 2.12. Determination of absolute declination (**a**) and absolute right ascension (**b**)

Here φ is the instantaneous polar altitude, which is determined by measurements of the zenith distances of circumpolar stars close to the pole at upper and lower culmination z_u and z_l, respectively. We have $\varphi = 90° - \frac{1}{2}(z_l - z_u)$.

The *measurement of absolute right ascension* can be equated to the determination of the sidereal time interval between the meridian transit of the vernal point ($\alpha = 0$) and the meridian transit of the star under consideration. For the determination of the vernal point, the ascending node of the solar path over the equator, observations of the sun are obviously necessary. The observation of the meridian transit of the vernal point can be referred back to the observation of the meridian transit of the sun (mid-point of the disc): through absolute measurements of the declination of the sun, the right ascension of the sun $\alpha_\odot$ can be obtained, whereupon already the distance of *one* observable object from the vernal point is given. Observations during the solstices, the sun's turning points, give the maximal values of $|\delta_\odot|$, which are equal to the obliquity of the ecliptic ε. Accordingly, from observations of the declination of the sun in the neighbourhood of the equinoxes (Fig. 2.12b), the sun's right ascension can now be calculated, since $\sin \alpha_\odot = \tan \delta_\odot / \tan \varepsilon$. The right ascension of the star α_* now follows from the difference between the sidereal time readings U_* and $U_\odot$ for the meridian transits of the star and the sun: $\alpha_* - \alpha_\odot = U_* - U_\odot$.

Derivation of a Fundamental System of Absolute Positions and Proper Motions: Absolute determinations of α and δ are available for a rather large collection of stars and for rather widely differing epochs — reaching back in part for about a century and of course with lower accuracy then than the more recent measurements. For stellar astronomy the significance of these observations lies chiefly in that their comparison can furnish changes in star positions over time as a result of the motion of the stars through space. For this, however, it is necessary to transform all the coordinate statements to the same reference system, and hence to a Normal Equinox. These transformations of the coordinate systems at the epoch of the observations to a different point in time represent the real problem of Fundamental Astrometry, which proves to be very complicated.

If the stars were ideal fixed points on the sphere, then absolute observations of α and δ for a sufficiently large number of epochs t_1, t_2, ... would enable one to determine all the constants of their behaviour from the known secular (proportional to time) and periodic variations of α and δ through precession, nutation, aberration and parallax. Because of the proper motions, however, the positions of the individual stars themselves vary in an unpredictable way. The problem of Fundamental Astrometry therefore, is not soluble without some assumptions. The method followed up to now postulates Newtonian mechanics for the calculation of the influence of planetary precession on the position of the vernal equinox: the calculation of the orbits of the planets and of the moon with allowance for the mutual disturbances provides the coordinates of these bodies in relation to an inertial system. This gives in particular the desired positional variation of the orbit of the earth, and so of the ecliptic, relative to an inertial system as a result of planetary precession. Position determinations of the planets and of the moon can now lead to the specification of this inertial system.

The constant of lunisolar precession and the constants of nutation are however obtained from star observations (because until now a calculation of these constants from the mass distribution in the interior of the earth was still not possible with the desired accuracy). Accordingly, assumptions on the statistical behaviour of the proper motions of the stars are essential. The spatial motions of the stars have proved to be not fully random, containing as they do a certain systematic fraction (Sect. 3.4.2).

The constant of aberration is always calculated from the velocity of light and earth's orbital velocity (on the basis of the size of the Astronomical Unit).

The approximation to an inertial system obtained in this way is called the Fundamental Astronomical Coordinate System. For its practical specification it is essential to have absolute measurements of the positions (α, δ) of a rather large number of stars distributed over the whole sphere, as well as the planets and the moon, at a rather large number of epochs $t_1, t_2, \ldots$ (in practice at least several decades). These provide the following results:

1. The constants of the transformation equations for passing over from absolute positions (α, δ), related to the true equinox at the observation epoch, to a coordinate system at rest, centred at the centre of gravity of the solar system, with a specified position of the equator and of the vernal point, or in short: for transformation of the positions to a fixed equinox t_0.

2. A so-called fundamental system of absolute positions and absolute proper motions of the observed stars for the equinox t_0.

These data are compiled in the so-called Fundamental Catalogues.

The first Fundamental Catalogue was arranged by F.W. Bessel and published in 1818 under the title Fundamenta Astronomiae. There followed other star catalogues of this type, improved correspondingly to the current state of the observational material. Since 1964 the volume in current use is the Fourth Fundamental Catalogue (usual abbreviation: FK 4) produced by the Astronomisches Rechen-Institut in Heidelberg by Fricke et al. (1963). Its successor, FK 5,

No.	Name	Mag.	Sp.	α	$\dfrac{d\alpha}{dT}$	$\dfrac{1}{2}\dfrac{d^2\alpha}{dT^2}$	μ	$\dfrac{d\mu}{dT}$	Ep. (α)	$m(\alpha)$	$m(\mu)$
697	θ CrA	4.69	G 5	18^h 29^m 55^s.960 / 18 31 43.030	+ 428^s.365 / + 428.194	− 0^s.331 / − 0.354	+ 0^s.277 / + 0.278	+0^s.002 / +0.002	18.13	5.9	27
700	Grb 2655 Dra	5.84	K o	18 32 10.421 / 18 30 57.735	− 290.105 / − 291.379	− 2.594 / − 2.500	− 0.182 / − 0.183	−0.002 / −0.002	21.90	6.8	31
1483	Grb 2603 Lyr	6.66	A o	18 32 22.814 / 18 33 5.187	+ 169.480 / + 169.506	+ 0.053 / + 0.051	− 0.002 / − 0.002	−0.002 / −0.002	32.53	3.0	15
1482	α Sct	4.06	K o	18 32 29.108 / 18 33 50.719	+ 326.449 / + 326.439	− 0.016 / − 0.021	− 0.130 / − 0.125	+0.020 / +0.020	14.60	1.4	7
1484	+9°3783 Oph	5.40	F 2	18 34 4.632 / 18 35 16.170	+ 286.142 / + 286.163	+ 0.041 / + 0.039	− 0.062 / − 0.060	+0.008 / +0.008	31.05	2.9	17
1485	83 G. Sgr	5.80	A 5	18 34 54.772 / 18 36 24.563	+ 359.196 / + 359.129	− 0.129 / − 0.138	+ 0.003 / + 0.004	+0.005 / +0.005	17.19	3.1	14
699	α Lyr	0.14	A o	18 35 14.655 / 18 36 5.436	+ 203.114 / + 203.137	+ 0.047 / + 0.046	+ 1.708 / + 1.701	−0.028 / −0.028	07.55	1.3	5
701	Grb 2640 Dra	6.00	A 3	18 36 3.843 / 18 36 8.495	+ 18.714 / + 18.501	− 0.426 / − 0.428	+ 0.210 / + 0.203	−0.030 / −0.030	23.24	3.7	18
698	ζ Pav	4.10	K o	18 37 12.297 / 18 40 7.381	+ 700.992 / + 699.663	− 2.560 / − 2.756	+ 0.158 / + 0.183	+0.099 / +0.098	07.23	9.3	46
1486	δ Sct	4.74 var.	F o	18 39 32.109 / 18 40 54.220	+ 328.460 / + 328.425	− 0.068 / − 0.073	+ 0.044 / + 0.044	0.000 / 0.000	18.76	1.5	8
702	ε Sct	5.09	G 5	18 40 47.819 / 18 42 9.492	+ 326.710 / + 326.675	− 0.068 / − 0.073	+ 0.137 / + 0.137	−0.001 / −0.001	21.66	1.7	10

EQUINOX AND EPOCH 1950.0 AND 1975.0

No.	δ	$\dfrac{d\delta}{dT}$	$\dfrac{1}{2}\dfrac{d^2\delta}{dT^2}$	μ'	$\dfrac{d\mu'}{dT}$	Ep. (δ)	$m(\delta)$	$m(\mu')$	GC	N30
697	−42°21' 2".24 / −42 19 55.57	+ 258".96 / + 274.41	+30".92 / +30.86	− 2".06 / − 2.05	+0".04 / +0.04	16.68	6.3	27	25313	4115
700	+77 30 34.35 / +77 31 43.16	+ 280.51 / + 269.98	−21.01 / −21.11	+ 0.07 / + 0.06	−0.03 / −0.03	17.14	3.2	13	25372	4124
1483	+46 10 44.06 / +46 11 55.75	+ 283.70 / + 289.78	+12.17 / +12.16	+ 1.47 / + 1.47	0.00 / 0.00	25.43	3.7	18	25379	4126
1482	− 8 16 50.48 / − 8 15 46.03	+ 251.95 / + 263.68	+23.48 / +23.46	− 31.19 / − 31.19	−0.02 / −0.02	13.93	2.2	12	25385	4129
1484	+ 9 4 53.84 / + 9 6 6.13	+ 284.04 / + 294.31	+20.56 / +20.54	− 12.88 / − 12.88	−0.01 / −0.01	28.99	4.7	23	25422	4133
1485	−21 26 27.08 / −21 25 11.13	+ 297.36 / + 310.26	+25.81 / +25.77	− 6.78 / − 6.78	0.00 / 0.00	19.86	5.0	21	25450	4136
699	+38 44 9.68 / +38 45 34.47	+ 335.48 / + 342.82	+14.68 / +14.67	+ 28.47 / + 28.53	+0.24 / +0.24	00.46	1.8	5	25466	4138
701	+65 26 37.71 / +65 27 58.37	+ 322.32 / + 322.96	+ 1.30 / + 1.28	+ 8.23 / + 8.24	+0.03 / +0.03	16.53	3.0	15	25491	4141
698	−71 28 28.15 / −71 27 7.96	+ 308.18 / + 333.30	+50.36 / +50.14	− 15.76 / − 15.75	+0.02 / +0.03	12.15	4.7	22	25522	4143
1486	− 9 6 7.26 / − 9 4 39.77	+ 344.09 / + 355.84	+23.51 / +23.48	+ 0.06 / + 0.06	+0.01 / +0.01	16.02	2.3	13	25580	4146
702	− 8 19 34.55 / − 8 18 4.18	+ 355.64 / + 367.31	+23.37 / +23.33	+ 0.74 / + 0.74	+0.02 / +0.02	16.11	2.9	14	25610	4152

Fig. 2.13. A reproduction from FK 4 [Fricke et al. (1963)]: the upper halves of two consecutive pages, the first of which deals with the right ascension, and the second with the declination. The stars are arranged in ascending order of right ascension in the usual way. The first column contains the number of the fundamental star, followed by the name, apparent visual magnitude and spectral type. α and δ are actually given for equinox and epoch 1950.0 and for 1975.0 (below); μ_α and μ_δ per century are denoted here by μ and μ', $d\alpha/dT$, $\frac{1}{2}d^2\alpha/dT^2$ and the corresponding derivatives for δ are the changes per century through precession and proper motion. Ep(α), Ep(δ) are the mean epochs of the underlying individual observations of α and δ: for example 30.21 means 1930.21. $m(\alpha)$, $m(\mu)$ and so on are the mean errors (standard deviations) of the individual quantities shown in brackets. The two last columns GC and N30 contain the numbers of the stars in Boss's (1936; 1937) "General Catalogue" and Morgan's (1952) "Catalogue of 5268 Standard Stars", respectively, whose positions refer to an epoch lying at about 1930

is in preparation. FK 4 contains 1535 stars with apparent magnitudes down to $m = 7$; equinox and epoch are 1950.0 and also 1975.0. The internal accuracy, for example around $\delta = +150°$, is for positions $\Delta\alpha \cos \delta = \pm0\overset{s}{.}008$, $\Delta\delta = \pm0\overset{''}{.}09$ and for proper motions (per century) $\Delta\mu_\alpha \cos \delta = \pm0\overset{s}{.}01$, $\Delta\mu_\delta = \pm0\overset{''}{.}05$. Two pages from FK 4 are reproduced in part in Fig. 2.13.

For a much larger number of stars the positions were determined by differential measurements against fundamental stars. A Fundamental Catalogue arising in this way is the General Catalogue (GC) by Boss (1936; 1937) with 33 342 stars down to $m = 9$. Of the results of similar undertakings, using differential measurements on photographic plates in certain zones of declination, the so-called Zone Catalogues of the Astronomische Gesellschaft AGK 1 (Epoch 1880), AGK 2 (1930) and AGK 3 (1975) should be mentioned. These include over 180 000 stars down to $m = 9$, and partially to $m = 13$, for which moreover absolute proper motions could be derived.

Another extensive work is the star catalogue of the Smithsonian Astrophysical Observatory, known for short as the "SAO Catalog" (1966). It achieves a combination of various earlier catalogues. In four volumes it gives positions and proper motions for 258 997 stars in the system of FK 4 (down to about the limit of visual brightness $V = 10$); V is given for 99% of these stars and also spectral type for 80%.

A compilation of catalogues of positions and proper motions is contained in Voigt (1965) and Schaifers and Voigt (1982).

2.1.5 Extra-Galactically Based Reference System, Radioastrometry

The proper motions of the stars had already at an early stage suggested the idea of using very far distant extra-galactic stellar systems as representing an absolute coordinate system. Already Herschel (1785) and Laplace (1797) saw in the "nebulae" the appropriate reference objects for that purpose.

The brighter galaxies show a diffuse structure and only very faint objects are sufficiently pointlike to provide a satisfactorily precise measure of position. It is only in the last decade – at Lick Observatory and at the Observatory of Pulkowo – that measurement on photographic plates of proper motion relative to galaxies has been successful. Absolute proper motions have thereby been achieved directly. The limiting magnitude in these programmes amounted to only about $m_B = 17$, so that not sufficient starlike objects can be included. It was therefore not yet possible by this method to determine a system of positions of galaxies with the high accuracy today achieved by the traditional methods described in Sect. 2.1.4.

This goal is on the other hand already attainable for compact extragalactic radio sources: the modern technique of radiointerferometry makes possible the measurement of absolute positions in relation to the celestial equator with extraordinarily great precision. Since the programme of measurement can include such optical stars as are simultaneously radio emitters, the resulting extragalactic reference system can be related to the Fundamental Coordinate System.

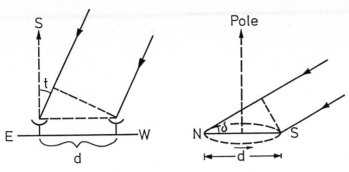

Fig. 2.14. *Left:* Interferometric observation of an object with $\delta = 0°$. *Right:* The baseline d rotates daily in a plane perpendicular to the polar axis. When $\delta \neq 0$ the effective baseline is therefore $d \cos \delta$; if the object is at the pole, the phase difference is zero

The principle of the measurement of δ and α may be explained with an interferometer consisting of two elements, whose baseline of length d lies exactly in the east-west direction (Fig. 2.14). Between the wave trains arriving at the two antennae there exists a phase difference Φ, which obviously varies with the daily motion of the source, and thus with its hour angle t, and furthermore depends on the declination δ of the source. We have

$$\Phi(t) = \frac{2\pi}{\lambda} d \cos \delta \sin t$$

where λ denotes the wavelength. The measured values show a sine-function dependence on t, the amplitude of which gives the declination. If the right ascension of only one radiosource is known, this enables the right ascension of the other sources to be determined from the hour angle. In the first successful measurements of this type (1972) with the 5 km basis length interferometer at Cambridge, England, the object was the well known variable β Persei = Algol recorded in FK 4, which is also a radio emitter.

Further measurements were carried out with very long baseline interferometers, in which the individual antennae are no longer connected by cables, and in the extreme case not even by radio (Very Long Baseline Interferometry, often abbreviated to VLBI, above all with intercontinental baselines). The resultant accuracy of the individual position determination of extragalactic sources is about $\pm 0''\!.05$ and so already surpasses the classical absolute measurements with optical instruments − this number must not be directly compared with the accuracy given in Sect. 2.1.4 for the positions of FK 4 stars, which rest on observations over a long period of time! Comparisons made up to now between the classical Fundamental Coordinate System and the radioastrometric reference system have given good agreement (to about $\pm 0''\!.1$) and allow one to expect that the FK 4 system already coincides very closely with an extragalactic inertial system.

2.2 Distances

The methods for deriving the distances of stars of the Milky Way system can essentially be classified as geometric-kinematic or photometric. The photometric procedures assume that one already knows the distances of a sufficient number of stars of each type that one proposes to study (secondary methods). Most of the geometric-kinematic procedures, on the other hand, are primary in character, since they manage without the knowledge of distances of other stars, and fundamental basic quantities (for example, the velocity of the sun) can themselves be obtained from geometric-kinematic investigations.

2.2.1 Primary Methods

Trigonometric Parallax: The annual parallactic motion, explained in Sect. 2.1.2 and described in the equatorial system by (2.15), can be observed for the nearest stars. By measurement of the parallax p the distance r [pc] $= 1/p''$ can be derived.

Even for the nearest star, Proxima Centauri, the parallax is only $0''.76$, so that for nearby stars the distances can only be obtained with errors of a few per cent if the amplitude of the annual movement can be measured with a systematic accuracy of about $\pm 0''.01$. Whilst not possible by absolute determinations of α and δ, this can be achieved by measurements of position relative to a number of apparently neighbouring, but very much more distant, stars which show no measurable parallactic variation, in other words faint stars with particularly small proper motions.

The observational material for the derivation of these "trigonometric" parallaxes consists of photographic plates, taken at appropriate intervals throughout the year and generally obtained using long focal length telescopes. The average error of good determinations of the actual position of the parallactic ellipse relative to the mean of the reference stars (Fig. 2.15) is about $\pm 0''.01$,

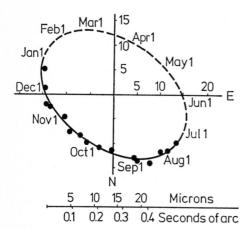

Fig. 2.15. Observational results for the parallactic motion of the M-dwarf Ross 248 which is near the sun. The points represent the mean values of numerous individual measurements which were carried out in the years 1937–1946 at the same observatory. Result for the parallax determination: $p = 0''.318$. [After Van de Kamp (1956)]

and can be reduced still further by enhancing the number of plates. If one requires that the uncertainty of the trigonometric parallax be less than 20%, then one is restricted to values of $p \gtrsim 0\rlap{.}{''}05$ and the region of space covered is given by $r \lesssim 20\,\text{pc}$.

Most of the results are compiled in the "General Catalogue of Trigonometric Stellar Parallaxes" edited by Jenkins (1952) and its associated supplements: data for 5822 stars in the 1952 edition and for about 600 additional stars in the 1963 supplement. Only for about a third of the listed stars are the parallax values determined with acceptable accuracy. Of the trigonometric parallaxes now known for about 7000 stars only about 5% of the errors are shown as less than 10%, which implies values of $p > 0\rlap{.}{''}05$.

Moving Cluster Parallaxes: By a moving cluster or moving group one understands a group of stars whose velocities through space relative to the sun are approximately parallel and of equal magnitude. Examples are open star clusters, as perhaps the Hyades and the Praesepe. In studying the data on proper motions and radial velocities one also notices scattered but neighbouring groups of stars with uniform motion, whose appearance on the sphere would not otherwise suggest a close relationship. The noteworthy examples are the Taurus group, to which the Hyades belong (Fig. 2.16), the Ursa Major group including the so-called Sirius group and the Scorpio-Centaurus group, members of which appear scattered on a region of the sphere of about $30° \times 60°$. These moving groups make possible, in the following manner, the determination of the distances of their members and, so far as this is meaningful, of their mean distance.

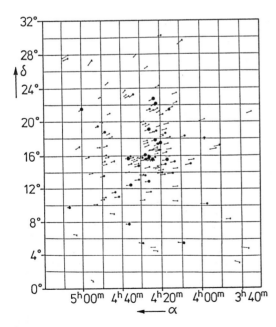

Fig. 2.16. The Taurus moving group with the Hyades. The individual motions of the stars are indicated by darts. [After Van Bueren (1952)]

49

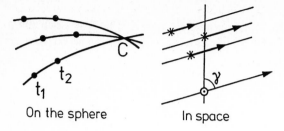

On the sphere   In space

Fig. 2.17. Determination of the Convergence Point of a moving cluster

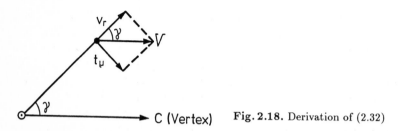

Fig. 2.18. Derivation of (2.32)

    If one plots the positions of each member of the group at two different moments of time t_1, t_2, and the great circles which are determined by the directions (position angles) of the proper motions, then these circles will intersect approximately at a point of the sphere, the Convergence Point (C in Fig. 2.17, left-hand side). If one has obtained the coordinates of this point then the destination point or vertex of the star motions is obviously known (Fig. 2.17, righthand side) and one can calculate the angles between this direction and the line of sight to each individual member of the group. If the radial velocities of these stars, v_r, are also known, then one can calculate their tangential velocities, $t_\mu = v_r \tan \gamma$ (Fig. 2.18). On the other hand these are also given by $t_\mu = K r \mu''$ [see Sect. 2.1.3, (2.24)]. From γ, v_r and the value of the individual proper motion μ'' of a member of the group one can therefore derive its distance:

$$r = \frac{v_r \tan \gamma}{K \mu''} \ . \tag{2.32}$$

    For a satisfactory accuracy of determination of the Convergence Point, and hence of the individual angles γ, it is obviously important that the members of the moving-cluster be not concentrated on only a small area of the sphere. Favourable cases are therefore represented by the already mentioned vastly extensive moving groups. The number of stars involved should naturally be as large as possible. If all the members of the group have practically the same space velocity, one needs in principle that the radial velocity v_r of only one of these stars be known in addition: v_r, γ and the position angle of the proper motion determine the space velocity vector V!

Table 2.1. Galactic coordinates of the Convergence Point l, b and the space velocity V (relative to the solar system) together with the mean distance r and extent D for three important moving groups. N is an estimate of the number of members of the group

Name	Convergence point		V	r	D	N
	l	b	[km s^{-1}]	[pc]	[pc]	
Taurus group (Hyades)	208°	+ 4°	43	46	20	170
Ursa Major group	9	−28	15	100	150	150
Scorpio-Centaurus group	210	−25	23	170	80 × 200	60

For a distance $r = 100\,\mathrm{pc}$ and a tangential velocity $t_\mu = 10\,\mathrm{km\,s^{-1}}$ one obtains a proper motion of magnitude $\mu'' = t_\mu/4.74\,r \approx 0.''02$ per year. The domain of the Convergence Point method therefore covers a few hundred pc. Results for the Convergence Points, space velocities and mean distances for the three moving groups already mentioned are given in Table 2.1.

The values for the mean distances of the U Ma- and Sco-Cen groups serve only as crude orders of magnitude; the value for the Tau group applies to the centre of the Hyades group and indicates a mean error of only about ±4 pc.

The *Distance of the Hyades* possesses the significance of a "fundamental" measure of cosmic distances, for this quantity forms the most important connecting link between the geometric procedures for determining distances and the essentially longer range photometric procedures described below under Sect. 2.2.2. Besides the moving cluster parallax method other ways of determining the distance of the Hyades have also been followed: determinations of the trigonometric parallaxes of individual members of the group, derivation of the dynamical parallaxes of binary stars of the group (see below) and the fitting of the Hyades main sequence to the theoretical Zero Age Main Sequence obtained from the computations of stellar models. More recent results show agreement within the limits of error and give for the group centre the mean value (see References)

$$r = 46 \pm 2\,\mathrm{pc}\ .$$

Statistical Parallaxes: The moving cluster method is applicable only to a very limited extent, especially only those star types are covered which are found in moving groups. A more general method for determining distance is one using the parallactic motion of the stars as a result of the motion of the sun through space. As with the trigonometric parallax the major semi-axis of the earth's orbit serves as a baseline, in this case we use the path of the sun through space during a specified time interval. Because of the superposition of the peculiar velocities of individual stars, however, one can in this way derive only the *mean* parallax of a group of stars, and moreover certain assumptions concerning the velocity distribution of these stars are necessary. The method is nevertheless very useful, for example in the determination of the absolute magnitudes of certain star types (RR Lyrae Stars, Cepheids, sub-dwarfs, etc.).

51

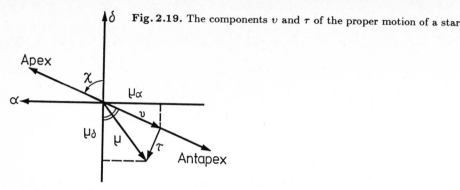

Fig. 2.19. The components v and τ of the proper motion of a star

The starting point is the resolution of the proper motion of each selected star into two components (Fig. 2.19): one component v (Greek upsilon), which lies in the great circle through the star, the apex and the antapex of the sun's motion, and one perpendicular component τ (Greek tau). The component v consists of the whole parallactic motion of the star together with the component of its peculiar motion along that great circle; τ is just the component of the peculiar motion of the star perpendicular to the parallactic motion. If χ denotes the position angle of the direction towards the apex, and μ_α, μ_δ stand for $\mu''_\alpha = 15\,\mu^s_\alpha$ cos δ, μ''_δ, so that v and τ are expressed in seconds of arc per year, then we obtain

$$v = \mu_\alpha \sin \chi + \mu_\delta \cos \chi$$
$$\tau = -\mu_\alpha \cos \chi + \mu_\delta \sin \chi \; . \tag{2.33}$$

Figure 2.19 shows an example in which the resultant motion has the same sign as the parallactic motion directed towards the antapex: i.e. $\mu_\alpha < 0$, $\mu_\delta < 0$ and hence $v < 0$. For the calculation of χ and of the angle between the star and the apex η (see Fig. 2.20) one obtains the system of equations

$$\sin \eta \cos \chi \; = \cos D_\odot \sin \delta \cos(\alpha - A_\odot) - \sin D_\odot \cos \delta$$
$$\sin \eta \sin \chi \; = \cos D_\odot \sin(\alpha - A_\odot)$$
$$\cos \eta \qquad = \cos D_\odot \cos \delta \cos(\alpha - A_\odot) + \sin D_\odot \sin \delta \; . \tag{2.34}$$

Here (α, δ) and $(A_\odot, D_\odot)$ denote the equatorial coordinates of the star and of the solar apex. The proper motion components v and τ correspond to the tangential velocity components $[\mathrm{km\,s}^{-1}]$

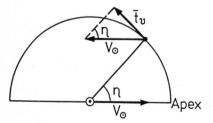

Fig. 2.20. The concept of the secular parallax

$$t_v = Krv \ ,$$

$$t_\tau = Kr\tau \quad (K = 4.738) \ . \tag{2.35}$$

We treat first the case of a group of stars which is distributed over a *small region of the sphere*. Solving (2.35) for v and τ and forming the mean values $\overline{v}$ and $\overline{|\tau|}$ (since $\overline{\tau} = 0$!) yields, on the understanding that parallax $p = 1/r$ and that the tangential velocities t_v and t_τ are statistically independent:

$$\overline{v} = \frac{1}{K}\overline{p} \cdot \overline{t_v} \tag{2.36}$$

$$\overline{|\tau|} = \frac{1}{K}\overline{p} \cdot \overline{|t_\tau|} \ . \tag{2.37}$$

Both equations allow the derivation of the *mean parallax* $\overline{p}$ of the star group. In the application of (2.36) the sun's velocity $S_\odot$ and the coordinates of the apex must be known. Then, under the assumption that the peculiar parts in $\overline{t_v}$ cancel out,

$$\overline{t_v} = \pm S_\odot \sin \eta \quad \text{(sign as in } \overline{v}) \ , \tag{2.38}$$

where η is the angle between the centre of the region under consideration and the apex, to be calculated from (2.34) (Fig. 2.20). The mean value $\overline{p}$ to be obtained in this way from (2.36) will be called the "*secular parallax*"; the same term is of course also used for the annual parallactic shift of a star with $\eta = 90°$ to the sun's motion: $p_s = (1/K)pS_\odot$.

Equation (2.37) enables one to deduce $\overline{p}$, provided that $\overline{|t_\tau|}$ can be determined. This can be done with the help of the radial velocity data on the assumption that the velocity distribution is invariant in space. Then $\overline{|t_\tau|}$ can be equated to the mean magnitude of the radial velocities $\overline{|v_r|}$ of stars in a small field at $90°$ angle from the region under consideration, and therefore at the pole of the half great circle depicted in Fig. 2.20. With isotropic velocity distribution, of course, $\overline{|t_\tau|}$ can be set equal to the mean value $\overline{|v_r|}$ for stars of the region under consideration itself.

In the case of a large region of the sphere, (2.35) with (2.38) and (2.37) can be extended, provided that the mean parallax of the stars under consideration is the same in all parts of the region. For star types which belong to the galactic disc, and whose spatial density distribution in a plane parallel to the plane of symmetry of the galaxy is approximately constant, one can assume that the above provision is met in a sufficiently narrow latitudinal zone $(b, b + \Delta b)$ of the galaxy. For each selected star therein one can then write

$$v = \pm \left(\frac{1}{K}S_\odot \sin \eta\right)p + \Delta v \ . \tag{2.39}$$

Here Δv represents the contribution of the peculiar motion. The coefficient of p varies with the position of the star, and will be treated as known. If the Δv have the character of a random error, then treatment by the method of least squares leads to the result:

$$\bar{p} = \frac{K}{S_\odot} \frac{\Sigma v \sin \eta}{\Sigma \sin^2 \eta} \, . \tag{2.40}$$

When using (2.37) we again need to know a suitable value for $\overline{|t_\tau|}$.

The procedure described takes it for granted that the velocity distribution of the stars under consideration is independent of the distance r. For r greater than about 200 pc this assumption can no longer be sustained, since then the differential rotation of the galaxy (Sect. 3.4.2) causes noticeable differences between the mean motions of the nearer and the more distant stars, which depend on the galactic coordinates l, b. For star groups at these great distances one must therefore first, before deriving v and τ from (2.33), subtract from the observed values of μ_α and μ_δ the contributions from the systematic differential rotation [Sect. 3.4.1, (3.65 and 71)], then invert the system (2.30).

It should be noticed that the mean parallax $\bar{p}$ of a group of stars, obtained as the mean value of the reciprocal distances, is in general not the same as the reciprocal of the mean distance $\bar{r}$: using $F(r)$ to denote the distribution of the distances of the stars we find that

$$\bar{p} = \overline{\left(\frac{1}{r}\right)} = \int \frac{1}{r} F(r) dr \neq \frac{1}{\bar{r}} = \frac{1}{\int r F(r) dr} \, .$$

To obtain $\bar{r}$ from $\bar{p}$ requires knowledge of the distribution $F(r)$ or of $f(p) = F(1/p)$.

Let us consider groups of stars of the rotating galactic disc which are so distant that the effect of the differential rotation on radial velocities v_r can be obtained with acceptable accuracy from the average of the measured v_r, in spite of this accuracy being affected by peculiar velocity contributions. By taking into consideration the latitude dependence, one has for small fields on the sphere approximately [Sect. 3.4.1, (3.64) and p. 136]

$$\bar{v}_r = A\bar{r} \sin 2\bar{l} \cos^2 \bar{b} \, .$$

We thus have a geometric-kinematical procedure which leads to a mean distance — it has of course been named "Rotation Parallax". For r greater than about 2 kpc one has to fall back on the general relation (3.59). Knowledge of the rotational constant A, and for $r>2$ kpc of the deviations from the formula (3.64), form the basis of the method. The advantages lie in the great range and in the freedom from the influence of interstellar extinction. By this method in association with photometric data the absolute magnitudes of certain types of variables can be derived.

Determination of Distances of Visual Binaries: Here the most favourable but rare case occurs with a system of large orbital major semi-axis for which one can determine not only the angular size but also, through spectroscopic measurements of the radial velocities, the linear size. For a circular orbit one simply has $2\pi a = vP$ with $v = $ orbital velocity and $P = $ period. Then the distance as well as the parallax of the system are given by the clear relation

$$\frac{1}{r[\text{pc}]} = p'' = \frac{a''}{a[\text{AU}]} \; . \tag{2.41}$$

If only a'' is known, one calculates $a[\text{AU}]$ from Kepler's Third Law and obtains

$$p'' = \frac{a''}{[P^2(\mathcal{M}_1 + \mathcal{M}_2)]^{1/3}} \; , \tag{2.42}$$

where the period P is expressed in years and the sum of the masses of both components $\mathcal{M}_1 + \mathcal{M}_2$ in solar masses. Since $\mathcal{M}_1 + \mathcal{M}_2$ appears under a cube root, comparatively good results can be obtained even with a gross approximation for these quantities. The value of p'' obtained from (2.42) will be called the "dynamical parallax". The relatively small spread of star masses about the value of the solar mass $(=1)$ prompts the approximate assumption $\mathcal{M}_1 + \mathcal{M}_2 = 2$ ("hypothetical parallax"). It is often possible, however, to make certain assertions concerning the spectral type and luminosity class of the components, so that with the aid of the mass-luminosity relationship one can estimate $\mathcal{M}_1$ and $\mathcal{M}_2$. The method then assumes a secondary character, in so far as the determination of the relationship between star type and luminosity depends upon distance determinations already made.

2.2.2 Secondary Methods

Photometric Determination of Distance: Between the apparent magnitude m, the absolute magnitude M, the interstellar extinction ("absorption") A of radiation on the way to the observer − all three quantities referring to the same spectral region − and the distance r of a star, there exists the relation

$$m - M = -5 + 5 \log r + A \; . \tag{2.43}$$

m and A (see below) can be measured by photometric means. If one knows M, then r is easily calculated. For normal stars this method has therefore acquired a very practical importance, since there are close relationships between M and several spectral criteria. Once these are known, close bounds can be put on M by means of comparatively easily determined characteristics of the stellar spectrum. The values obtained for r will be called *spectrophotometric distances,* although in the older literature the term "spectroscopic parallax" is often to be found (for $1/r$).

To derive of the interrelationship between the luminosity criteria and the absolute magnitude M, the "calibration" of the luminosity criteria, one uses those stars for which the distance has been determined either directly or indirectly by primary methods. Besides trigonometric parallaxes, above all the other methods, the distance determinations of open star clusters described below as well were important. Since all the members of a cluster are at practically the same distance it follows that, if one knows the distance parameter $m-M$ of a star cluster, then its colour-magnitude diagram (m vs. colour-index) provides immediately the absolute magnitude of every star type contained in the cluster. The accepted values for the M_V of the several MK-spectral types were to a large extent obtained in this way. To obtain the most precise determination possible of the absolute magnitude of a star type, one often refers these days to a continuous, quantitative spectral classification of the stars, on the basis of multi-colour photometry and photometric measurements of particular spectral characteristics, such as, for example, the Hβ-Index: $\beta = m_\beta(30\,\text{Å}) - m_\beta(150\,\text{Å})$, where $m_\beta(\Delta\lambda)$ is the magnitude measured through an interference filter centered on the hydrogen line Hβ with half-width $\Delta\lambda$. For $\Delta\lambda\sim30\,\text{Å}$ one records mainly radiation from the line, for $\Delta\lambda\sim150\,\text{Å}$ mainly from the continuum. Therefore β is a measure of the strength of the Hβ line.

For stars with periodically variable brightness (Cepheids, RR Lyrae stars, etc.) quantitative relationships between mean absolute magnitude and period of brightness variation have been established (Sect. 3.3.2); for other types of stars approximate values are known for the extreme or mean values of absolute magnitudes, e.g. for Novae (at maximum) which usually show a larger scatter. After the identification of a star with one of these types its absolute magnitude can therefore be given with more or less accuracy.

The *interstellar extinction* can be determined with the help of the empirical relationship $A_V = R\,E_{B-V}$, where $E_{B-V} = (B-V) - (B-V)_0$ denotes the colour excess and as a rule R is set at $R = 3$ [Sect. 4.1.1, especially (4.12ff.)]. $(B-V)$ is the observed colour index; $(B-V)_0$ is the colour index of the star under consideration in the absence of interstellar extinction, its so-called intrinsic colour. $(B-V)_0$ can be given immediately if the MK spectral type of the star is known; other ways of obtaining $(B-V)_0$ arise from the quantitative classification using multi-colour photometry.

The *range of photometric determination of distances* for single stars for a fixed limiting magnitude m^* depends strongly on the absolute magnitude of the star under consideration and the increasing effect of interstellar extinction with increasing distance. Near the galactic plane one can on average set $A_V(r) = 1.5 \times 10^{-3} r$ [mag] in the visual region of the spectrum, where r is expressed in pc. As a guide, Table 2.2 presents range values r_* calculated using this expression in (2.43).

The *accuracy* of photometric distance determinations depends on the errors in the absolute magnitude and in the interstellar extinction. From (2.43) one readily gets (for $\Delta m = 0$)

$$\frac{|\Delta r|}{r} \approx \tfrac{1}{2}(\Delta M + \Delta A) . \tag{2.44}$$

An error in the absolute magnitude of $\Delta M = \pm 0^{\text{m}}\!5$ introduces already an uncertainty in the distance of 25%!

56

Table 2.2. Distance range values r_* from the photometric method of distance determination for stars at low galactic latitude with various visual absolute magnitudes M_V. m_V^* is the visual apparent magnitude of the faintest observable stars (for the assumption concerning the interstellar extinction see text)

m_V^* M_V	$r_*[pc]$ 10	15	20
-5	2200	4500	7150
0	650	2200	4500
$+5$	95	650	2200
$+10$	10	95	650
$+15$	1	10	95

Distance Determination of Star Clusters: If one knows the distance of *one* star cluster, this enables one to determine the distances of other clusters of the same type (open or globular) with the help of their colour-magnitude diagrams (CMD). We limit ourselves in this section to open clusters; the application of the procedure to globular clusters will be discussed in Sect. 3.3.2.

First of all let us assume for simplicity that no interstellar extinction is present. For the cluster of known distance we can take the M_V-$(B-V)$ diagram as given, it having been constructed on the basis of measurements of apparent magnitudes $m_V = V$ and $m_B = B$. For another cluster of unknown distance only the m_V-$(B-V)$ diagram (CMD) is available. The distance of this second cluster is deduced from the vertical displacement of its CMD from the main sequence of the first, as shown in Fig. 2.21. The distance parameter of the second cluster $m_V - M_V$ is obtained simply from the difference between the two ordinate scales.

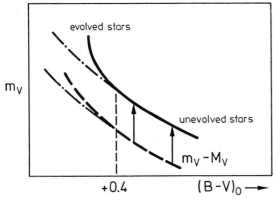

Fig. 2.21. The photometric distance determination of open clusters. The plain curve shows the main sequence of the cluster of known distance in the M_V-$(B-V)_0$ diagram. In the region $(B-V)_0 < 0.4$ this curve deviates from the main sequence of the unevolved stars ("Zero Age Main Sequence"), which is here shown chain-dotted

The deviation from the main sequence alters with the age of the star cluster. The lower portion, roughly $(B - V)_0 > 0^m 4$, corresponds to spectral types from F5 and usually remains unchanged, since these stars stay on the main sequence for at least a few 10^9 years.

When one has the CMD for both star clusters, then the vertical displacement needed to achieve coincidence of the lower portions of both main sequences gives the difference between the distance parameters of the clusters

$$\Delta(m_V - M_V) = (m_V - M_V)_2 - (m_V - M_V)_1$$
$$= 5 \log r_2 - 5 \log r_1 . \tag{2.45}$$

Example: Between the Praesepe and the Hyades one finds that $\Delta(m_V - M_V) = +3.0$. Since for the Hyades $(m_V - M_V)_1 = +3.3$, corresponding to $r_1 = 46$ pc, it follows that for the Praesepe $(m_V - M_V)_2 = +6.3$ or $r_2 = 180$ pc. In both cases one may set $A_V \approx 0$.

A difficulty arises if one wishes to determine the distances of young open clusters, in which very early spectral types occur, with the aim to calibrate the luminosity criteria shown by O- and B-stars. These clusters are relatively rare, and the nearest examples are already at such great distances that their main sequences below $(B - V)_0 \approx +0.4$ can no longer be defined with the accuracy desired ($r = 1$ pc implies that when $M_V = +5^m$ and $A_V = 1^m 5$ then $m_V = 16^m 5$!). A direct relation to the Hyades, in which no O- and B-stars occur, is not possible. One therefore advances step by step, using as the intermediate step the Pleiades, which extend down to $(B - V)_0 \approx -0.1$. The course of the Pleiades main sequence in the region $(B-V)_0 \approx +0^m 1 \ldots +0^m 4$, below the clearly marked turnoff point from the main sequence of unevolved stars, can now be used as the starting basis for the determination of distances of significantly younger star clusters with O-stars, like perhaps the double cluster h and χ Persei. The main sequence fitting procedure described simultaneously furnishes in an empirical way the main sequence of stars of zero age ("Zero Age Main Sequence", often abbreviated to ZAMS). The outcome has been tested and validated against the early types contained in the Scorpio-Centaurus moving group.

The interstellar extinction alters the CMD of a star cluster in that for each individual star the ordinate m_V undergoes a shift on account of the total A_V, and the abscissa $B - V$ a shift because of the colour excess E_{B-V}. Since in many cases all members of the cluster can be reckoned to undergo a uniform interstellar attenuation of the light it can be treated as steady shifts of the "whole diagram" in *both* coordinate directions, which can be easily corrected by the known value of E_{B-V} and hence also the known $A_V = 3 E_{B-V}$ (see above). After that the procedure described for distance determination may be applied. If in particular only the abscissae $(B - V)_0$ have been corrected in the CMD of the two star clusters to be compared, then the difference $(A_V)_2 - (A_V)_1$ has to be added in the right-hand side of (2.45).

Table 2.3. Standard main sequence (ZAMS) in the two colour diagrams of the UBV- and of the RGU-systems. [After Becker and Fenkart (1971)]

M_V	$B - V$	$U - B$	M_G	$G - R$	$U - G$
-3	-0.30	-1.12	-3	$+0.02$	$+0.08$
-2	-0.25	-0.91	-2	$+0.08$	$+0.28$
-1	-0.20	-0.69	-1	$+0.13$	$+0.51$
0	-0.14	-0.47	0	$+0.19$	$+0.75$
$+1$	-0.06	-0.21	$+1$	$+0.25$	$+1.04$
$+2$	-0.10	$+0.08$	$+2$	$+0.41$	$+1.33$
$+3$	$+0.31$	$+0.02$	$+3$	$+0.62$	$+1.33$
$+4$	$+0.47$	-0.02	$+4$	$+0.81$	$+1.27$
$+5$	$+0.63$	$+0.12$	$+5$	$+0.98$	$+1.39$

The determination of E_{B-V} is simple if the MK spectral types are known for a number of cluster stars (see above). In most cases, however, one has to proceed differently. An important method is based on a three-colour-photometry of the star cluster. The starting point is provided by the two-colour-diagrams (TCD). In the case of a UBV-photometry $U - B$ (as ordinate) is then plotted against $B - V$. If no interstellar extinction is present, the main sequence stars fall on a certain S-shaped curve. Table 2.3 displays the numerical relationship for the UBV- and for the RGU-system ($U - G$ replacing $U - B$, $G - R$ replacing $B - V$). Interstellar extinction brings about for each star a downward shift in the ordinate by the quantity $E_{U-B} = A_U - A_B$ and a shift to the right in the abscissa by $E_{B-V} = A_B - A_V$. The observed colour indices accordingly differ from the intrinsic values as follows:

$$U - B = (U - B)_0 + E_{U-B} \ , \quad B - V = (B - V)_0 + E_{B-V} \ . \tag{2.46}$$

The ratio of both colour excesses is however practically constant [see Sect. 4.1.1, (4.9f)]:

$$E_{U-B}/E_{B-V} = 0.72 + 0.05 \, E_{B-V} \ . \tag{2.47}$$

The interstellar extinction accordingly produces a parallel translation of all the field points in a constant direction along a straight "reddening line" (Fig. 2.22).

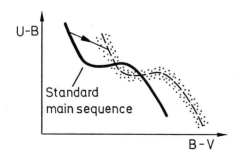

Fig. 2.22. The effect of uniform interstellar extinction on the distribution of main sequence stars of an open cluster in the two-colour-diagram (schematic). The *unbroken curve*, denoted Standard Main Sequence, describes the case of vanishing interstellar extinction

Then E_{U-B} and E_{B-V} can be obtained by shifting the observed main sequence back against the colour shift until coincidence is achieved with the standard main stream for unreddened stars.

Two further secondary methods for distance determination of lesser accuracy will now finally be briefly mentioned: (1) The angular diameter of a star cluster of a certain type, perhaps a sub-class of open clusters, adjudged by relative concentration, can serve for distance determination if the true diameter of the cluster type is already known for several examples. As a geometric method it has the advantage of being independent of interstellar extinction. (2) The strength of the interstellar absorption lines in the stellar spectrum (Sect. 4.1.4) can be used in low galactic latitudes as a rough measure of the distance of the star. For example one finds that on average, with considerable statistical scatter, for the Ca II-K line: $r[\text{kpc}] \approx 3.0 \, A_\lambda(K)$, and for the Na I-$D$ line: $r[\text{kpc}] \approx 1.2 \, [A_\lambda(D_1) + A_\lambda(D_2)]$. Here A_λ is the equivalent width of the line, measured in Å.

3. Structure and Kinematics of the Stellar System

3.1 Apparent Distribution of the Stars

3.1.1 General Star Counts, Integrated Starlight

For the *quantitative description* of the distribution of the stars on the sphere (galactic coordinates l, b) and by apparent magnitude, one deals with either the star numbers

$A(m|l, b)\Delta m$ = number of stars in the interval of apparent magnitude $m - \frac{1}{2}\Delta m \ldots m + \frac{1}{2}\Delta m$, which are to be found in the square degree centred on (l, b)

or the corresponding cumulative star numbers per square degree up to the apparent magnitude m :

$$N(m|l, b) = \int_{-\infty}^{m} A(m|l, b)dm \ .$$

(3.1)

The *observational data,* from which star numbers are obtained on a large scale, consist of so-called *"Durchmusterungen" or surveys,* whose results are published in extensive catalogues. These contain the positions and apparent magnitudes of all stars for the whole sky or for selected fields, up to a certain limiting magnitude. The star coordinates are specified there, not to the highest possible precision, but accurately enough to enable each star to be identified (in α about $\pm 1^s$, in δ about $\pm 0'.1$).

An example of such an undertaking is the "Bonner Durchmusterung" (BD), carried out by the University Observatory of Bonn, for the northern sky between $\delta = +90°$ and $\delta = -2°$ (published 1859/62). This was later complemented by the "Südliche Bonner Durchmusterung" (SD) down to $\delta = -23°$ and the "Cordoba Durchmusterung" (CD or CoD) carried out in Argentina for $\delta = -22°$ to $\delta = -90°$. The limiting magnitude was $m_V \approx 10$. BD and SD include around 458 000 stars, and CD about 589 000. The magnitude values are based on visual estimates.

Even today the names of the stars in these old catalogues are still often used: The stars in each $1°$ width declination zone − in ascending order of right ascension – were simply numbered. Example: α Lyrae = Vega lies between $\delta = +38°$ and $\delta = +39°$ and gets the title BD + 38°3238. Instead of BD, SD or CD, recent English publications mostly use DM (Durchmusterung). Another useful identification is the number of the star in the Henry Draper Catalogue by Cannon and Pickering (1918–1924; 1925–1949), the result of a spectral survey extending over the whole sky, which however includes only all stars up to $m_V \approx 8$. This list gives Vega as HD 172 167 (the numbers increase with the right ascension).

Other large observation programmes have provided locations and photographic magnitudes in part for stars up to magnitude 11 or are still in progress (see also the references at the end of Sect. 2.1.4). The advance to ever fainter magnitudes, however, swells the number of stars so much that a restriction to random samples is necessary. For example, the estimated total of all stars up to photographic magnitude $m_{pg} = 13$ is already 5×10^6, and up to $m_{pg} = 20$ it is about 10^9. Accordingly star-counts down to very faint stars were carried out in 206 "Selected Areas" evenly distributed over the whole sky – sometimes called after their originator, J.C. Kapteyn (1906). The sizes of the areas used lay between $15' \times 15'$ (in low galactic latitudes) and $80' \times 80'$ (in high galactic latitudes). Large undertakings were the "Harvard-Groningen Durchmusterung" (1918/24) of all 206 areas (251 000 stars, limiting magnitude $m_{pg} = 11$ to 16) and the "Mount Wilson Catalogue of Photographic Magnitudes" (1930) of 139 Selected Areas between $\delta = +90°$ and $\delta = -15°$ (68 000 stars, limiting magnitude $m_{pg} = 13$ to 18.5). The magnitude values of these old catalogues are subject almost throughout to appreciable systematic errors, which can amount to a few tenths of a magnitude. The limited success of classical stellar statistics described in the Introduction has had the result that this type of programme has not been repeated or extended. There are available, however, a considerable number of individual investigations of selected areas, in which not only magnitudes, often in two or more colours, but also spectral types or equivalent parameters have been determined.

In order to give an impression of the *general numbers of stars* under discussion, Table 3.1 lists the cumulative numbers of stars per square degree $\overline{N}(m|b)$ averaged over all galactic longitudes, as a function of galactic latitude.

The increase in concentration near the galactic equator with decreasing apparent brightness illustrated by these numbers, especially the ratios between the star numbers at the galactic pole and at the galactic equator given at the foot of the table, can be explained qualitatively in a simple way by the assumption of a plane parallel stratum of stars about the galactic equator (Fig. 3.1). If the sun lies in the plane of symmetry, then far fewer stars appear near $b = 90°$ with increasing range of observations, characterised by the effective limiting magnitude, than appear near $b = 0°$. The result is a concentration of stars near the galactic equator rapidly increasing with m.

Integrated Starlight: From the star numbers $A(m|l, b)$ the portion of the surface brightness of the Milky Way supplied by the stars can be obtained by sum-

Table 3.1. Abstract of the results of general cumulative star numbers per square degree, averaged over all galactic longitudes ($m = m_{\mathrm{pg}}$)

b	$\overline{N}(m\vert b)$		
	$m = 6$	$m = 12$	$m = 18$
± 90	0.039	10.2	617
70	0.042	11.5	692
50	0.048	14.8	1000
30	0.059	23.5	2512
20	0.071	31.6	4678
10	0.089	43.6	9332
0	0.166	89.1	20900
$\overline{N}(m\vert 90°)/\overline{N}(m\vert 0°)$	1:4	1:9	1:34

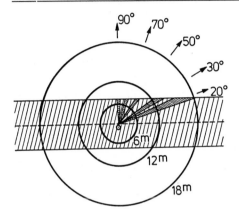

Fig. 3.1. Interpretation of the observed dependence of the star-counts on the galactic latitude by the assumption of a plane parallel stratification of stars (schematic: star density decreases upwards and downwards within the disc)

mation. The radiation flux from a star of magnitude 10, spread over a square degree ($\square°$) on the sphere, often serves as a unit for this purpose, so this gives the dimension of the intensity, written as $1\,S_{10}/\square°$. The value of this unit in terms of absolute intensity naturally depends on the colour system taken as the basis of the apparent magnitude. Thus, for example, for visual magnitudes at $\lambda = 5500\,\text{Å}$, $1\,S_{10}(V)/\square° = 1.2 \times 10^{-9}\,\mathrm{erg\,cm^{-2}s^{-1}\,sterad^{-1}\,\text{Å}^{-1}}$ approximately, whilst for B-magnitudes at $\lambda = 4400\,\text{Å}$, $1\,S_{10}\,(B)/\square° = 2 \times 10^{-9}\,\mathrm{erg\,cm^{-2}s^{-1}\,sterad^{-1}\,\text{Å}^{-1}}$.[1]

One star of the m-th magnitude per square degree contributes to the surface brightness (intensity), in units of $S_{10}/\square°$,

$$I_m = \frac{S_m}{S_{10}} = 10^{-0.4(m-10)} \quad .$$

---

[1] The variation of the absolute value of the unit $1\,S_{10}$ with variable wavelength λ is identical with the distribution of the extra-terrestrial energy flux Φ_λ in the spectrum of an A0V-star with $m_\lambda = 10$, since the magnitude is a (logarithmic) measure for the radiation flux Φ_λ of the star under consideration in units of the flux Φ_λ of an A0V-star with $m_\lambda = 0$.

All the stars inside a square degree with the mid-point (l, b) therefore contribute to the surface magnitude

$$I^*(l,b) = \sum_{m=-\infty}^{+\infty} 10^{-0.4(m-10)} A(m|l,b)[S_{10}/\square°] \; . \tag{3.2}$$

The results for the integrated starlight $I^*(l, b)$ spread over the whole sphere are still largely based on the large old photometric survey catalogues, and accordingly refer to the visual or classical photographic B-spectral regions. Before one attempts to use these numbers, the systematic errors of the old magnitude system should be corrected. Comparison with direct measurements of the surface brightness, however, shows that the contribution of the faint stars is not taken into account with satisfactory accuracy (Sect. 3.1.3). For the region of the galactic equator the values are often significantly too low on that account. For this reason Table 3.2 gives the values $I_B^*(l, b)$ obtained for a few galactic longitudes, but only for the higher galactic latitudes.

Table 3.2. Surface brightness of the integrated starlight $I_B^*(l, b)$ in the B-spectral region. Unit: $1 S_{10}(B)/\square°$. (After Sharov and Lipaeva, 1973)

l \\ b	$-80°$	$-40°$	$-20°$	$20°$	$40°$	$80°$
0°	11	29	100	40	20	10
100	14	19	40	32	22	10
200	15	15	25	40	20	10
300	11	24	33	49	20	10

Summations of the starlight for spectral regions in the ultraviolet give distributions with very strong concentrations at the galactic equator. Typical individual values are given in Chap. 4, Table 4.4. Below 2000 Å practically only the O- and B-stars make appreciable contributions.

3.1.2 Apparent Distribution of the Individual Star Types

The concentration at the galactic equator is very different for stars of different physical types. O- and B-stars as well as open star clusters show a strong measure of preference for the immediate neighbourhood of the galactic great circle (Fig. 3.2). Later spectral types are as a rule less concentrated in the Milky Way than early types. Table 3.3 illustrates this by means of the star numbers $\overline{N}(m|b)$ per square degree up to limiting magnitude $m_{pg} \approx 8.5$ for the individual spectral classes. The decrease in the galactic concentration from B to G must of course be partially attributed to the fact that the luminosity, and hence the range of the observations, first of all becomes smaller on passing over to the later types (cf. Fig. 3.1). For K and M the Main Sequence stars with $m_{pg} < 8.5$ are indeed all located within short distances, but the range is

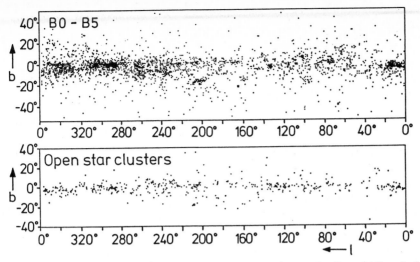

Fig. 3.2. Apparent distribution of the B0- to B5-stars (from Charlier, 1926) and of the open star clusters (from Collinder, 1931)

Table 3.3. Star numbers $\overline{N}(8\overset{m}{.}5|b)$ averaged over all galactic longitudes for the different spectral types as a function of the galactic latitude b. (Excerpt from Nort, 1950)

b	$\overline{N}(8\overset{m}{.}5|b)$					
	B	A	F	G	K	M
85°	0.00	0.18	0.30	0.20	0.38	0.03
45°	0.02	0.19	0.27	0.18	0.44	0.04
25°	0.07	0.37	0.33	0.25	0.62	0.05
5°	0.54	0.72	0.37	0.28	0.70	0.06
−5°	0.71	0.75	0.35	0.26	0.74	0.06
−25°	0.13	0.40	0.36	0.27	0.56	0.04
−45°	0.04	0.21	0.32	0.26	0.42	0.04
−85°	0.02	0.12	0.25	0.26	0.40	0.03

materially greater for the strongly luminous giants of both these types, which are included here. Therefore at least for these objects a qualitatively lesser concentration at the galactic plane can be deduced. The concentration of the globular star clusters at the galactic equator is even weaker.

For the variables and the other special types there are also significant differences with regard to the distribution over galactic latitudes. For example the concentration at the galactic equator is for the Type I Cepheids very strong, for Type II Cepheids (W Virginis stars), long period variables of late type (Mira etc.) or central stars of planetary nebulae distinctly less strong, and for RR Lyrae stars only as weak as for globular star clusters, in which these types of variable also occur.

The apparent *distribution with galactic longitude* also shows pronounced differences for the different objects. Characteristic of the globular clusters, the individually occurring RR Lyrae stars and also the remaining star types only moderately concentrated at the galactic equator, is a strong increase in their frequency in the neighbourhood of the direction of the galactic centre $l \approx 0°$. For the O- and B-stars, on the other hand, a concentration in the direction of the centre is scarcely noticeable, but they exhibit a very uneven distribution, with a tendency to the formation of large star groups, so-called star associations (see also Sect. 3.3.1).

If one considers only B-stars which are apparently bright, and therefore relatively close, for example $m_B < 8$, then it is evident that the great circle of strongest concentration is noticeably inclined to the galactic equator. Also the general star numbers $N(m|l, b)$ produce as the position of greatest apparent star density at the specified limiting magnitude m an approximate great circle, which on passing to brighter limiting magnitudes deviates increasingly from the currently accepted galactic equator: for $m_B \approx 8$ the inclination is about $10°$, for $m_B \approx 4$ nearly $20°$. This phenomenon was described for the southern sky by John Herschel as long ago as 1847, and it was investigated in general by B.A. Gould in 1874. Since then one speaks of Gould's Belt of bright stars, in which local peculiarities of the spatial distribution of nearby stars of spectral types O, B and A are reflected.

3.1.3 Distribution of the Surface Brightness of the Milky Way in the Visible and in the Infrared

Visible Radiation: The directly measured radiation energy in a specified region of the spectrum per unit of area on the sphere − free from the influence of atmospheric extinction − consists of several components:

1. direct starlight
2. starlight scattered from interstellar dust particles ("Diffuse Galactic Light")
3. sunlight scattered from interplanetary dust particles (Zodiacal Light)
4. luminosity of the upper atmosphere (Airglow)
5. light from sources (1) to (4) scattered in the earth's atmosphere by molecules and vapour droplets.

The contributions of the galactic emission nebulae (Sect. 4.2.1) and of the extra-galactic star systems are negligible compared with the direct starlight in the ordinary broadband optical spectral regions (e.g. U, B, V). Isolated peaks in the surface brightness distribution produced by a few nearby very bright stars are appropriately eliminated. To obtain the pure Milky Way brightness (1) and (2), it is nessary to ascertain and remove the contributions (3) to (5) from the measured results. For (3) and (4) this is essentially possible, since these components exhibit distributions oriented to the ecliptic and the sun in the first case, and the horizon in the second. The component (1) should in the ideal case agree with the integrated starlight discussed in Sect. 3.1.1.

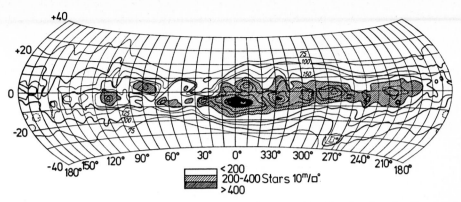

Fig. 3.3. Isophotes of the brightness distribution in the Milky Way for the visual spectral region. Unit of surface brightness: 1 star magnitude 10. per square degree. (From Elsässer and Haug, 1960)

The result of a photoelectric surface photometry (components 1 + 2) of the whole Milky Way for the V-spectral region is shown in Fig. 3.3 – we cannot go any further into the measurements and their reduction. The angular resolution amounted here to about 3°. A corresponding chart has been obtained by the same authors for the blue spectral region. As well as other similar observational results for the B- and U-regions, there is available an older photographic surface photometry of the northern and southern Milky Way (see Literature Summary), which has the advantage of relatively higher angular resolution (about $\frac{1}{2}$ degree) and accordingly also records small scale fluctuations of the brightness distribution.

The contour charts clearly show a systematic increase of the surface brightness towards the direction of the galactic centre; the greatest brightness is reached at a maximum lying a few degrees south of the galactic equator. Moreover it appears that the brightnesses in the northern Milky Way ($l \approx 30° \dots 210°$) are on average significantly lower than in the southern Milky Way. We notice also the appearance of a number of distinct brightness maxima, for example in the directions $l \approx 73°$ (Cygnus) and $l \approx 287°$ (Carina), the interpretation of which will be discussed in Sect. 3.3.4.

Infrared Radiation: Close to the galactic equator the background brightness of the Milky Way is modified to a considerable degree in the visible region by the effect of clouds of interstellar dust (Sect. 4.1.1). In particular the light from the inner region of the galaxy is strongly attenuated by interstellar extinction. This effect is lessened with increasing wavelength, however, and already at $2.2\,\mu$m (K-region) for example it is only about 20% of the value for the V-region. By means of surface photometry in the near infrared, therefore, the galactic distribution of stars, which still radiate strongly in this spectral region, can be examined even in the neighbourhood of the galactic plane to great distances. In particular it becomes possible to measure the emission of the stellar

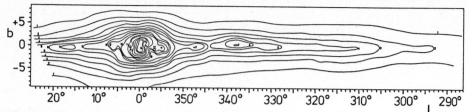

Fig. 3.4. Distribution of surface brightness in the southern Milky Way at $\lambda = 2.4\,\mu$m. The outermost contour corresponds to a brightness of $1 \times 10^{-10}\,\mathrm{W\,cm^{-2}\,sterad^{-1}\,\mu m^{-1}}$; the intensity difference between neighbouring contours also corresponds to one such unit. (From Hayakawa et al., 1979)

concentration in the galactic kernel, which in the visible is barely detectable. As an example of a recent result, Fig. 3.4 shows a brightness contour chart of the southern Milky Way for $\lambda = 2.4\,\mu$m. Here the central region of the galaxy shows up clearly. The contour corresponding to half of the maximum brightness value is approximately an ellipse with semi-axes of 5° in galactic longitude

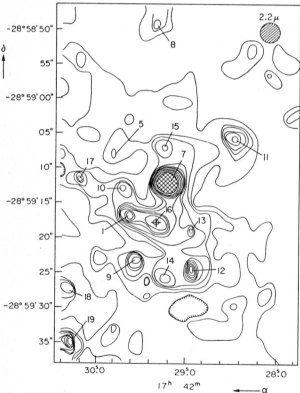

Fig. 3.5. Brightness contour chart of the galactic centre for $\lambda = 2.2\,\mu$m, obtained with the 5 m. Hale telescope on Mt. Palomar. The angular resolution amounts to $2''5$ (shaded circular area near the upper right corner). The cross (+) marks the galactic centre (explanation in text). The change in radiation flux from one contour to the next is $2.5 \times 10^{-18}\,\mathrm{W\,m^{-2}\,Hz^{-1}\,sterad^{-1}}$. (By permission of Becklin and Neugebauer, 1975)

and 3° in galactic latitude. Significantly longer wave infrared radiation was also observed in this region.

Observations of the kernel region with higher angular resolution in different spectral bands in the near infrared show that within a distance from the centre of only about $20''$ (corresponding to about 1 pc!) there is a concentration of numerous individual sources. An example is given in Fig. 3.5. It is assumed that this radiation comes from giant stars of late spectral type − very bright individual objects or whole star clusters. Noteworthy is the outstanding object No. 7: in the double hatched region there should be 35 additional contours! The centre point of source No. 16 lies at $\alpha = 17^{\mathrm{h}}42^{\mathrm{m}}29\overset{\mathrm{s}}{.}3\pm0\overset{\mathrm{s}}{.}15$, $\delta = -28°59'18''\pm3''$ (1950.0) and accordingly lies near to the kernel of the radiosource "Sagittarius A West" and hence at the same time near to an "ultracompact" radiosource (see Sect. 4.2.2). In this region is presumably the actual centre of our galaxy.

3.2 The Local Galactic Star Field

In this section we discuss the derivation of the spatial distribution and the motions of the stars for the region around the sun which till now has been covered by the methods of stellar statistics. The extent of this depends strongly on the luminosity of the star types actually being considered, but close to the galactic plane even in favourable circumstances barely extends more than 2 kpc.

3.2.1 Methods of Stellar Statistics

Distribution Functions: The goal of the study of our stellar system cannot consist in establishing the position and velocity coordinates of every individual star. One seeks here to obtain statistical statements, namely the distributions of the attributes of the stars, such as position, velocity, luminosity, etc. Very extensive information is already contained in the absolute frequency distributions $N(\boldsymbol{r}, \boldsymbol{v}, \boldsymbol{S})$ of the position vector $\boldsymbol{r}(x, y, z)$, the velocity vector $\boldsymbol{v}(U, V, W)$ and the star type $\boldsymbol{S}$ (characterised by the spectral type Sp and absolute magnitude M) with the attribute

$$N(\boldsymbol{r}, \boldsymbol{v}, \boldsymbol{S})\Delta\boldsymbol{r}\,\Delta\boldsymbol{v}\,\Delta\boldsymbol{S} \tag{3.3}$$

= number of stars in volume element $x \ldots x + \Delta x$, $y \ldots y + \Delta y$, $z \ldots z + \Delta z$ with velocity components in the intervals $U \ldots U + \Delta U$, $V \ldots V + \Delta V$, $W \ldots W + \Delta W$, with spectral types and absolute magnitudes in the intervals $Sp \ldots Sp + \Delta Sp$, $M \ldots M + \Delta M$, respectively.

The formulation of the problem of stellar statistics is directed primarily at the density distribution and the velocity distribution, which are obtained from $N(\boldsymbol{r}, \boldsymbol{v}, \boldsymbol{S})$ by integration over the velocity $\boldsymbol{v}$ and over the position vector $\boldsymbol{r}$, respectively. One defines

$$\int N(\boldsymbol{r}, \boldsymbol{v}, \boldsymbol{S})d\boldsymbol{v} = D(\boldsymbol{r}, \boldsymbol{S}) = D(\boldsymbol{r}) \cdot \varphi(\boldsymbol{S}|\boldsymbol{r}) \qquad \text{with} \qquad (3.4)$$

$$D(\boldsymbol{r}) = \int D(\boldsymbol{r}, \boldsymbol{S})d\boldsymbol{S}$$

$$\qquad = \text{ spatial number density of the stars(“} \textit{star density} \text{”)} \qquad (3.5)$$

and

$$\varphi(\boldsymbol{S}|\boldsymbol{r}) \quad = \quad \text{fraction of the stars in a finite volume around the position}$$
$$\boldsymbol{r}, \text{ which have the attribute } \boldsymbol{S} \text{ (referred to an interval}$$
$$\varDelta \boldsymbol{S} = 1) \qquad (3.6)$$

In the jargon of probability theory, this is the conditional distribution of $\boldsymbol{S}$ (the condition is the presence of the attribute $\boldsymbol{r}$). By summation over the spectral types one obtains from $\varphi(\boldsymbol{S}|\boldsymbol{r})$ the distribution of the absolute magnitude $\varphi(M|\boldsymbol{r})$, which is also called the normalised luminosity function: by definition

$$\int\limits_{-\infty}^{+\infty} \varphi(M|\boldsymbol{r})dM = 1 \ . \qquad (3.7)$$

Integration over space to obtain the pure velocity distribution is meaningful only over a small region of the star system, since the velocity field of the stars systematically changes its character, not only with distance from the galactic centre, but also with increasing distance from the galactic plane. One therefore studies the relative velocity distribution of the stars which are located in a finite volume around the position $\boldsymbol{r}$ and also belong to a specific type $\boldsymbol{S}$:

$$\psi(\boldsymbol{v}|\boldsymbol{r}, \boldsymbol{S}) = \frac{N(\boldsymbol{r}, \boldsymbol{v}, \boldsymbol{S})}{D(\boldsymbol{r}, \boldsymbol{S})} \ . \qquad (3.8)$$

Fundamental Equation for the Density Distribution: The first problem of classical stellar statistics is to deduce the density distribution $D(\boldsymbol{r}) = D(r, l, b)$ along the line of sight in the direction (l, b) from the functions $A(m|l, b)$ or $N(m|l, b)$ which have themselves been obtained from the star counts (see Sect. 3.1.1). The currently fixed variables l and b will be omitted in what follows. In the simple ideal case of uniform density distribution $D(r) = D_0$ and equal absolute magnitude of all stars $M = M_0$ and a total absence of interstellar extinction, $N(m)$ can easily be obtained: then $N(m) {\sim} r_m^3$, with r_m denoting the distance of stars with apparent magnitude m, and the radiation flux of a star in the spectral region under consideration is $\varPhi_m {\sim} r_m^{-2}$ at the position of the observer. Elimination of r_m leads to $N(m) {\sim} \varPhi_m^{-3/2}$, where $\varPhi_m {\sim} 10^{-0.4m}$ from the definition of m. Accordingly one obtains

$$N(m) {\sim} 10^{0.6m} \quad \text{or} \qquad (3.9)$$

$$\log N(m) = 0.6m + \text{const.} \qquad (3.10)$$

The cumulative star numbers would increase by a factor $N(m+1)/N(m) = 10^{0.6} \approx 4$ for each increase of one magnitude. The numbers in Table 3.1 however correspond to a factor between 2.5 and 3. The assumptions cannot therefore be valid. In accordance with the qualitative interpretation given in Sect. 3.1.1 of the dependence of star numbers on the galactic latitude, we assume that the star density $D(r)$ in directions of higher latitudes decreases with increasing distance; in low latitudes the interstellar extinction − in all directions of the galactic belt − simulates a density decrease!

We now turn to the general case of arbitrary density distribution $D(r)$ and arbitrary luminosity function $\varphi(M|r)$. At distance r the number density of stars with absolute magnitude in the interval $M \ldots M + dM$ is

$$D(r,M)dM = D(r)\varphi(M|r)dM \ .$$

If we express M in terms of m and r using (2.43):

$$M = m + 5 - 5\log r - \Delta m(r) \ ,$$

where $\Delta m(r)$ denotes the interstellar extinction in the spectral region under consideration up to the distance r, then we have the number density of the stars with apparent magnitudes in the interval $m \ldots m + dm$ for the distance r. In the solid angle ω (Fig. 3.6) the number of stars in the distance interval $r \ldots r + dr$ with apparent magnitudes m to $m + dm$ is

$$D(r)\varphi(m + 5 - 5\log r - \Delta m(r)|r)\omega r^2 dr \, dm \ .$$

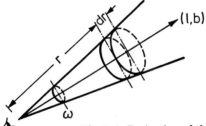

Fig. 3.6. Derivation of the Fundamental Equation of stellar statistics (3.11)

Integration with respect to the distance then gives the number of all stars with apparent magnitudes between $m - \frac{1}{2}$ and $m + \frac{1}{2}$ in the solid angle under consideration as:

$$A(m) = \omega \int_0^\infty D(r)\varphi(m + 5 - 5\log r - \Delta m(r)|r)r^2 dr \ . \tag{3.11}$$

This expression is often called the *Fundamental Equation of Stellar Statistics*. (The use of the usual notation A for the interstellar extinction would cause this letter to appear here with two differential meanings). The solid angle corresponding to a square degree ($1\square°$) is given by

$$\omega = \left(\frac{2\pi}{360}\right)^2 = 3.046 \times 10^{-4} \ .$$

If $\varphi(M\,r)$ can be taken as known, then (3.11) represents an integral equation of the first kind for $D(r)$ (for determination of $\Delta m(r)$, see Sect. 4.1.1, (4.12ff.), and also Sect. 5.4.1).

The mathematical character of the integral (3.11) becomes especially clear if one puts it into the form of a convolution integral by introducing the distance parameter $\varrho = -5 + 5\log r$ as new integration variable. If we first in the interest of simplicity neglect the interstellar extinction $\Delta m(r)$ and assume $\varphi(M|r)$ to be independent of r, we then obtain

$$r = 10^{(\varrho+5)/5} = e^{(\varrho+5)/5\,\mathrm{Mod}} \quad (\mathrm{Mod} = \log e) \tag{3.12}$$

and with

$$\Delta(\varrho) = \omega e^{3(\varrho+5)/5\,\mathrm{Mod}} \frac{1}{5\,\mathrm{Mod}} D(e^{(\varrho+5)/5\,\mathrm{Mod}}) \tag{3.13}$$

it follows that

$$A(m) = \int_{-\infty}^{+\infty} \Delta(\varrho)\varphi(m-\varrho)d\varrho \ . \tag{3.14}$$

When $D(r) = D_0 = \text{const.}$ and $M = M_0 = \text{const.}$, so that $\varphi(M) = \delta(M - M_0) = \text{deltafunction}$, which vanishes except when $M = M_0$ (limiting case of the Gauss Distribution when the scatter $\sigma \to 0$), one obtains $A(m) = \Delta(m - M_0) \sim \exp(3m/5\,\mathrm{Mod}) = 10^{0.6m}$. The integration over m as in (3.1) produces the same functional form for $N(m)$ as we had already found by an elementary method in (3.9).

Equation (3.14) clearly means that $A(m)$ results from the "smoothing" of the variation of $\Delta(\varrho)$. Table 3.6 shows that the general luminosity function $\varphi(M)$ extends over an interval of about 20 magnitudes in M so that $A(m)$ represents a weighted mean value of the modified star density $\Delta(\varrho)$ over the interval $\varrho = m - 10$ to $m + 10$, corresponding to the distance ratio $r_2/r_1 = 10^4$. The structure of the density profile $\Delta(\varrho)$ or $D(r)$, across a spiral arm intersected by the line of sight, will therefore not be recoverable from the star numbers $A(m)$.

The classic period of stellar statistics is characterised above all by great efforts to invert the integral equation (3.11) with the general luminosity function, either analytically (H. v. Seeliger, K. Schwarzschild et al.), or numerically ("Kapteyn's scheme"). It became clear only rather belatedly that this problem is in practice insoluble, so the extensive investigations carried out led to density

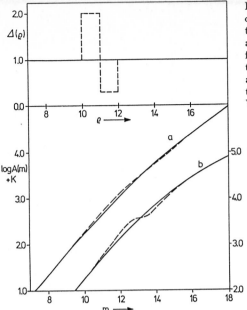

distributions of little practical use. We demonstrate the situation by means of two examples (Fig. 3.7). The star density was first specified to have the constant profile $D(r) = 1$, and then to make two deviations, taking the value $D(r) = 2$ between $r = 720$ and $1000\,\mathrm{pc}$ and the value $D(r) = 0.3$ between $r = 1000$ and 1350.

In both cases a homogeneous interstellar extinction $\Delta m(r) = 1 \times 10^{-3} r$ (r in pc) was assumed. With the very broad general luminosity function $\varphi(M)$ – as discussed in Sect. 3.2.2 – the function $A(m)$ assumed the profiles indicated by the letter a in the lower half of Fig. 3.7. The effect of the pronounced variations of the star density $D(r)$ or $\Delta(\varrho)$ was in practice scarcely detectable in $A(m)$. On the other hand if one gives $\varphi(M)$ a Gauss distribution

$$\varphi(M) = \frac{1}{\sigma_M \sqrt{2\pi}} e^{-(M-M_0)^2/2\sigma_M^2} \tag{3.15}$$

with the mean value of absolute magnitude $M_0 = +2.0$ and standard deviation $\sigma_M = \pm 1.0$, which is approximately true for the Main Sequence star groups with spectral types A5 to F0, then one obtains the curves marked b. Here the density variations do indeed come through better, but the conclusions on the density profile would still, by reason of the ever present statistical variations on $A(m)$, turn out to be rather inexact.

In agreement with the results of detailed model computations we arrive at the conclusion that the inversion of the integral equations (3.11) or (3.14) to obtain the star density distribution $D(r)$ is possible in at best satisfactory fashion if the scatter σ_M of the absolute magnitudes of the star types under consideration is less than about one magnitude.

In modern investigations one has therefore dealt with star counts for star types each with the narrowest possible scatter band of luminosity, usually narrow spectral groups (Index s) with known distribution of absolute magnitudes $\varphi_s(M)$ – approximated by a Gauss distribution. Equation (3.11) is then applied for the corresponding functions $A_s(m)$, $\varphi_s(M)$ and $D_s(r)$.

If one substitutes multi-colour photometry for the spectral classification, then it is possible to progress to substantially fainter stars and one can obtain the $A(m)$ for small intervals of absolute magnitude $M - \frac{1}{2}\Delta M \ldots M + \frac{1}{2}\Delta M$. Let us mention as an example a method developed by W. Becker in 1962, which starts from a three-colour photometry: for each specified interval of apparent magnitude $m - \frac{1}{2}\Delta m \ldots m + \frac{1}{2}\Delta m$ (for example $m = m_V$) a "fractionated" two-colour diagram (see Fig. 2.22) is produced. By shifting the diagram in the direction of the colour axes up to the Standard Main Sequence, one can first eliminate the influence of the interstellar extinction. Inasmuch as the two intrinsic colours permit an unambiguous determination of the absolute magnitudes, the numbers $A(m, M)$ can now be obtained from direct enumerations of the stars which are to be found in the interval $M - \frac{1}{2}\Delta M \ldots M + \frac{1}{2}\Delta M$. For a specified M the density is now given simply by

$$D_M(r) = \frac{\sum\limits_{m} A(m, M)}{V(m, M)} \,, \tag{3.16}$$

where $V(m, M)$ denotes the volume defined by the limits of the magnitude interval $m - \frac{1}{2}\Delta m$ and $m + \frac{1}{2}\Delta m$ under consideration.

One can say in general that authentic density distributions can be obtained only if the required *solution of the fundamental equation* (3.11) or (3.14) assumes only the character of a *correction* of the distributions $A(m, M)$ or $A_s(m)$ for the remaining scatter of the absolute magnitudes about M or the mean value of the spectral group $\overline{M}_s$. For this, various methods can be applied. Favourite these days is the approximation of the integral equation by a linear system of equations with the density values as unknowns, and their solution by the usual methods. Regarding the exact formulation for the numerical computation and also for other methods the literature listed in the Appendix should be consulted.

Description of the Velocity Distribution: The observations lead first of all to the peculiar velocities of the stars (Sect. 2.1.3). The distribution defined in (3.8) can be directly deduced from this only for velocities relative to the local reference system. For the stars in the neighbourhood of the sun one obtains from this the "scatter" of the individual velocities about the mean flow as a result of the galactic rotation. In the case of a "random" distribution of the peculiar velocities of the stars, one would obtain for each component of the space velocity $\dot{\xi}$, $\dot{\eta}$, $\dot{\zeta}$ in the galactically oriented coordinate system (see Sect. 2.1.3) one and the same Gauss distribution about the respective mean value $\overline{\dot{\xi}}$, $\overline{\dot{\eta}}$, $\overline{\dot{\zeta}}$, with the squared deviation $\Sigma^2 = \overline{(\dot{\xi} - \overline{\dot{\xi}})^2} = \overline{(\dot{\eta} - \overline{\dot{\eta}})^2} = \overline{(\dot{\zeta} - \overline{\dot{\zeta}})^2}$.

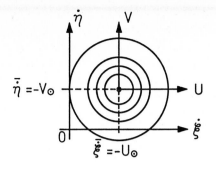

Fig. 3.8. Illustration of the velocity distribution

Figure 3.8 shows for this case the projection of the distribution on the $(\dot\xi, \dot\eta)$ plane of the velocity space, indicated by a few lines of equal projected frequency. With the mean point (centroid) of the point cloud in velocity space as the new origin: $U = \dot\xi - \bar{\dot\xi}$, $V = \dot\eta - \bar{\dot\eta}$, $W = \dot\zeta - \bar{\dot\zeta}$ we obtain the special three-dimensional Gaussian distribution

$$\psi_0(U, V, W | S) = \frac{1}{(2\pi)^{3/2} \Sigma^3} \exp\left[-\frac{1}{2}\left(\frac{U^2 + V^2 + W^2}{\Sigma^2}\right)\right] \tag{3.17}$$

where the subscript $_0$ denotes reference to the neighbourhood of the sun.

The observational data have shown that the peculiar velocities of the stars are not "randomly" distributed. For velocities which are not too high (about $<60 \,\mathrm{km\,s}^{-1}$), however, the surfaces of equal point density in the velocity space can be approximately represented by tri-axial ellipsoids, and the point density profiles in the directions of the main axes can be approximated by Gaussian distributions with different dispersions. Accordingly one often chooses for the starting point the general three-dimensional Normal Distribution (Karl Schwarzschild, 1907):

$$\psi_0(u_1, u_2, u_3 | S) = f(u_1, u_2, u_3) = C \exp\left[-\tfrac{1}{2}Q(u_1, u_2, u_3)\right] \tag{3.18}$$

with the quadratic form

$$Q(u_1, u_2, u_3) = \sum_{i,k=1}^{3} a_{ik} u_i u_k \ . \tag{3.19}$$

We have here denoted the peculiar velocity components U, V, W by the convenient notation u_1, u_2, u_3. The constant C is found by carrying out the normalisation integration

$$\int\limits_{-\infty}^{+\infty} \int \int f(u_1, u_2, u_3) du_1 \, du_2 \, du_3 = 1 \quad \text{whence}$$

$$C = \frac{\sqrt{A}}{(2\pi)^{3/2}} \ . \tag{3.20}$$

Here A denotes the value of the determinant of the a_{ik}. Since the terms $a_{ik}u_iu_k$ and $a_{ki}u_ku_i$ can also be grouped to form $(a_{ik}+a_{ki})u_iu_k$, we can assume without loss of generality that $a_{ki} = a_{ik}$.

The purpose of the analysis is the determination of the directions (l_i, b_i) of the three principal axes of the *velocity ellipsoid* and the three velocity dispersions in these directions $\Sigma_1^2, \Sigma_2^2, \Sigma_3^2$. In the reference system given by the principal axes $f(u_1, u_2, u_3)$ transforms into

$$F(u_1', u_2', u_3') = \frac{1}{(2\pi)^{3/2}\Sigma_1\Sigma_2\Sigma_3}\exp\left\{-\frac{1}{2}\left(\frac{u_1^2}{\Sigma_1^2} + \frac{u_2^2}{\Sigma_2^2} + \frac{u_3^2}{\Sigma_3^2}\right)\right\}. \qquad (3.21)$$

One calls the surface defined by

$$Q(u_1', u_2', u_3') \equiv \frac{u_1'^2}{\Sigma_1^2} + \frac{u_2'^2}{\Sigma_2^2} + \frac{u_3'^2}{\Sigma_3^2} = 1 \qquad (3.22)$$

the dispersion ellipsoid. The transformation to the principal axes system calls for the appropriate rotation of the coordinate system and hence the linear transformation

$$u_j' = \sum_{l=1}^{3} \beta_{jl}u_l \quad (j = 1, 2, 3)$$

where $\beta_{jl} = \cos(u_j', u_l)$ denotes the direction cosine between the u_j'-axis and the u_l-axis. In this principal axes transformation, formally treated in mathematical text-books, the matrix a_{ik} is usually considered as given. In the present application, however, it is appropriate to take instead the elements of the variance matrix

$$s_{ik} = \int\!\!\!\int\limits_{-\infty}^{+\infty}\!\!\!\int u_iu_kf(u_1, u_2, u_3)du_1\,du_2\,du_3 = \overline{u_iu_k} \qquad (3.23)$$

and hence the directly calculable mean values $\overline{u^2}, \overline{uv}, \ldots \overline{U^2}, \overline{UV}, \ldots$, respectively, over all the selected stars as the starting point. One can show that the matrices a_{ik} and s_{ik} are mutually inverse and $\Sigma_1, \Sigma_2, \Sigma_3$, as roots of the cubic equation for Σ (Secular Equation),

$$\Delta \equiv \begin{vmatrix} s_{11} - \Sigma^2 & s_{12} & s_{13} \\ s_{21} & s_{22} - \Sigma^2 & s_{23} \\ s_{31} & s_{32} & s_{33} - \Sigma^2 \end{vmatrix} = 0 \qquad (3.24)$$

can be calculated. For the β_{jl}, by which the directions of the principal axes are determined, we have a linear system of equations, the solution of which gives

$$\beta_{jl} = \pm\frac{\Delta_{jl,j}}{\sqrt{\Delta_{j1,j}^2 + \Delta_{j2,j}^2 + \Delta_{j3,j}^2}}. \qquad (3.25)$$

Here $\Delta_{jl,j}$ denotes the sub-determinant of Δ which is conjugate to the index pair j, l, evaluated with $\Sigma = \Sigma_j$. If the components u_1, u_2, u_3, for example, refer to the equatorial system, then the coordinates of the principal axis system (α_j, δ_j) are

$$\tan \alpha_j = \beta_{j2}/\beta_{j1} \quad , \quad \sin \delta_j = \beta_{j3} \quad (j = 1, 2, 3) \ . \tag{3.26}$$

Methods for Deriving the Luminosity Function: The enumeration of the stars with known absolute magnitudes inside a certain volume, the direct way of obtaining the luminosity function, produces approximate completeness, for objects of low absolute brightness, only in the immediate neighbourhood of the sun, r less than about 20 pc. With relatively small numbers of stars involved, most reliable results have been obtained up to now by this method.

One possible way of obtaining the luminosity function for larger and also more distant volumes is offered by the procedures described above for *determining density for narrow spectral or luminosity groups*. If the mean absolute magnitudes M_{0s}, the dispersions σ_{Ms}, and the star densities $D_s(r)$ are available for a sufficient number of individual spectral groups (index s), then $\varphi(M|r)$ follows:

$$D(r)\varphi(M|r) = \sum_s D_s(r)\frac{1}{\sigma_{Ms}\sqrt{2\pi}}\exp\left[-\frac{(M-M_{0s})^2}{2\sigma_{Ms}^2}\right] \tag{3.27}$$

where $D(r) = \Sigma D_s(r)$. In a similar way one can use the densities $D_M(r)$ obtained from (3.16) with a photometric classification basis. Since the star numbers $A_s(m)$ or $A(m, M)$ taken as a basis up to now extend only to limiting magnitudes, usually appreciably below $m_V \sim 20$, conclusions on $\varphi(M, r)$ can be reached in this way only to medium luminosities (see Sect. 2.2.2, Table 2.2!).

So far the profile of the luminosity function for faint absolute magnitudes, M_V greater than about $+10$, can be obtained only for a region restricted to relatively short distances. To extend this volume as far as possible, and to include as many stars as possible, procedures have long been employed which start from the extensive data on measured proper motions and parallaxes: the proper motion can also be used on its own as a distance criterion. These observational data, however, give systematic preference to stars with easily detectable and hence large proper motions. Because of the incompleteness with regard to stars with small proper motions, one has to apply a statistical correction – usually considerable.

P.J. van Rhijn's method uses the catalogue of trigonometric parallaxes. Stars with small proper motions are there systematically under-represented. One considers first of all those stars with proper motions $\mu > \mu_0$ (the limit μ_0 depends on m: for $m = 5 \ldots 9$, μ_0 is about $0\rlap{.}''2$/year for $m = 9 \ldots 15$, $\mu_0 = 0\rlap{.}''5$/year approximately). The frequency distribution $G'(m, p'|\mu > \mu_0)$ uses observed parallaxes $p' = p + \Delta p$ with observational errors Δp. The distribution $G(m, p|\mu > \mu_0)$, corrected for the errors Δp, is obtained by inversion of the convolution:

$$G'(p') = \int\limits_{-\infty}^{+\infty} G(p)W(p' - p)dp$$

with $W(\Delta p) =$ Gaussian distribution with mean value $\overline{\Delta p} = 0$ and standard deviation σ_p ($=$ average of the mean errors of parallax from the internal accuracy). For this purpose $G(p)$ is expanded for example in a Taylor series.

The incompleteness factor $k(p; \mu_0)$, which converts $G(m, p|\mu > \mu_0)$ into $G(m, p)$ can obviously be defined by

$$\frac{1}{k} = \frac{G(m, p|\mu > \mu_0)}{G(m, p)} = \iint\limits_{\mu > \mu_0} \Psi(t_\alpha, t_\delta)dt_\alpha \, dt_\delta \;, \tag{3.28}$$

where $\Psi(t_\alpha, t_\delta)$ is the distribution of the components of the tangential velocity. The integration region is given by $t = \sqrt{t_\alpha^2 + t_\delta^2} = K\mu/p > K\mu_0/p = t_0(p)$ (Fig. 3.9). Since stars of the entire sphere are included, $\Psi(t_\alpha, t_\delta)$ is regarded as circularly symmetric and one writes

$$\Psi(t_\alpha, t_\delta) = \int \Psi(t_\alpha, t_\delta, v_r)dv_r$$

$$= \frac{1}{2\pi \Sigma^2}\exp\left\{ -\frac{(t_\alpha - \overline{t}_\alpha)^2 + (t_\delta - \overline{t}_\delta)^2}{2\Sigma^2} \right\} \tag{3.29}$$

where Σ is taken to be the mean value of the velocity dispersions appearing in (3.21); the mean values $\overline{t}_\alpha$, $\overline{t}_\delta$, which because of the sun's motion are non-zero, are determined by the magnitude of the sun's velocity $S_\odot$ and the angles η, χ introduced in (2.33) and (2.38), which depend upon the star position (α, δ): $\overline{t}_\alpha = S_\odot \sin \eta \sin \chi$, $\overline{t}_\delta = S_\odot \sin \eta \cos \chi$. For example, for $\Sigma = 30 \,\mathrm{km\,s^{-1}}$ and $S_\odot = 25 \,\mathrm{km\,s^{-1}}$ one finds on average over the whole sphere the values $k(p; \mu_0)$ given in Table 3.4. At a distance of $r = 50 \,\mathrm{pc}$, and for the limiting proper motion $\theta_0 = 0''.5/\mathrm{year}$ corresponding to $m \geq 9$, only about 1.2 per thousand stars is included!

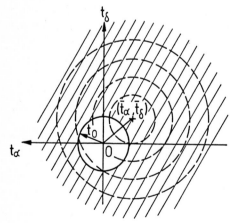

Fig. 3.9. Calculation of the integral on the right-hand side of (3.28). The integration extends over the region outside the circle about the origin with radius $t_0(p) = K\mu_0/p$ (shaded)

Table 3.4. Typical example for the values of the incompleteness factor $k(p; \mu_0)$ defined in (3.28). Explanation in text. (From Trumpler and Weaver, 1953)

p	r [pc]	$k(p; \mu_0)$	
		$\mu_0 = 0''5$	$\mu_0 = 0''2$
$0''02$	50	830	2.7
0.05	20	2.7	1.18
0.10	10	1.28	1.04
0.20	5	1.06	1.01

Once $G(m,p)$ has been established, the substitution $m = M - 5 - 5\log p$ gives a frequency distribution $F(M,p)$, the integration or summation of which over p gives the absolute values of the luminosity function:

$$D_0\varphi_0(M) = \frac{1}{V_0} \int_{p_0}^{\infty} F(M,p)dp .$$ (3.30)

Here in the ideal case $p_0 = 1/r_0$ is the smallest parallax still included and V_0 is the corresponding volume. In actual fact, $F(M,p)$ can be determined with sufficient reliability only inside the limited region $(p_1 \ldots p_2)$, whose position and extent depends on M, so that the integration should be carried out only over this interval and the corresponding volume used instead of V_0.

Other classic methods for the determination of the general luminosity function choose proper motion star-counts and catalogues as starting point, and by counting obtain from them the frequency function $A(m, \mu)$. Following a method developed by W.J. Luyten specially for large μ ($>0''5/\text{year}$), and hence nearby stars, ("Method of the mean absolute magnitudes"), one can deduce the conditional distribution $\Psi(M|m, \mu)$ and then obtain

$$F(m, \mu, M) = A(m, \mu)\Psi(M|m, \mu) .$$ (3.31)

By integration with respect to μ one then gets a distribution $G(m, M)$ – in general after correction for the restriction $(\mu > \mu_0)$ by a factor $k(m, M; \mu_0)$ – and by integration over m and division by V_0 finally $D_0\varphi_0(M)$. $\Psi(M|m, \mu)$ can (for large μ) be found in a similar way. Analogously to the relation between M, m and p, we set

$$H = m + 5 + 5\log \mu .$$ (3.32)

It then appears for stars of known absolute magnitudes, that the mean value $\overline{M}$ over the stars with a specified value of H varies linearly with H (linear regression):

$$\overline{M}(H) = \alpha + \beta H$$ (3.33)

and moreover that, approximately,

$$\Psi(M|m,\mu) = \Psi(M|H) = \frac{1}{\sigma\sqrt{2\pi}}\exp\left\{-\frac{[M-\overline{M}(H)]^2}{2\sigma^2}\right\} \tag{3.34}$$

where σ is independent of H. If α, β and σ are found for the stars of known M, then the distribution $\Psi(M|m,\mu)$ calculated from (3.32) to (3.34) can be applied to the $A(m,\mu)$ from the proper motion catalogue.

3.2.2 Luminosity Function and Spatial Distribution of the Stars

Stars Near the Sun: We consider the region of space $r\lesssim 20$ pc, the only region for which trigonometric parallaxes can be inferred with errors less than about 20% (Sect. 2.2.1), and whose stars can be included with the greatest degree of completeness. For this "neighbourhood" the star density D_0 and the absolute luminosity distribution $D_0\varphi_0(M)$ can be obtained by direct counting and division by the volume V_0 under consideration.

The available observational data for stars of this "nearest neighbourhood of the sun" are collected in a catalogue produced by Gliese (1969) at the Astronomisches Rechen-Institut of Heidelberg, whose second edition includes the known data up to 1968. For 1529 individual stars, double and multiple systems, in total 1890 stars with parallaxes $p \geq 0''045$, this catalogue provides: α, δ, μ_α, μ_δ, p and in most cases also v_r, $\dot{\xi}$, $\dot{\eta}$, $\dot{\zeta}$ (there denoted by U, V, W), spectral type, visual magnitude V, $B-V$, $U-B$ (also here in part $R-I$) and M_V. The interstellar extinction here plays no role. Of these stars 1277 are at distances $r \leq 20$ pc. An extension of the catalogue to the space $r \leq 25$ pc was compiled at the Royal Greenwich Observatory and was published in 1970. The data obtained since 1969 have been published in a special list (Gliese and Jahreiss, 1979). In Table 3.5 the essential data for the 25 nearest stars are collected.

The more recent programmes for discovering nearby stars and determining their trigonometric parallaxes concentrate above all on the white and red dwarfs, which are themselves by no means yet fully discovered in this neighbourhood, to determine their frequency and the profile of the luminosity function for $M_V > +10$. In order to identify these objects, one is searching for the faintest possible stars — in extreme cases down to $m_V \sim 21$ — with large proper motions (greater than about $0''2$/year). Also included here are, of course, intrinsically brighter stars with high space velocities and at distances $r > 20$ pc, which are later eliminated.

The star medley of the nearest neighbourhood of the sun shows the following composition by spectral and luminosity classes: at least 77% of the stars are late Main Sequence types K and above all M, hence red dwarfs. On the other hand there is here not a single O- or B-star — a star of this unusual type so near would be one of the brightest stars in the sky! The giants (III, IV) make up less than 1%, the fraction of sub-dwarfs may be about 4%. The middle Main Sequence stars of types A, F, G constitute about 8%, and the

Table 3.5. The 25 stars nearest to the sun. The first column gives the number of the star in Gliese's catalogue, μ denotes the magnitude of the proper motion, m_V the apparent, and M_V the absolute visual magnitude, p the parallax, and ξ, η, ζ are the components of the space velocity relative to the sun defined in Sect. 2.1.3: ξ-axis $l = 0°$, $b = 0°$, η-axis $l = 90°$, $b = 0°$, ζ-axis $b = +90°$. (Data from Gliese, 1969)

No.	Name[a]		α_{1950} [h m s]	δ_{1950} [° ']	μ ['/year]	m_V [mag]	Spectral type	p[b] [']	M_V [mag]	ξ	η	ζ [km s^{-1}]
15	BD+43°44	A	0 15 31	+43 44.4	2.90	8.07	M1Ve	0.282	10.32	−49	−11	− 4
		B				11.04	M6Ve		13.29	−52	− 5	− 6
65	L 726−8	A	1 36 25	−18 12.7	3.36	12.45	dM5.5e	0.367	15.27	−44	−20	−19
	= UV Cet	B				12.95	dM5.5e		15.8	−45	−20	−22
71	τ Cet	A	1 41 45	−16 12.0	1.92	3.50	G8Vp	0.277	5.72	+18	+29	+12
244	α CMa	A	6 42 57	−16 38.8	1.33	− 1.46	A1V	0.377	1.42	+15	0	−11
		B				8.68	A		11.56			
280	α CMi	A	7 36 41	+ 5 21.3	1.25	0.37	F5VI–V	0.285	2.64	+ 5	− 9	−18
		B				10.7	F		13.0			
406	Wolf 359		10 54 6	+ 7 19.2	4.70	13.53	dM8e	0.426	16.68	−26	−44	−18
411	BD+36°2147		11 0 37	+36 18.3	4.78	7.50	M2Ve	0.397	10.49	+46	−53	−74
447	Ross 128		11 45 9	+ 1 6.0	1.38	11.10	dM5	0.301	13.50	+18	− 4	−17
551	Proxima Cen		14 26 19	−62 28.1	3.85	11.05	dM5e	0.761	15.45	−25	− 3	+14
559	α Cen	A	14 36 11	−60 37.8	3.68	− 0.01	G2V	0.743	4.35	−29	+ 1	+14
		B				1.33	K0V		5.69			
699	Barnards star		17 55 23	+ 4 33.3	10.34	9.54	M5V	0.552	13.25	−91	−55	−19
725	BD+59°1915	A	18 42 12	+59 33.3	2.29	8.90	dM4	0.282	11.15	−26	−12	+26
		B	18 42 13	+59 33.0	2.27	9.69	dM5		11.94	−24	− 3	+31
729	AC−24°2833−183	A	18 46 45	−23 53.5	0.72	10.6	dM4.5e	0.345	13.3	− 6	+ 1	− 9
820	61 Cyg	A	21 4 40	+38 30.0	5.22	5.22	K5Ve	0.296	7.58	−90	−53	− 8
		B				6.03	K7Ve		8.39			
845	ε Ind		21 59 33	−56 59.6	4.70	4.68	K5Ve	0.291	7.00	−77	−38	+ 4
866	L 789−6		22 35 45	−15 35.6	3.26	12.18	dM7e	0.305	14.60	−67	− 2	+41
887	CoD−36°15693		23 2 39	−36 8.5	6.90	7.36	M2Ve	0.279	9.59	−102	−15	−57
905	Ross 248		23 39 26	+43 55.2	1.60	12.29	dM6e	0.318	14.80	+34	−77	+ 1

[a] BD = Bonner Durchmusterung, CoD = Cordoba Durchmusterung, L = Luyten's Catalogue, AC = Astrographical Catalogue.

[b] The errors in parallax are about $\pm 0''.010$ throughout

white dwarfs about 10%. These numbers should be regarded only as a guide. Precise allowance for the selection effect could still increase the contribution of intrinsically faint white and red dwarfs.

Also remarkable is the fraction of double and multiple stars: if one counts up the components in the region of near completeness, $r \leq 5$ pc, it appears that 59% of all stars near the sun are members of double or multiple systems! The ratio of the number of these systems to the total of all the true single stars plus the systems amounts to 0.39.

The star density D_0 for the data of Gliese's catalogue shows a systematic decrease when one increases the region of the count from $r \leq 4$ pc progressively to $r \leq 20$ pc. The following numbers give the ratio of the densities inside the successive spherical shells to the density for $r \leq 4$ pc:

Distance interval [pc]	0...4	4...5	5...10	10...15	15...20
$D_0/D_0(r \leq 4 \text{ pc})$	1.00	0.76	0.48	0.36	0.28

The reason is the growing incompleteness of the catalogue with increasing distance. Completeness is assumed up to $m_V \approx 8.5$. Inside 20 pc all the stars with $M_V \approx 7$ or brighter are included. At the same time the stars of low absolute brightness — in actuality particularly frequent — down to $M_V \approx 10.5$ should still be completely accounted for up to 4 pc. Even for $r \leq 4$ pc the stars with $M_V \gtrsim 11$ will be only partially included. The extrapolation of the absolute star densities to a fully inclusive limiting value $(r \to 0)$ is still relatively uncertain and leads currently to the conclusion:

$$D_0 \approx 0.10 \quad \text{to} \quad 0.15 \, \text{stars pc}^{-3} \, ,$$

corresponding to 3000 to 4000 stars within 20 pc. We therefore know only the smaller part of all stars in this volume! As a standard value one can reckon a number of about 100 stars per 1000 pc^3. The 25 stars of Table 3.5 are located inside a sphere of radius $r = 3.7$ pc, corresponding to $p = 0''.27$, and accordingly lead to about 120 stars per 1000 pc^3.

Luminosity Function (LF): The absolute frequencies $D_0 \varphi_0(M_V) V_0$ for the spherical volume with $r = 20$ pc are shown in Fig. 3.10. The incompleteness of the data was here allowed for to a certain degree by successive decrease of the underlying volume in going to ever fainter absolute magnitudes. The most noteworthy feature of the profile is that the maximum occurs at $M_V \approx +14$.

The summation gives $D_0 V_0 \approx 3600$ stars. The three intrinsically brightest of these stars are the giants (III) Aldebaran (α Tau A), Arcturus (α Boo) and Capella (α Aur A) with $M_V = -0.6, -0.24$ and $+0.09$. Of the (incompletely known) intrinsically faint stars, BD $+ 4°4048$ B, the component of a brighter star, is the faintest, with $M_V = +18.57$; of the currently known single stars, the faintest has a brightness $M_V = +17.0$.

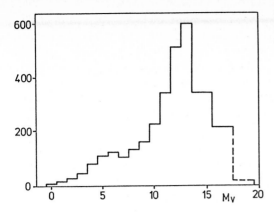

Fig. 3.10. Absolute values of the luminosity function $D_0\varphi_0(M_V)V_0$ of the stars near the sun. (After Jahreiss, 1974)

With the help of the empirical mass-luminosity relationship $\mathcal{M} = \mathcal{M}(M_V)$ the mean mass density of stellar material in the region near the sun $\overline{\varrho}_0 = D_0\sum\varphi_0(M_V)\mathcal{M}(M_V)$, expressed in solar masses $\mathcal{M}_\odot$ per cubic parsec, can be calculated from $D_0\varphi_0(M_V)$. With the profile of Fig. 3.10 we derive:

$$\overline{\varrho}_0 = 0.046\,\mathcal{M}_\odot\mathrm{pc}^{-3}\hat{=}3\times10^{-24}\mathrm{g\,cm}^{-3}\ .$$

This corresponds to about 2 hydrogen atomic masses per cubic centimetre. Since $\varphi_0(M_V)$ is possibly still incomplete for $M_V\gtrsim +13$, this value could still be somewhat too low (see Sect. 2.1.4).

We turn now to the applications of the stellar statistical methods described in Sect. 3.2.1, first of all to the luminosity function and subsequently to the density distribution in the most widespread possible region in the neighbourhood of the sun. Of the older results for the luminosity function we produce only the profile derived in 1936 by van Rhijn by the method of trigonometric parallaxes (see Fig. 3.11). Since trigonometric parallaxes can be regarded as reliable

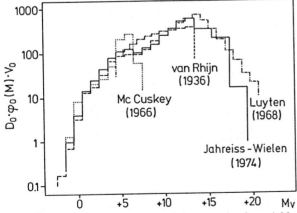

Fig. 3.11. Luminosity function of the stars in the neighbourhood of the sun from various investigations. The continuous line corresponds to the profile shown in Fig. 3.10 for stars near the sun ($r\lesssim20\,\mathrm{kpc}$). Ordinate: number of stars in the interval $M_V - \frac{1}{2}\ldots M_V + \frac{1}{2}$ in the volume of a sphere with radius 20 pc. (From Jahreiss, 1974)

only when they are greater than $0''.01$, the volume included is fixed at about $r \lesssim 100\,\mathrm{pc}$. The extensive recent results for the region of faint absolute magnitudes were obtained by Luyten with the method of mean absolute magnitude. The initial data for an investigation of 1968 were the proper motions of about 4000 stars with $\mu > 0''.5$/year – about half with $m_{\mathrm{pg}} = 15\ldots21$ – which Luyten had found on over a hundred plates taken with the Palomar-Schmidt telescope. About 15% of these stars are closer than 10 pc, but a significant fraction were located beyond $r = 20\,\mathrm{pc}$.

The procedures starting from (3.27) or (3.16), based on the derivation of star densities $D_S(r)$ for narrow spectral groups or of star numbers for narrow luminosity groups $A(m, M)$, have enabled us to make assertions on the luminosity function at great distances and in various directions. The first way was originally described by S.W. McCuskey in 1956: first of all a total of about 18 000 stars to a limiting brightness $m_{\mathrm{pg}} = 12$ were classified on objective-prism photographs of 11 fields of the northern Milky Way at $b \approx 0°$, taken with the 60 cm Schmidt telescope of the Warney and Swasey Observatory in Cleveland (Ohio). M_{0s} and σ_{MS} were derived for spectral groups with $\sigma_{MS} \leq 0''.9$, and thereafter the star densities $D_S(r)$ were obtained from the star numbers $A_S(m)$ (we shall go into this further). Then (3.27) gave the luminosity function for each of the fields. The specified limiting magnitude has the effect that the conclusions extend at most to $M_{\mathrm{pg}} \approx +7$ and hence the distance coverage is $r \approx 100\ldots400\,\mathrm{pc}$. More recently these investigations have been extended to further fields.

The results of van Rhijn, Luyten, McCuskey (mean profile for 9 fields) and the profile for the stars near the sun ($r \leq 20\,\mathrm{pc}$) are compared with one another in Fig. 3.11. If one overlooks the discrepancies which arise in each case over about the last two magnitudes from the end, the agreement can be regarded as very satisfactory. The true maximum appears to be bracketed by the profiles of Luyten and Jahreiss-Wielen ($r \leq 20\,\mathrm{pc}$): it is located at about $M_V = +13$ to $+14$. The luminosity function of the stars near the sun is given in Table 3.6 numerically, together with a frequently used "Standard Profile" based on the results of van Rhijn, McCuskey and Luyten.

Between the results for various fields of the Milky Way at great distances one finds in many cases differences up to a factor of four, which may be only partly real. At distances of about 1 kpc and over, one is already including within galactic latitudes $|b| \approx 6°$ stars at distances from the galactic plane $|z| > 100\,\mathrm{pc}$, so that the result is modified by the dependence of the luminosity function upon z (see below).

The fourth and fifth columns of Table 3.6 contain the *volume emission* or "luminosity density" in the V-spectral region $\varepsilon_V(M_V)$ and the *mass density* $\overline{\varrho}(M_V)$, respectively, of the star field in the solar neighbourhood, in each case for the interval $M_V - \frac{1}{2}\ldots M_V + \frac{1}{2}$. $\varepsilon_V(M_V)$ is given by multiplication of $D_0\varphi_0(M_V)$ by the luminosity ratio $L_V/L_V^\odot = 10^{-0.4(M_V - M_V^\odot)}$, in which the value $M_V^\odot = +4.77$ has been used. The emission is thereby expressed in solar luminosities $L_V^\odot$. $\overline{\varrho}(M_V)$ follows by multiplication of $D_0\varphi_0(M_V)$ by $\mathcal{M}(M_V)/\mathcal{M}_\odot$

Table 3.6. Luminosity function $D_0\varphi_0(M_V)$ for $r\leq 20$ pc after Jahreiss, and also a mean profile from the results of van Rhijn, McCuskey and Luyten, which covers roughly the region $r\lesssim 100$ pc. For this profile we also give the emission of the star field per unit volume in the V-region $\varepsilon_V(M_V)$, expressed in solar luminosities $L_V^\odot$, and the mass density $\bar\varrho_0$ in solar masses $\mathcal{M}_\odot$, for the interval of absolute magnitude $M_V - \frac{1}{2}\ldots M_V + \frac{1}{2}$. The reference volume is always $10^4\,\mathrm{pc}^3$

M_V	$D_0\varphi_0(M_V)$ $[10^{-4}\,\mathrm{pc}^{-3}\,\mathrm{mag}^{-1}]$		$\varepsilon_V(M_V)$ $[10^{-4}\,L_V^\odot\,\mathrm{mag}^{-1}\,\mathrm{pc}^{-3}]$	$\bar\varrho_0(M_V)$ $[10^{-4}\,\mathcal{M}_\odot\,\mathrm{mag}^{-1}\,\mathrm{pc}^{-3}]$
	$r\leq 20$ pc	$r\lesssim 100$ pc		
-6		1×10^{-4}	2	5×10^{-3}
-5		6×10^{-4}	5	2×10^{-2}
-4		2.9×10^{-3}	9	6×10^{-2}
-3		1.3×10^{-2}	17	1.7×10^{-1}
-2		5×10^{-2}	26	0.5
-1		0.25	51	1.6
0	1	1	81	4
1	4	3	97	10
2	7	6	77	12
3	13	12	61	18
4	23	19	39	23
5	32	34	28	37
6	36	42	14	38
7	30	35	4	26
8	39	42	2	26
9	47	54	1	29
10	73	78	1	34
11	102	98		35
12	153	107		34
13	178	117		28
14	102	129		23
15	102	125		20
16	64	120		15
17	64	107		9
18	>5	83		6
19	>5	50		4
20		30		2
21		13		1
22		5		1
Total	>1080	1310	515	437

Total over $\varepsilon_{pg}(M_{pg})$: 777

from the mass-luminosity relationship. As one can see, the total volume emission in the V-region — and hence the surface brightness of the Milky Way I_V — is contributed mainly by the numerically sparse stars of considerable intrinsic brightness, whilst the total stellar mass density is supplied mainly by the numerous intrinsically faint stars. The reason for this is the relatively slow decrease in mass in going from intrinsically bright stars to intrinsically faint stars $\mathcal{M}\sim L^{1/3}$ approximately. The total emitted light $\mathcal{L}_V = \sum \varepsilon_V(M_V)$ in the neighbourhood of the sun turns out to be about $0.5\,L_V^\odot$ per cubic parsec.

A characteristic quantity for a star medley is the total *mass-luminosity ratio* of the volume under consideration. For the star field of the solar neighbourhood we obtain from the numbers at the foot of Table 3.6

$$\frac{\mathcal{M}/\mathcal{M}_\odot}{\mathcal{L}_V/\mathcal{L}_V^\odot} = 0.85 \ , \quad \frac{\mathcal{M}/\mathcal{M}_\odot}{\mathcal{L}_{pg}/\mathcal{L}_{pg}^\odot} = 0.56$$

so that in the visual region it approximates to a solar mass per solar luminosity. With the data for the stars near the sun ($r \leq 20\,\mathrm{pc}$) the values are insignificantly different, for example 0.7 for the photographic region. Let us quote for comparison the ratio $(\mathcal{M}/\mathcal{M}_\odot)/(L_V/L_V^\odot)$ for individual stars of the types A0V, G0V and M0V: the values obtained are, roughly, 0.1, 1 and 20.

Of interest for comparison with other galaxies are the luminosity function and especially the mass-luminosity ratio in a cylinder perpendicular to the galactic plane — the latter quantity can be directly estimated for extragalactic star systems. From the results discussed further below on the spatial distribution of the stars, the star density falls off significantly less steeply with distance z from the galactic plane for the intrinsically faint Main Sequence stars than for the bright early types (Table 3.7). Accordingly the luminosity function changes with increasing z in the sense that the fraction of the inherently fainter stars increases in comparison with that of the inherently bright stars: $\varphi(M|z)$ shows for $|z| > 100\,\mathrm{pc}$ a notably steeper climb with M than $\varphi_0(M)$. For the cylinder perpendicular to the galactic plane, therefore, one gets a higher mass-luminosity ratio, by about a factor of two, than for $|z| \lesssim 100\,\mathrm{pc}$.

Table 3.7. Relative density distribution $D(z)$ perpendicular to the galactic plane in the direction of the galactic north pole for various normal star types, expressed as a percentage of the respective value of $D(z)$ when $z = 50\,\mathrm{pc}$. (From Upgren, 1962, 1963; Sturch and Helfer, 1972)

z[pc]	Main sequence (V)				Giants (III)			
	A0...A5	A7...F2	F5...F8	G0...G2	G8...G9	K0...K2	M	
50	100	100	100	100	100	100	100	
100	51	23	53	97	100	59	100	
200	8	6	23	80	100	32	93	
400	1	1	8	50	80	11	30	
600	0.3	0.4	–	–	55	5	13	
800	0.2	0.2	–	–	40	4	10	
1000	0.1	0.1	–	–	30	2	7	

From $D_0\varphi_0(M_V)$ we can, by means of the mass-luminosity relationship $\mathcal{M} = \mathcal{M}(M_V)$, obtain the frequency distribution of the star masses, the so-called *mass-function*:

$$F(\mathcal{M}) = D_0\varphi_0(M_V)\frac{dM_V}{d\mathcal{M}} \ . \tag{3.35}$$

$F(\mathcal{M})$ falls off very steeply with increasing mass above about $\mathcal{M} = 0.5\mathcal{M}_\odot$, on average roughly proportional to $\mathcal{M}^{-4}$; for $\mathcal{M}<0.5\mathcal{M}_\odot$ the decrease is appreciably flatter. We shall return to this later.

Initial Luminosity Function: The luminosity function is of importance also for the theory of star formation. The stars of the vicinity of the sun show a mix of objects of various ages. Corresponding to their relatively short stay on the Main Sequence $\tau_{\mathrm{MS}}(M_V)$, the inherently bright B0V stars must, for example, be younger than about 2×10^7 years. For Main Sequence stars with $M_V>5^m$, on the other hand, $\tau_{\mathrm{MS}}\gtrsim T =$ age of the galaxy $\approx 10^{10}$ years ($\approx$ age of the oldest globular clusters): all mass-poor and inherently faint stars with $M_V> + 5^m$ originating in the past are still extant, whereas more massive and inherently brighter Main Sequence stars have an age of only up to $\tau_{\mathrm{MS}}(M_V) -$ the remainder of these objects have become giants or white dwarfs already.

One defines the "initial luminosity function" of the newly created stars on the Zero-Age Main Sequence $\psi(M_V)$, and also the star formation rate $s(t) =$ number of stars which come into existence at time t per unit of time and of volume. The number of stars with absolute magnitudes in the interval $M_V\ldots M_V+dM_V$ which are created in the time interval $t\ldots t+dt$ ("birth rate") is therefore given by $\psi(M_V)dM_V s(t)dt$. Integration from $t = 0$ up to the present time $t = T$ gives all the stars ever created in the interval $M_V\ldots M_V + dM_V$. Of these, however, only those stars created after the time $t = T - \tau_{\mathrm{MS}}(M_V)$ are still extant, so that

$$D_0\varphi_{\mathrm{MS}}(M_V) = \psi(M_V) \int\limits_{T-\tau_{\mathrm{MS}}(M_V)}^{T} s(t)dt .$$ (3.36)

We wish to consider here only the case of constant star formation rate $s(t) = s_0$. Then for $M_V<+5$ we have directly $D_0\varphi_{\mathrm{MS}}(M_V) = \psi(M_V)s_0\tau_{\mathrm{MS}}(M_V)$ and for $M_V> + 5$, since $\tau_{\mathrm{MS}} = T$, $D_0\varphi_{\mathrm{MS}}(M_V) = \psi(M_V)s_0T$. Since all stars with $M_V> + 5$ are still extant, for this region $\varphi_{\mathrm{MS}}(M_V) = \psi(M_V)$, and so $s_0 = D_0/T$. We thus find that

$$\varphi_{\mathrm{MS}}(M_V) = \psi(M_V)\cdot\frac{\tau_{\mathrm{MS}}(M_V)}{T}$$ (3.37)

with $\tau_{\mathrm{MS}}(M_V)/T\leq 1$, i.e. $\varphi_{\mathrm{MS}}(M_V)\leq\psi(M_V)$: the intrinsically bright, shortlived star types created before the time $t = T - \tau_{\mathrm{MS}}(M_V)$ are missing.

With the observed luminosity function under the restriction to Main Sequence stars $\varphi_{\mathrm{MS}}(M_V)$, and the values for $\tau_{\mathrm{MS}}(M_V)$ and $\tau_{\mathrm{MS}}(Sp)$ from the theory of star evolution the initial luminosity function $\psi(M_V)$ can be derived from (3.37). Whilst using the distribution $\varphi_{\mathrm{MS}}(M_V)$ it should be borne in mind that a portion of the stars continually created near the galactic plane moves to a greater distance z in course of time through gravitational interaction with other stars and interstellar clouds. According $\varphi_{\mathrm{MS}}(M_V)$ refers to a cylinder perpendicular to the galactic plane. The result for $\psi(M_V)$ is shown in Fig. 3.12 together with $\varphi_{\mathrm{MS}}(M_V)$.

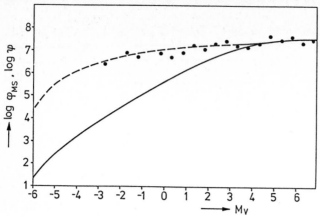

Fig. 3.12. Initial luminosity function $\psi(M_V)$ (- - -) and the luminosity function $\varphi_{MS}(M_V)$ for the Main Sequence stars (——). The points reproduce the profile for the Pleiades. The normalisation factor is chosen so that the two profiles coincide for $M_V > 5$

In young open star clusters there should be a distribution of absolute magnitudes which approximately coincides with $\psi(M_V)$. This is actually the case and confirms the usefulness of the relation (3.37). As an example Fig. 3.12 shows the luminosity function obtained by simple counting for the Pleiades, whose age is at most about 10^8 years. The assumption of a constant star formation rate throughout cannot of course be taken as established. It is simply the simplest assumption with which the observed facts can be explained within their framework of error limits. The density of non-stellar material was probably higher at the start of the development of the galaxy than now, so that there was then also a higher star formation rate. After a few 10^9 years, however, it seems that a roughly constant star formation rate had been established.

Since the mass function $F(\mathcal{M})$ described above refers to a mix of various old stars, one defines analogously to $\psi(M_V)$ an initial mass function $F_0(\mathcal{M})$. As a larger fraction of massive objects is still among the young stars, $F_0(\mathcal{M})$ falls off less steeply in the region $\mathcal{M} > 1 \mathcal{M}_\odot$ than $F(\mathcal{M})$ – corresponding to the behaviour of $\psi(M_V)$ compared with $\varphi(M_V)$ (Fig. 3.12). One usually takes $\log \mathcal{M}$ as independent variable and by means of

$$D_0 \psi(M_V) dM_V = \xi_0(\mathcal{M}) d\ln\mathcal{M} = F_0(\mathcal{M}) d\mathcal{M} \tag{3.38}$$

introduces the function $\xi_0(\mathcal{M}) = F_0(\mathcal{M})\mathcal{M}$. $\xi_0(\mathcal{M})d\mathcal{M}$ is the mass of all the young stars contained in unit volume with individual masses in the interval $\mathcal{M} \ldots \mathcal{M} + d\mathcal{M}$. E.E. Salpeter found in 1955 that, for $\mathcal{M} \gtrsim 2 \mathcal{M}_\odot$, $\xi_0(\mathcal{M}) \sim \mathcal{M}^{-1.35}$. The comparison with the mass function $\xi(\mathcal{M}) = F(\mathcal{M})\mathcal{M}$ observed nowadays is shown schematically in Fig. 3.13. For $\mathcal{M} < 0.5 \mathcal{M}_\odot$ the common profile of $\xi(\mathcal{M})$ and $\xi_0(\mathcal{M})$ is still not sufficiently well determined.

For questions of star formation from interstellar matter (Sect. 6.3.4) it is of interest to express the *magnitude of the star formation rate* in mass units

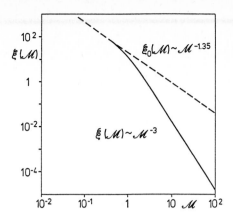

Fig. 3.13. Observed and initial mass functions (schematic). Abscissa: star masses in units of the solar mass on a logarithmic scale. The ordinate values give the mass contribution per 1000 pc^3 of the stars with individual masses in the interval $\mathcal{M} \ldots \mathcal{M} + \Delta\mathcal{M}$ of width $\Delta\mathcal{M} = 1\mathcal{M}_\odot$

per unit of time and per unit of volume. Restricting consideration merely to stars with $M_V \gtrsim +5$, and hence $\mathcal{M} \lesssim 1\mathcal{M}_\odot$, this quantity is obtained simply by analogy to $s_0 = D_0/T$ as

$$\frac{d\overline{\varrho}_0}{dt} = \frac{\overline{\varrho}_0(M_V \gtrsim 5)}{T} = \frac{1}{T}\int\limits_5^\infty D_0\varphi(M_V)\mathcal{M}(M_V)dM_V \ .$$

The last column of Table 3.6, combined with $T = 10^{10}$ years, gives for the right-hand side of this equation about $4 \times 10^{12}\mathcal{M}_\odot\mathrm{pc}^{-3}$ per year. Taking account of the massive but infrequent stars with $M_V < 5$ raises the mass density only by about 10 to 30 per cent. As the lowest value of the total star formation rate in the vicinity of the sun we therefore obtain

$$\frac{d\overline{\varrho}_0}{dt} \approx 5 \times 10^{-12}\mathcal{M}_\odot\mathrm{pc}^{-3}\mathrm{year}^{-1} \ .$$

Star Density Distribution in the Galactic Plane: For the present we exclude O- and early B-stars, since the number density of these scarce objects is too small for the application of the statistical methods described in Sect. 3.2.1; they will be treated in Sect. 3.3.1. For the reasons explained (Sect. 3.2.1), the only investigations appropriate for the derivation of density distributions for the main body of normal stars are those in which a partition has been made into narrow spectral and luminosity groups and the relevant profile of interstellar extinction $\Delta m(r)$ has been taken into account. This requirement has been satisfied in essence only by the results of the last three decades. The great expenditure of effort means that usually only small individual fields – and there often only the distribution of a narrow type group of stars – have been investigated. For years past completed programmes, in which a larger part of the Milky Way is systematically covered by numerous random sample fields, are correspondingly scarce. It amounts to this, that the range of density determinations depends strongly upon the luminosity of the star type considered (Sect. 2.2.2, Table 2.2). Accordingly, in spite of a large number of individual investigations, we have today still no fully verified and generally accepted detailed chart of the star density distribution in the galactic plane up to distances of one, and even more so, several kiloparsecs, whether for the individual middle and late spectral groups or even for the totality of all these common types. We restrict ourselves therefore to discussion of examples, namely the partial results from two large programmes: (1) work on the basis of spectral classification which has already been mentioned in the discussion of the luminosity function, and (2) an investigation on the basis of a three-colour photometry [Sect. 3.2.1, (3.16)].

For the first case we take the determination of the density function $D_s(r)$ from the star number profile $A_s(m_{pg})$ for early A-stars in the directions of 11 Milky Way fields between $l \approx 40°$ and $l \approx 210°$. The size of field lies in each case between 10 and 20 square degrees, the limiting magnitude in the older investigations (S.W. McCuskey, 1956) described here first of all was $m_{pg} \approx 12$, and later $m_V \approx 13$. On the basis of luminosity determinations by primary methods the values obtained [see (3.15)] for the spectral groups B8...A0 and A2...A5 were $M_0 = +0.5$ and $+2.0$, respectively, and $\sigma_M = \pm 0.8$ (both groups). The profile of the important interstellar extinction $\Delta m_{pg}(r)$ was individually derived for each field: from the colour index CI of an adequate number of stars for which, besides m_{pg}, the magnitudes in a further spectral region were available it follows that $m_{pg} = KE$ with the colour excess $E = CI - (CI)_0$ and $(CI)_0$ denotes the intrinsic colour (by spectral type); the constant K is known from the law of interstellar reddening (Sect. 4.1.1). A number of numerical methods were used for the solution of the fundamental equation (3.11). The resulting density distributions in the galactic plane are shown in Fig. 3.14 for both spectral groups.

A frequently occurring feature of such results, which is also detectable in Fig. 3.14, consists in a preferential extension of the star concentration oriented

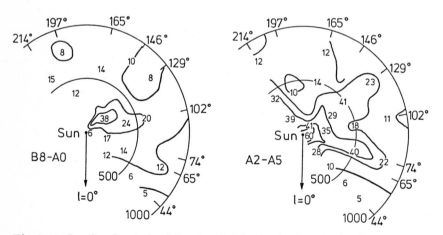

Fig. 3.14. Results of a study of the star distribution in the galactic plane $l = 40°...210°$, $r \lesssim 1$ kpc for the two spectral groups B8...A0 and A2...A5. The density values are quoted in units of 10^5 stars pc^{-3}. The galactic longitudes of the 9 selected fields are indicated outside. (From McCuskey, 1965)

radially from the sun. These shapes are usually not real but are produced by errors (scatter) in the values used for the absolute magnitudes and the interstellar extinction. Thus for example the radial extension of the concentration of A-stars at $l \approx 130°$ (Cassiopeia − Perseus) − which also appears for F0...F5 and G-giants − was discounted to a considerable degree, as more recent investigations were based upon more exact UBV- and $uvby$-magnitudes and absolute magnitudes on the basis of MK-classifications and measurements of the Hβ-index. The maximum of the star density lies at a distance $r \approx 250\,\mathrm{pc}$ (at $l \approx 133°$). The "Perseus-Arm" (Sect. 3.3.1) at $r \approx 2\,\mathrm{kpc}$ appearing in the distribution of the O- and B-stars is possibly indicated in more recent results for B8...A0 stars. The star concentration around $l \approx 70°$ (Cygnus) visible in Fig. 3.14 can be regarded as a part of the "local arm" (see Sect. 3.3.1).

As an example of the application of the method of quantitative photometric classification of the stars by a multi-colour photometry, we consider the results from W. Becker's Basle school for a small field of 0.14 square degrees near the galactic centre in Sagittarius (centre point of the field: $l = 1.1°$, $b = -1.1°$). The basic data consisted of the magnitudes in the three spectral regions R, G, U (red, yellow, ultraviolet analogous to V, B, and U) − with m_R, m_G, m_U replaced for simplicity in the usual way by R, G, U. The magnitudes were determined by photographic photometry on plates from the 122 cm Schmidt telescope at Mt. Palomar Observatory, for all stars of the field up to a limiting magnitude $G = 16.5$, altogether 2220 stars. The most important advantage of the exclusive application of the integral magnitudes is the relatively long range.

The "fractionated" two-colour diagrams, in which in this case $U-G$ is plotted against $G-R$, were obtained for various regions of the apparent magnitude G, in each case for a specific interval in G. These were then applied to the derivation of the interstellar extinction Δm_G in that G interval. From the shift of the field point compared with the Main Sequence or Giant Branch for non-discoloured stars, the colour excess E_{G-R} can be determined for different intervals of the absolute magnitude M_G (displacement along the reddening line). The interstellar extinction in this colour system is given by $\Delta m_G = 2.7 E_{G-R}$. The distance r is determined by G, M_G and m_G, so that $\Delta m_G(r)$ then follows. The result for E_{G-R} as a function of the distance parameter is reproduced in Fig. 3.15. After a weak, or indeed barely verifiable, "foreground extinction"

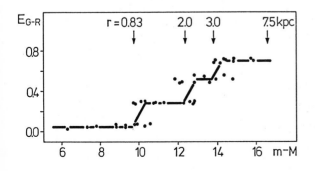

Fig. 3.15. Reddening E_{G-R} as a function of the distance module $m - M = G - M_G$ for a small field near the direction of the galactic centre. (From Gschwind, 1975)

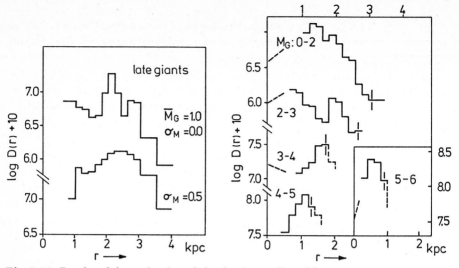

Fig. 3.16. Results of determination of the density profile $D(r)$ for Main Sequence stars in different intervals of absolute magnitude M_G and for late giants by the method of three-colour photometry for a small field near the direction of the galactic centre. The completeness limit is indicated by vertical strokes. In the lower curve in the left-hand diagram, the effect of a scatter of the absolute magnitudes $\sigma_M = 0.5$ is taken into account. Unit for $D(r)$: number of stars pc^{-3}. (From Gschwind, 1975)

one finds three clouds at distances $r = 0.83$, 2.0 and 3.0 kpc with extinction values $\Delta m_G = 0.65$, 0.65 and 0.5 mag. The occurrences near the distances of the clouds, of points at equal abscissae with two different values of E_{G-R} is caused by errors in $m-M$.

The star density profile $D(r)$ for the Main Sequence stars or the giants, in each case in a specific interval of the absolute magnitude M_G, is obtained from (3.16) in the manner described there. An extract of the results is shown in Fig. 3.16. Disregarding here the detail, which because of the random sample character of these profiles should not be accredited with too much significance, one can make the following observation: in each case, starting from the value D_0 for the region near the sun, the densities generally increase to a maximum value which is reached at between 1 and 2 kpc; thereafter the densities decrease. This finding is in accord with the idea that, at a distance of about 1.5 kpc from the nearby "local arm", the next inner spiral arm of the Milky Way system is located (Sect. 3.3.1). The course of the curves in Fig. 3.16 accordingly represents a section of this "Sagittarius Arm".

A general problem in density determinations lies in that the completeness of the data, especially for the fainter stars, decreases rapidly with distance. Thereby a fall in density is frequently simulated long before the boundary of the distance region under consideration is reached.

Distribution Perpendicular to the Galactic Plane: Because of the strong concentration of interstellar dust at the galactic plane, predominantly at distances $|z| < 100$ pc (Sect. 5.4.1), the star numbers $A(m|l, b)$ for medium and high galactic latitudes b are far less affected by the influence of interstellar extinction than at $b \approx 0°$. The decrease in star density with increasing z for individual star types is therefore relatively well known.

Already the apparent distribution of the different spectral groups on the sphere (Table 3.3) allows the conclusion of very different concentrations near the galactic plane. The application of the methods of density analysis described above confirmed this and produced the relative density profiles displayed in summary in Table 3.7.

It is noticeable above all that the density decrease in going from early to middle Main Sequence stars is distinctly flatter; this tendency appears to continue up to the late dwarfs, although the range here is only small. The giants show throughout a significantly weaker density gradient than A- and F-stars. A comprehensive discussion of the distribution of the various objects in the Milky Way system perpendicular to the galactic plane follows in Sect. 3.5.1.

3.2.3 Motions of the Stars in the Solar Vicinity

The determination of the individual space velocity of a star requires knowledge of the proper motion (PM), radial velocity (RV) and distance (Sect. 2.1.3). The PM is known for a relatively large number of stars, the RV as well is known for only a portion of these stars, and the distances can be determined for only a sub-group of these. Space velocities can be derived only for this last, relatively small number of stars, which does not provide a representative sample for that purpose (selection for large PM, preference for intrinsically bright stars, etc.). Even in the 1131 "nearby" stars in Gliese's (1969) catalogue with known μ_α, μ_δ, v_r and r the intrinsically faint objects with large space velocities are over-represented, so that the nearby M-dwarfs with small space velocities are underestimated. Fortunately important information on the kinematics of a group of stars can also be obtained from the RV alone, and with certain limitations also from the PM, namely evidence on the solar motion relative to the star group and on the velocity ellipsoid. However let us first of all consider the data in Gliese's catalogue.

Kinematics of the Stars near the Sun with $r \leq 20$ pc: We consider results for the kinematic behaviour of the Main Sequence spectral groups A2...A6, A7...F1, F2...F6, F7...G3 and K...M. The last group consists of the stars of a spectral survey of red dwarfs carried out by the McCormick Observatory, in which no selection was made for large PM. The star groups are of various mean ages, since these field stars may indeed have originated at all possible times, but according to spectral type the age must lie below a certain maximum value (lifetime on the Main Sequence). One obtains mean ages, which increase from 4×10^8 years for the A2...A6 group up to 5×10^9 years for K...M.

First of all the mean values of the velocity components in the galactically oriented reference system $\bar{\xi}$, $\bar{\eta}$, $\bar{\zeta}$ (see Sect. 2.1.3) have been calculated for each

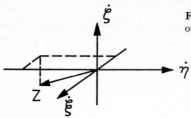

Fig. 3.17. Location of the centroid (Z) in velocity space of the red dwarf stars in the neighbourhood of the sun

star group, whose field point in velocity space is denoted as the centroid of the motions of these stars. For example, for the last group, the red dwarfs, one gets the results illustrated in Fig. 3.17, $\bar{\dot{\xi}} = -5\pm3\,\mathrm{km\,s}^{-1}$, $\bar{\dot{\eta}} = -19\pm2\,\mathrm{km\,s}^{-1}$, $\bar{\dot{\zeta}} = -7\pm2\,\mathrm{km\,s}^{-1}$. With reversed signs these indicate the components of the solar velocity relative to these stars $U_\odot$, $V_\odot$, $W_\odot$ in the directions $l = 0°$ and $b = 0°$, $l = 90°$ and $b = 0°$, and $b = +90°$. The magnitude of the velocity vector in this case is about $21\,\mathrm{km\,s}^{-1}$ and its direction points to $L_\odot = 76°\pm7°$, $B_\odot = +18°\pm4°$. Table 3.8 displays the components of the solar velocity obtained in relation to each of the five spectral groups. The systematic increase of $V_\odot$ in going from early to late spectral types is noticeable here.

Table 3.8. Components $U_\odot$, $V_\odot$, $W_\odot$ of the solar velocity and magnitude of the total velocity dispersion Σ (all quantities in $\mathrm{km\,s}^{-1}$) for different spectral groups of the stars in the neighbourhood of the sun (from Jahreiss, 1974)

Component	A2...A6	A7...F1	F2...F6	F7...G3	K...M
$U_\odot$	2	11	12	16	5
$V_\odot$	6	6	10	21	19
$W_\odot$	8	6	11	6	7
Σ	20	19	32	47	50

The distribution of the peculiar velocities

$$U = \dot{\xi} - \bar{\dot{\xi}}, \quad V = \dot{\eta} - \bar{\dot{\eta}}, \quad W = \dot{\zeta} - \bar{\dot{\zeta}}$$

of the stars near the sun is illustrated in Fig. 3.18 by the example of the results for the group K...M. The projections of the velocity data on the coordinate planes are in each case roughly schematised by three contours of equal frequency. Qualitatively similar results are also obtained for the other spectral groups. We summarise the implications of Fig. 3.18 in the following statements:

1. the surfaces of equal frequency can be approximately represented by triaxial ellipsoids;
2. the longest axis lies close to the line joining the galactic centre to the anti-centre ($l = 0°$, $180°$, respectively, $b = 0°$), but when this coincidence is not exact, one speaks of a "vertex deviation";
3. the two shorter axes are nearly equal;

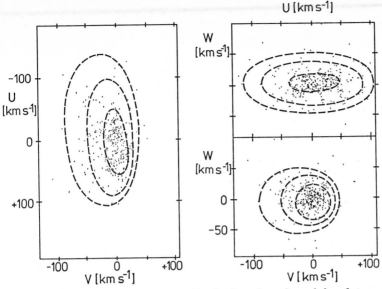

Fig. 3.18. Projections of the velocity distribution of nearby red dwarf stars. Explanation in text

4.   the V-component shows an asymmetry, which consists of an increased number of stars with large negative values of V ("asymmetry of large peculiar velocities").

Moreover, the values of the standard deviation of velocity $\Sigma = [\Sigma_U^2 + \Sigma_V^2 + \Sigma_W^2]^{1/2}$ (Sect. 3.2.1) displayed at the foot of Table 3.8, like $V_\odot$, show a systematic increase for the late spectral types, and hence with increasing age of the stars, a matter to which we shall return.

Motion of the Sun on the Basis of PM and RV: The right-hand equations (2.26), with $\dot{x} = u - u_\odot$, $\dot{y} = v - v_\odot$, $\dot{z} = w - w_\odot$, according to (2.28) and (2.24), lead to

$$K\mu_\alpha''/p = t_\alpha = -\gamma_{11}u_\odot - \gamma_{21}v_\odot - \gamma_{31}w_\odot + t_{\alpha p}$$
$$K\mu_\delta''/p = t_\delta = -\gamma_{12}u_\odot - \gamma_{22}v_\odot - \gamma_{32}w_\odot + t_{\delta p}$$
$$v_r = -\gamma_{13}u_\odot - \gamma_{23}v_\odot - \gamma_{33}w_\odot + v_{rp} \ . \tag{3.39}$$

These three equations hold for each star with certain known coefficients γ_{ik}. Here we have in each case combined the terms arising from the peculiar velocity (u, v, w) with the property $\bar{u} = \bar{v} = \bar{w} = 0$ and denoted them by $t_{\alpha p}$, $t_{\delta p}$ and v_{rp}. We shall treat these quantities as random errors with mean values $\bar{t}_{\alpha p} = \bar{t}_{\delta p} = \bar{v}_{rp} = 0$.

The copious data on proper motions permits relatively good determinations of the direction of the solar motion, the *solar apex*, in relation to certain categories of the stars. For this purpose just the first two equations (3.39) are available. Under the assumption of spatial invariance of the velocity distribution, one can by the method of least squares determine the three unknowns

$$\frac{\bar{p}}{K}u_{\odot}\ , \quad \frac{\bar{p}}{K}v_{\odot}\ , \quad \frac{\bar{p}}{K}w_{\odot}$$

where $\bar{p}$ denotes a suitably chosen mean value of the parallaxes of the stars under consideration. Equation (2.29) also gives the equatorial coordinates of the apex $A_{\odot}$, $D_{\odot}$; the magnitude of the solar velocity $S_{\odot}$, however, remains open, so long as $\bar{p}$ is unknown. Since $\bar{p}$ must depend on the galactic latitude, the zones of galactic latitude have to handled separately.

Because of its clarity and simplicity, let us now sketch a rough approximate procedure for calculating $A_{\odot}$ and $D_{\odot}$: if all the stars were fixed in space, so that the solar motion alone gave rise to the proper motions, then stars on the celestial equator would have to show the behaviour of their PM components μ_{α} that is sketched in Fig. 3.19: at the right ascensions of the apex and the anti-apex μ_{α} will pass through zero and change sign. Accordingly $A_{\odot}$ can be deduced simply by counting up the signs of μ_{α} in a star catalogue for stars near the equator, for example with $|\delta|<40°$, and noting the difference $N_{+}-N_{-}$ between the numbers of stars with $\mu_{\alpha}>0$ and with $\mu_{\alpha}<0$ for each whole hour of α. An example is shown in Fig. 3.20. Due to actual peculiar motions one obtains not a rectangular form of distribution, but one can readily infer that $A_{\odot} \approx 18^{\mathrm{h}}$. By an analogous discussion of the signs of the PM components μ_{δ} of stars near a circle through the celestial pole and the equatorial points with $\alpha = A_{\odot}$ and $\alpha = A_{\odot} + 12^{\mathrm{h}}$ respectively, one can now find $D_{\odot}$ also ($\mu_{\delta} = 0$, change of sign); in fact $D_{\odot} \approx +30°$.

For stars with known *radial velocities* the least squares solution of the third equation (3.39) leads directly to $u_{\odot}$, $v_{\odot}$, $w_{\odot}$ and hence not only $A_{\odot}$ and $D_{\odot}$ but also $S_{\odot}$ – without knowledge of distances. For the B-stars in the region $r \lesssim 500$ pc an improvement in the representation is obtained if a supplementary

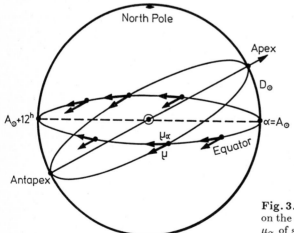

Fig. 3.19. Effect of the solar motion on the component of the proper motion μ_{α} of stars along the celestial equator

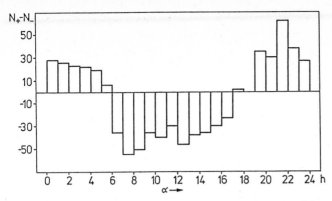

Fig. 3.20. Sign frequency $N_+ - N_-$ of the component μ_α of the proper motion for the stars of the fourth Fundamental Catalogue (FK 4) with $|\delta| < 40°$. Explanation in text

additive constant is introduced. This so-called *K-term* turns out to be positive and of a few $\mathrm{km\,s^{-1}}$, corresponding to a weak expansion of this star group ("K-effect").

Results for the solar motion have been derived from many categories of the stars, selected from various viewpoints, and they show significant differences from one another. The reason for this is that stars of different physical structure, different chemical constitution and different ages have in general different kinematic attributes. The value of many older investigations of mixtures of kinematically homogeneous groups, such as all bright stars with $m < 6$, is now therefore very limited. Accordingly the solar motion relative to the main mass of the brighter stars contained in the PM- and RV-catalogues (Main Sequence A...G, giants, supergiants) now serves simply as a conventional approximate value, generally called the "Standard Apex", with the equatorial coordinates $A_\odot = 18^\mathrm{h} = 270°$, $D_\odot = +130°$, a position in the constellation Hercules, and $S_\odot = 20\,\mathrm{km\,s^{-1}}$.

Relative to different star groups, each of a definite spectral and luminosity class, just as with the results for the region $r \lesssim 20\,\mathrm{pc}$, one often finds significantly different values for the apex coordinates and the solar velocity. A summary of the results is presented in Table 3.9. These numbers give only a rough overview of the kinematics of the stars, since it has to be remembered that the stars of a certain spectral and luminosity group frequently still do not form a homogeneous kinematic group.

In passing from early to late type Main Sequence stars one finds a systematic increase of $L_\odot$ and $S_\odot$, corresponding to an increase of the component $V_\odot$, which has already been discussed for the stars in the sun's vicinity (Table 3.8). Thus the later, on the average older types show, in contrast to the generally younger A-stars, an *"asymmetric drift"* directed towards the negative V-axis (against the galactic rotation). In Sect. 6.1.4 it is shown that this drift can be readily interpreted as the result of deviations of star paths from co-planar circular orbits about the galactic centre. For the same reason the velocity dispersions Σ_U, Σ_V, Σ_W also shown in Table 3.8 (explained further below) show

97

Table 3.9. Solar motion in galactic coordinates ($L_\odot$, $B_\odot$, $S_\odot$), velocity dispersions (Σ_U, Σ_V, Σ_W) and galactic longitude of the vertex direction l_1 (= direction of the major axis of the velocity ellipsoid) for different spectral and luminosity groups (from Delhaye, 1965)

Star types	$L_\odot$	$B_\odot$	$S_\odot$	Σ_U	Σ_V	Σ_W	l_1
			[km s^{-1}]	[km s^{-1}]			
Main Sequence							
A0...F3	43°	+33°	14	20	9	9	19°
F4...F8	51	+23	17	27	17	17	13
F9...G1	56	+15	26	26	18	20	2
G2...G7	70	+11	24	32	17	15	14
G8...K2	55	+23	20	28	16	11	3
K3...K6	68	+14	25	35	20	16	11
K8...M2	68	+25	17	32	21	19	8
M3...M6	64	+23	23	31	23	16	353
Giants							
A	41	+30	21	22	13	9	27
F	43	+19	29	28	15	9	14
G0...G8	57	+28	15	26	18	15	12
G9...K5	62	+18	21	31	21	17	18
M	76	+18	20	31	23	16	7
Supergiants							
F...M	56	+25	16	13	9	7	18

a systematic increase from early to late types. It therefore suggests the idea of defining a distinct local reference system in which one extrapolates to $\Sigma_U = 0$ the empirical relationship between Σ_U and $V_\odot$. This would then represent the limiting case for the motion of all stars under consideration with equal velocities on a circular orbit. One must however expect that the centroid of the A-stars already provides a good approximation to a local reference system *dynamically* defined in this way. Relative to the A-stars, the K-giants and the M-dwarfs with distances up to about 100 pc, the solar motion in a galactically oriented UVW-system (J. Delhaye 1965) turns out to be

$$U_\odot = +9\,\mathrm{km\,s}^{-1}\ ,\quad V_\odot = +11\,\mathrm{km\,s}^{-1}\ , W_\odot = +6\,\mathrm{km\,s}^{-1}\ , \tag{3.40}$$

corresponding to $L_\odot = 51°$, $B_\odot = +23°$, $S_\odot = 15.4\,\mathrm{km\,s}^{-1}$ (Fig. 3.21). This

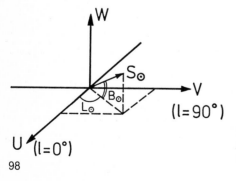

Fig. 3.21. Location of the vector of solar motion in the galactically oriented coordinate system

result is today generally known as *Basic solar motion*. It is a good approximation to the *peculiar motion of the sun relative to the circular orbit motion of the galactic rotation* at the position of the sun (more in Sect. 6.1.4) and defines the Local Standard of Rest (LSR) currently accepted as representative. Relative to the OB-stars and the classic Cepheids at $r \lesssim 1\,\mathrm{kpc}$ one gets nearly the same values.

Velocity Ellipsoid: We discuss first of all the establishment of propositions on the basis of *radial velocities,* since this route leads to the goal directly and without knowledge of the distances of the stars. The assumption of an ellipsoidal velocity distribution of the peculiar velocities in the form of a three-dimensional normal distribution means that for a fixed direction (α, δ) the distributions $\Psi(t_\alpha, t_\delta, v_\mathrm{r} | \alpha, \delta)$ also, and in particular moreover

$$\Psi_\mathrm{r}(v_\mathrm{r} | \alpha, \delta) = \int\limits_{-\infty}^{+\infty}\!\!\!\int \Psi(t_\alpha, t_\delta, v_\mathrm{r} | \alpha, \delta) dt_\alpha dt_\delta$$

are normal distributions. We can accordingly set

$$\Psi_\mathrm{r}(v_\mathrm{r} | \alpha, \delta) = \frac{1}{\sigma_\mathrm{r}\sqrt{2\pi}} \exp\left\{-\frac{(v_\mathrm{r} - \overline{v}_\mathrm{r})^2}{2\sigma_\mathrm{r}^2}\right\} ,$$

where $\overline{v}_\mathrm{r}$ and $\sigma_\mathrm{r}^2 = \overline{(v_\mathrm{r} - \overline{v}_\mathrm{r})^2}$ depend on α and δ. For a small area on the sphere with centre (α, δ) we have from (2.26)

$$\begin{aligned} v_\mathrm{r} - \overline{v}_\mathrm{r} &= \gamma_{13}\dot{x} + \gamma_{23}\dot{y} + \gamma_{33}\dot{z} - (\gamma_{13}\overline{\dot{x}} + \gamma_{23}\overline{\dot{y}} + \gamma_{33}\overline{\dot{z}}) \\ &= \gamma_{13}u + \gamma_{23}v + \gamma_{33}w \end{aligned}$$

and also (averaging over all included stars in the area)

$$\begin{aligned} \sigma_\mathrm{r}^2 = \overline{(v_\mathrm{r} - \overline{v}_\mathrm{r})^2} &= \gamma_{13}^2 s_{11} + \gamma_{23}^2 s_{22} + \gamma_{33}^2 s_{33} \\ &\quad + 2\gamma_{13}\gamma_{23} s_{12} + 2\gamma_{23}\gamma_{33} s_{23} + 2\gamma_{13}\gamma_{33} s_{13} \ . \end{aligned} \qquad (3.41)$$

The s_{ik} are identical with the elements of the variance matrix of the peculiar velocities $(u = u_1, v = u_2, w = u_3)$ introduced in Sect. 3.2.1, (3.23). If $\sigma_\mathrm{r}^2(\alpha, \delta)$ has been determined for many small fields in different directions, then one can obtain the six elements $s_{ik} = s_{ki}$ under the assumption of spatial invariance of the velocity distribution $\psi(u, v, w)$ from a system of equations of type (3.41) with known coefficients γ_{ik} by the method of least squares. From the s_{ik} then follow the standard deviations Σ_j in the three main axes and their directions (α_j, δ_j) or (l_j, b_j) [Sect. 3.2.1, (3.24) to (3.26)].

  If one starts from *proper motions,* then it turns out to be appropriate not to form the deviations referred to the respective mean values, corresponding to (3.41), but to form $\overline{t_\alpha^2}$, $\overline{t_\delta^2}$ and $\overline{t_\alpha t_\delta}$ directly, because then the always problematical mean value of the square of the parallax $\overline{p^2}$ appears simply as a factor.

One obtains, for example,

$$\overline{t_\alpha^2} = \gamma_{11}^2 s_{11}' + \gamma_{21}^2 s_{22}' + \gamma_{31}^2 s_{33}' + 2\gamma_{11}\gamma_{21} s_{12}'$$
$$+ 2\gamma_{21}\gamma_{31} s_{23}' + 2\gamma_{11}\gamma_{31} s_{13}' \qquad \text{with} \qquad (3.42)$$

$$s_{11}' = \overline{\dot{x}^2} = \overline{(u - u_\odot)^2} = \overline{u^2} - u_\odot^2 = s_{11} - u_\odot^2$$
$$s_{12}' = \overline{\dot{x}\dot{y}} = \overline{(u - u_\odot)(v - v_\odot)} = s_{12} + u_\odot v_\odot$$

and so on.

Using (2.26), the moments of the PM components $\overline{(\mu_\alpha'')^2}$, $\overline{(\mu_\delta'')^2}$ and $\overline{(\mu_\alpha''\mu_\delta'')}$ are obtained as expressions of the type (3.42), but with the factor $p^2(b)/K$ on the right-hand side. Without knowledge of this factor one can find only the directions of the axes of the velocity ellipsoid and the ratios of the dispersions $\Sigma_1 : \Sigma_2 : \Sigma_3$, and hence determine the shape of the ellipsoid. One can obtain the "absolute scale" of the Σ_j by including the RV data. This method was first used by C.V.L. Charlier in 1924 in treating the PM of 4182 stars with $m{<}6$ from Boss's Preliminary General Catalogue.

The derivation of the velocity distribution of intrinsically brighter stars, which can be included from a relatively large region of space, requires the elimination of the effect of differential galactic rotation dealt with in Sect. 3.4.2 – with high claims for accuracy already for distances $r{>}100$ pc. This means that the vectorial difference between the local circular orbit velocity at the position of the sun and the circular orbit velocity at the position of the star must be subtracted from the peculiar velocities (U, V, W) obtained. The resulting peculiar velocities (U^*, V^*, W^*) are then always *relative to the local standard of rest for the position of the star*. One gets (cf. Fig. 3.43, p. 133)

$$U^* = U - \Theta(R)\eta R^{-1}$$
$$V^* = V - \Theta(R)(R_0 - \xi)R^{-1} + \Theta_0 \qquad (3.43)$$
$$W^* = W \ .$$

Here $R = \sqrt{(R_0 - \xi)^2 + \eta^2}$ and R_0 are the distances of the star and the sun from the galactic centre, $\Theta(R)$ and Θ_0 are the corresponding circular orbit velocities and $\xi = r \cos l \cos b$, $\eta = r \sin l \cos b$ the rectangular galactic coordinates of the star (ξ-direction: $l = 0°$, $b = 0°$).

The existing *results* for the velocity ellipsoid of the normal stars with spectral types A...M for larger regions around the sun confirm the findings for $r{\lesssim}20$ pc (Fig. 3.18), according to which the principal axes approximately coincide with the directions of the galactically oriented UVW-system. In Table 3.9 the dispersion Σ_U, Σ_V and Σ_W are therefore given *in the coordinate axes*. The first one, Σ_U, is always significantly greater than $\Sigma_V \approx \Sigma_W$. The galactic longitude l_1 of the major axis of the ellipsoid (last column) characterises the so-called vertex deviation (from the direction of the galactic centre). The uncertainties of the dispersion values lie between $\pm 1\,\text{km s}^{-1}$ (for A-stars) and

$\pm 4\,\mathrm{km\,s^{-1}}$ (for the late dwarfs). In accordance with Table 3.8 for the stars near the sun, there is a systematic increase of the Σ-values in going from early to late spectral types clearly shown, but the ratios Σ_U/Σ_V and Σ_U/Σ_W only change to a lesser degree.

For the *OB-stars* it should be noted that these objects frequently occur grouped in associations with a small internal motion ($<3\,\mathrm{km\,s^{-1}}$ in the components). As entities the OB-associations have mean velocities of the order of $\Sigma \approx \pm 10\,\mathrm{km\,s^{-1}}$ relative to the local standard of rest. Similarly small velocity scatters are found for the *classical Cepheids,* which on average are somewhat older than the OB-stars: an investigation on the basis of radial velocities, proper motions and photometric distances for the region of space $r \lesssim 1\,\mathrm{kpc}$ gives (from Wielen, 1973):

$$\Sigma_U = \pm 8\,\mathrm{km\,s^{-1}}\ , \quad \Sigma_V = \pm 7\,\mathrm{km\,s^{-1}}\ , \quad \Sigma_W = \pm 5\,\mathrm{km\,s^{-1}}\ .$$

All these relatively young objects evidently still show, in essentials, the kinematic attributes of the interstellar medium from which they were created, namely, about $\pm 6\,\mathrm{km\,s^{-1}}$ (Sect. 4.1.5).

Local Moving Groups: The derivation of space velocities for numerous stars in the solar neighbourhood has led to the realisation that, in addition to the typical star clusters, stars widely scattered on the sphere can also constitute a moving group with nearly equal velocity vectors for all members. Thus numerous stars in Taurus have practically the same velocity components as the Hyades, namely $(U, V, W) = (-30, -5, +5)$, each in $\mathrm{km\,s^{-1}}$ and referred to the local centroid, defined in (3.40). Another moving group is composed of stars in the Ursa Major region, including five of the stars of the Plough, and a group of stars around Sirius, including itself. The velocity components of the Ursa Major-Sirius group are $(U, V, W) = (+23, +13, -3)$. In the UV-plane of velocity space these two moving groups show maximum densities which occur either in the first quadrant ($U>0$, $V>0$) or in the third quadrant ($U<0$, $V<0$) and thereby simulate a vertex deviation. Since, however, significant vertex deviations also appear for star types whose space velocities do not belong to a group − an example is provided by the M-dwarfs with emission lines − the values of l_1 presented in Table 3.9 must in the main have another cause (see Sect. 6.1.4).

Another local motion phenomenon makes its appearance in the previously mentioned K-effect of the brighter B-stars. It is the apparent distribution of these objects that causes Gould's Belt (Sect. 3.1.2).

Kinematic Groups and Ages of the Stars: Stars of middle or late type sorted by the usual spectral criteria of the same spectral and luminosity classes can differ considerably in age (long survival time on the Main Sequence!). Moreover significant differences have been found from star to star in the relative abundances of the higher elements from carbon onwards ("metals") in their atmospheres. Older stars must, as a result of repeated gravitational interactions with other stars and with massive interstellar clouds, have different kinematic attributes from young objects; theoretical considerations lead us to expect greater velocity scattering. For very old stars, which were created at an early stage of the galaxy, a truly particular kinematics is to be expected.

These qualitative considerations are confirmed by the observations: the increase in the velocity dispersions Σ_i towards late spectral types shown in Table 3.9 indicates that in general larger Σ_i are shown by older star groups; the relatively short-lived A-stars should therefore form an approximately homogeneous kinematic group. Moreover it appears that spectral criteria characterising *metal content,* namely the strength of the metal lines, are in general correlated with the kinematic behaviour. Consider, for example, groups of F-stars with in each case a certain mean metal deficit compared with the solar atmosphere, measured by the quantity $[M/H] = \log(M/H) - \log(M/H)_\odot$. Values of the velocity dispersion Σ_W are found which increase from about $\Sigma_W = 20\,\mathrm{km\,s}^{-1}$ to $\Sigma_W \approx 60\,\mathrm{km\,s}^{-1}$ as the relative metal abundance decreases from $[M/H] \approx 0$ to $[M/H] = -1.5$. An extreme case is represented by the especially metal-poor sub-dwarfs, with $[M/H] \approx -2$, whose kinematics we shall go into further below. On the basis of the magnitudes of the UBV- or $uvby$-system, the metal deficiency is frequently characterised by the ultraviolet excess $\delta(U - B) = |(U - B) - (U - B)_0|$ or the metal line index $m_1 = (v - b) - (b - y)$.

As a pure criterion of age, the presence and the strength of the Ca II lines H and K have been established for K- and M-dwarfs. Table 3.10 shows the relationship between the line strength (HK) on the one hand and $V_\odot$, Σ_U and $\Sigma = [\Sigma_U^2 + \Sigma_V^2 + \Sigma_W^2]^{1/2}$ on the other hand. Here the line strength is given in an arbitrary scale, which ranges from HK $= +8$ (very strong) to HK $= -5$ (very weak or no emission). As may be seen, both the velocity scatter and the "asymmetric drift", determined by $V_\odot$ (see above), increase monotonically with decreasing Ca II emission. The last column gives the average "kinematic" age, which was obtained from the relationship between Σ-values of Table 3.8 and the average ages given on p. 93. The possibility of an independent test is provided by the Hyades. For the late dwarf stars of this star cluster the value

Table 3.10. Kinematic attributes ($V_\odot = V$-component of the solar motion, $\Sigma_U =$ dispersion of the U-component, $\Sigma =$ total velocity dispersion) and age τ of around 180 K- and M-dwarf stars as functions of the strength (HK) of the emissions in the Ca II lines H and K (in an arbitrary scale). (From Jahreiss, 1974 and Wielen, 1977)

HK	$V_\odot$ [km s^{-1}]	Σ_U [km s^{-1}]	Σ [km s^{-1}]	τ [years]
+8...	+3 12	19	26	3×10^8
	+2 15	24	34	1×10^9
	+1 21	32	46	3×10^9
	0 19	40	53	5×10^9
	−1 41	56	82	7×10^9
−2...	−5 27	66	92	9×10^9

$HK = +3$ was observed. The theoretically determined evolutionary age of this group is found to be about 8×10^8 years and lies close to the value predicted by Table 3.10.

Asymmetry of Large Peculiar Velocities: The representation of the surfaces of equal point density in velocity space as ellipsoids is satisfactory only for stars with velocity magnitudes smaller than about $60 \, \text{km} \, \text{s}^{-1}$. This was already noticeable in Fig. 3.18. If one also includes stars with velocities greater than $60 \, \text{km} \, \text{s}^{-1}$, known as *high velocity stars,* then the distribution shows a strong asymmetry, which is illustrated by Fig. 3.22. Here U and V are referred to the local centroid. In the direction of the positive V-axis ($l = 90°$) there are no large velocities, the occupancy decreases rapidly to nil at $V \approx +60 \, \text{km} \, \text{s}^{-1}$; in the opposite direction, the negative V-axis, velocities up to around $V = -500 \, \text{km} \, \text{s}^{-1}$ occur, and the occupancy decreases only slowly. The centre point of the enveloping circle O lies at $V \approx -200... - 250 \, \text{km} \, \text{s}^{-1}$. In the W-V plane the distribution with regard to high velocities appears similar and the corresponding enveloping circle has the same centre point on the V-axis.

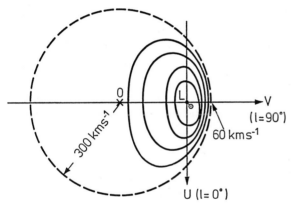

Fig. 3.22. Distribution of the components U and V of the peculiar velocities with inclusion of the high velocity stars, depicted by contours of equal density of occupancy (schematic). L indicates the origin of the local standard of rest. Explanation in text

This finding has the following significance, at the moment qualitative only: high velocity stars frequently have high velocity W-components perpendicular to the galactic plane. These stars therefore move on strongly inclined galactic orbits and must − in contrast to the normal stars with small peculiar velocity − occupy a roughly spherical or only slightly flattened volume, whose centre point is the galactic centre of mass. It therefore has to be accepted that the totality of all stars with high velocities has no, or only a weak, rotational motion as an entity about the galactic centre, and that the centre point 0 of the boundary circle depicted in Fig. 3.22 is at rest relative to the centre. Then the local standard of rest L must move with a velocity $\Theta \approx 200...250 \, \text{km} \, \text{s}^{-1}$ in the direction $l = 90°$ [for notation see Sect. 6.1.4, (6.36)], and we conclude

103

further that the normal stars near the sun orbit the galactic centre to a first approximation in circular paths parallel to the galactic plane with this velocity. The sharp boundary of the V-components at $+60\,\mathrm{km\,s^{-1}}$ can in this scenario be identified with attainment of the escape velocity from the system $\Theta_E = \Theta_0\sqrt{2}$ ("parabolic velocity") (J.H. Oort, 1927; W. Fricke, 1948), since for the given circular orbit velocity Θ_0 it follows within the error limits that $\Theta_E \approx \Theta_0 + 60 \approx 300\,\mathrm{km\,s^{-1}}$.

In general it is remarkable here that the study of the motions of the stars in a relatively constricted neighbourhood of the sun leads to conclusions on the large scale kinematics and dynamics of the whole galaxy. Also the alignment of the major axis of the velocity ellipsoid of normal stars with the brightest regions of the Milky Way in Sagittarius can be seen as an indication that the controlling centre of force for the star motions is to be sought far outside the sun's neighbourhood in that direction (Sect. 6.1.4).

A direct confirmation of the motion of the local standard of rest in the direction $l \approx 90°$, $b \approx 0°$ has been derived from the radial velocities of extra-galactic stellar systems. The result $\Theta_0 = 200\ldots300\,\mathrm{km\,s^{-1}}$ is, however, relatively imprecise.

The stars with large peculiar velocities are not a homogeneous group. They have in common, however, the spectral characteristic of abnormally weak metal lines: Quantitative analysis gives the metal deficit $[M/H] \approx -1\ldots-2$. A sub-group consists of the so-called *normal high velocity stars,* which in the Hertzsprung-Russell Diagram (HRD) lie in the region of the Main Sequence or, less frequently, on the Giant branch. These objects represent a very small fraction of the generality of Main Sequence stars in B- and A-stars, but a considerably greater fraction in the late types. One concludes from this that we are dealing here with older stars.

Another high velocity star group consists of the *sub-dwarfs,* which are to be found about 1^m below the Main Sequence in the HRD. Sub-dwarfs show the normal hydrogen lines corresponding to spectral types F...K, but the metal lines shown strongly by middle Main Sequence stars of small velocity, and for example the strong Ca I $\lambda\,4227\,\text{Å}$ of late types, are extraordinarily weak or quite undetectable, so the spectra are similar to Type A. In the statistical relationship between the kinematic attributes and the metal deficit, characterised by the ultraviolet excess $\delta(U\text{-}B)$, the sub-dwarfs are clearly distinguished from the normal stars. Figure 3.23 shows this on the basis of velocity magnitude perpendicular to the galactic plane $|W|$ and the value $\delta(U\text{-}B)$ for a representative group of normal F- and G-stars with $m_V < 5.5$ and for 128 sub-dwarfs (cf. also Table 3.11).

Also the long period variables of Mira type, the short period RR Lyrae stars and the globular star clusters belong to the objects of large peculiar velocity relative to the local standard of rest. The kinematics of the last two sub-groups will be treated in (3.43).

Results for the mean value of the V-component relative to the local reference system ("asymmetric drift") and the velocity scatters Σ_W and $\Sigma =$

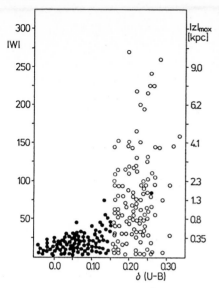

Fig. 3.23. Correlation between the magnitude of the velocity component perpendicular to the galactic plane $|W|$ in $\mathrm{km\,s^{-1}}$ and the ultraviolet excess $\delta(U - B)$ for subdwarfs (*open circles*) and normal Main Sequence stars of the same spectral type (F...G). The right-hand ordinate scale relates to the maximum distance $|z|$ in kpc which a star with the velocity component $|W|$ can attain. (From Eggen, 1970)

Table 3.11. Mean value of the V-component (direction $l = 90°$, $b = 0°$) relative to the local standard of rest as well as the velocity dispersions Σ_W and Σ for different sub-groups of the stars with large peculiar velocities. ΔS is the Preston Spectral Index. Its magnitude is correlated with the metal deficit. (After Fricke, 1949; Eggen, 1970; Smak, 1966; Plaut, 1963; Van Herk, 1965; Kinman, 1959, and Woltjer, 1975)

Sub-group	Number of objects used	$\overline{V}$ $[\mathrm{km\,s^{-1}}]$	Σ_W $[\mathrm{km\,s^{-1}}]$	Σ $[\mathrm{km\,s^{-1}}]$
Normal high velocity stars	598	$- 53$	40	120
Sub-dwarfs $\delta(U - B) > 0.15$	128	-230	90	260
Mira stars with periods				
$P = 140 \ldots 200$ days	68	-100	60	140
$P = 200 \ldots 250$ days	94	$- 60$	50	80
$P = 250 \ldots 300$ days	112	$- 30$	30	80
$P = 300 \ldots 350$ days	127	$- 20$	35	60
$P = 350 \ldots 400$ days	59	$- 10$	20	60
RR-Lyrae stars with				
$P = 0.1^d \ldots 0.4^d$ $\}\Delta S < 5$	19	-20	30	90
$P = 0.4^d \ldots 0.6^d$	26	$- 90$	80	170
$P = 0.2^d \ldots 0.4^d$	9	-130	80	120
$P = 0.4^d \ldots 0.6^d$ $\}\Delta S > 5$	35	-140	85	200
$P = 0.6^d \ldots 0.8^d$	20	-250	80	220
Globular star clusters	70	-170	~ 100	~ 200

$[\Sigma_U^2 + \Sigma_V^2 + \Sigma_W^2]^{1/2}$ are summarised for the known sub-groups in Table 3.11. For the variables under consideration, partition into groups, each with approximately the same period, reveals a significant dependence of the kinematic data on the period: the stars in a certain period interval do indeed form a kinematically homogeneous subgroup. The Preston Spectral Index is defined by $\Delta S = 10 \, [Sp(\mathrm{H}) - Sp(\mathrm{Ca\,II})]$, where $Sp(\mathrm{H})$ and $Sp(\mathrm{Ca\,II})$ are the spectral types derived from the Balmer lines of hydrogen and the $\mathrm{Ca\,II}$ K-line (whole class difference $= 1$) respectively.

3.3 Large Scale Distribution of the Stars

The foregoing results for the spatial density distribution of the star types A...M, which form the main body of the stars in the galactic disc in the neighbourhood of the sun, do not extend far enough to provide a picture of the large scale structure of the whole stellar system. Near the galactic plane a further extension in the visible and ultraviolet regions is possible only within the restriction to intrinsically very bright stellar objects, because of the strong interstellar extinction (cf. Sect. 2.2.2, Table 2.2). Above all these are the star types O, B, Supergiants of type A and later, and Cepheids of type I, which according to their distribution on the sphere are always strongly concentrated at the galactic plane. These objects form the flattest sub-system of the Galaxy.

The differences found between different star types in the solar neighbourhood, with regard to the concentration at the galactic plane and the kinematic attributes, show that there is a whole series of sub-systems with different extents perpendicular to the galactic plane. A case of relatively slight galactic concentration consists of the globular star clusters, which have already been discussed in the Introductory Survey from the point of view of their appearance on the sphere. Most members of a sub-system which is so nearly spherical offer the advantage that only a relatively short length of the line of sight lies within the light-absorbing galactic dust layer. So long as a sufficiently bright object is involved, about $M_V \lesssim 0$, it is therefore possible to study the large scale structure of the "galactic halo", and thereby to obtain evidence on the extent of the whole stellar system.

3.3.1 OB Stars, Young Open Star Clusters and OB Associations

OB Stars: The pinpointing of certain star types of great intrinsic brightness proceeds in two stages. The first stage consists in the identification of objects of the type in question with the help of appropriate available surveys of the sky. The second stage concerns the data necessary for determination of the distance. The photometric method is used, and for this the absolute magnitude M_V of each star and the interstellar extinction $\Delta m_V = A_V$ have to be derived.

The *identification* of stars of the Harvard spectral types O...B9 is effected with the help of objective-prism photographs, as they were used for the determination of the spectral types given in the Henry Draper catalogue of the Harvard Observatory for all stars with $\delta > -25°$ up to $m_B \approx 8.5$. The fraction of the O- and B-stars contained in this catalogue is roughly 1%. Additional stars of these types up to about $m_B = 12$ are contained in the "Henry Draper Extensions" for special areas. The procedure is limited to $m_B \lesssim 13$, however, because there is too much superposition of spectra for fainter stars in the star-rich fields of the Milky Way. Up to close to this limit is covered by, for example, the so-called "Luminous Stars Catalogues" of the northern and of the southern Milky Way and similar products of major spectral surveys for intrinsically very bright stars (O, B, AI...GI) with the Schmidt telescopes at the Hamburg Observatory, at the Warner and Swasey Observatory in Cleveland (Ohio) and for the southern region by the University of Michigan and the Heidelberg Observatories in Chile and South Africa, respectively. These catalogues contain altogether well over 11 000 O- and early B-stars, which at first were assessed only roughly in three sub-classes: OB, OB$^-$ and OB$^+$. The criteria for this "OB" classification, originated by W.W. Morgan (1951), are the strengths of the Balmer lines, of the Balmer jump and of the Ca II K-line; the dispersion amounted to about 600 Å/mm for the Hγ-line (4340 Å). The magnitudes of the OB stars were at first only estimated on the objective-prism plates.

For the second stage extensive *photometric observations* must be carried out, to provide accurate apparent magnitudes, for example U, B, V, and quantitative luminosity criteria, such as the Hβ-index, for all the stars in the survey catalogue. The derivation of the relationship between estimated or photometric luminosity criteria and the absolute magnitude M_V itself, poses a special problem in the case of the OB stars, since the nearest of these relatively scarce objects is already too far away for measurement of the trigonometric parallax. The method of moving cluster parallaxes provides a direct route only for a few early types (in the Scorpio-Centaurus group), whilst in the motion group with by far the best determined distance, the Hyades, there are no O- or B-stars. The goal has been achieved by the indirect photometric procedure already described in Sect. 2.2.2 for the distance determination of young open star clusters with OB stars on the basis of the colour-magnitude diagram. The M_V of the early MK spectral classes were first derived in this way. In corresponding manner the relationship between M_V and, for example, the Hβ-index (β, for definition see Sect. 2.2.2) can be obtained, or one picks out those stars for which not only β but also the MK-type is known, and so takes as basis the already derived relationship between MK-type and M_V (Fig. 3.24). The uncertainty of M_V thus obtained from β is today in general about $\pm 0^{\text{m}}5$.

MK spectral types (or a quantitative photometric classification) are also needed for the derivation of the individual interstellar extinction A_V from the colour excess $E_{B-V} = (B - V) - (B - V)_0$ [Sect. 4.1.1, (4.12ff.)], since its connection with the intrinsic colour $(B - V)_0$ is known. Otherwise A_V has to be interpolated from existing results for neighbouring objects.

From V, M_V and A_V the distance of the star can finally be calculated. The results of an investigation carried out in the manner described, for the

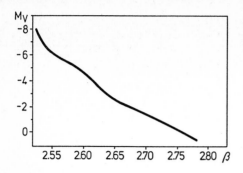

Fig. 3.24. The relationship between the absolute magnitude M_V and the Hβ-index. (From Neckel and Klare, 1980)

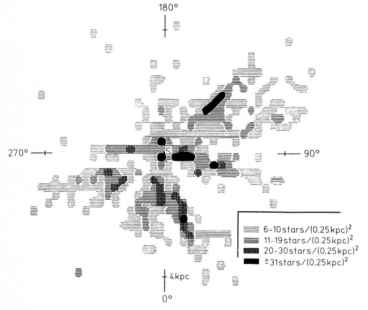

Fig. 3.25. Distribution of about 6200 OB-stars in the galactic plane. The position of the sun is indicated by the letter S. (From Klare and Neckel, 1967)

distribution of the OB stars, projected on to the galactic plane, are presented in Fig. 3.25. The range amounts to about 4 kpc, so that large scale structure should already be indicated. Although from (2.44), just because of the uncertainty of M_V to about $\pm 0^m.5$, distance errors of about 25% must in general be expected, which leads to a corresponding smearing and broadening of the fine structure actually present, the following conclusion is clearly apparent: the OB stars in the region covered are concentrated primarily on three roughly parallel "arms", of which the middle one, the so-called "local arm", includes the sun. If one looks for comparison at photographs of typical spiral galaxies, then it appears plausible to assume that these structural elements belong to the spiral arms of our Milky Way system, and perhaps also partially form bridges between the spiral arms. One must also bear in mind that the spiral arms of the extra-galactic systems by no means have a smooth profile throughout, but on the contrary often show significant irregularities.

The distribution of the OB stars perpendicular to the galactic plane (distance z) can be approximately represented by the expression

$$D(z) = D_0 \left[1 + \left(\frac{z}{H} \right)^2 \right]^{-1} .$$

For the region $r \lesssim 2\,\mathrm{kpc}$ one obtains the values $D_0 \approx 5 \times 10^{-5}$ stars pc^{-3} (cf. Table 3.6) and $H \approx 50\,\mathrm{pc}$. The concentration of OB stars at the galactic plane is therefore extraordinarily strong. One often refers to the mean value of distance from the galactic plane $\overline{|z|}$. For the OB stars $\overline{|z|} \approx 65\,\mathrm{pc}$.

Young Open Star Clusters and OB Associations: These objects because of their intrinsically bright members are also observable to distances of several kiloparsecs. For the open clusters it is a matter of a loose assemblage of at most several hundred stars, in which the star density is about 10 or 100 times the star density in the neighbourhood. For example the star clusters h and χ in Perseus each contain around 300 stars (to $m_V = 15^{\mathrm{m}}$) and have a diameter of about 15 pc. In many young open clusters, i.e. in objects with O- and early B-stars, interstellar matter can be detected.

The naming of a star cluster − apart from objects known since ancient times − usually consists of the (abbreviated) title of the catalogue in which it is introduced, or the (possibly abbreviated) name of the catalogue editor, and its catalogue number. Examples: M 11 = NGC 6705, IC 4665, Haf 16, King 22. Here M stands for the name Messier, NGC for "New General Catalogue" and IC for the extension of the NGC known as the "Index Catalogue"; Haf is the abbreviation for the name Haffner. The total number of known open clusters is about 10^3.

Associations are loose assemblages of stars of a narrow spectral or type group, in which the partial density of these objects alone is greater than in the neighbourhood − and less than the total star density in the neighbourhood. The diameters of associations are on average about a factor of 10 greater than the open clusters; the number of members lies between about 100 and 1000 stars. In addition to the OB stars, similar assemblages are also formed by the T Tauri stars found in the pre-Main Sequence development, the so-called T-associations, whose members however do not attain the great intrinsic brightness of OB stars. Stellar associations are always connected with clouds of interstellar gas and dust, or embedded in them. Associations of late B- and A-stars, in which reflection nebulae (Sect. 4.1.3) occur, are called R-associations.

The naming of stellar associations was at first taken from the (abbreviated) name of the constellation prefixed by a Roman numeral, e.g. I Per, II Per, etc. In the meantime another nomenclature has been introduced, which indicates also the type of association. The following examples may serve as illustrations: Per OB1 (= I Per), Per OB2 (= II Per); Cyg T1; CMa R1. Frequently in the region of an OB association open star clusters are to be found, for example h and χ Per in Per OB1. The number of known associations is around 10^2.

The notion of star associations was conceived by W. A. Ambarzumjan (1947), who also recognised the cosmological significance of these objects. Whilst the mutual gravitation can produce long-term stability in star clusters, the extended associations should disintegrate through differential motions within about 10^7 years. The approximately spherical shape therefore carries the implication that this involves an expanding group of newly created stars. The predicted expansion can be confirmed in a few OB associations on the basis of the observed proper motions and radial velocities ($\approx 10\,\mathrm{km\,s^{-1}}$). The ("expansion"-)age of these objects is thus found to be of order 10^6 years. The observed general frequency of B-stars can be explained by the supposition that all these stars arise from OB associations.

The basis for the derivation of the spatial distribution of the open star clusters and the OB associations is the CMD method described in Sect. 2.2.2. More recent results for the distribution on the galactic plane of young open star clusters containing spectral types earlier than about B3 are shown in Fig. 3.26

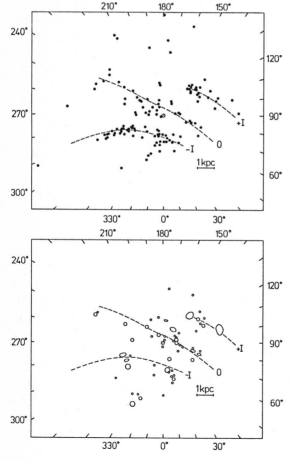

Fig. 3.26. *Above*: Distribution of young open star clusters. (From Vogt and Moffat, 1975). *Below*: Distribution of associations. (From Humphreys, 1976)

110

(top); the corresponding result for associations containing super-giants is given in Fig. 3.26 (bottom).

If it is remembered here again that distance errors of about 25% can occur thoughout, the reality of the three arm structure already noticed in Fig. 3.25 can scarcely be doubted. Usual labels are: *"Perseus Arm"* for the "outer" grouping, only clearly marked for $l<150°$, *"Local Arm"* for the middle one and *"Sagittarius Arm"* or even "Sagittarius-Carina Arm" for the "inner" one, lying nearer the galactic centre. For the local arm one also occasionally comes across the older name "Orion Arm", which emphasises its branching in the direction $l \approx 210°$. Today one often uses the unambiguous abbreviated labels +I (Perseus Arm) 0 and −I. An additional arm −II ("Norma-Centaurus Arm") lying further inside is suggested, but is not established by these data alone. The three confirmed structure elements were first detected in the distribution of OB associations in 1953 (W.W. Morgan, A.E. Whitford and A.D. Code).

If one plots similarly the points for *all* open star clusters for which usable distance determinations are available (around 200 objects), thus including also the older clusters with earliest spectral types later than B3, then the arm structure disappears. The same happens if one considers just the distribution of the older open clusters. The arms are therefore a characteristic of the distribution of the *younger* stars! According to the theory of stellar evolution, with OB-stars and young open clusters one is dealing with objects which must be younger than about 10^7 years. On the other hand their velocities relative to a local standard of rest are of order $10 \, \mathrm{km \, s^{-1}}$, so that since their formation they can have traversed paths of only about $10 \, \mathrm{km \, s^{-1}} \times 10^7$ years $\times \, 3 \times 10^7 \, \mathrm{s/year} = 3 \times 10^{15} \, \mathrm{km} = 100$ pc. The young stars must therefore have been formed in the arms in which we find them today.

The distribution of the open star clusters perpendicular to the galactic plane is very similar to that of the OB-stars. We get $\overline{|z|} \approx 60 \, \mathrm{pc}$ with only a relatively narrow difference between the young and the older clusters.

3.3.2 Globular Clusters, RR Lyrae Stars and Cepheids

Globular clusters are spherical star systems with strong central condensation. They have diameters up to 50 pc and contain 10^4 to 10^7 stars. The nearest globular clusters are already so distant that Main-Sequence stars of the solar type have apparent magnitudes of $m_V \approx 19^m$. Therefore the construction of their CMD requires observations with large telescopes. A schematic representation of a typical CMD is shown in the accompanying Fig. 3.27. Its interpretation leads to the conclusion that we are dealing here with the oldest objects (ca. 15×10^9 years) of the Milky Way system. The lower branch of the CMD turns out to be identical with the lower part of the Main-Sequence. The more massive stars of the upper Main-Sequence have evolved more rapidly and moved toward the region of the red giants. Part of them constitute now a *horizontal branch* which includes the short-periodic RR Lyrae variables (between $B−V = 0^m2$ and 0^m4). H. Shapley and H.B. Sawyer divided the globular clusters into 12

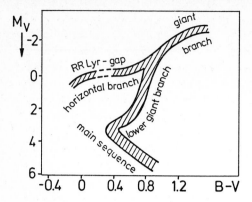

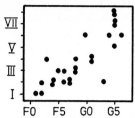

Fig. 3.27. Schematic colour-magnitude diagram of globular clusters

Fig. 3.28. Correlation between Morgan's metallicity classes and the integral spectral types of globular clusters

classes of decreasing star concentration I...XII. The number of known globular clusters in the Milky Way system is 129 (Kukarkin's general catalogue, 1975).

Just as for individual stars, one can determine "integral" spectral types for whole globular clusters. One finds types between A5 and G8. W.W. Morgan has therefore divided the integral spectra of the clusters into eight classes I...VIII, with increasing strength of the metal lines (not to be confused with the concentration classes!). Here the Metallicity Class I is defined by the weakest metal lines and Class VIII by the strongest. Class I corresponds to a metal abundance (relative to hydrogen) of about 1/100 of that of normal stars, for example the sun, whilst Class VIII attains roughly these normal values. For the globular cluster M3, for example, we have type F7 II. The integral spectral type and the metal content of the globular clusters are however not independent of one another (Fig. 3.28).

With decreasing metal content the Giant branch in the CMD deviates towards the "top", and hence towards greater brightness, by a space of about three magnitudes. A remarkable relationship exists between the intrinsic colour for the point in the CMD where the Giant branch and the horizontal branch meet, $(B-V)_0^*$, and the metal deficit relative to the solar atmosphere $[M/H] = \log(M/H) - \log(M/H)_\odot$ introduced in Sect. 3.2.3. A recent investigation (by Butler, 1975) finds that

$$[M/H] = 7.2(B - V)_0^* - 7.0 \ .$$

The RR *Lyrae stars* are here involved in two ways: (1) examples occurring in globular clusters ("cluster variables") are of fundamental importance for the photometric distance determination of these objects, (2) isolated occurring "field stars" of this variable type form a characteristic galactic subsystem with similar characteristics to the system of globular clusters. The distribution of the *Cepheids* will be discussed in this section, since the absolute magnitudes of these variables have to be derived largely by the same methods that are applied for the likewise pulsating RR Lyrae stars.

Absolute Magnitudes of RR Lyrae Stars and Cepheids: These quantities, which are fundamental for long-range distance determinations, can be obtained by three methods: (1) application of the method of statistical parallaxes (Sect. 2.2.1), (2) finding objects of this type in star clusters or moving groups with known distances and (3) determining the star radius $\overline{R}$ for the mean magnitude from the light curve and the radial velocity curve, and calculating the visual luminosity $\overline{L}_V = 4\pi\overline{R}^2\overline{\Phi}_V$ with the mean visual radiation flux at the star surface $\overline{\Phi}_V$.

The first method is specially appropriate for *RR Lyrae stars* outside globular clusters, because the centroid of these variables has a high velocity relative to the sun ($\overline{V}$ in Table 3.11). These stars, however, form no kinematically homogeneous groups and therefore must first of all be classified according to their real kinematic properties. Accordingly the object numbers of individual subgroups fall and, because of their large velocity dispersions (Σ_U, Σ_V, Σ_W), also the accuracy of parallax determinations. For the derivation of M_V itself there is of course still the interstellar extinction to allow for, whose uncertainty however remains in general relatively small for higher galactic latitudes.

The results of more recent investigations lie between $M_V = +0.4$ and $M_V = +1.3$, the mean error amounting to at least ±0.5. To what extent M_V depends on the chemical composition of RR Lyrae stars, and whether field stars and cluster stars have the same M_V, are still open questions.

The second method can be followed not only for RR Lyrae stars in sufficiently near globular clusters, but also for variables of this type in loose star groups with uniform motion. In the first case the distance of the star cluster has been obtained following the procedure explained in Sect. 2.2.2, by adjusting the Main Sequence of its CMD to the course obtained from the theory of metal-poor stars. From RR Lyrae stars in six globular clusters the values obtained in this way lie between $M_V = 0.0$ and $+0.6$. In the second case one can apply the purely empirical method of moving cluster parallaxes (Sect. 2.2.1) for determination of the distance of each individual member covered by the group − even the prototype RR Lyrae itself belongs to such a group. An investigation of this type with allowance for interstellar extinction also gave values of M_V between 0.0 and +0.6.

For the derivation of the absolute magnitudes of the *classical Cepheids* (Type I), and hence the absolute determination of the relationship between M_V and the period P, the method of photometric distance determination of open star clusters has already achieved fundamental importance. A dozen of these variables with $P = 3^d\ldots41^d$, which belong to open star clusters or associations − four of them in the double cluster h and χ Persei − led to the relations (from Tammann, 1970)

$$M_V = -2.469 - 3.534\log P + 2.647(\overline{B} - \overline{V})_0 \quad \text{and}$$
$$M_V = -1.700 - 3.382\log P + 1.834(\overline{U} - \overline{B})_0 \;, \tag{3.44}$$

where P is expressed in days and $\overline{B}$ and $\overline{V}$ denote the mean magnitude of the star.

The derivation of M_V for RR Lyrae stars, as for Cepheids, by the third method mentioned, from the star radius $\overline{R}$ and the radiation flux $\overline{\Phi}_V$ at the star surface, requires a relatively exact determination of these two quantities. For this one uses a relationship between Φ_V and the colour index $(B-V)_0$ or the spectral type, which itself is based essentially on the direct determination of star diameter with the intensity interferometer. An application to RR Lyrae for example gave $M_V = +0.5$; δ Cephei gave the value $M_V = -3.6$, whilst (3.44) for the period of this star $P = 5\overset{d}{.}37$ in combination with its mean colour index $(\overline{B-V})_0 = +0.6$ gives: $M_V = -3.5$. In many cases the agreement between the results of the two methods is not so good.

Distance Determination for Globular Clusters: The nearest of these objects is already too far away for the application of the moving cluster method. We have, however, the procedure of fitting of the lower Main Sequence of the CMD to the Zero Age Main Sequence (Sect. 2.2.2). Since the brightest stars still lying on the Main Sequence have themselves apparent magnitudes $m_V \approx 19$ in the nearest globular clusters, this is possible in only a few cases. Because of the low metal content of globular cluster stars, we must use here the correspondingly modified Standard Main Sequence as the basis.

Significantly longer range is attained by the photometric determination of distance by means of the horizontal branch in the CMD of the cluster, whose position in the $M_V - (B-V)$ diagram can be fixed on the basis of the absolute magnitudes of the RR Lyrae stars. This method is nowadays applicable to more than half of the known globular clusters. In about twenty clusters the individual RR Lyrae stars can be detected and used for derivation of the distance module.

If the clusters are so distant that only the upper part of the Giant branch still lies above the magnitude limit, then the mean apparent magnitude of the 25 brightest cluster stars can be used as a distance indicator, in that one assigns the corresponding mean absolute magnitude to be as it is for nearer clusters with well known distances: $\overline{M}_B(25) \approx -0.4$ (Basis: $M_V = +0.6$ for RR Lyrae stars). More exactly, there is a dependence on the absolute total magnitude of the cluster M_B^* (from Woltjer, 1975):

$$\overline{M}_B(25) = 3.0 + 0.46 M_B^* .$$

After the distance has first been determined assuming that $\overline{M}_B(25) = -0.4$, one can obtain an approximate value for M_B^* and then get a better value for $\overline{M}_B(25)$ from the above relationship.

For the remaining globular clusters, one tries to estimate the distance from the angular diameter. If α denotes the angular diameter in arc minutes, inside which 90% of the light originates, and one wants the distance r in kpc, then according to Woltjer (1975) for the globular clusters with known distances $\overline{r \cdot \alpha} \approx 80$ on average, with a scatter of about $\pm 30\%$; the corresponding mean linear diameter of the cluster is around 25 pc. This relation yields an approximate value for r from α.

Apart from the last purely geometric procedure, the methods described produce first of all the apparent distance module of the cluster, which still has to be adjusted for the influence of interstellar extinction: $m-M = (m-M)_0 + A$ with $(m-M)_0 = -5 + 5\log r$. One proceeds in a similar way as for individual stars: there is a relationship between the integrated intrinsic colour index for the cluster $(B-V)_0$ and the integrated spectral type Sp. From this one can obtain the intrinsic colour index from Sp and hence the colour excess for the whole cluster $E_{B-V} = (B-V) - (B-V)_0$, and finally the visual interstellar extinction $A_V \approx 3 \cdot E_{B-V}$.

For the derivation of the relation between the integrated quantity $(B-V)_0$ and Sp the two-colour diagram for the integrated values of U, B, V can be used. If one plots the clusters of a narrow spectral group, it turns out that the points lie on a straight "reddening line", $E_{U-B}/E_{B-V} = \text{const}$. Two examples are shown in Fig. 3.29. The clusters with the smallest $B-V$ are taken to be unreddened – as a rule objects in the higher galactic latitudes – from which the colour index $(B-V)_0$ of the spectral group is found. The mean relationship obtained between $(B-V)_0$ and Sp is linear in the region covered, F2...G5, and is fixed approximately by the two points F2: $(B-V)_0 = 0.60$, G5: $(B-V)_0 = 0.82$.

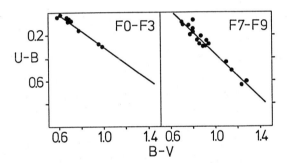

Fig. 3.29. Reddening lines in the two-colour diagram for globular clusters with integrated spectral types F0...F3 and F7...F9. (From Racine, 1973)

The E_{B-V} of the globular clusters shows a characteristic dependence on galactic latitude b, which can be approximately represented in the mean by the equation $E_{B-V} = 0.06 \cdot \operatorname{cosec} b - 0.06$. For the determination of the interstellar extinction for a number of very faint clusters, for which no integrated spectral types are available, we have to fall back on this relationship (see also Sect. 4.1.1).

Distribution of Globular Clusters: Figure 3.30 shows the results of a more recent investigation. Here the origin of coordinates is the sun; the x-axis points in the direction of the galactic centre $l = 0°$. The basis of the distance determination here was the value $M_V(\text{RR Lyr}) = +0.6$.

A typical globular cluster has an absolute total magnitude $M_V \approx -8$ and at a distance $r = 20\,\text{kpc}$ with $A_V = 0$ would still show as an easily observable object with magnitude $V \approx 8.5$. The globular clusters of our galaxy can

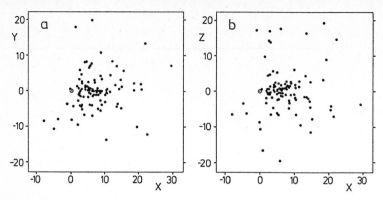

Fig. 3.30a,b. Distribution of 110 globular clusters (**a**) projected on to the galactic plane and (**b**) projected on to the perpendicular plane which contains the position of the sun and the galactic centre. Origin of the coordinate system is the sun (⊙). Unit of length: 1 kpc. (By permission of Harris, 1976)

therefore only escape detection through the presence of excessive interstellar extinction near the galactic plane. When one studies the distribution in the xz-plane (Fig. 3.30b), it is obviously the case that for $|z|<2$ kpc the number of globular clusters found decreases sharply with increasing $x>5$ kpc, whereas for $|z|>2$ kpc a sharp fall in the number of objects summed over z first sets in when $x \approx 20$ kpc. At least in the region $|z|<2$ kpc, therefore, the observational material is incomplete. Overall calculations of the extent of the total population of globular clusters in the Milky Way system on the basis of these results must leave at least 200 objects to be accounted for.

For the determination of *the distance R_0 of the centre* of the globular cluster system, which we identify with the centre of the galaxy, one has to restrict oneself to consideration of objects with $|z|>2$ kpc. For this, the underlying data of Fig. 3.30 gives in this way

$$\overline{x} = R_0 = 8.5 \pm 1.5 \text{ kpc} .$$

The relatively large error includes the uncertainty of the absolute magnitude of the RR Lyrae stars by almost $\pm 1^m 0$.

The question arises, whether globular clusters with significantly different metal abundances show also a different spatial distribution. Since metal abundance and integral spectral type are correlated with one another (Fig. 3.28), the answer can be found by consideration of sub-groups of the globular clusters with certain integrated spectral types. The results are shown in Fig. 3.31: clusters of spectral group F, with high metal deficits, show only a very weak concentration at the galactic plane and form a large, nearly spherical, halo. The distribution of the G types is quite different. These objects, with small metal deficit, are almost without exception to be found near the galactic plane. In Fig. 3.31 they appear to form a flat disc with smaller radius than that of the sub-system of "halo clusters" of spectral type F. If one studies the distribution on the sphere, it is seen that about 75% of the globular clusters of type G lie in-

116

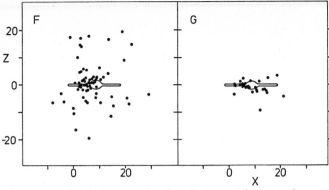

Fig. 3.31. Distribution of globular clusters with spectral type F ("metal poor") and G ("metal rich") in the $z - x$ plane. Origin is the position of the sun. Unit of length: 1 kpc. The galactic centre is taken to be at $x \approx 8.5$ kpc. (By permission of Harris, 1976)

side a circle of 10° radius about the direction of the centre of the galaxy. These objects would therefore occupy a cigar-shaped volume, whose major axis points towards the sun! That is very improbable. One must therefore assume that the globular clusters of type G are more strongly concentrated at the galactic centre than the F types and so occupy a smaller, slightly flattened volume. The cigar shape of the distribution must appear because the observation of almost all these objects undergoes interstellar extinction to a significant extent near the galactic plane, so that large distance errors (x-coordinate!) result. Summarising, one can say that near the galactic centre there are globular clusters with low, medium and "high" metal content in roughly equal numbers. At greater distances from the centre one finds first of all no more globular clusters with high metal content, and finally at greater distances only the metal poor objects.

Globular Clusters as Stellar Systems: On the basis of the distances obtained, the diameters, the distribution of the absolute magnitudes of the cluster stars and, in combination with other data, the masses as well as the mass-luminosity ratio of the globular clusters can be derived.

The mean *diameter* (inside which 90% of the light originates) has already been given as around 25 pc. The scatter about this value is relatively narrow at about 30%; large deviations are shown by only a few examples, including ω Centauri, for which the diameter is about 150 pc. The central star density in globular clusters is in general about a factor 10^3 higher than the number density of the star field of the solar neighbourhood, and so is about 10^2 stars pc^{-3}.

The *luminosity function (LF) of the globular cluster stars* $\Phi(M_V)$, usually defined as the number of stars of the cluster in the interval $M_V - \frac{1}{2} \ldots M_V + \frac{1}{2}$, is given in principle by enumeration according to apparent magnitude, together with the well known distance parameter. For the nearest globular clusters, such as M3 or M92, one has a distance parameter $m_V - M_V \approx 15$, so that with the currently attainable limiting magnitude $m_V \approx 23$, the region $M_V \lesssim +8$ is ac-

117

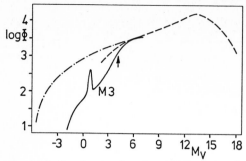

Fig. 3.32. Luminosity function $\Phi(M)$ of the globular cluster M3 = NGC 5272 (—). Shown for comparison are the LF of the stars in the neighbourhood of the sun (– – –; arbitrarily fitted at $M_V = +6$) and the initial LF (– • –). The arrow indicates the location of the "knee" in the CMD

cessible. Figure 3.32 shows the essential profile of results obtained for these relatively close objects, for example M3, in comparison with the LF for the stars in the neighbourhood of the sun. The steep rise of $\Phi(M_V)$ for $M_V < +6$ has superimposed on it, slightly above $M_V = 0$, a peak which is produced by the stars of the horizontal branch. Above $M_V \approx +6$, and traceable so far only up to about $M_V = +8$, more recent investigations show that the rise does not continue: possibly the curve is then already starting to fall. Theoretical arguments for this have been put forward: if a globular cluster has passed the galactic plane on its strongly inclined galactic orbit, the associated "gravitational shock" could cause the mass-poor Main Sequence stars of the cluster with $M_V \gtrsim +10$ and $\mathcal{M} \lesssim 0.4 \mathcal{M}_\odot$ to be left behind.

We enquire next into the contribution of the stars of various absolute magnitudes to the total visual luminosity of a globular cluster $\mathcal{L}$. For this we shall consider the ratio

$$\frac{\mathcal{L}(M < M_0)}{\mathcal{L}} = \frac{\displaystyle\sum_{M=-\infty}^{M_0} L(M)\Phi(M)}{\displaystyle\sum_{M=-\infty}^{+\infty} L(M)\Phi(M)} , \qquad (3.45)$$

where the subscript V has been dropped for the sake of brevity and $L(M)$ denotes the luminosity of a star with the absolute magnitude M. Since $\Phi(M)$ is known only for $M \lesssim +8$, normalisation of this function is not possible and one therefore determines the denominator $\mathcal{L}$ from the directly observed integral absolute total visual magnitude.

The numerical values for the ratio $\mathcal{L}(M < M_0)/\mathcal{L}$ for M3 are assembled in Table 3.12. The total visual luminosity of the globular cluster is evidently produced predominantly by the relatively small number of intrinsically bright stars with $M_V < +1$ – just as with the stars of the galactic disc in the neighbourhood of the sun (cf. Table 3.6). It is a different matter for the corresponding contributions of the stars with $M < M_0$ to the total mass of the globular cluster $\mathcal{M}$:

118

Table 3.12. The relative contributions of the stars with visual absolute magnitudes $M<M_0$ to the total luminosity and to the total mass ($\mathcal{M} = 2.5 \times 10^5\,\mathcal{M}_\odot$) of the globular cluster M 3. (From Sandage, 1958)

M_0	$\mathcal{L}(M<M_0)/\mathcal{L}$	$\mathcal{M}(M<M_0)/\mathcal{M}$
−1	0.35	0.00
+1	0.68	0.01
3	0.83	0.03
5	0.96	0.14

$$\frac{\mathcal{M}(M<M_0)}{\mathcal{M}} = \frac{\sum\limits_{M=-\infty}^{M_0} \mathcal{M}(M)\Phi(M)}{\sum\limits_{M=-\infty}^{+\infty} \mathcal{M}(M)\Phi(M)} \,, \tag{3.46}$$

where $\mathcal{M}(M)$ denotes the mass of a star with visual magnitude M from the mass-luminosity relationship. The third column of Table 3.12 shows that the chief contribution to the total mass $\mathcal{M}$ comes from the intrinsically faint and unmassive members of the clusters. (Compare with Table 3.6).

The *total mass*, standing in the denominator of (3.46), can itself be determined by a dynamical method from the scatter of the star velocities and the radius of the globular cluster. Under the assumption of statistical equilibrium of the gravitating stellar masses, the whole cluster satisfies the energy equation

$$2E_{\text{kin}} + E_{\text{pot}} = 0 \,. \tag{3.47}$$

For the total kinetic energy of the star motions relative to the centre of mass of the cluster we can set approximately

$$E_{\text{kin}} \approx \tfrac{1}{2}\mathcal{M}\overline{v^2} \tag{3.48}$$

with $\overline{v^2}$ = square of the velocity scatter of the cluster members, whilst for the total potential energy of the cluster stars we have

$$E_{\text{pot}} = -\sum_{i,k} G\frac{\mathcal{M}_i\mathcal{M}_k}{r_{ik}} \approx -G\frac{\mathcal{M}^2}{R} \tag{3.49}$$

(r_{ik} is the mutual distance separating stars of the cluster with individual masses $\mathcal{M}_i$, $\mathcal{M}_k$, G is the gravitational constant, R is the effective radius of the cluster). Then it follows that

$$\mathcal{M} \approx \frac{R\overline{v^2}}{G} \,. \tag{3.50}$$

The procedure can be further refined by assuming a model for the globular cluster that is consistent with the observed distribution of star numbers and surface brightness over the cluster. The velocity scatter $\sqrt{\overline{v^2}}$ is obtained from an analysis of the line profile of the integrated spectrum, for example from comparison with the artificially broadened lines of appropriately selected individual stars. A recent investigation of ten globular clusters gave values $\sqrt{\overline{v^2}}$ between $7.5\,\mathrm{km\,s^{-1}}$ and $19\,\mathrm{km\,s^{-1}}$. The resulting mass values cover the range

$$\mathcal{M} \approx 2 \times 10^5 \ldots 1 \times 10^6 \, \mathcal{M}_\odot \ .$$

For the average cluster diameter $2R = 25\,\mathrm{pc}$ given above, and the round value $\sqrt{\overline{v^2}} = 10\,\mathrm{km\,s^{-1}}$, (3.50) gives $\mathcal{M} \approx 3 \times 10^5 \, \mathcal{M}_\odot$. The resulting *mass-luminosity ratio of globular clusters*

$$\frac{\mathcal{M}/\mathcal{M}_\odot}{\mathcal{L}_V/L_V^\odot} \approx 1 \ldots 3$$

therefore lies somewhat higher than in the volume of the galactic disc near the sun (3.22). The reason is the absence of the intrinsically bright stars of the upper Main Sequence from the globular clusters.

In young open star clusters with massive O- and B-stars one finds the opposite: because of their high contribution to $\mathcal{L}_V (L \sim \mathcal{M}^3)$, the mass-luminosity ratio is considerably lower than unity.

Isolated RR Lyrae Stars and Distance of the Galactic Centre: RR Lyrae variables occurring outside globular clusters have on average an apparent magnitude of $m_B \approx 16$ at a distance of $10\,\mathrm{kpc}$ and for negligible stellar extinction. Accordingly *one* requirement is met for obtaining the large scale distribution of these stars in the galaxy. Observations on extra-galactic spiral systems have shown that the RR Lyrae stars occurring in isolation are strongly concentrated around the system centre. If this is also the case in our galaxy, these objects are especially well suited for the determination of the distance of the sun from the centre R_0. The difficulty arises here, however, that has already been discussed in connection with globular clusters of spectral type G, that objects which are strongly concentrated about the centre appear above all at low galactic latitudes, and hence in a region of high interstellar extinction.

Because of the cloud structure of the interstellar medium there are, even at low galactic latitudes, a few "windows", in which A_V is less then about two magnitudes up to distances of more than $10\,\mathrm{kpc}$. The "*galactic window*" nearest to the centre is situated at $l = 0°9$, $b = -3°9$; the line of sight passes within $700\,\mathrm{pc}$ of the centre. It is a quarter of a square degree in size and contains the globular cluster NGC 6522, whose distance is about $6.5\,\mathrm{kpc}$ and whose reddening E_{B-V} was found to be 0.50. In 1951 W. Baade found in this field, on plates taken by the $2.5\,\mathrm{m}$ reflector at the Mt. Wilson Observatory, altogether

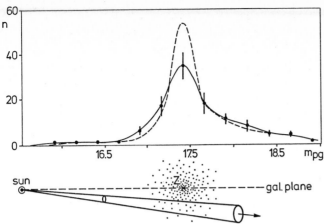

Fig. 3.33. Number of RR Lyrae stars as a function of the apparent magnitude $n(m_{pg})$ in intervals of width $0\overset{m}{.}25$, for a small field of high transparency at $l = 0\overset{\circ}{.}9$, $b = -3\overset{\circ}{.}9$. Vertical strokes indicate the size of the mean errors. The *dashed curve* is given by correction of the observed profile for random errors in m_{pg}, assuming a scatter $\sigma(m_{pg}) = \pm 0\overset{m}{.}26$. (From Oort and Plaut, 1975). The lower part of the picture explains the location of the conical field of view

76 RR Lyrae stars up to a limiting magnitude $m_{pg} \approx 20^m$. The behaviour of the numbers $n(m_{pg})$ with increasing m_{pg} enabled Baade not only to establish the strong concentration of the RR Lyrae stars towards the galactic centre, but also to determine the distance R_0 to the centre. Figure 3.33 shows the distribution $n(m_{pg})$ in this field of a total of 110 RR Lyrae stars from a recent study by Oort and Plaut (1975), in which similar profiles were also derived for five other fields with $|b|<30°$, $l<4°$ and high interstellar transparency.

In order to derive the true density distribution of the RR Lyrae stars along the line of sight, one needs the mean absolute magnitude of these variables and the interstellar extinction. The latter can always be obtained for a large number of RR Lyrae stars from the observed $B - V$ and the known intrinsic colour $(B - V)_0$ of these variables. For Baade's field the value found in this way was $A_{pg} = 1.8$, which corresponds closely to the reddening of NGC 6522 given above (Determination of the ratio A_{pg}/A_V see Sect. 4.1.1). If one associates the most frequently occurring apparent magnitude in Baade's field (Fig. 3.33) $m_{pg}(\text{Max})=17.3$ with the distance of the galactic centre $r = R_0$, then assuming that $M_{pg}(\text{RR Lyr})=+0.7$ – corresponding to $M_V(\text{RR Lyr})=+0.5$ – and $A_{pg}=1.8$ we obtain the value $R_0 = 9.1$ kpc. As mean value of all the fields studied, the uncertainty of the critical estimate is

$$R_0 = 9\pm 1\,\text{kpc}$$

in good agreement with the corresponding result for the centre of the globular cluster system.

If one interprets the profiles $n(m_{pg})$ to mean that the surfaces of equal number density of the RR Lyrae stars are ellipsoids of rotation with equal axis

121

ratios c/a, then we have nearly spherical symmetry with c/a not less than 0.8. For the decrease of density with increasing distance from the centre R, we obtain approximately $D(R) \sim R^{-3}$.

Distribution of the Classical Cepheids: High luminosity and a good method for determining M_V − from the period P of the light source using (3.44) with $\Delta M_V \approx \pm 0^m.2$ − make the Cepheids of Type I appear especially appropriate for studies of the large scale structure of the galaxy. Since they occur preferentially near the galactic plane, however, there is, as for the OB stars, a significant effect from interstellar extinction. The reddening E_{B-V} − and hence also $A_V \approx 3E_{B-V}$ − can be calculated from P and the observed colour indices $\overline{B} - \overline{V}$ and $\overline{U} - \overline{B}$ (after Tammann, 1970)

$$E_{B-V} = -0.652 - 0.129 \log P + 2.243(\overline{B} - \overline{V}) - 1.554(\overline{U} - \overline{B}) \quad . \quad (3.51)$$

This formula follows from (3.44), if one inserts therein:

$$(\overline{B} - \overline{V})_0 = \overline{B} - \overline{V} - E_{B-V} \; ,$$
$$(\overline{U} - \overline{B})_0 = \overline{U} - \overline{B} - E_{U-B} \quad \text{with} \quad E_{U-B} = 0.8 E_{B-V}$$

[cf. Sect. 4.1.1, (4.9f.)], the two right-hand sides being set equal and the resulting equation solved for E_{B-V}. Then from the first equation (3.44) for M_V we can also obtain $(\overline{B} - \overline{V})_0$ from the observed colour index $\overline{B} - \overline{V}$.

Distances can be derived for well over 200 classical Cepheids on the basis of photoelectrically measured magnitudes. The spatial distribution of these variables shows a strong concentration towards the galactic plane, with $\overline{|z|} \approx 70\,\text{pc}$. Throughout the galactic plane in the distance range considered, $r \lesssim 5\,\text{kpc}$, there appears to be a randomly fluctuating number density devoid of any clearly marked impressed structure. For the intrinsically brightest, long period Cepheids, however, a distribution is to be expected like that which one finds for the OB

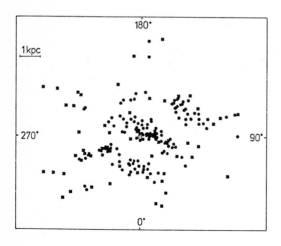

Fig. 3.34.
Distribution of young Cepheids with $P > 11^d.25$ in the galactic plane (*squares*). For comparison are shown the young star clusters and H II regions also (*circles*). (From Tammann, 1970, by courtesy of the International Astronomical Union)

stars and the young open star clusters: from the theory of star evolution these massive objects reach the pulsation phase relatively quickly; for $P > 10^d$, corresponding to $M_V < -4^m$, the mean age turns out to be $\tau < 3 \times 10^7$ years. The Type I Cepheids occurring in open star clusters of known age have even made possible the derivation of a corresponding empirical relationship between age and period: $\log \tau$ [10^7 years] $\approx 1.16 - 0.651 \log P$ [days]. The intrinsically brightest Cepheids must have arisen from short-lived OB stars and should therefore show essentially the same spatial distribution.

Figure 3.34 shows the distribution of 52 classical Cepheids with $P > 11^d 25$, whose distances ($\lesssim 5$ kpc) can be determined with a random error of only about $\pm 10\%$. A preference for the three arms $-I$, 0 and $+I$ (and possibly an inner arm $-II$) is at least suggested (cf. Fig. 3.26).

3.3.3 Stars in the Galactic Halo

Globular clusters and isolated RR Lyrae stars are not the only objects which occur at great distances from the galatic plane and hence belong to the "Halo" of our galaxy. The bulk of spheroidally distributed objects is according to present knowledge provided by "normal" stars of middle and late spectral type with metal deficient atmospheres, and the sub-dwarfs, of whom this attribute is also characteristic.

They are stars of the same classes which in the solar neighbourhood emerged as "high velocity stars" and were often distinguished by large velocity components W perpendicular to the galactic plane.

Since the spatial distribution of these objects largely follows from observations at high galactic latitudes, the influence of interstellar extinction is usually small. Whilst the faintly luminous sub-dwarfs can nevertheless be included only up to short distances, the brightness of Main-Sequence stars with, say, $M_V \approx +5$, for a limiting magnitude of $m_V \approx 20$, can be measured for objects as far as $r \approx 10$ kpc. The derivation of the density profile to such great distances is therefore possible, using methods which are based on multicolour photometry in sufficiently broad spectral regions (Sect. 3.2.1).

The advantage of this method, first developed by W. Becker, lies in that the stars with little metal content can be relatively easily recognised, since the metal deficit manifests itself in a superfluity in the shortwave part of the spectrum, and hence leads to an "ultraviolet excess". Figure 3.35 illustrates this in the light of the two-colour diagram for Becker's RGU-system. Compared with the UBV-system this has the advantage that the differences in the metal abundances affect almost exclusively the U-magnitude, so that the deficit appears in a larger and therefore better detectable ultraviolet excess $\delta(U - G) = (U - G) - (U - G)_0$; in the two-colour-diagram it produces a vertical shift upwards for all stars which show a metal deficit.

The coordination of absolute magnitude M_G with the Main Sequence profile for metal deficient stars can be obtained by means of computations of model

123

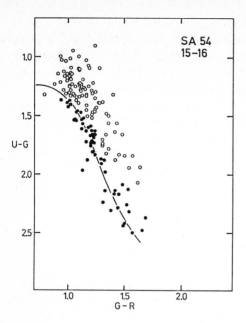

Fig. 3.35. Fractionated two-colour diagram of the stars with apparent magnitude $G = 15\ldots16$ in a field of $2.5\,\square^\circ$ size at $l = 199^\circ$, $b = +59^\circ$ (Selected Area 54). The curve indicates the behaviour of the normal Main Sequence (cf. Table 2.3). The stars with ultraviolet excess $\delta(U-G)$ ($\circ\,\circ\,\circ$) stand out clearly from the Main Sequence stars with normal metallicity. (From Fenkart, 1968)

atmospheres: a point on the normal Main Sequence corresponding to a certain absolute magnitude M_G experiences a shift in a certain direction upon transition to lower metal abundance. The separation of the giants cannot be gone into here.

W. Becker and his fellow workers have applied the procedures sketched in Sect. 3.2.1, (3.16) to the determination of the density profile $D(r)$ in an increasing number of small test fields which already exceeds ten, which lie roughly on a great circle, perpendicular to the galactic equator and passing through the direction of the galactic centre. This "Halo Programme" is in a way the counterpart of the studies of the density distribution in the galactic plane in selected Milky Way fields (Sect. 3.2.2). In each Halo test field separate profiles of $D(r)$ were derived for the metal rich "disc stars" and for the metal deficient "halo stars" with ultra-violet excess. Figure 3.36 shows a typical result for Main Sequence stars with absolute magnitudes $M_G = +3^m \ldots + 8^m$. Intrinsically bright stars are extremely rare in the halo.

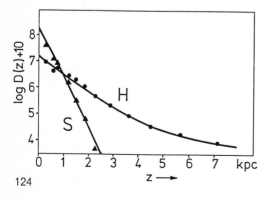

Fig. 3.36. Density profiles perpendicular to the galactic plane for metal-poor stars with ultraviolet excess, which evidently belong to the galactic halo (H), and for stars with normal metallicity (S), which are to be found only in the galactic disc. These results are based on observations in a field at the galactic north pole. Abscissa: distance from the galactic plane z in kpc. Ordinate: $\log D(z) + 10$, where $D(z)$ is referred to $1\,\mathrm{pc}^3$. (From Fenkart, 1967)

The general results can be summarised in somewhat the following way. In the neighbourhood of the sun, hence in the galactic plane, the metal deficient main sequence and giant stars amount to a numerical fraction of about one per thousand to one per cent, according to the selected limit for metal deficiency, of the metal rich stars of this type. These are the normal "high velocity stars", whose numerical density thus appears to be lower than that of the sub-dwarfs according to the counts for $r \lesssim 20$ pc discussed in Sect. 3.2.2. With increasing distance from the galactic plane the average metal deficit increases. At a distance of about $|z| > 1.5$ kpc the density of disc stars falls below the density of halo stars. An approximation of the profiles displayed in Fig. 3.36 by the expression $D(|z|) = D_0 \exp(-|z|/z_0)$ produces $z_0 = \overline{|z|}. \approx 300$ pc for the disc stars, but $\overline{|z|} \approx 700$ pc for the metal deficient halo objects (for $|z| > 1$ kpc).

If one forms the luminosity function with the density values existing for each narrow M_G interval, then with increasing distance $|z|$ we find a decrease in the fraction of intrinsically bright stars with $M_G < +5^{\mathrm{m}}$, which is more strongly marked for the halo stars.

For the *local mass density of the halo stars* in the galactic plane in the locality of the sun we have values around $1 \times 10^{-3} \mathcal{M}_\odot \mathrm{pc}^{-3}$, which are much smaller than the mass density of the disc stars (Sect. 3.2.2). The mass ratio of halo to disc stars, inside a cylinder standing perpendicular to the galactic plane (from $z = -\infty$ to $+\infty$), lies on the other hand between about 0.1 and 0.25.

3.3.4 Surface Brightness of the Milky Way and Galactic Structure, Stellar Emission of the Central Region

"Surface Brightness" and Volume Emission: The observational results described in Sect. 3.1.3 refer to the distribution of the intensity $I_\lambda(l, b)$, hence to the incident radiation energy at the wavelength λ under consideration, per unit of surface, time, solid angle and wavelength, as a function of the galactic coordinates l and b. So far as the interstellar extinction permits, the spatial distribution of the sources emitting at the wavelength under consideration finds expression in $I_\lambda(l, b)$. The surface area of an extended emitting region viewed inside a fixed solid angle increases proportionally to the square of the viewing distance. The attenuation of radiation proportionally to the square of distance travelled on its way to the observer is therefore precisely compensated. In comparison with studies based on magnitude measurements of individual stars, there is accordingly a significant gain in range, in that this is limited only by interstellar extinction.

The connection between the intensity I_λ and the volume emission $\varepsilon_\lambda =$ energy emitted per unit of volume, time, solid angle and wavelength, in the absence of interstellar extinction, is as follows (Fig. 3.37): the contribution of the volume element $dF\, dr$ at a distance r to the energy flowing across the element dF' at the position of the observer is given by

$$dI_\lambda dF' d\omega = \varepsilon_\lambda(r) dF\, dr\, d\omega' \quad .$$

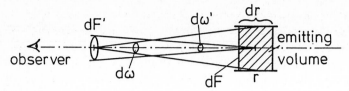

Fig. 3.37. Derivation of the connection between intensity and volume emission

Here the solid angle element $d\omega = dF/r^2$ and $d\omega = dF'/r^2$, so that dF and dF' can be eliminated. One obtains the following functional dependence on distance:

$$dI_\lambda = \varepsilon_\lambda(r)dr \quad .$$

If the radiation emitted at r is attenuated on the way to the observer by the factor $\exp[-\tau_\lambda(r)]$, with the optical depth $\tau_\lambda(r)$, then integrating along the line of sight, the intensity becomes

$$I_\lambda = \int_0^\infty \varepsilon_\lambda(r)e^{-\tau_\lambda(r)}dr \quad . \tag{3.52}$$

This equation is valid for all wavelengths. For the various bands of the optical region (e.g., U, B, V, R, I, ...) τ is determined simply by the effect of the interstellar dust. Using the definition of the magnitude scale, we can express τ by the interstellar extinction A in magnitudes:

$$e^{-\tau(r)} = 10^{-0.4A(r)} \quad . \tag{3.53}$$

If the observed intensity is expressed in stars with $m_V = 10^m$ per square degree — we take the visual region V as an example — then it is appropriate to choose a corresponding unit for the volume emission ε_V, for example 1 star with $M_V = +5$ (approximately corresponding to the sun) per pc^3. In this case (3.52) for a fixed direction transforms into

$$I_V[\text{stars } 10^m/\square^\circ] = 3.05 \times 10^3 \int_0^\infty \varepsilon_V(r) \cdot 10^{-0.4A_V(r)}dr$$

(*r*-scale in pc) . $\tag{3.54}$

The influence of the interstellar extinction decreases with increasing wavelength (see Sect. 4.1.1). Above about $10\,\mu$m the transparency factor (3.53) is practically equal to unity, so that observations lead directly to the integrated volume emission at the wavelength under consideration. Here, of course, not only the stars but also the thermal emission of the interstellar dust contribute to the observed magnitude. However, even observations in the near infra-red allow penetration of the interstellar dust and study of the assemblies of stars in the central region of the galaxy, even at $b = 0°$. The volume emission ε of stellar origin so derived naturally relates primarily to the numerous cool red stars of late spectral type, whose contribution to the surface brightness of the Milky Way is of less importance in the visible region: there the contribution of early types becomes vital. Since these occur predominantly in the spiral arms, the galactic spiral structure should become evident in the visible region in spite of interstellar extinction.

Interpretation of Observations in the Visible: We discuss first of all the distribution of brightness in galactic longitude on the basis of the results presented in Sect. 3.1.3. On account of the strong concentration of light-absorbing interstellar dust at the galactic plane it would not be meaningful to consider $I(l,b)$ for $b = 0°$. We therefore investigate the profile along the "crest line" of brightness peaks, which in general occurs where $b \neq 0$, and which is displayed in Fig. 3.38. Qualitatively one can distinguish three components:

1. a systematic climb in brightness towards the direction of the centre, with a strong maximum at $l \approx 0°$,
2. clearly marked peaks of brightness, which lie fairly symmetrically about the direction of the centre,
3. "microfluctuations" (in observations with higher angular resolution) which occur throughout.

The first component can be broadly identified with the contribution of the galactic disc, whose volume emission obviously reaches a maximum at the central region. Most of the brightness peaks of the second component can be

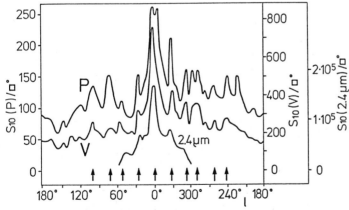

Fig. 3.38. Profiles of the surface brightness of the Milky Way along the "crest lines" in the visual region (V) and in the so-called international (photographic) blue region (P), from Elsässer and Haug, 1960), and also for the circumcentral region in the infrared at $\lambda = 2.4\,\mu\text{m}$, from Hayakawa et al. (1979). The maxima listed in Table 3.13 are indicated by arrows. Abscissa: galactic longitude; ordinates: intensity in units of 1 A0V-star 10th magnitude at each of the wavelengths considered, per square degree

127

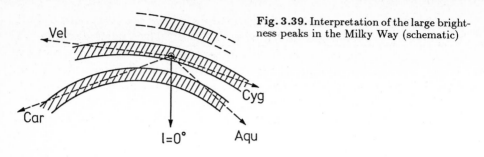

Fig. 3.39. Interpretation of the large brightness peaks in the Milky Way (schematic)

interpreted either as contributions from spiral arms for which the line of sight runs tangentially to the mid-line of the arm (Fig. 3.39), or as star clouds in spiral arms. A few of the large maxima may be "windows" between the interstellar clouds.

A summary of the brightness maxima and their significance is given in Table 3.13. (We cannot here go into the detailed justification by studies of the interstellar extinction in these directions, but further arguments will be introduced in Sect. 5.4.4, which are based on results for the distribution of interstellar gas.) The "microfluctuations" arise through the cloud-like structure of the interstellar medium, which we go into in Sects. 4.1.1 and 5.4.1.

Table 3.13. Location and interpretation of large brightness maxima of the Milky Way, which are observed in the visual *and* blue spectral regions

Brightness maximum l	Constellation	Interpretation
26°	Scutum	Star cloud
49°	Aquila	Tangency to the Sagittarius arm (−I)
73°	Cygnus	Tangential direction to the local arm (0)
101°	Lacerta	} Produced by weak local
242°	Puppis	} interstellar extinction
266°	Vela	Tangential direction to the local arm
287°	Carina	{ Tangential direction to the { Sagittarius arm (−I)
307°	Centaurus	} Star clouds or inner spiral arm structure
332°	Norma	

For a quantitative representation of the observed distributions $I(l, b)$ in the visual and in the blue spectral regions by (3.54) one can separate the galactic distribution of the volume emission into two parts: $\varepsilon = \varepsilon^{(d)} + \varepsilon^{(s)}$, where $\varepsilon^{(d)}$ denotes the contribution of the rotationally symmetric disc and $\varepsilon^{(s)}$ that of the spiral arms. In a study of this sort by Neckel (1968), for example, the following expression was chosen:

$$\varepsilon^{(d)}(R, z) = \frac{\varepsilon_0^{(d)}(R)}{1 + \left(\frac{z}{h(R)}\right)^2} \quad , \tag{3.55}$$

where R denotes the distance from the galactic centre and z the distance from the galactic plane; $h(R)$ represents the thickness of the disc as a function of R.

Because the knowledge of the overall spiral arm structure of our galaxy is too uncertain, including all the irregularities to be expected from photographs of other spiral galaxies, the quantitative discussion of the brightness maxima in $I(l,b)$ must in essence be restricted to the regions of Cygnus, Vela and Carina. For this it is sufficient to approximate $\varepsilon^{(s)}$ by an appropriate model for the local arm (0) and the next inner arm (−I), where the dependence on z can be represented in a similar manner to (3.55). Finally the interstellar extinction $A(r; l, b)$ in the visual and in the blue spectral regions has to be obtained by special investigations, which we shall discuss in Sects. 4.1.1 and 5.4.1.

Figure 3.40 shows results for the inferred radial profiles for the stellar emission $\varepsilon_0^{(d)}(R)$ at $z = 0$ and the effective half-thickness $h(R)$ of the galactic disc. The steep ascent of $\varepsilon_0^{(d)}$ towards the galactic centre follows necessarily from the observed increase in brightness when approaching the direction $l = 0°$. In spite of the strong interstellar extinction, the surface brightness of the Milky Way is not determined only by local conditions, but in fact provides information on the large scale structure of the galaxy.

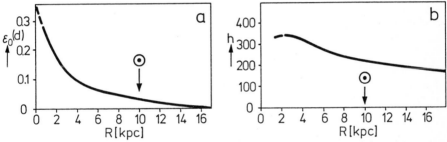

Fig. 3.40a,b. Model of the visual stellar emission of the galactic disc: (**a**) radial profile of the volume emission [stars with $M_V = 5^m$ per pc^3] in the galactic plane, (**b**) "half-thickness" h[pc] of the disc. (From Neckel, 1968)

In the neighbourhood of the sun the volume emission has the components $\varepsilon_0^{(d)}(R_0) = 0.039$ and $\varepsilon_0^{(s)}(R_0) = 0.025$, each expressed in stars with $M = 5^m$ per pc^3. The sum of these two numbers should agree approximately with the values obtained from Table 3.16 by summation over $\varepsilon(M) \cdot 10^4$, since the unit $L_\odot$ used there corresponds approximately to $M = 5^m$. This is seen to be so. One finds moreover that the contribution to the blue spectral region, of the spiral arm component $\varepsilon_0^{(s)}(R_0)$ introduced here, is largely explicable by the emission of the OB stars mentioned in Sect. 3.3.1.

The overall brightness of the galactic disc in the visual turns out to be $2 \times 10^{10} L_V^\odot$, corresponding to $M_V = -21^m$; for the integrated colour index an investigation gave $(B - V)_0 = +0^m.61$. The primary contribution here is the inner region of the system, compared with which the young objects of the spiral arms are not important. Similar values are obtained, for example, for the nearest large spiral galaxy, the Andromeda Nebula.

The Near Infrared: In addition to the brightness profile in the visible region, Fig. 3.38 shows the variation of the surface brightness at 2.4 μm along the galactic equator in the neighbourhood of the direction of the centre, which follows from the contour chart shown in Fig. 3.4. Here the climb towards the maximum at $l = 0°$ is, in accordance with expectation, still more strongly marked than in the visible; on the other hand the similarity of the two curves, in the neighbouring maxima also, cannot be overlooked.

The steep falling off in brightness of the infrared curve symmetrically about $l = 0°$ has the interpretation that the star distribution in the galactic disc inside $R \lesssim 2$ kpc distance from the centre attains a high density peak, which governs the brightness profile within $|\Delta l| \approx 10°$ from the direction of the centre, whilst further out the density falls off less strongly with increasing R. This attribute of the radial density profile corresponds with the result of an analysis of observations in the visible region, shown in Fig. 3.40.

To be able to make further quantitative statements based on infrared measurements on *the central star concentration* requires knowledge of the remaining interstellar extinction A_λ in the direction of the centre for the wavelengths under consideration. Since the true (unreddened) energy distribution of the galactic kernel region is of course not known, one makes do with assumptions, such as, for example, that the spectrum of the kernel corresponds to the kernels of other comparable spiral galaxies whose centres can be observed, being comparatively free from the influence of extinction because of the strong inclination of their principal plane to the line of sight.

That the star content of the galactic kernel is not very different from that in the Andromeda Galaxy M 31 is also suggested by observations of the integrated spectrum in the galactic "window" around the globular cluster NGC 6522 (Sect. 3.3.2), which is possible even in the blue and violet regions. They show, in agreement with the results for extra-galactic systems, a spectral type of about K0 with strong metal lines and especially CN-bands.

If one now compares this "true" energy distribution with that observed in the galactic kernel region at various infrared wavelengths, assuming the "normal" reddening rule for $\lambda = 2.4\,\mu\text{m}(K)$ given in Sect. 4.1.1, one finds an extinction $A_K = 3^\text{m} \ldots 4^\text{m}$; this corresponds to a visual extinction A_V of $20^\text{m} \ldots 30^\text{m}$. The measurements available up to now do not permit a more precise statement. However, on the "crest line" of the visual brightness distribution near $l \approx 0°$, for which profile $b{\neq}0$, one would assume a significantly smaller value, since otherwise the central region would not be so clearly marked in $I_V(l, b)$.

By integration over the spectrum in the near infrared, free from the influence of interstellar extinction, one finds the total emission from a central volume, which subtends an angle of 2° and therefore has a diameter of about 300 pc, to be about 10^{10} solar luminosities. The principal source of this emission is presumed to be intrinsically bright stars, hence giants, of middle and late spectral types (see also Sect. 3.5.2); the total mass of these stars would then amount to about $10^9 \mathcal{M}_\odot$.

A second component of the emission from the galactic central region, with about the same extent as the source described in the 2.4 μm chart, was detected by observations in the far infrared between 75 and 125 μm. The spectrum of this radiation can be represented approximately by the Planck curve for a temperature $T = 30$ K. It is an obvious inference that this is a question of warm radiation from a great aggregate of interstellar dust in this region of space (cf. Sects. 4.2.4 and 5.3.3). The energy source for the warming of the dust clouds may be the short wave radiation of the stars of the central region. For the same volume as that mentioned in the above statement on the emission from the central star concentration, the dust component produces a few times 10^8 solar luminosities. Of the total radiation from the central star concentration, a few per cent – namely the short wave portion – must therefore be absorbed by dust clouds in the immediate vicinity of these stars and re-emitted in the form of long wave infrared radiation. Most of the (long wave) stellar radiation is only scattered by these solid particles and finally leaves the central region.

The infrared sources observed at $\lambda = 2.2 \,\mu$m in the innermost central region with $R \lesssim 1.5$ pc (Sect. 3.1.3, Fig. 3.5) are presumably intrinsically very bright individual stars of late spectral type or whole star clusters. For the luminosity of these objects one expects values of the order of $10^4 \, L_\odot$: for Source No. 7 in Fig. 3.5 even $10^5 \, L_\odot$. At $\lambda = 10 \,\mu$m the high resolution contour chart already looks quite different: only Source No. 7 still survives, whilst other sources have appeared. Probably here one is seeing in essence the thermal emission of the dust concentrations, warmed up by the newly formed stars contained therein.

3.4 Large Scale Motion of the Stars

The kinematics of objects near the sun already allows statements on the rotation of the galactic disc relative to the stars with low metal content moving on strongly inclined galactic orbits (Sect. 3.2.3: Asymmetry of Large Peculiar Velocities). An independent approach to the galactic rotation and its behaviour with increasing distance from the centre of the system is made possible by the kinematic data from far distant stars of the galactic disc. In what follows we shall first of all give a formal derivation of the effects to be expected in the RV and PM. Further conclusions on the large scale kinematics of the star system are suggested by observations of halo objects such as globular clusters and RR Lyrae stars.

3.4.1 Shear, Rotation and Dilatation of the Velocity Field

Within larger regions of space, r greater than about 100 pc, the field of peculiar velocities may no longer be considered as, in the statistical sense, spatially

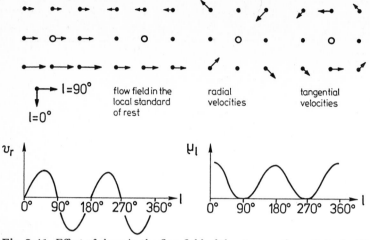

$l=90°$ $l=0°$ flow field in the local standard of rest radial velocities tangential velocities

v_r

0° 90° 180° 270° 360° l

μ_l

0° 90° 180° 270° 360° l

Fig. 3.41. Effect of shear in the flow field of the stars on the radial velocities and the proper motions

homogeneous: non-rigid rotation of the stellar system about the centre lying in the direction $l = 0°$ gives the flow pattern shear (Fig. 3.41, left); in addition, with any rotation all velocity vectors acquire spin relative to an inertial system. Moreover, dilatation (expansion or contraction) of the stellar system may be present, which provides further inhomogeneity. We therefore consider first of all the effects of such anomalies of the flow pattern from statistical homogeneity on the radial velocities and the proper motions.

Qualitative Considerations: Figure 3.41 illustrates the case of pure shear in a straight-line parallel flow directed towards $l = 90°$ in the galactic plane with a linear gradient. Upper left is a detail from the basic flow pattern. The origin of the local standard of rest (position of the sun) is indicated by an open circle. The two upper right diagrams show the resolution into radial and tangential components v_r and t_l, respectively. The left-hand lower portion of Fig. 3.41 shows the expected dependence of radial velocity on galactic longitude l for stars at equal distances r. The amplitude of the resulting double wave about the mean value over all longitudes $\bar{v}_r = 0$ is obviously proportional to the distance r. Finally, on the right is shown the (distance independent) proper motion component $\mu_l = t_l / Kr (> 0)$ as a function of l.

If the shear is the result of non-rigid rotation about a relatively distant centre lying in the direction $l = 0°$, then the flow pattern is spinning relative to the fundamental coordinate system fixed in space. A spin has no effect on the radial velocities, but it does produce a constant additive term for the proper motions, illustrated by Fig. 3.42. If ω denotes the angular velocity of the rotational motion of the origin of the local standard of rest (circular orbit!), then the change in the proper motion in unit time ($\Delta t = 1$) in the given sense of rotation is expressed by $\Delta\mu_l = -\omega$, so that the double wave of proper motion μ_l sketched in Fig. 3.41 is shifted bodily in the negative ordinate direction.

Fig. 3.42. Effect of rotation in the flow field on the proper motions

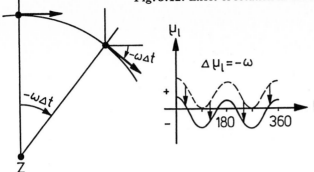

An isotropic expansion or contraction of the stellar system in the region of space under consideration must obviously give an additive term to the radial velocities, which leaves the proper motions unchanged.

Quantitative Treatment: First of all we consider stars which move in the galactic plane with velocity $\Theta = \Theta(R)$ in circular orbits with radius R about the galactic centre. The angular velocity of the orbital motion will be denoted by $\omega(R)$. For the distance of the sun from the centre $R = R_0$ we shall introduce the abbreviations: $\Theta(R_0) = \Theta_0$ and $\omega(R_0) = \omega_0$. We seek to ascertain what radial velocities and proper motions would be observed in the local standard of rest, if no further motions were superposed.

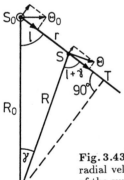

Fig. 3.43. Derivation of the effect of the differential galactic rotation on the radial velocities and the proper motions. S_0, S and Z indicate the positions of the sun, the star under consideration and the galactic centre

From Fig. 3.43 we have:

$$v_{\mathrm{r}} = \Theta \sin(l + \gamma) - \Theta_0 \sin l \tag{3.56}$$

$$t_l = \Theta \cos(l + \gamma) - \Theta_0 \cos l \quad . \tag{3.57}$$

In order to eliminate $\sin(l + \gamma)$ we use the sine rule for the triangle ZSS_0 :

133

$$\frac{R}{R_0} = \frac{\sin l}{\sin [180^\circ - (l + \gamma)]} = \frac{\sin l}{\sin(l + \gamma)} \quad ;$$

$\cos(l + \gamma)$ is given by the relation between the projections of $\overline{ZS_0}$ and $\overline{ZS}$ on the line of sight to the star:

$$\overline{S_0 T} = R_0 \cos l = R \cos(l + \gamma) + r \quad .$$

Then, from (3.56)

$$v_r = \Theta \frac{R_0}{R} \sin l - \theta_0 \sin l = R_0 \left[\frac{\Theta}{R} - \frac{\Theta_0}{R_0} \right] \sin l \quad \text{or} \tag{3.58}$$

$$v_r = R_0 [\omega(R) - \omega_0] \sin l \tag{3.59}$$

and, from (3.57) [definition of K, see Sect. 2.1.3, (2.24) and (2.25)]:

$$t_l = K r \mu_l = R_0 [\omega(R) - \omega_0] \cos l - r\omega(R) \quad . \tag{3.60}$$

In order to calculate $v_r(r, l)$ and $t_l(r, l)$ in terms of the rotation function $\omega = \omega(R)$ and the known value R_0 one has still to express R in terms of r by means of the cosine rule

$$R^2 = R_0^2 + r^2 - 2 R_0 r \cos l \quad . \tag{3.61}$$

In the *close neighbourhood of the sun*, $r \ll R_0$, and therefore $r \ll R$, we can substitute the useful approximation

$$\omega(R) - \omega_0 \approx (R - R_0)\omega_0' \quad \text{with} \quad \omega_0' = \left(\frac{d\omega}{dR} \right)_{R_0} \tag{3.62}$$

and also obtain from (3.61) the simple approximate expression for $R - R_0$:

$$R = R_0 \sqrt{1 - \left(\frac{r}{R_0} \right)^2 - 2 \frac{r}{R_0} \cos l} \approx R_0 \left[1 - \frac{r}{R_0} \cos l \right] = R_0 - r \cos l \quad ,$$

whence

$$R - R_0 \approx -r \cos l \quad . \tag{3.63}$$

Introducing (3.62) and (3.63) into (3.59) and (3.60), and using the trigonometric formulae

$$\cos l \sin l = \tfrac{1}{2} \sin 2l \ , \quad \cos^2 l = \tfrac{1}{2}(1 + \cos 2l)$$

(and neglecting r^2 in comparison with $r R_0$), we obtain the required results:

$$v_r = Ar \sin 2l \tag{3.64}$$

$$K\mu_l = A \cos 2l + B \quad . \tag{3.65}$$

Here we have introduced the so-called *Oort Constants*:

$$A = -\tfrac{1}{2}\omega_0' R_0$$
$$B = -\tfrac{1}{2}\omega_0' R_0 - \omega_0 \tag{3.66}$$

with the property

$$A - B = \omega_0 \quad . \tag{3.67}$$

Usually A and B are given in $\mathrm{km\,s^{-1}kpc^{-1}}$, and correspondingly r in kpc. When describing μ_l'' one uses $P = A/47.4$ and $Q = B/47.4$ instead of A and B, with the dimensions arc seconds per century. The linearised expressions for the effects of the differential galactic rotation satisfy the imposed accuracy requirements when r is less than or equal to about $R_0/10$.

An additional *dilatation* of the star system can be described by $\dot{R} = \varepsilon \cdot R$, where perhaps $\varepsilon = \varepsilon(R)$; $\varepsilon > 0$ indicates expansion. It introduces an additive term into the radial velocity (Fig. 3.44)

$$\Delta v_r = [-\dot{R}\cos(l + \alpha)] - [-\dot{R}_0 \cos l] \quad .$$

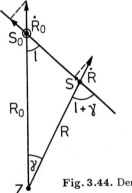

Fig. 3.44. Derivation of the effect of a dilatation

In complete analogy with the transition from (3.57) to (3.60), apart from a sign, one obtains

$$\Delta v_r = -R_0[\varepsilon(R) - \varepsilon_0]\cos l + r\varepsilon(R) \quad , \tag{3.68}$$

where $\varepsilon_0 = \varepsilon(R_0) = (\dot{R}/R)_{R_0}$. Linearisation for $r \ll R_0$ with the substitution $\varepsilon(R) - \varepsilon_0 \approx (R - R_0)\varepsilon_0'$, and using (3.63), leads to

$$\Delta v_{\rm r} = r \left[2C \left(\frac{r}{R_0} - \cos l \right) \cos l + \varepsilon_0 \right] \quad \text{where} \tag{3.69}$$

$$C = -\tfrac{1}{2} \varepsilon_0' R_0 \quad . \tag{3.70}$$

In the case $\varepsilon = \varepsilon_0 = $ const. we have $\Delta v_{\rm r} = \varepsilon_0 r$ (>0 for expansion).

Consideration of non-zero galactic latitudes for the stars under study (orbital motion parallel to the galactic plane!) introduces a factor $\cos b$ into (3.59) and a factor $\cos^2 b$ into (3.64); the proper motion in longitude $\mu_l'' = \dot{l} \cos b$ remains unchanged, but we now have a component μ_b'', given by

$$K\mu_b'' = -\tfrac{1}{2} A \sin 2b \sin 2l \tag{3.71}$$

[see, for example, Trumpler and Weaver (1953), p. 562].

3.4.2 Rotation of the Galactic Disc

We turn now to observational data from which the differential rotation of the star field of the galactic disc can be deduced. Two examples are shown in Fig. 3.45: the variation of the RV with galactic longitude for the Cepheids and for early B stars in certain distance intervals around $\bar{r} = 2.3$ and 2.0 kpc, respectively. The continuous curves have been calculated from (3.64) with appropriately chosen amplitude factors Ar (terms of higher order make only minor deviations from this at these distances).

Although nearby stars of this type with $r \lesssim 100$ pc, and hence $Ar \approx 0$, show scarcely more than "randomly" distributed peculiar velocities in their RV, the relatively far distant objects presented in Fig. 3.45 show the dominating influence of large scale non-rigid rotation. The velocities referred in each case to the local standard of rest at the position of these stars show only a relatively small scatter about the profile arising from the shear of the velocity field.

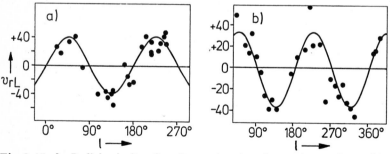

Fig. 3.45a,b. Radial velocity after discounting the solar motion $v_{rL} [\text{km s}^{-1}]$ as a function of the galactic longitude: (a) for classical Cepheids with mean distance $\bar{r} = 2.3$ kpc (from Joy, 1939) and (b) for early type B-stars with $\bar{r} = 2.0$ kpc (from Feast and Thackeray, 1958)

Similar results are available for the components of the PM in the galactic longitude μ_l, which can be found using (3.65). Here the distance does not appear, which means that the Oort Constants A and B can in principle be determined solely from the μ_l. Only for sufficiently far distant stars, however, does the contribution of the peculiar motion to the observed PM become sufficiently strong that the pure rotation effect can be obtained with adequate accuracy. The spin of the velocity field can be found from the PM alone: $A - B = \omega_0$ is the angular velocity of the spin at the position of the sun.

Determination of the Oort Constants A and B from the Observations: For this purpose one represents the individual observed values, relative to the sun, of v_r, μ_l, μ_b of each of the selected objects by means of the equations

$$v_r = (v_r)_\odot + (v_r)_{\text{rot}} + (v_r)_{\text{pec}} + K$$
$$\mu_l = (\mu_l)_\odot + (\mu_l)_{\text{rot}} + (\mu_l)_{\text{pec}} + \Delta\mu_l \qquad (3.72)$$
$$\mu_b = (\mu_b)_\odot + (\mu_b)_{\text{rot}} + (\mu_b)_{\text{pec}} + \Delta\mu_b \quad .$$

Here the contributions of the solar motion relative to the local standard of rest (at the position of the sun), the differential galactic rotation, and the peculiar motion of the star (relative to the local standard of rest at its position) are denoted by the subscripts $\odot$, rot and pec, respectively; K represents the dilatation term ("K-term") denoted in Sect. 3.4.1 by Δv_r.

The correction terms $\Delta\mu_l$, $\Delta\mu_b$ arise from the fact that the proper motions available up to now have been derived under the assumption of the old Newcomb value of the precession constants, which are still based upon 19th century observations of absolute star positions. If the Newcomb values for the lunisolar precession ψ and the planetary precession χ are both in error by the quantities $\Delta\psi$ and $\Delta\chi$, the proper motions obtained using (2.5) and (2.6) in Sect. 2.1.2 will both be in error by the quantities

$$\Delta\mu_\alpha^s = \Delta m + \Delta n \sin\alpha \tan\delta \quad \text{with} \quad \Delta m = \Delta\psi \cos\varepsilon - \Delta\chi$$
$$\Delta\mu_\delta'' = \Delta n \cos\alpha \qquad\qquad \text{with} \quad \Delta n = \Delta\psi \sin\varepsilon \quad . \qquad (3.73)$$

The corresponding quantities $\Delta\mu_l$ and $\Delta\mu_b$ in galactic coordinates are given by (2.30) in Sect. 2.1.3. In studies of proper motion, however, one generally uses the second and third equations (3.72) in the equatorial system.

The system of equations of type (3.72), set up for a number of appropriate stars, contains as unknowns the Oort Constants A and B, the dilatation constants ε_0 and C, the precession corrections $\Delta\psi$ and $\Delta\chi$, which occur only in the proper motion part, and one can also treat the components of the solar motion as unknowns. If the data are available for sufficiently far distant stars, one may even try to determine the second derivative ω_0'' in the expansion of the rotation function

$$\omega(R) = \omega_0 + \omega_0'(R - R_0) + \tfrac{1}{2}\omega_0''(R - R_0)^2 \quad . \qquad (3.74)$$

In the solution of the system it is assumed that the contribution of the peculiar motion vanishes in the average.

The *proper motion data* for 512 stars of the FK 4 and a supplement (see Sect. 2.1.4) give $\Delta\psi = +1''\,10\pm0''\,15$ (per century), and so lead to the new precession constant given in Sect. 2.1.2 – $\Delta\chi$ was found by using new and more accurate planetary masses in the calculation of χ. The fundamental proper motions yielded (from Fricke, 1977) the Oort Constants

$$P = A/47.4 = 0''\,33\pm0''\,08/\text{cent.} \quad \text{or} \quad A = 15.6\pm2.8\,\text{km s}^{-1}\text{kpc}^{-1}$$
$$Q = B/47.4 = -0''\,23\pm0''\,06/\text{cent.} \quad \text{or} \quad B = -10.9\pm2.8\,\text{km s}^{1}\text{kpc}^{-1} \quad .$$

The corresponding evaluation of proper motion data from 166 179 stars of the AGK 3 (spectral types B3 ... M) led to the same correction $\Delta\psi$ and

$$A = +16.1\pm1.9\,\text{km s}^{-1}\text{kpc}^{-1} \,, \quad B = -9.0\pm1.9\,\text{km s}^{-1}\text{kpc}^{-1} \quad .$$

Determinations of A from *radial velocity data* in conjunction with carefully calculated distances have been carried out in more recent times for, amongst others, OB stars, classical Cepheids and open star clusters with distances up to a few kpc. The values obtained lie between 15 and 19 km s^{-1}kpc^{-1}. Attempts to obtain also the second derivative of the rotation function (3.74) led to $w_0'' \approx +0.3$ to $+0.6\,\text{km s}^{-1}\text{kpc}^{-2}$. In practice this means that $w(R)$ falls off somewhat more slowly for R greater than about 10 kpc than the linear profile $w(R) = w_0 + w_0'(R - R_0)$ with $w_0' = -2A/R_0 < 0$.

The *K term* is always found to have small (positive or negative) values up to about 4 km s^{-1}. In so far as the quantities found usually exceed their mean error, we are presumably seeing – at least as regards the influence of stars with $r \lesssim 500\,\text{pc}$ – predominantly the effects of local systematic anomalies in the general galactic rotation (region of Gould's Belt, see Sect. 3.2.3: local moving groups). A general expansion of the galaxy, like that revealed by the kinematics of the interstellar gas in the central region of the galaxy, could however be occurring throughout the star field with a local expansion rate of only $\varepsilon_0 < 4\,\text{km s}^{-1}\text{kpc}^{-1}$.

Rotation Constants and Distance of the Sun from the Galactic Centre: The state of knowledge at the beginning of the sixties led in 1963 to a recommendation of the IAU for the general adoption of the round figures $A = +15$, $B = -10$, and hence $w_0 = 25\,\text{km s}^{-1}\text{kpc}^{-1}$ and the value $R_0 = 10\,\text{kpc}$. For the linear velocity of the local standard of rest, these figures give $\Theta_0 = w_0 R_0 = 250\,\text{km s}^{-1}$.

The more recent results for R_0, A and B reviewed here, together with further results, have prompted a new discussion of the question of the best values.

Today the best approximation for *Oort's Constants* still appears to be:

$$A \approx +15\,\text{km s}^{-1}\text{kpc}^{-1} \,, \quad B \approx -10\,\text{km s}^{-1}\text{kpc}^{-1} \quad . \tag{3.75}$$

From this it follows that the *angular velocity at the position of the sun* is

$$\omega_0 = A - B \approx 25.0\,\mathrm{km\,s^{-1}kpc^{-1}} \qquad \text{or} \tag{3.76}$$

$$\omega_0/47.4 \approx 0\rlap{.}{''}53 \ \text{per century}$$

and, as the *orbital period of the local standard of rest*,

$$T_0 = \frac{360°}{(\omega_0/47.4)} \approx 2.5 \times 10^8 \ \text{years} \quad . \tag{3.77}$$

The results for the central distance of the sun obtained from the distributions of the globular clusters and the RR Lyrae stars led in 1985 to a recommendation of the IAU for the use of the value

$$R_0 = 8.5\,\mathrm{kpc} \quad , \tag{3.78}$$

and − in good agreement with the accepted negative vertex velocity of the extreme halo objects − the *circular orbit velocity*

$$\Theta_0 = \omega_0 R_0 \approx 220\,\mathrm{km\,s^{-1}} \quad . \tag{3.79}$$

The linear *velocity gradient* at $R = R_0$ is:

$$\left(\frac{d\Theta}{dR}\right)_{R_0} = \left(\frac{d}{dR}\omega R\right)_{R_0} = \omega_0' R_0 + \omega_0$$

$$= -(A+B) \approx -5\,\mathrm{km\,s^{-1}kpc^{-1}} \quad . \tag{3.80}$$

As the direct consequence of these numbers we immediately estimate the total mass $\mathcal{M}$ (stars + interstellar matter) inside $R = R_0$ whose effect is concentrated at the centre, by equating the gravitational acceleration $G\mathcal{M}/R_0^2$ to the centripetal acceleration Θ_0^2/R_0. We accordingly obtain

$$\mathcal{M} = \frac{\Theta_0^2 R_0}{G} \approx 2 \times 10^{44}\mathrm{g} \approx 1 \times 10^{11}\,\mathcal{M}_\odot \quad . \tag{3.81}$$

3.4.3 Objects of the Galactic Halo

Globular Clusters: These objects introduce a new aspect of the large scale kinematics of the galactic system. Here we have only radial velocities, whose accuracy usually only amounts to $\pm20\,\mathrm{km\,s^{-1}}$: proper motions are not available for globular clusters. It is appropriate to relate the radial velocities directly to the local standard of rest defined in (3.40): an object with the galactic coordinates (l, b) which is at rest in the local standard of rest (peculiar velocity = 0) has, by virtue of the solar motion, a radial velocity

$$(v_r)_\odot = -U_\odot \cos l \cos b - V_\odot \sin l \cos b - W_\odot \sin b \quad . \tag{3.82}$$

This expression follows directly by transforming the third equation (3.39) to galactic coordinates using γ_{ik} from (2.27), and averaging ($\overline{v}_{rp} = 0$). The radial velocity of a globular cluster in the local standard of rest is then

$$v_{rL} = v_r - (v_r)_\odot \quad . \tag{3.83}$$

If one now sets

$$v_{rL} = \overline{U} \cos l \cos b + \overline{V} \sin l \cos b + \overline{W} \sin b + v_{rLp} \quad , \tag{3.84}$$

then the centroid of the globular cluster velocities $(\overline{U}, \overline{V}, \overline{W})$ relative to the local standard of rest can be obtained by equating $\overline{v}_{rLp}$ to zero. Here U, V, W denote the velocity components in the galactically oriented system (U-component in the direction $l = 0°$, $b = 0°$, V-component in the direction $l = 90°$, $b = 0°$). An investigation based on radial velocity data for 70 globular clusters gave

$$\overline{U} \approx 0 \ , \quad \overline{V} \approx -170\,\mathrm{km\,s}^{-1} \ , \quad \overline{W} \approx 0 \quad . \tag{3.85}$$

The uncertainty of this result amounted to about $\pm 30\,\mathrm{km\,s}^{-1}$. The scatters of the components U, V, W about the mean values (3.85) are in each case of the order of $\pm 100\,\mathrm{km\,s}^{-1}$. The solution of the equation system (3.84) with the inclusion of an additive constant K ("K term") produced no significant indication of an expansion ($K>0$) or a contraction ($K<0$).

According to the interpretation expounded in Sect. 3.2.3, this result means that the globular clusters take part in the rotation of the galactic disc either weakly or not at all. This can also be proved in another way: if a globular cluster rotates parallel to the galactic plane with the angular velocity $\omega(R)$, then there arises a component of the radial velocity which according to (3.59) and the indication at the end of Sect. 3.4.1 is given by

$$(v_r)_{\mathrm{rot}} = R_0[\omega(R) - \omega_0] \sin l \cos b \quad . \tag{3.86}$$

Here R_0 denotes the distance of the sun from the centre and ω_0 the angular velocity of the local standard of rest at the position of the sun. The factor $\cos b$ takes account of the finite galactic latitude of the cluster.

Let us take an average of the radial velocities, corrected for solar motion, of a large number of clusters in a restricted field with centre point at (l, b), whose distances R from the rotation axis of the galactic disc lie in a narrow interval. Only the differential effect of the rotation with reference to the local standard of rest should, on average, be left over: $\overline{v}_{rL} \approx (v_r)_{\mathrm{rot}}$. If one sets a value between 200 and 250 $\mathrm{km\,s}^{-1}$ (Sects. 3.2.3, 3.4.2) for the orbital velocity $R_0\omega_0 = \Theta_0$ of the local standard of rest, then $R_0\omega(R)$ for the globular clusters in the R-interval under consideration can be estimated from (3.86). Hence one finds that the rotational velocity of the globular cluster system as a whole, near $R \approx R_0$, can only be about $50 \, \mathrm{km \, s^{-1}}$ or less.

This result is consistent with (3.85) and $\Theta_0 = \omega_0 R_0 = 200 \ldots 250 \, \mathrm{km \, s^{-1}}$ within the limits of error, and it corresponds to the small flattening of the system (Sect. 3.3.2). The globular clusters obviously move in orbits around the centre whose inclinations to the galactic plane are essentially randomly distributed. The scatter of the velocity components lying along the line of sight $v_{rL} \approx - \Theta_0 \sin l \cos b + v_{rLp}$ (since $R_0 \omega(R) \approx 0$ and $\overline{U} \approx \overline{W} \approx 0$) was estimated to be about $\pm 90 \, \mathrm{km \, s^{-1}}$ for the G types strongly concentrated around the centre and $\pm 130 \, \mathrm{km \, s^{-1}}$ for the "halo clusters".

RR Lyrae Stars: For well over a hundred of these variables both the radial velocities (RV) and the proper motions (PM) are available. The limiting magnitude is about $m_{pg} = 14$, so that (for $M_{pg} \approx + 0.5 \ldots + 1.0$) distances from the sun extend up to a maximum of about $4 \, \mathrm{kpc}$. As already described in Sect. 3.3.2, statistical parallaxes can first of all be derived for kinematically homogeneous sub-groups of this material, and hence the absolute magnitude of the RR Lyrae stars can be determined. With these we can then obtain individual photometric distances, which together with the radial velocities and proper motions lead to the space velocities of the individual stars (Sect. 2.1.3).

If one is interested only in the velocity mean values (centroid) and dispersions, the statistical parallaxes can be used directly. Results of this sort are displayed in Table 3.11. The individual groups differ in length of period P and metal deficit, measured by the Preston Spectral Index ΔS. RR Lyrae stars with $P < 0^{\mathrm{d}}\!4$ show a smaller metal deficit than the longer period stars of this type. Although the uncertainties in the numbers may amount to 20 to 30%, there can be no doubt that not only the magnitude of the asymmetric drift $-\overline{V}$ but also the velocity dispersion increase significantly with the period and with the metal deficit. This is clearly shown by two sharply differentiated main groups, which are characterised by $P < 0^{\mathrm{d}}\!4$, $\Delta S < 5$ and $P > 0^{\mathrm{d}}\!4$, $\Delta S \gtrsim 5$. The first group lags only slightly behind the rotation of the galactic disc, whilst the second group on average shows very similar kinematic data to the globular clusters of the halo type.

3.5 General Summary, Stellar Populations

3.5.1 Sub-Systems

From the results shown up to now, stellar objects of different types, such as for example Main Sequence stars of spectral class F or, say, RR Lyrae stars with Preston Spectral Index $\Delta S > 5$, evince in general different spatial distributions and different kinematic attributes. If one looks for a one-dimensional classification, then a possible parameter might be the asymmetric drift, or the degree of participation in the galactic rotation, with which the velocity disper-

Table 3.14. Sub-systems with different kinematics and different spatial distribution, ordered according to the magnitude of the asymmetric drift $\bar{V}$. Explanation in text

Objects		$\bar{V}$ [km s⁻¹]	Σ_W [km s⁻¹]	$\overline{\lvert z \rvert}$ [pc]	$\left(\dfrac{\partial \log D}{\partial \log R}\right)_{R_0}$	D_0 [1/1000 pc³]	ρ_0 [$M_\odot$/1000 pc³]	ΣD
O...B2(OB) ($M_V \lesssim -3$)		0	<10	65	0	5×10^{-4}	10^{-2}	10^4
Cepheids Type I		0	5	70	0	1×10^{-4}	5×10^{-4}	3×10^4
Open clusters	O...B6	0	6	50	0	4×10^{-4}	0.1	3×10^4
	B7...F			80				
Main Sequence (V)	A	0	5	90	0	0.5	38	10^{11}
	F	0	15	130	–	2.5		
	G	−10	20	180	−1...−2	96		
	K			270				
	M			270				
Giants (III, IV)	G	−10	15	400	−1.5	1	1	
	K			270	−2			
	M			290				
White dwarfs		−10	20	(270)	−2	8	7	10^{10}
Planetary nebulae		−10	–	140	−1.5	5×10^{-5}	5×10^{-5}	5×10^4
Long period variables (Mira stars)	$P = 350...400$	−10	20	250	−1	10^{-3}	10^{-3}	10^6
	$300...350$	−20	35	300	−2			
	$250...300$	−30	30	400				
RR Lyrae stars	$\Delta S < 5$ $P < 0^d4$	−50	30	(400)	−2	10^{-5}	10^{-5}	5×10^4
Novae		–	(20)	(400)	−2	–	–	25...50 p.a.
Long period variables	$P = 200...250$	−60	50	600	−3	10^{-3}	10^{-3}	10^6
	$140...200$	−100	60	900				
RR Lyrae stars	$\Delta S \geqq 5$ $P > 0^d4$	−200	70...90	(2000)*	−3*	10^{-5}	10^{-5}	5×10^4
Sub-dwarfs	$\delta(U - B) > 0.15$	−230	90	2000*	−3.5*	1	1	10^{10}
Globular clusters	$Sp > G1$	−200	100*	2000*	−4.5*	10^{-8}	1×10^{-3}	2×10^2
	Halo-Type ($\leqq$ G1)			6000*	−3.4*			

sion and also the spatial distribution are correlated directly. One thus arrives at B. Lindblad's concept, already mentioned in the Introductory Survey, of a galaxy composed of sub-systems. The possibility of progressive refinement of the classification is in practice limited, and one ends up with a scheme − optionally in more or less detail − such as is displayed in the summary Table 3.14.

Table 3.14 characterises the kinematics of the sub-systems not only by the asymmetric drift $\overline{V}$ ($= V$-component of the centroid of the velocities relative to the local standard of rest), but also by the velocity dispersion perpendicular to the galactic plane Σ_W. The spatial distribution is indicated in each case by the mean value of the distance from the galactic plane $\overline{|z|}$ and by the logarithmic radial gradient of the number density D: $\partial \log D / \partial \log R$ is in the case of a power law $D(R) \approx R^n$ simply equal to the exponent n: its magnitude characterises the degree of concentration about the galactic centre. The quantities Σ_W, $\overline{|z|}$, the density gradient and the values of the number density D_0 and the mass density ϱ_0, which are also given, are normally quoted for the position of the sun ($R \approx R_0$). If the numbers are marked with an asterisk (*), however, then they apply to the whole sub-system, for example with the globular clusters with integrated spectral type earlier than G 1, which occur almost exclusively in the region $R < R_0$. The last column contains under the heading ΣD rough estimates of the overall total number of members of the sub-system.

3.5.2 Stellar Populations and Evolution of the Galaxy

The present day concept of stellar populations is dominated by the relationship between the chemical and the kinematic attributes of the different classes of stellar objects; the chemical composition is further correlated − to a high degree − with the mean age. If one enquires into those objects which exhibit almost the same kinetics (and distribution), and it turns out that this can be so for quite different types, such as, for example, globular clusters and RR Lyrae stars occurring in isolation with periods greater than $0\overset{d}{.}4$, then they often prove to have the same chemical composition and also the same mean age. Accordingly stellar populations can be defined as groups of stellar objects with the same characteristics relating to kinematics, distribution, chemical composition and mean age.

Classification of Populations: The boundary between different stellar populations involves a certain arbitrariness. The "working classification" evolved in 1957 and mentioned in the Introduction is displayed in Table 3.15. The five populations introduced there are characterised by the magnitude of distance from the galactic plane z, the velocity dispersion Σ_W in the z-direction, the concentration about the galactic centre (qualitative), the metal content in the stellar atmospheres relative to that of the solar atmosphere ($= 1$) and the age of the objects.

143

Table 3.15. Classification scheme for stellar populations. Explanation in text

	Halo-Population II	Intermediate Population II	Disc-population	Older Population I	Extreme Population I
Typical objects	Globular clusters (Halo type)	Normal high velocity stars ($\lvert W\rvert >30\ \mathrm{km\,s^{-1}}$)	F- to M-stars with "weak metal lines"	A-stars F- to K-stars with "strong metal lines"	OB-stars I Cepheids Supergiants
	RR Lyr stars with $P>0\overset{\mathrm{d}}{.}4$	Long period variables with $P<250^{\mathrm{d}}$	Planetary nebulae	K- to M-dwarfs with emission lines	Open clusters with early B-stars
	Sub-dwarfs		RR Lyr stars with $P<0\overset{\mathrm{d}}{.}4$		T Tau-stars
			Novae		
$\overline{\lvert z\rvert}$ [pc]	2000	700	400	150	70
Σw [$\mathrm{km\,s^{-1}}$]	$80\ldots100$	$40\ldots60$	$20\ldots40$	$10\ldots20$	$5\ldots10$
Concentration towards centre	strong	strong	strong/ moderate	weak	none
Metallicity	$0.01\ldots0.1$	$0.01\ldots0.1$	$0.03\ldots0.3$	$0.3\ldots1$	1
Age [years]	$\sim 10^{10}$	$\sim 10^{10}$	$2\times10^{9}\ldots10^{10}$	$5\times10^{8}\ldots5\times10^{9}$	$<5\times10^{8}$

The Extreme Population I contains the youngest stellar objects, which show "normal" abundances of the heavy elements from carbon onwards ("metals") relative to hydrogen, are very strongly concentrated at the galactic plane, and there exhibit a random distribution in the broad lines of a spiral structure. The same features are shown by the interstellar medium. The Extreme Population II or Halo Population II on the other hand contains the oldest known stellar objects of the Milky Way. In the atmospheres of these stars the metal abundances are about two orders of magnitude lower than in the Extreme Population I and the spatial distribution is nearly spherical symmetric with strong concentration around the centre but at the same time extending to a very large halo.

The main body of the galactic disc, including the central bulge, with spectral types A...M (Main Sequence, Giants) falls between these two extremes. Starting with the still relatively short-lived A stars, one finds here – according to the results mentioned in Sect. 3.2.2 (kinematics) and 3.2.4 (kinematic groups and ages) – a smooth transition from the normal metal abundances ("stars with strong metal lines") to the metal deficits below 1/10 of this value ("stars with weak metal lines"). The other characteristics vary in corresponding fashion: mean ages as well as velocity scatter and distribution perpendicular to the galactic plane. The partition of the intervening region presented in Table 3.15, into an Older Population I, still with largely normal metal abundance, a Disc Population embracing the main mass of the galactic stars, and an Intermediate Population II, must therefore be seen as a relatively gross approximation.

Baade originally proposed that the star concentration in the galactic central bulge consisted essentially of objects in Population II, similar to the stars of the globular clusters of the halo. Later it became clear that metal rich stars contribute far more to the volume emission there than metal deficient stars. For example, in Baade's high transparency "window" at NGC 6522 (see p. 120) about twice as many M giants as RR Lyrae stars were found near the centre, thus making the former the dominant star types, which do not occur in the globular clusters of the halo. The spectrum of the integrated light of this field for the blue spectral region, which stems predominantly from objects of the central region, shows relatively strong metal lines and robust CN bands. The spectrum of the kernel region of the Andromeda Galaxy has a similar appearance. The predominant stellar population of the galactic central region, according to these and other observations, has roughly the spectral characteristics of the totality of stars of the old open star cluster NGC 188. Stars with weak CN bands, but strong metal lines, however, can also be detected, such as occur in metal rich globular clusters, and hence typical Population II objects.

Populations and Evolution of the Galaxy: The significance of the population schema lies primarily in that, in combination with our knowledge of the evolution of individual stars, it leads almost inevitably to a certain scenario for the evolution of the whole Milky Way system, which represents at least a good working hypothesis. One supposes first that the chemical elements, with the exceptions of hydrogen, the greater part of helium – which was probably already produced in the "Big Bang" – and the light elements boron, beryllium and lithium, were formed in the interiors of the stars, and in later unstable evolutionary phases were expelled again into the interstellar medium. A progressive enrichment of interstellar matter with "metals" accordingly ensued. The stars formed from this medium, originally consisting only of hydrogen and helium, will therefore show a higher metal abundance from generation to generation. One may assume that the chemical composition of the atmosphere of a star as a rule still tallies with that of the interstellar medium at the time of the star's formation. The atmospheres of the oldest known objects, namely the members of Halo Population II, should therefore have the lowest metal abundances – as observed.

One is further led to the conclusion that at the time of the formation of Halo Population II, about 10^{10} years ago, a roughly spherical turbulent gas cloud must have existed. This proto-galaxy, at first very extended, contracted because of its great mass under the effect of its own gravitational forces. The energy released was largely thermalised and radiated away. One may assume that the total original angular momentum of the cloud was not by chance exactly equal to zero, so its conservation during the further process of collapse must inevitably lead to a flattening of the system. The gas not yet condensed into stars distributed itself into a rotating disc, in which an equilibrium was established between the gravitational force and the centrifugal force (Sect. 6.3.4). The stars formed in these late phases should therefore from generation to generation exhibit an ever decreasing scatter of velocity perpendicular to the disc and an ever increasing metal content (Disc Population). Outside the central region the remaining gas finally concentrated itself, under the influence of its own gravitation, very strongly on the plane of symmetry. The stars formed here have the attributes of the Extreme Population I.

Outlook: If one seeks to develop the fundamental ideas outlined here into a detailed picture of the evolution of the galaxy, then numerous problems arise in confronting the observations, which up to now can be only partially clarified, and whose difficulties have prompted certain modifications of the population schema of Table 3.15. First of all it appears that age and chemical composition are not correlated with one another in a simple fashion. It must be surprising that the solar atmosphere has the same elemental abundances as the young stars of the Extreme Population I. The chemical composition of the interstellar matter cannot therefore have changed much since the formation of the sun, and hence in about the last 4×10^9 years. The oldest metal rich open star clusters have still higher ages, for example 6×10^9 years for NGC 188, which do not differ excessively from the ages of the globular clusters. As an escape one can assume that the original collapse of the proto-galaxy extended over a period of several 10^9 years: the "halo globular clusters" most deficient in metal were formed first and therefore occupy the largest volume. The globular clusters with higher metal content were formed later, as the collapse was well under way, in a smaller region of space around the galactic centre. The evolution of the disc must have proceeded for more than 10^9 years after that before the formation of stars could begin in it. In the kernel region this evidently still continues, as the great H II regions existing there (Sect. 4.2.2) make dramatically clear.

The increase of the velocity scatter of the stars of the galactic disc with their age (Table 3.10) need not lead necessarily and exclusively to the later evolution of the galactic collapse. Local fluctuations of the gravitational field produce a "diffusion of stellar orbits", which is linked with an increase in velocity scatter in the course of time (Sect. 6.1.4).

Still largely unclarified is the question of the stellar objects of the first generation, which consisted only of hydrogen and helium, and in whose interiors the metals of the Halo Population II were formed. Other unsolved questions have been thrown up by, for example, the discovery of "super metal rich stars", in whose atmospheres the metal abundance appears to attain a multiple of the solar value – one is not dealing here with objects like the *Ap*- and *Am*-stars on whose surfaces anomalous elemental abundances are produced, probably through some special separation mechanism.

4. Interstellar Phenomena

4.1 The Generally Distributed Medium

Clearly demarcated strong concentrations of interstellar matter, such as the dense molecular clouds and the hot H II regions, are subjects for research as individual items. On the other hand, the far from homogeneous medium in the remaining vast spaces between such prominent features makes us aware of its existence initially only through integral phenomena: the distributed matter extends over a great length of the line of sight. A typical example is the interstellar extinction of starlight. Evidence pertaining to the material at a definite position cannot be obtained directly; regions of higher optical density, which may be called "clouds", are first distinguishable by determination of the absorption of the light from stars at various distances, or by star-counts in the direction under consideration. With other integral phenomena, such as the interstellar absorption lines in stellar spectra, and also the 21 cm line radiation from the universally distributed hydrogen, a localisation of the causative matter is possible only if the spatial velocity field is known.

4.1.1 Interstellar Extinction

Galactic Absorption Band, Nebula-Free Zone: The phenomenon of the interstellar extinction of starlight is particularly conspicuous in wide-angle photographs of the Milky Way (Fig. 1.6, p. 10). The star numbers $N(m|l, b)$ are severely reduced in a dark band, lying almost symmetrically about the galactic equator, with irregular breadth fluctuating between $10°$ and $40°$. Convincing proof that this does not involve a genuine dearth of stars, but attentuation of the light, is provided by the counts, mentioned in Chap. 1, of extra-galactic stellar systems, known in the older literature as (extra-galactic) "nebulae", but usually today called galaxies. In 1934, using photographic observations in 1283 selected fields with the 2.5 m telescope at the Mount Wilson Observatory, E. Hubble found the following relationship for the general (averaged over all l) decrease in the numbers of galaxies down to limiting magnitude $m_{pg} = 20$ per square degree with decreasing galactic latitude b:

$$\log \overline{N}(20^{\mathrm{m}}|b) = 2.115 - 0.15 \operatorname{cosec}|b|. \tag{4.1}$$

This is valid for $|b|$ greater than about $15°$. At lower latitudes there is an almost complete "zone of avoidance", generally coinciding with the "galactic absorption band" described above (Fig. 1.6).

For the explanation of this recently reinvestigated and basically confirmed result, one assumes a plain parallel layer of light-attenuating matter in the region of the galactic plane: extra-galactic light, starting with intensity I_0 at galactic latitude b, experiences attentuation on its journey to the observer (in the galactic plane) in accordance with (cf. Fig. 4.1)

$$I = I_0\, e^{-\tau} \quad \text{with} \quad \tau = \frac{\tau_0}{\sin b} = \tau_0 \cosec b \quad . \tag{4.2}$$

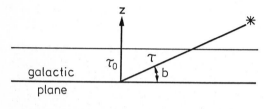

Fig. 4.1. Calculation of the extinction by a plane parallel layer

Here τ_0 denotes the optical semi-thickness of the layer perpendicular to the galactic plane ($b = \pm 90°$). Expressed in magnitudes the extinction is therefore

$$\Delta m = -2.5 \log \frac{I}{I_0} = 2.5 \log e \cdot \tau = 1.086\tau \quad \text{or} \tag{4.3}$$

$$\Delta m(b) = 1.086\tau_0 \cosec b = \Delta m_0 \cosec b \quad . \tag{4.4}$$

Let us assume first of all that the spatial number density of galaxies is constant and no light attenuation takes place. Then [see Sect. 3.2.1, (3.10)] $\log \overline{N}(m|b) = 0.6\,m + \text{const}$. The galactic extinction, however, causes a displacement of this straight line in the $\log \overline{N}(m|b) - m$ diagram towards higher values of m by the amount $\Delta m(b)$, so that for a fixed m the number of galaxies is reduced:

$$\log N(m|b) = 0.6m - 0.6 \cdot \Delta m_0 \cosec b + \text{const}.$$

Comparison with the empirical result (4.1) yields $0.6 \cdot \Delta m_0 = 0.15$, and hence a mean optical half-thickness in magnitudes of $\Delta m_0(\text{pg}) = 0^{\text{m}}25$. Recent galaxy countings carried out at the Lick Observatory gave the factor 0.24 instead of 0.15 in (4.1), from which it follows that $\Delta m_0(\text{pg}) = 0^{\text{m}}4$.

The precise interpretation of the photographic observations, however, must take into account that for an extended source with a fall off in brightness at the edge the integral brightness is more strongly attenuated than the intensity of a point source inside a galaxy: the edge regions fall increasingly below the detection limit, and the galaxy becomes apparently smaller. On the other hand, since

150

it turns out that the (visual) extinction in a radius of 30° around the galactic North pole is only about $0^m\!.06$, and within 30° of the South pole about $0^m\!.09$ (Appenzeller, 1975), the sun is fortuitously located in a region of particularly low dust density, and a value for the *average* optical half-thickness of the layer Δm_0 can be found by this route only by a relatively inexact extrapolation. Satisfactory results were first obtained by discussion of photoelectrically measured reddening (Sect. 5.4.1).

Dark Clouds and Brightness Fluctuations of the Milky Way: If one considers the galactic absorption band in celestial photographs in detail, one notices numerous more or less clearly demarcated regions with distinctly reduced numbers of stars, even to an almost complete absence of stars. These are the dark clouds, of which Fig. 1.5 and 4.2 give examples. The shapes of

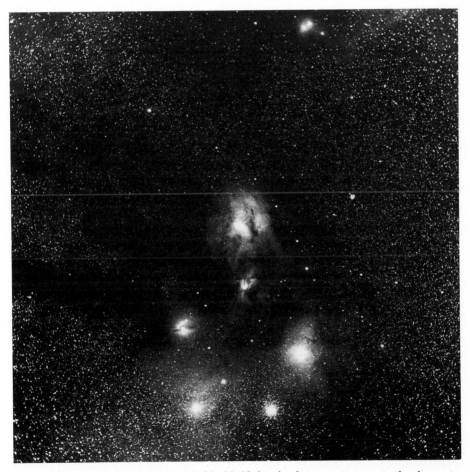

Fig. 4.2. Dark cloud at the star ϱ Ophiuchi. If the cloud were not present, the dense star field (right and lower left) would extend uninterrupted over the whole picture. ϱ Ophiuchi itself is in the bright nebula above, to whose luminosity it makes an essential contribution. The two bright objects near the lower edge of the picture are: Antares on the left, the great globular cluster M4 = NGC6121 on the right. (From Barnard, 1927)

the dark clouds are usually irregular, often elongated and filamentary. In many regions there is a tendency to similar orientation of several such "filaments". That this phenomenon must actually be accounted for by the assumption of light-absorbing "clouds", is most easily demonstrated by the general plotting of cumulative star numbers $N(m)$ against m for the dark region and for the undisturbed neighbourhood (Wolf Diagram, see Sect. 4.2.5). Further proof is provided by the discovery of particularly marked reddening of the starlight (see below) and the evidence of a weak scattered light from the dark cloud (see Sect. 4.1.3).

The strong fluctuations of the surface brightness and also of the star numbers $N(m)$ from place to place in the Milky Way are predominantly caused by the irregular, cloudy distribution of the light-attenuating matter. About two thirds of the mean square variation of the brightness is caused by the typical dark clouds recognisable as individual objects; one third consists of small scale fluctuations which are caused by "cloudlets" with extinction contribution below the detection threshold $\Delta m \approx 0\overset{m}{.}3 \ldots 0\overset{m}{.}4$ for individual clouds. The quantitative statistical interpretation of the brightness fluctuations by a stochastic model of the spatial distribution and the extinction contributions of the clouds (Sect. 5.4.1) cannot be discussed any further here (see Neckel, 1968, Peters, 1970, and Mattila, 1978).

Interstellar Reddening: The interstellar extinction is dependent on the wavelength. If one compares the measured energy distributions in the spectra of two stars of the same MK-spectral type, then as a rule there is only a constant ratio between the two radiation fluxes if both stars are sufficiently near, closer than about 50 pc, say. If at least one of the two stars is significantly further away, then one always finds — especially at low galactic latitudes — that the difference of the two spectral profiles, expressed as the ratio of the two radiation fluxes, is not constant, but increases with decreasing wavelength: the extinction increases in the shorter wavelengths, and so always causes a reddening of the light from a star.

Accordingly, measurements of integral magnitudes for stars at greater distances produce colour indices (formed in the sense of shortwave minus longwave) which are usually greater than the corresponding intrinsic colour of the spectral type of these stars, i.e., we obtain positive *colour excesses*, for example

$$E_{B-V} = (B-V) - (B-V)_0 > 0 \quad , \tag{4.5}$$

where $(B-V)_0$ denotes the intrinsic colour for the spectral type under consideration. For stars at nearly the same distance the interstellar reddening decreases on average with increasing galactic latitude, as one would expect for a nearly plane parallel absorption layer. For stars in the conspicuous dark clouds particularly high values are reached. For stars near the galactic equator at a distance $r = 1\,\text{kpc}$ one finds in general that E_{B-V} is about $0.3 \ldots 0.6$. For objects outside the layer at a galactic latitude b one finds on average that

$$E_{\text{B}-\text{V}} \approx 0.06 \operatorname{cosec} |b| - 0.06 \quad . \tag{4.6}$$

We can also express the colour excess by the "absorption values" for each of the effective wavelengths, usually denoted by $A_\lambda (= \Delta m_\lambda)$, and we obtain by rearrangement of the right-hand side of (4.5)

$$E_{\text{B}-\text{V}} = (B - B_0) - (V - V_0) = A_{\text{B}} - A_{\text{V}} \quad .$$

In order to derive the form of the reddening law it is sufficient in principle to observe two stars of the same MK-spectral type, and therefore with the same true spectral energy distribution (and equal luminosity), which must simply be at different distances (and in almost the same direction), so as to be exposed to different interstellar extinction. The difference of the apparent magnitudes of the two stars at a fixed wavelength λ is then given by

$$m_\lambda^{(1)} - m_\lambda^{(2)} = K_1 + A_\lambda^{(1)} - A_\lambda^{(2)} \quad , \tag{4.7}$$

where K_1 denotes a constant, independent of wavelength, determined only by the difference of the distances. The absorption values $A_\lambda^{(i)}$ $(i = 1, 2)$ can be expressed in terms of the respective optical thicknesses according to (4.3)

$$\tau_\lambda^{(i)} = \int_0^{r_i} \kappa_\lambda^{(i)}(r_i') dr_i' = \int_0^{r_i} k_\lambda \varrho_i(r_i') dr_i' \quad . \tag{4.8}$$

Here we have introduced the mass absorption coefficient k_λ; $\varrho(r_i')$ describes the profile of mass density of the light absorbing matter along the line of sight to the star i. We shall assume that the absorption has the same wavelength dependency in both directions, and hence the same k_λ. Then, substituting (4.8) in (4.7), using (4.3), we obtain:

$$\Delta m_\lambda = m_\lambda^{(1)} - m_\lambda^{(2)} = K_1 + 1.086 k_\lambda \left[\int_0^{r_1} \varrho(r_1') dr_1' - \int_0^{r_2} \varrho(r_2') dr_2' \right]$$

$$= K_1 + K_2 k_\lambda \quad . \tag{4.9}$$

Accordingly for each selected pair of stars we obtain k_λ only to within a constant factor and an additive constant. From the observed magnitude differences (using suffices V and B for the corresponding effective wavelengths) one now forms the quantity usually called "the normalised extinction":

$$F(\lambda) = \frac{\Delta m_\lambda - \Delta m_{\text{V}}}{\Delta m_{\text{B}} - \Delta m_{\text{V}}} = \frac{k_\lambda - k_{\text{V}}}{k_{\text{B}} - k_{\text{V}}} \quad , \tag{4.10}$$

which is independent of the distances of the stars and the two density profiles. If k_λ has the same wavelength dependency everywhere, then the same profile $F(\lambda)$ will be obtained for pairs of stars in the most varied directions. This is of course also true in particular if one of the two stars, $i = 2$, say, is quite unreddened $(A_\lambda^{(2)} = 0)$. Then it follows that

$$F(\lambda) = \frac{A_\lambda^{(1)} - A_V^{(1)}}{A_B^{(1)} - A_V^{(1)}} = \frac{E_{\lambda-V}^{(1)}}{E_{B-V}^{(1)}} \ . \tag{4.11}$$

The measurements available today of pairs of stars of the same spectral type extend from about 1000 Å in the ultraviolet right into the far infrared. Results for $F(\lambda)$ are shown in Fig. 4.3, where it should be noticed that the argument is chosen to be the reciprocal of the wavelength. For the visible region one then obtains an almost linear profile, since $F(\lambda) \sim \lambda^{-1}$ there approximately, and so $k_\lambda \sim \lambda^{-1}$ also. The definition of the normalised extinction (4.11) is such that all $F(\lambda)$ curves must pass through the two points $F(V) = 0$ and $F(B) = +1$. At $\lambda^{-1} = 2.3\,\mu\mathrm{m}^{-1}$, corresponding to $\lambda \approx 4400$ Å, the curve shows a slight bend, often called the "knee". In the UV-region, first explored by extra-terrestrial observations from rockets, and above all from the Orbiting Astronomical Observatory OAO-2, a prominent structural feature appears: a "hump" at $\lambda^{-1} = 4.6\,\mu\mathrm{m}^{-1}$, corresponding to $\lambda \approx 2200$ Å, with an associated minimum, followed by a relatively steep climb for $\lambda^{-1} > 7\,\mu\mathrm{m}^{-1}$, and so for $\lambda < 1400$ Å.

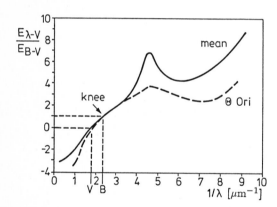

Fig. 4.3. Interstellar reddening curves. (—) mean profile, (- - -) results for the two stars Θ^1 and Θ^2 Orionis

In contrast to the absorption spectra of the stellar atmospheres with their numerous Fraunhofer lines, the observed profiles of the interstellar extinction coefficients exhibit no strongly marked lines or bands in the visible or UV which would permit direct inferences concerning the absorbing matter. We know of only a number of weak, so-called diffuse interstellar lines or bands, to which we shall return later. An important result, therefore, was the discovery of strong *absorption bands in the infrared* at $\lambda \approx 10\,\mu\mathrm{m}$, between about 8 and 12 $\mu\mathrm{m}$ (Fig. 4.4), and also at $\lambda \approx 20\,\mu\mathrm{m}$. Since normal stars are too faint at these wavelengths, these observations are based on infrared stars, which occur in regions of high interstellar extinction. These are primarily objects in the direction of the galactic central region and infrared sources such as the Becklin-

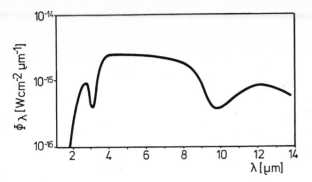

Fig. 4.4. Spectrum of the infrared star (BN Object see Sect. 4.2.4) called after Becklin and Neugebauer with characteristic interstellar absorption bands at 3.1 μm and at 9.7 μm

Neugebauer object in the Orion nebula. For these and a few other infared stars there also appears an absorption band at $\lambda = 3.1\,\mu\mathrm{m}$ (see Fig. 4.4), which can probably be ascribed to H_2O ice and which had long been sought. The $10\,\mu\mathrm{m}$ band is often interpreted as absorption due to silicates. An unequivocal explanation of this, as of the other strong absorptions in the infrared, cannot so far be given.

In the far infrared the mean profile of $F(\lambda)$ approaches the asymptotic value $F(\lambda = \infty) \approx -3$. The assumption based on what follows, of small solid particles as the cause of the interstellar extinction, would suggest vanishing values of A_λ for $\lambda \to \infty$. Equation (4.11) then gives $F(\infty) = -A_V/E_{B-V}$, and so

$$\frac{A_V}{E_{B-V}} \approx 3 \quad . \tag{4.12}$$

The significance of this very important result for stellar astronomy lies in that it enables us to obtain the visual absorption A_V from the colour excess $E_{B-V} = (B-V)-(B-V)_0$ alone. The above mentioned span of normal values for E_{B-V} at $r = 1\,\mathrm{kpc}$ and $b \approx 0°$ corresponds accordingly to $A_V = 1\ldots2\,\mathrm{mag\,kpc}^{-1}$. The ratio A_V/E_{B-V} is usually denoted in the literature by R. The present day best value is $R = 3.1\pm0.1$.

We can of course therefore infer A_λ for every other wavelength also. From the definition (4.11) it follows that

$$\frac{A_\lambda}{A_V} = 1 + \frac{E_{B-V}}{A_V}F(\lambda) \quad .$$

The *mean* behaviour of this ratio with increasing wavelength is given in Table 4.1. In particular, one finds that $A_B/A_V \approx 4/3$.

For several stars investigated, comparison with a practically unreddened star of the same spectral type gives relatively large deviations from the behaviour of $F(\lambda)$ usually found. As a typical example, Fig. 4.3 shows the reddening curve for ϑ^1 Ori and ϑ^2 Ori in the Orion Nebula: in the ultraviolet the hump and also the rise towards short wavelengths are considerably less pronounced; at the same time the size of the limit value $|F(\infty)| = A_V/E_{B-V} = R$ is higher, at about 4 to 6. Both aberrations appear to be a common property

155

Table 4.1. Mean observed interstellar extinction as a function of wavelength, expressed by the ratio A_λ/A_V. (After Schild, 1977; Becklin et al., 1974; Nandy et al., 1975)

λ [μm]	A_λ/A_V	λ [μm]	A_λ/A_V
0.1000	4.60	0.5000	1.16
0.1250	3.10	0.5480	1.00
0.1375	2.73	0.5840	0.91
0.1590	2.57	0.6050	0.88
0.1830	2.53	0.6436	0.83
0.2000	2.80	0.7100	0.71
0.2080	2.98	0.7550	0.68
0.2140	3.09	0.8090	0.62
0.2190	3.10	0.8446	0.58
0.2230	3.07	0.871	0.54
0.2360	2.68	0.970	0.47
0.2500	2.39	1.061	0.40
0.2740	2.03	1.087	0.38
0.3200	1.74	1.25	0.30
0.3400	1.64	2.2	0.15
0.3636	1.56	3.4	0.10
0.4036	1.43	4.9	0.05
0.4255	1.35	8.7	0.01
0.4566	1.26	10.0	0.01

of the extinction in the immediate neighbourhood of hot young stars with H II regions. The increase in $|F(\infty)|$ can also in isolated cases be simulated by circumstellar emission.

The troughs in the infrared absorption bands, expressed by the maximal optical thickness τ (Sect. 5.1.1), are in general the more strongly marked, the higher A_V is. For the 10 μm band the prevailing values in thick interstellar clouds are

$$\frac{A_V}{\tau_{10}} \approx 10\ldots40 \quad .$$

Diffuse Interstellar Bands: In stellar spectra there frequently occur diffuse absorption lines or bands which can be identified neither as groups of lines nor as molecular bands of the stellar gas. The strongest and therefore the best known of these structures is the diffuse band at λ 4430 Å (Fig. 4.5). In many stars it attains a breadth of 20 Å and more, with a central trough of only about 15% of the neighbouring continuum, and an equivalent width of 3 Å and more. Most of the total of about 40 known bands between 4400 and 7000 Å, and a few newly discovered bands in the near infrared, are however weaker. Relatively strong ones are the structures at 4765 Å, 4880 Å, 5780 Å, 5800 Å, 6180 Å and 6280 Å. One also finds considerably broader bands with $\Delta\lambda \approx 20\ldots200$, so that one can perhaps speak of a smooth transition to structures at about the 2200 Å hump in the reddening curve.

The strengths (equivalent widths) of the carefully investigated diffuse bands of the visible region − as also of the 2200 Å hump − have as a rule proved to

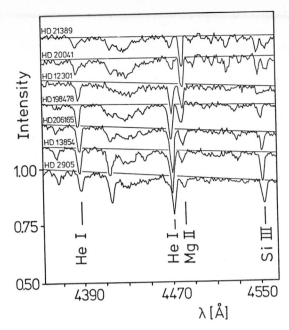

Fig. 4.5. Relative spectral intensity profiles for seven stars in the region of the diffuse interstellar band at 4430 Å. (After Danks and Lambert, 1975)

be well correlated with the interstellar extinction. For example, the equivalent width of the band at $\lambda\,4430\,\text{Å}$ approximately satisfies the relation $A_{4430}[\text{Å}] \approx 3 \cdot E_{\text{B}-\text{V}}[\text{mag}]$. The total spectrum of the diffuse bands generally increases with the interstellar extinction.

Interpretation of Extinction by Solid Particles.[1] The multitude of possible causes of the interstellar extinction can be narrowed down relatively simply to small solid particles with dimensions of a few wavelengths of light down to large molecules. For this, we rely not only on the general behaviour of the reddening curve between the near infrared and the near ultraviolet but also on the evidence of Oort's limiting value on the *maximal mass density* in interstellar space (cf. Sect. 6.1.5). For all components of the interstellar matter together, near the galactic plane in the wider neighbourhood of the sun, this can amount, at the most, to about $3 \times 10^{-24}\,\mathrm{g\,cm^{-3}}$, corresponding to about 2 H atoms in $1\,\mathrm{cm^3}$. The production of a visual extinction $A_{\text{V}} = 1\,\mathrm{mag}$ at $1\,\mathrm{kpc}$ by continuous absorption or scattering processes with electrons, atoms or molecules would lead to a mass density exceeding this limit value by a few powers of ten. Thus, for example, the assumption of scattering by free electrons would require a few hundred electrons in $1\,\mathrm{cm^3}$, which would imply the presence of a much greater number of atoms and ions. For Rayleigh scattering by atoms or molecules the number density required would be about $10^4\,\mathrm{cm^{-3}}$. In addition to this, the required wavelength dependency would not result either from electron scattering

---

[1] We deviate here briefly from the consistent account of the mere phenomena, for the advantage in the subsequent sections of this chapter of being able to make use of the knowledge that the interstellar extinction is caused by dust particles (Sect. 5.3).

(independent of λ) or from Rayleigh scattering ($\sim \lambda^{-4}$). Special atoms, ions or molecules (e.g. H^- ions) with continuous absorption, similar at least in part to the reddening curve, have very low relative frequency compared with the other atoms and therefore also lead to far too high mass densities.

If one considers the alternative to gas particles of small solid "dust particles", then the requirement of Oort's limiting mass value can easily be fulfilled. We shall assume for the sake of simplicity that all the particles are spheres with the same radius a. The "extinction coefficient" for 1 cm path length with a density of N_s particles in $1\,\mathrm{cm}^3$ is then equal to the effective screening by $1\,\mathrm{cm}^3$ of this dust medium (more precise argument follows in Sect. 5.3.1):

$$\kappa_\lambda = \frac{|\Delta I_\lambda|}{I_\lambda} = \pi a^2 Q_{\mathrm{ext}} N_S \quad . \tag{4.13}$$

Here we have introduced the efficiency factor Q_{ext}, in general dependent on λ. When $a \gg \lambda$, Q_{ext} is not equal to one, but two: in addition to the geometrically removed light, there is a contribution of equal magnitude by diffraction in a ring system, which is not received by the observer (Babinet's Principle). For the extinction amount we have

$$A_\lambda = 1.086\tau_\lambda = 1.086\pi a^2 Q_{\mathrm{ext}}(\lambda)\overline{N}_s r \quad . \tag{4.14}$$

With particle radius $a \approx \lambda$, Q_{ext} is of order unity in the visible region for all materials. An average visual extinction $A_V = 1$ mag for $r = 1\,\mathrm{kpc} = 3 \times 10^{21}$ cm, with $a = \lambda_V = 5 \times 10^{-5}$ cm, enables one to infer that $\overline{N}_s = (\pi a^2 r)^{-1} = 4 \times 10^{-14}$ particles cm^{-3}. With a density in the particles themselves of $1\,\mathrm{g\,cm}^{-3}$ this result corresponds to a mass density of the interstellar dust medium of $(4\pi/3)a^3\overline{N}_s = 2 \times 10^{-26}\mathrm{g\,cm}^{-3}$. The average mass density of the dust required accordingly amounts to less than 1% of Oort's limiting value and is still compatible with a gas-dust mass ratio of about 100:1.

Another strong argument for small solid particles is provided by the observed wavelength dependency. This is easily understood between the near infrared and the near ultraviolet: the theory of light scattering by small spheres shows that for the region $a \approx \lambda$ we have $Q_{\mathrm{ext}}(\lambda) \sim \lambda^{-1}$ (see Sect. 5.3.1). When $a \ll \lambda$, however, $Q_{\mathrm{ext}} \ll 1$ since the interaction with the radiation field is then small − like a dipole antenna whose length $\ll \lambda$!

Taking the simplest approach, one can say that, for a given mass of any solid material, the maximal optical extinction effect is achieved if it is divided into very small particles of the order of size of the wavelength of light. If one rolls out a $1\,\mathrm{cm}^3$ cube into a sheet of 10^{-5} cm thickness, one can cover $10^5\,\mathrm{cm}^2 = 10\,\mathrm{m}^2$; the sheet will then provide 10^{15} little cubes of 10^{-5}cm side length. Compared with the original cube of $1\,\mathrm{cm}^3$, the extinction effect is increased by a factor 10^5! The corresponding thing is true for light scattering (smoke signals!). Further sub-division into the region $a \ll \lambda$, however, achieves no further increase in effectiveness, because then, as already stated, Q_{ext} becomes much less than unity. Additional arguments for dust particles as the cause of interstellar extinction come from the observations of interstellar polarisation and the direct evidence of the starlight scattered from the particles (see Sects. 4.1.2 and 4.1.3).

4.1.2 Interstellar Polarisation

Description of Polarised Radiation: We consider first of all the simple electromagnetic wave. We set up a rectangular coordinate system in the plane containing the position of the observer and perpendicular to the direction of propagation, with ordinate direction through the unit vector e_2 tangential to the hour circle (in the direction of the celestial north pole), the abscissa along the unit vector e_1 perpendicular thereto. At a fixed position (the observer) we can then write the components of the electric vector E lying in the plane under consideration as

$$E_1 = E_{10} \sin(2\pi\nu t - \varepsilon_1)$$
$$E_2 = E_{20} \sin(2\pi\nu t - \varepsilon_2) \quad .$$

In addition to the frequency ν four constants occur here (ε_1, ε_2 denote the phase), of which only three are, however, mutually independent: with varying time the end point of the vector E describes an ellipse (elimination of $2\pi\nu t\,!$), which is characterised by the lengths of its two semi-axes and the position angle θ of the major axis (Fig. 4.6). One is accordingly taking the wave as having general elliptical polarisation. The axis ratio of the ellipse is often characterised by the angle β, defined by

$$\frac{\text{small axis}}{\text{large axis}} = \tan \beta \quad .$$

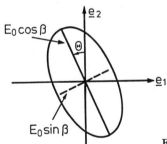

Fig. 4.6. Description of polarised radiation $E_0^2 = E_{10}^2 + E_{20}^2$

For $\beta = 0°$ we have wholly linear polarisation: the oscillation is then only in the fixed direction θ; for $\beta = 45°$ the ellipse becomes a circle, the polarisation is completely circular.

Instead of the four parameters E_{10}, E_{20}, ε_1, ε_2 one often introduces the *Stokes Parameters* I, Q, U, V, which are defined by the left-hand parts of the following equations:

159

$$I = E_{10}^2 + E_{20}^2 = \sqrt{Q^2 + U^2 + V^2} \tag{4.15}$$

$$Q = E_{20}^2 - E_{10}^2 = I \cos 2\beta \, \cos 2\theta \tag{4.16}$$

$$U = -2E_{10}E_{20} \, \cos(\varepsilon_2 - \varepsilon_1) = I \cos 2\beta \, \sin 2\theta \tag{4.17}$$

$$V = 2E_{10}E_{20} \, \sin(\varepsilon_2 - \varepsilon_1) = I \sin 2\beta \quad . \tag{4.18}$$

The Stokes Parameter I is simply the intensity. The right-hand part of (4.15), which is easy to verify, shows once again that only three independent parameters arise. I, $Q^2 + U^2 = I^2 \cos^2 2\beta$ and V are clearly invariant under changes of the coordinate system, whereas Q and U vary with θ under a rotation.

The observed radiation from a cosmic source always consists of a superposition of very many simple waves of this sort. The essential advantage of the introduction of the Stokes Parameters consists in that with weak polarisation the values of I, Q, U, V for the individual waves are simply additive. For natural light the phase differences $\varepsilon_2 - \varepsilon_1$ and the amplitude ratios E_{20}/E_{10} of the individual waves are not mutually correlated, and as a result the totals $Q = U = V = O$. If there is correlation of the individual waves, then in general a partial elliptic polarisation results. The resulting Stokes Parameters I, Q, U, V can be resolved into two parts: (1) an unpolarised part with $I_1 \neq 0$, $Q_1 = U_1 = V_1 = O$ and (2) a completely polarised part with in general non-zero values I_2, Q_2, U_2, V_2. The *degree of linear polarisation* of the resulting radiation is defined as

$$P = \frac{\sqrt{Q_2^2 + U_2^2}}{I_1 + I_2} = \frac{\sqrt{Q^2 + U^2}}{I}$$

and the quantity $e = |V_2|/(I_1 + I_2) = |V|/I$ is called the *ellipticity*. If the second part is wholly linearly polarised (direction θ), then it follows that $V_2 = V = O$ and $I_2^2 = Q_2^2 + U_2^2 = Q^2 + U^2$. The smallest intensity $I_{\min}$, found for the direction $\theta + 90°$ with a rotatable analyser, is in this case identical with I_1, whilst the largest intensity $I_{\max}$, occurring in the polarisation direction θ, is equal to $I_1 + I_2$. For partial linear polarisation of the radiation one accordingly obtains the following relations, which are important in practice,

$$P = \frac{I_2}{I_1 + I_2} = \frac{I_{\max} - I_{\min}}{I_{\max} + I_{\min}} \tag{4.19}$$

and, with $I = I_2$ in (4.16) and (4.17),

$$Q = (I_{\max} - I_{\min}) \cos 2\theta = PI \cos 2\theta \tag{4.20}$$

$$U = (I_{\max} - I_{\min}) \sin 2\theta = PI \sin 2\theta \tag{4.21}$$

with $I = I_{\max} + I_{\min}$.

Sometimes the *degree of polarisation* is also expressed in *magnitudes* by the definition:

$$\Delta m_{\mathrm{p}} = -2.5 \log(I_{\mathrm{min}}/I_{\mathrm{max}}) \quad . \tag{4.22}$$

For the small degrees of polarisation occurring in interstellar polarisation $P \ll 1$ with $I_{\mathrm{max}} \approx I_{\mathrm{min}}$ it follows from (4.19) that $P \approx [1 - (I_{\mathrm{min}}/I_{\mathrm{max}})]/2$, and so $I_{\mathrm{min}}/I_{\mathrm{max}} \approx 1-2P$. Thereby (4.22) can be rewritten as $\Delta m_{\mathrm{p}} = -1.086 \ln(I_{\mathrm{min}}/I_{\mathrm{max}}) \approx +1.086 \times 2P$ or

$$\Delta m_{\mathrm{p}} \approx 2.172 P \quad . \tag{4.23}$$

Observational Results: Systematic polarisation measurements have so far been carried out for about $10\,000$ stars in the northern and southern skies, involving different spectral types, degrees of polarisation P and polarisation direction θ. The degree of linear polarisation observed is usually less than 2%. The errors of measurement in typical investigations for stars brighter than 7^{m} are at most $\Delta P = 0.0005$ or 0.05%, and increase for stars of magnitude 9 to about 0.25%. The errors in direction of polarisation for $P > 0.01$ are below $\Delta \theta = 5°$. The most important results of the observations so far, which also in particular reveal the interstellar origin of the polarisation, are as follows:

1) The degree of polarisation P or Δm_{p} is correlated with the interstellar extinction measured, for example, by $E_{\mathrm{B-V}}$ or A_{V}. Degrees of polarisation above about 1% are nearly always connected with reddening of the stars — on the other hand reddened stars do not always show polarisation. The ratio $\Delta m_{\mathrm{p}}/A_{\mathrm{V}}$ is on average about 0.03 and the highest value reached in the individual case is 0.065. One can say that the extinction coefficients of the interstellar dust medium for light vibrating parallel to and perpendicular to the polarisation direction differ at most by about 6%.

2) The greatest degrees of polarisation occur overwhelmingly in the neighbourhood of the galactic equator. The polarisation directions of the light from individual stars show strong mutual correlation in many regions of the sphere (see Chap. 1, Fig. 1.11). Examples are the Milky Way regions around $l \approx 140°$ (Perseus-Cassiopeia) and $l = 320°$ (Centaurus-Norma) with almost uniform orientation of the polarisation parallel to the galactic equator. An almost random distribution of polarisation directions is found in the regions around the galactic equator at $l \approx 80°$ (Cygnus-Cepheus) and $l \approx 260°$ (Puppis-Vela). At some places the directions show a strongly eddied structure. For stars in regions of elongated filamentary clouds one finds in some cases a correlation of the polarisation directions with the orientation of the "filaments".

3) Measurements on more than a hundred stars at different wavelengths between 3300 Å and $1\,\mu$m show only a relatively small variation of the degree

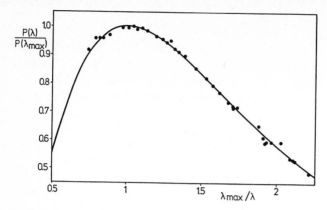

Fig. 4.7. Observed wavelength dependence of the interstellar polarisation of the starlight, *solid curve* according to (4.24). (After Serkowski, Mathewson, and Ford, 1975)

of polarisation P, which differs somewhat from star to star. On average $P(\lambda)$ shows a shallow profile with a maximum between 5000 and 7000 Å. A noticeable characteristic is that $P(\lambda)/P_{max}$, where P_{max} is the maximal value of P, attained at $\lambda = \lambda_{max}$, shows always nearly the same behaviour, as a function of λ_{max}/λ (Fig. 4.7). This relationship can be represented by the following expression (Serkowski, 1973):

$$\frac{P(\lambda)}{P_{max}} = \exp\{-1.15[\ln(\lambda_{max}/\lambda)]^2\} \quad . \tag{4.24}$$

λ_{max} varies between 0.45 and 0.8 μm and lies on average at 0.55 μm. Observations in the near infrared, however, argue for replacing the factor of 1.15 by 1.7 λ_{max} (Wilking et al., 1980). The connection with the reddening E_{B-V} is given by $P_{max} \lesssim 0.09 E_{B-V}$ [see also Sect. 5.3.2, (5.167)].

The facts contained in (1) and (2) admit of the following qualitative explanation: The interstellar polarisation arises from an anisotropy in the interstellar extinction of the starlight. This could be the result of a portion of the solid particles having an elongated, non-spherical form, so that the extinction differs for light waves vibrating in different directions. The linear polarisation caused by very many elongated particles on the line of sight would then require that the main structural axes of these particles are to a certain extent uniformly aligned in space. In Sect. 5.3.3 it is explained that a partial alignment may be the result of a spatial interstellar magnetic field, whose strength amounts to at least a few 10^{-6} Gauss $= 10^{-10}$ Tesla. With this mechanism, proposed by L. Davis and J.L. Greenstein in 1951, the polarisation direction – lying perpendicular to the major axes of the particles – is parallel to the magnetic field. Figure 1.11 can then be regarded as a representation of the magnetic field distribution projected on to the sphere. Together with the distances of the observed stars, this allows one to derive a model of the spatial development of the magnetic field itself. One obtains a field line formation parallel to the local spiral arm (see also Sect. 4.1.8).

For a number of stars one detects not only the linear polarisation but also a weak *circular polarisation* with strengths $|V|/I \approx (1 \ldots 3) \times 10^{-4}$. Here it is noticeable above all that V/I varies with the wavelength and also that it undergoes a change of sign near the wavelength of the maximum of linear polarisation $\lambda \approx \lambda_{max}$.

Finally it should be mentioned that, in addition to the interstellar part of the polarisation of the starlight, for some particular star types a weak "intrinsic polarisation" — usually variable, and therefore in most cases first detectable — can also be established. Thus one finds, in particular, for a number of eclipsing variables, including β Lyrae, a periodically oscillating polarisation, by which the original prediction of Chandrasekhar can be at least qualitatively confirmed (see Chap. 1).

4.1.3 Reflection Nebulae and Diffuse Galactic Light

With the extinction and partial polarisation of light from the stars by interstellar dust, a part of the radiation incident on the solid particles is absorbed and another part is simply scattered, and so diverted without loss in a different direction. The absorbed light causes an insignificant warming of the dust particles and thereby increases its thermal emission in the far infrared (in Sect. 5.3.3 it is explained that the temperature of the particles is in general below 20 K). The starlight scattered with a certain directional distribution produces a faint surface luminosity, whose intensity depends on the contribution of scattering to the extinction, and whose spectral profile corresponds to a superposition of the spectra of each of the stars making significant contributions.

In the *reflection nebulae* (about 500 are known, of which an example is given in Fig. 1.7) we evidently encounter specially favourable cases of this phenomenon. Here are located one, or even several, sufficiently bright stars in the direct proximity of a "cloud" of interstellar dust (more precisely, stars with spectral types later than B0, which are not able to ionise the hydrogen gas of the cloud and so produce an "H II region" dominating the dust-scattered light by strong line emissions). In Table 4.2 are collected some data for a selection of typical reflection nebulae, and also an indication of the illuminating stars.

Between the angular size of the nebula and the apparent magnitude m^* of the illuminating star there exists a linear relationship of the form

$$m^* = -k \log a + K \quad ,$$

where a is the angular distance from the star to the "edge" of the nebula in arc minutes. For the reflection nebulae, one finds that to within error limits $k = 5.0$, whilst the constant K depends on the particular detection limits achieved for surface brightness (Fig. 4.8). This result is compatible with the idea that the light scattering interstellar dust is distributed statistically homogeneously, so that the extent of the nebulae is in general determined only by the luminosity

Table 4.2. Data on selected bright reflection nebulae. The nebulae around the Pleiades stars Electra (17 Tau), Maja (20 Tau) and Merope (23 Tau) are to be seen in Fig.1.7. The unit of the quoted surface brightnesses in the V region is 1 star with $V = 10^m$ per square degree. $\Delta(B-V) = (B-V)_n - (B-V)_s$ denotes the difference in colour index $B-V$ between the nebula and the illuminating star. Regarding the naming of nebulae see Sect. 4.2.1

Name	α_{1950}	δ_{1950}	Size	Surface brightness $[S_{10}(V)/\square°]$	$\Delta(B-V)$	Illuminating star Sp	V	$B-V$
IC 348	3^h41^m	$+31°59'$	$10' \times 10'$	62	-0^m71	B5V	8^m53	$+0^m69$
Electra-Nebula	3^h42^m	$+23°57'$	$20' \times 16'$	430	-0^m19	B6III	3^m71	-0^m11
Maja-Nebula	3^h42^m	$+24°24'$	$30' \times 30'$	360	-0^m45	B7III	3^m88	-0^m07
Merope-Nebula	3^h43^m	$+23°43'$	$30' \times 30'$	520	-0^m11	B6IVn	4^m18	-0^m06
Cederblad 44	5^h19^m	$8°20'$	$3' \times 3'$	47	-0^m59	B1V	5^m77	-0^m13
NGC 2068 = M 78	5^h44^m	$0°00'$	$8' \times 6'$	750	-0^m37	B5	10^m49	$+0^m62$
Antares-Nebula	16^h26^m	$-26°20'$	$126' \times 78'$	43	-1^m19	M1Ib	0^m92	$+1^m84$
IC 1287	18^h29^m	$-10°50'$	$20' \times 10'$	68	-0^m49	B2V	5^m72	$+0^m24$
NGC 7023	21^h01^m	$+68°00'$	$10' \times 8'$	62	-0^m24	B5e	7^m39	$+0^m38$

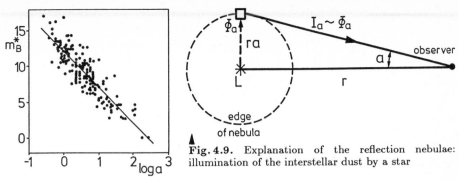

Fig. 4.8. Relation between the apparent magnitude of the illuminating star m_B^* and the logarithm of the greatest distance a [arc minutes] of the edge of the nebula from the star, measured on copies of the blue plates of the Palomar Sky Survey. This result is frequently known as the "Hubble Relation" after its discoverer. For the ruled straight line $m_B^* = -4.95 \log a + 12.01$. (After Dorschner and Gürtler, 1965)

of the star: a star with luminosity L and at distance r produces at the edge of the nebula at an angular distance a (in arc measure), the radiation flux $\Phi_a = L/4\pi(ra)^2$ (Fig. 4.9). If the same fraction of the incident radiation is reflected isotropically in all nebulae, then the observed surface brightness I_a (intensity), independent of distance, at the edge of the nebula is equal to Φ_a to within a constant factor. On the other hand, for all observed nebulae I_a is identical with the detection limit. Hence $\Phi_a = L/4\pi(ra)^2$ is a constant and we can write for the stellar radiation flux at the observer $\Phi^* = L/4\pi r^2 = \text{const.} \cdot a^2$, or

$$m^* = -2.5 \log \Phi^* = -5 \log a + K \quad . \tag{4.25}$$

The reflection nebulae are bluer as a rule than the illuminating stars: $\Delta(B - V) < 0$. This is to be expected on account of the increase of interstellar extinction towards shorter wavelengths, discussed in Sect. 4.1.1: the portion removed from the starlight makes its reappearance here as the scattered light becoming more intense in the blue. The nebula light moreover usually shows as a result a significant linear polarisation. The colour difference $\Delta(B - V)$ and the polarisation P vary with angular distance φ from the illuminating star. For the Merope Nebula (Pleiades), for example, $\Delta(B - V)$ varies between $\varphi = 0$ and $15'$ from -0.6 to $+0.1$ and $P(V)$ increases from 1% to 10%.

The name "reflection nebula" is applied to a number of phenomena; thus, for example, to objects with comet-like shape, which probably form a special group of interstellar objects. The (Hubble's) Nebula NGC 2261, illuminated by the variable star R Monocerotis, belongs to this class.

Extraterrestrial observations in the *ultraviolet* down to 1200 Å also show in the neighbourhood of early type stars reflection nebulae with the continuous spectra of these stars, for example over the greater part of the Orion constellation. We therefore arrive at the conclusion that the scattering power of interstellar dust – like the extinction – is very strong at these short waves.

Diffuse Galactic Light: The total interstellar dust producing the galactic absorption band causes a general scattered light, whose source is the whole galactic disc. This so-called Diffuse Galactic Light has been measured, both in the visual region and also in the short waves – by rocket and satellite observations into the far ultraviolet. Its contribution to the surface brightness of the Milky Way amounts from this to 20 to 50% of the integrated starlight, according to wavelength!

The determination of the absolute intensity of the Diffuse Galactic Light is difficult, since even in very small, practically star-free, areas of the Milky Way, in addition to a contribution from the undetected faintest stars there also occurs an appreciable brightness contribution from the Zodiacal Light, and for terrestrial measurements also the luminosity of the upper atmosphere (Airglow) as well as the light from all extraterrestrial sources scattered in the lower atmosphere. Zodiacal light and Airglow can be removed, since they – in contrast to the diffuse galactic light – are symmetrically distributed with respect to the ecliptic and to the horizon, respectively. Typical results for the individual components of the surface brightness of the Milky Way for wavelengths from 4250 Å to 1550 Å are presented in Table 4.3 (for the determination of the integrated starlight see Sect. 3.1.1). As a convenient measure of intensity one usually quotes the number of stars of magnitude 10 for the spectral region under consideration per square degree, which would produce the same energy flux per unit of solid angle. One writes, for example, for the intensity of the diffuse galactic light $I_{\mathrm{DGL}}(B) = 47\,S_{10}/\square°$ (cf. Sect. 3.1.1).

Table 4.3. Example of results for the various parts of the surface brightness of the sky near the galactic equator for $|b| \approx 1°$ and $l = 65° \ldots 145°$, reached on the basis of extraterrestrial measurements on board the Orbiting Astronomical Observatory, OAO-2. Unit: $1\,S_{10}(\lambda)/\square° = 1$ star of type A0V with $m_\lambda = 10^{\mathrm{m}}$ per square degree. (After Lillie and Witt, 1970)

λ [Å]	4250	3320	2460	2040	1550
Integrated starlight I^*	113	108	88	80	140
Contribution of the stars weaker than 14^{m}	53	33	7	5	5
Zodiacal light	41	31	3	13	2
Diffuse galactic light I_{DGL}	47	55	22	19	51
(I_{DGL}/I^*)	0.42	0.51	0.25	0.24	0.36

After what has been said it cannot be surprising that even the relatively strong concentrations of interstellar dust which appear as *dark clouds* have a faint surface luminosity, since they reflect a portion of the radiation from the stars outside the cloud. Thus it appears, for example, that even the darkest spots in the "Coalsack" (see Fig. 1.5) still show scattered light in the B-region with a brightness of about $50\,S_{10}/\square°$. This luminosity becomes especially noticeable with isolated clouds in higher galactic latitudes. A dark cloud at $b = +37°$ in the constellation Libra serves as an impressive example, which with a surface brightness of $25\,S_{10}/\square°$ in the visual is noticeable above the general diffuse galactic light, already very faint at these latitudes (Mattila, 1970).

4.1.4 Interstellar Absorption Lines in Stellar Spectra

In this and the following sub-sections we turn our attention to the integral phenomena of the interstellar gas. In optical astronomy the evidence for gas generally distributed between the stars of the galactic disc was based on observations of absorption lines in the visible part of stellar spectra, which differ in the following fundamental respects from the stellar lines:

1.　relatively narrow widths (Fig. 4.10);
2.　systematic increase in the line strengths, not only with the distance of the stars − for a fixed direction near the galactic plane − but also with the interstellar extinction;
3.　usually different Doppler shifts from those of normal Fraunhofer lines − in the case of double stars no participation in the periodic displacements ("stationary" lines).

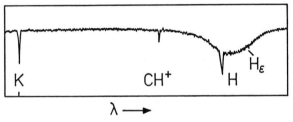

Fig. 4.10. Interstellar absorption lines Ca II-K λ 3933.66 Å, CH$^+$ λ 3957.74 Å, and Ca II-H λ 3968.47 Å in the spectrum of the star ζ Ophiuchi. The Ca II-H line lies inside the broad stellar hydrogen line $H\varepsilon$

The narrow line widths are a result of the extremely low density and pressure of the interstellar gas (no collision broadening!). The great widths of the diffuse bands mentioned in Sect. 4.1.1 on the other hand are readily explained if one takes into account that this phenomenon is caused by very small solid bodies (see also Sect. 5.3.2).

Since the wealth of Fraunhofer lines in middle and late spectral types makes the detection and measurement of interstellar absorption lines very difficult or quite impossible, the existing identifications are based in essence only on spectra of early types. This restriction is not serious, as only these hot stars have strong continua in the ultraviolet and the absorption lines of the elements most abundant in interstellar space, such as H, C, N, O etc., occur exclusively in these spectral regions, nowadays made accessible through rocket and satellite observations. In interstellar space far from stars practically all atoms, ions or molecules occur in their ground states: accordingly, nearly all the interstellar absorption lines observed in the optical and UV-regions are resonance lines.

Precisely for the important abundant elements, however, the first excited state is already so high above the ground state that the resulting lines usually have wavelengths less than 3000 Å. Interstellar gas with temperatures above 10^4 K shows its presence only by absorption lines in the ultraviolet spectral region. In the case of hydrogen the energy difference to the state $n = 2$ is about 10 eV and the corresponding Lyman α-line ($n = 1 \rightarrow 2$) is already in the far ultraviolet, with wavelength $\lambda = 1216$ Å.

Table 4.4. Interstellar lines in the region $\lambda = 3000 \ldots 8000$ Å

Lines of atoms or ions			Molecular lines		
Atom spectrum		λ [Å]	Molecule	λ [Å]	Transition
NaI	D_1	5895.92	CH	4300.31	$A^2\Delta - X^2\Pi(0,0)$
	D_2	5889.95		3890.23 ⎫	
		3302.94		3886.39 ⎬	$B^2\Sigma^- - X^2\Pi(0,0)$
		3302.34		3878.77 ⎭	
				3146.01 ⎫	
KI		7698.98		3143.15 ⎬	$C^2\Sigma^+ - X^2\Pi(0,0)$
		7664.91		3137.53 ⎭	
CaI	g	4226.73			
CaII	H	3968.47	CH⁺	4232.58	$A^1\Pi - X^1\Sigma^+(0,0)R(0)$
	K	3933.66		3957.74	$(1,0)R(0)$
				3745.33	$(2,0)R(0)$
TiII		3383.76		3579.02	$(3,0)R(0)$
		3241.98			
		3229.19			
		3072.97	CN	3874.61 ⎫	
				3875.77 ⎬	$B^2\Sigma^+ - X^2\Sigma^+(0,0)$
FeI		3859.91		3874.00 ⎭	
		3719.94			

Spectral Region $3000 \ldots 8000$ Å: In this region, not observable from the earth's surface, apart from the well known H and K lines of the CaII spectrum, and also the D lines of NaI and the g line at $\lambda\,4227$ Å of CaI, only about 10 more interstellar absorption lines of atoms or ions are known. Table 4.4 gives a list of the wavelengths. The very thoroughly researched spectrum of the O 9.5 star ζ Ophiuchi moreover shows additional, of course very faint, probably interstellar lines of LiI (6708 Å), BeI, AlI, KI (4047/44), Ti I, TiII, V I, Cr I, Mn I, Ni I, Cu I and Co I. Other well measured lines of interstellar origin in the visible region can be attributed to electronic transitions – coming from the lowest rotational levels of the ground state – of the molecular radicals CH (7 lines), CH⁺ (4) and CN (3) (for the basis see Appendix B). Table 4.4 gives the wavelengths and the transitions.

Ultraviolet Region: The most important objective of the search for interstellar absorption lines in the region $\lambda<3000\,\text{Å}$ by extra-terrestrial observations was above all the detection of interstellar hydrogen. The first successful observations of the Lyman α-line were made by a research group at Princeton Observatory (USA) by means of a rocket experiment in 1965. An objective grating spectrograph was used for the production of spectra of a number of hot stars of the Orion region. The results were soon confirmed by similar experiments by other groups of workers at the Goddard Space Flight Centre and at the Naval Research Laboratory. Further comprehensive observational data have been provided since 1969 by the NASA satellites of the Orbiting Astronomical Observatory series OAO-2 and OAO-3, called "Copernicus". The satellite Copernicus had been equipped by the Princeton group with an 80 cm telescope with Rowland grating spectrograph specially developed for measuring interstellar absorption lines. With this resolutions of 0.05 Å and 0.1 Å could be achieved in the regions 950–1450 Å and 1650–3000 Å, respectively, at a range for early B-stars of about $m_V = 5^m$. Below the Lyman boundary $\lambda = 912\,\text{Å}$ the strong continuous absorption of interstellar hydrogen (Lyman continuum) sets a limit to observation (down to about 100 Å the "range" $< 100\,\text{pc}$). Besides the first term of the Lyman series for hydrogen a large number of other interstellar atoms and ions of the elements C, N, O, Si, Mg, S, Ar, Mn, Fe, Ni, Cl, P etc. were identified. A selection is contained in Table 4.5.

Atom spectrum	λ [Å]	A_λ [mÅ]
HI	1215.67 (Ly α)	16700
CI	1328.83	52
	1277.25	74
	1193.03	54
	1139.79	42
CII	1334.53	189
NI	1200.71	133
NII	1134.17	119
	1083.99	125
OI	1302.17	201
	1039.23	108
MgI	2852.13	218
MgII	2795.53	312
SiII	1808.01	87
	1526.71	(190)
	1304.37	132
	1260.42	170
	1193.29	147
	1020.70	74
SiIII	1206.51	123
PII	1152.81	61
SII	1259.52	112
SIII	1190.21	67
ArI	1048.22	97
MnII	2605.70	111
	2576.11	138
FeII	2599.40	226
	2382.03	238
	1608.46	(85)
	1144.95	91

◄**Table 4.5.** Selection of strong lines of interstellar atoms and ions in the ultraviolet, measured by the UV spectrometer of the satellite "Copernicus" in the spectrum of the O 9.5 V star ζ Ophiuchi. λ denotes the laboratory wavelength, A_λ is the equivalent width. Here we took only the line from the lowest fine-structure state. (After Morton, 1975)

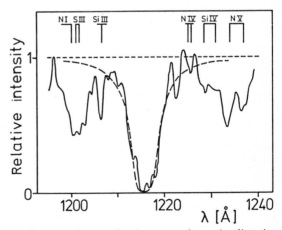

Fig. 4.11. Interstellar Lyman α absorption lines in the spetrum of the O 9.5 star ζ Orionis. The *broken lines* represent an optimally fitted theoretical profile and the assumed continuum. (By permission of Smith, 1972)

Because of the overwhelming abundance of hydrogen the interstellar Lyman α-line is very strong, attaining widths of 10 Å and more, although the profiles of this line are determined almost entirely by the natural line width (radiation broadening). The differentiation from the even broader stellar Lyman α-line is made possible by the fact that for the earliest spectral types O5...B1 the stellar line is only weakly marked and hence the interstellar component is dominant; from about B3 the stellar Lyman α-line predominates. Figure 4.11 shows an example of the measured profile of the interstellar line.

Of considerable consequence was the discovery of broad interstellar OVI-lines at 1031.9 and 1037.6 Å in many of the spectra of early star types studied (with strengths unrelated to the spectral type), which arise through transitions from the ground state. The existence of so highly ionised atoms requires very high temperatures. If one interprets the line width as the thermal Doppler effect, (cf. Sect. 5.2.5), then $T \approx 10^6$ K! It must therefore be accepted that a significant fraction of the volume of the galactic disc is filled with a very thin plasma of this temperature (cf. also Sect. 4.1.9: Cosmic x-rays; also Sect. 5.2.3). Other interstellar lines of a higher degree of ionization detected in the UV are Si IV λ 1400 Å and C IV λ 1550 Å.

UV observations have special significance in making available *evidence of molecular hydrogen*. The homonuclear H_2 molecule has no permanent electric dipole moment, so that (in the ground state) the vibration-rotation bands lying in the long-wave spectral region are forbidden and therefore very weak (quadrupole radiation; evidence in Orion Nebula see Sect. 4.2.4). A radiospectrum does not occur. There remain the allowed transitions between electronic states. Here the situation is similar to that for the H atom: all H_2 molecules find themselves in the ground state ($^1\Sigma_g^+$) and the first excited state ($^1\Sigma_u^+$) is already so high that the longest wave resonance absorption lines, the Lyman bands, fall in the region $\lambda \lesssim 1100$ Å. (Refer for the basis to Appendix B.) It is therefore a matter of transitions between vibration levels of the lower and upper electronic states (quantum numbers v'' and v'). Each $(v'' \to v')$ transition produces numerous lines as a result of splitting of each vibration level in a single rotation state (quantum numbers below and above: J'', J', respectively). These lines form several groups: the values of the difference $\Delta J = -1, 0, +1$ correspond to each of the line groups ("branches" of the rotation structure), which are denoted by the symbols $P(J'')$, $Q(J'')$ and $R(J'')$ ($J'' = 0, 1, 2, \ldots$). In the spectrum of the O7 star ξ Persei the interstellar H_2 Lyman bands $(v' \leftarrow v'') = (0-0), (1-0)$ to $(7-0)$ could be detected by 1970 rocket flights. Figure 4.12 shows extracts from the high resolution spectra obtained by the satellite Copernicus for ζ Oph and Φ^1 Ori with several lines of the $(1-0)$ Lyman band. It should moreover be mentioned that the Lyman bands of the isotope molecule HD were also observed.

Another interstellar molecule that can be detected through its band system in the extreme UV is carbon monoxide, CO. For this molecule also, important because of the relatively great abundance of its components and its high stability, it is a matter of transitions between electronic states. In ζ Oph, for example,

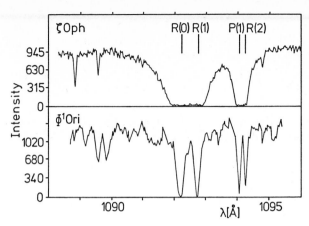

Fig. 4.12. Spectra of the stars ζ Ophiuchi (O 9.5 V, $E_{B-V} = 0.32$) and Φ^1 Orionis (B 0.5 IV–V, $E_{B-V} = 0.11$) in the region of the (1, 0) Lyman band of the H_2 molecule from observations with the ultraviolet telescope of the Copernicus Satellite. The star with the higher interstellar extinction shows essentially stronger H_2 lines. (By permission of Savage, Bohlin, Drake, and Budich, 1977)

11 individual lines of the CO molecule were observed with wavelengths between 1076 and 1447 Å.

Fine Structure and Doppler Displacements. The older observations of interstellar absorption lines in the visible region with relatively low dispersions (e.g. 30 Å/mm) and resolutions ($\lambda/\Delta\lambda < 3 \times 10^4$) resulted in proportionality of the equivalent width A_λ of the lines to the distances r of the stars studied, over a wide distance region. On the other hand the theory of the line absorption of a homogeneous layer in the case of small line broadening for great distances (= layer thicknesses) leads one to expect a saturation gradient: if the stellar continuum is completely absorbed (residual intensity $\approx$ 0) at the middle of the line, then A_λ must increase more slowly than in proportion to the "layer thickness" r (curve of growth, see Sect. 5.2.5).

However, so far as the gas takes part in the nonrigid rotation of the galaxy, interstellar atoms at different large distances will produce absorption lines with different Doppler displacements and so saturation will be impeded. For the galactic longitudes $l = 0°$, $90°$, $180°$ and $270°$, for each of which A_λ was found to be proportional to r, this effect, of course, does not operate. These considerations led O.C. Wilson and P.W. Merrill in 1937 to the conclusion that the interstellar gas must have a discrete cloud structure: each interstellar absorption line consisted of several unresolved sharp components, produced by individual clouds with different velocity components along the line of sight. The total strength of each line is then roughly proportional to the *number* of clouds encountered by the line of sight, which again increases linearly (on average) with distance.

171

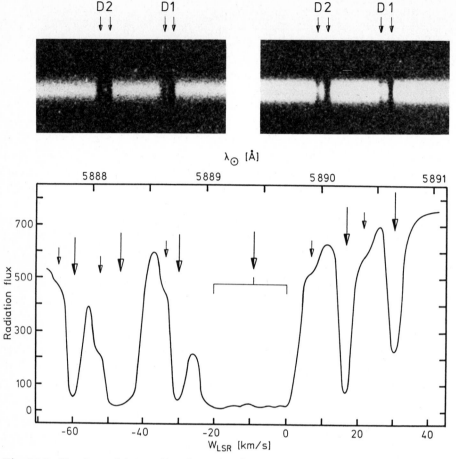

Fig. 4.13. Structure of interstellar absorption lines in the spectra of very distant stars. *Top:* Photographic spectrogram of the interstellar Na I lines D_1 and D_2 with a resolution of $7\,\mathrm{km\,s^{-1}}$ for the star HD 14134 (Per), left, and HD 12953 (Cas) at distances of 3.0 and 2.5 kpc, respectively. Each line is split into two relatively broad components with a separation distance in radial velocity of $30\,\mathrm{km\,s^{-1}}$. The stronger component is only slightly violet shifted and is caused by the local spiral arm, the weaker component is caused by the Perseus arm (cf. Fig. 4.14). Observations of very distant stars in these directions with very high spectral resolution show that each of the two components is again split into several individual lines. As an example is shown *below* the photoelectrically measured profile of the D_2 line in the spectrum of the star 6 Cas with a resolution of $2\,\mathrm{km\,s^{-1}}$. It was obtained with an interferometric spectrometer at the Coudé focus of the 2.2 m telescope of the Calar Alto Observatory. $\lambda_\odot$ = wavelength referred to the sun, W_{LSR} = radial velocity in the local standard of rest (Sect. 3.2.3). The broad trough around $W_{\mathrm{LSR}} = -10\,\mathrm{km\,s^{-1}}$ corresponds to the stronger component in the photographic spectrum. The line group from -30 to $-63\,\mathrm{km\,s^{-1}}$ arises in a distant spiral arm. The displacements of a few of the weaker lines, e.g. at $+17$ and $+30\,\mathrm{km\,s^{-1}}$, cannot be explained by the effect of the differential galactic rotation but correspond to high peculiar velocities relative to the local standard of rest at the position of this cloud. (After Münch)

The increase in spectral resolution – on modern interferometers up to $\lambda/\Delta\lambda$ greater than 3×10^5, corresponding to $1\,\mathrm{km\,s}^{-1}$ down to $0.5\,\mathrm{km\,s}^{-1}$ and less – later showed that actually for about 95% of 300 O- and B-stars studied, the interstellar Ca II- and Na I-lines could be resolved into several components, in many cases more than five. The splitting amounts correspond to velocities up to about $\pm 50\,\mathrm{km\,s}^{-1}$, an example being shown in Fig. 4.13. Simple lines are observed only for stars at higher latitudes or at distances $r < 200\,\mathrm{pc}$. In combination with the photometric distances of the stars – derived by consideration of the interstellar extinction – the statistics of the observational data show that near the galactic plane the line of sight encounters about 5 to 10 clouds in a distance of $1\,\mathrm{kpc}$. Here it is predominantly a matter of those weak concentrations, known as "small clouds" or "diffuse clouds", whose dust components produce the minor fluctuations in the Milky Way brightness mentioned in Sect. 4.1.1.

The *participation of the interstellar gas in the galactic rotation* is detected by the systematic Doppler displacement of the components of each interstellar absorption line, whose dependence on the galactic longitude l can on average be represented by (3.64) (after allowing for the solar motion!).

The amplitude of this double wave is proportional to the distance of the absorbing gas. In a portrayal of the radial velocities of the individual line components as a function of the galactic longitude, therefore, the spatial distribution of the gas also comes to light. A fine example of this is shown in Fig. 4.14. Here, for the longitude region from 90° to 180°, there is a split into two branches, which obviously correspond to two cloud complexes at different distances, namely the clouds of the local spiral arm with very small rotation effect, and the clouds of the Perseus arm at more than about 2 kpc distance.

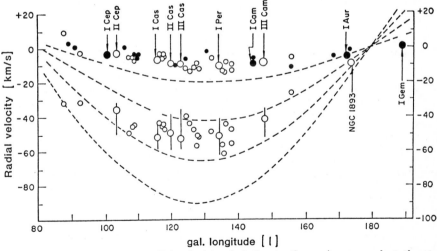

Fig. 4.14. Variation of the radial velocities (relative to the sun) measured at the strong components of the interstellar K and D lines, as a function of the galactic longitude, after an investigation by Münch (1965). Larger symbols refer to O-associations. For clouds at constant distances from the sun of $r = 1, 2, 3$ and $4\,\mathrm{kpc}$ (from *top* to *bottom*), which partake in the galactic rotation, the radial velocities must fall on the *broken curves*

The velocities of the clouds relative to the galactic rotation motion, on the other hand, are as a rule smaller than about $20\,\mathrm{km\,s^{-1}}$. Components – mostly faint – are however also observed implying peculiar velocities of $40\ldots100\,\mathrm{km\,s^{-1}}$ and more. Two examples are given in Fig. 4.13. Observations of distant galactic or extra-galactic sources show that the number of these *high velocity clouds* encountered by the line of sight at higher galactic latitudes increases with the distances of the target stars. Therefore these clouds must be also located at large distances from the galactic plane – in contrast to the typical diffuse clouds of low velocity. In the ultraviolet high velocity clouds are represented by spectral lines of singly-ionized C II, Si II, Mg II, Fe II etc. as well as of the more highly ionized C IV, Si IV and O VI. The temperatures of these clouds obviously cover a large range, perhaps from 10^3 to $10^6\,\mathrm{K}$ (further in Sects. 4.1.5, 5.2.3, 5.4.5).

4.1.5 The 21 cm Radio Line of the Interstellar Hydrogen

Transition, Line Width: In the foregoing sections it has already been indicated that in the interstellar space, far from very bright stars, all atoms and ions are normally found in their ground state. The ground state of the neutral hydrogen atom $1^2S_{1/2}$ (quantum numbers: $n = 1$, $l = n - 1 = 0$, $s = \frac{1}{2}$, $j = l + s = \frac{1}{2}$) consists of two hyperfine structure levels, which correspond to the parallel and the anti-parallel configuration of electron and nuclear spins s and i. For the total angular momentum $\boldsymbol{f} = \boldsymbol{j} + \boldsymbol{i}$, with $\boldsymbol{j} = \boldsymbol{l} + \boldsymbol{s} =$ total electron angular momentum, there are two values, which are characterised by the hyperfine structure quantum numbers $f = \frac{1}{2} + \frac{1}{2} = 1$ (parallel, upper state) and $f = \frac{1}{2} - \frac{1}{2} = 0$ (anti-parallel, lower state). For the statistical weights of the two levels $g = 2f + 1$ (further splitting of each level in an external magnetic field) we have the values $g_1 = 3$ and $g_0 = 1$.

The energy difference $E_1 - E_0$ between the two levels corresponds to a frequency $\nu_0 = (E_1 - E_0)/h$ and a wavelength $\lambda_0 = c/\nu_0$. Recent laboratory measurements, in agreement with theory, give

$$\nu_0 = 1420.40575\,\mathrm{MHz} \quad \text{or} \quad \lambda = 21.1049\,\mathrm{cm} \quad .$$

The selection rules of Quantum Theory allow only transitions between even and odd levels, i.e. even and odd values of the orbital angular momentum quantum number l. Since $l = 0$ for both levels, the spontaneous transition $f = 0 \rightarrow f = 1$ is forbidden, i.e. there is no electric dipole radiation. Magnetic dipole radiation is possible, however, with the very small transition probability $A_{10} = 2.87 \times 10^{-15}\,\mathrm{s^{-1}}$, corresponding to a very long lifetime for the upper "metastable" level $T = 1/A_{10} = 11 \times 10^6$ years.

The long lifetime corresponds according to the uncertainty principle to an extraordinarily narrow width of the upper level, so that an extremely thin line results – the lower level is infinitely thin, since there is no spontaneous transition $0 \rightarrow 1$. The natural line width in frequency measure is therefore $\Delta\nu_N = (2\pi T)^{-1} = 4.5 \times 10^{-16}\,\mathrm{Hz}$!

Collisions between the atoms shorten the lifetime of the upper level to the value $T_{col} = 1/Z$, where Z is the mean number of collisions per second, given by the product of the effective cross-section, the number density and the velocity of the atoms. The results given in Sect. 5.4.2 for the densities ($\approx 1\,\mathrm{cm}^{-3}$) and thermal motions ($\approx 1\,\mathrm{km\,s}^{-1}$) of the interstellar atoms, with an effective cross-section $\approx \pi a_1^2$, a_1 being the innermost Bohr radius, lead to a collision broadening of the line $\Delta\nu_{col} = (\pi T_{col})^{-1} \approx 10^{-11}$ Hz. Even this value still corresponds to a very narrow line width. On the other hand thermal and macroscopic motions with velocity v of only $1\,\mathrm{km\,s}^{-1}$ already lead to Doppler displacements $\Delta\nu_D = \nu_0 v/c \approx 5 \times 10^3$ Hz $= 5$ kHz. It can therefore be said that the line profiles of the 21 cm line of the interstellar hydrogen are completely determined by the thermal and above all the macroscopic motions of the emitting neutral hydrogen atoms.

Measurable Quantities of Radio Astronomy: We first of all comment on a few of the ideas applied in radioastronomy. If one directs a radiotelescope at a cosmic radio source with the intensity distribution $I_\nu(\vartheta, \varphi)[\mathrm{W\,m}^{-2}\mathrm{Hz}^{-1}$ sterad$^{-1}]$, where $\vartheta = 0$ (z-axis) is the direction of maximal signal, then the power received by the antenna $P_\nu[\mathrm{W\,Hz}^{-1}]$ is given by

$$P_\nu = \tfrac{1}{2}A \int \int I_\nu(\vartheta, \varphi) f(\vartheta, \varphi) d\Omega \qquad (4.26)$$

with $d\Omega = \sin\vartheta\, d\vartheta\, d\varphi$. The factor $\tfrac{1}{2}$ appears because the feed receives only the radiation of one polarisation direction. We denote by A the effective antenna area (to think in the $x - y$ plane) and $f(\vartheta, \varphi)$ is the directivity pattern of the antenna (Fig. 4.15).

If we think of the whole situation as being inside a closed space, whose walls have the temperature T, then I_ν is given in thermodynamic equilibrium by the Planck function, which in the radio region ($h\nu/kT \ll 1$) transforms into the Rayleigh-Jeans rule: $I_\nu = 2(\nu^2/c^2)kT$. Because of isotropy it then follows that the received power $P_\nu = \tfrac{1}{2}AI_\nu\Omega_A$ and the *effective solid angle* of the antenna

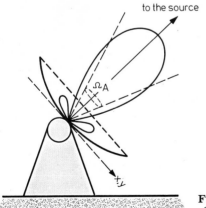

Fig. 4.15. Definition of the directivity pattern and effective solid angle of an antenna

175

$$\Omega_A = \int\int f(\vartheta, \varphi)d\Omega \quad .$$

In complete analogy with the well known optical formula for the theoretical angular resolving capability $\alpha \approx \lambda/D$, for the antenna of the radiotelescope $\Omega_A = \lambda^2/A$ – this must here be accepted without proof. Accordingly we have $P_\nu = kT$. In accordance with this result one generally associates with the radiation power P_ν of a cosmic radio source a temperature by the definition

$$P_\nu = kT_A \quad .$$

T_A is known as the *"antenna temperature"* and is thus defined as the temperature of a black body radiator in which the radiotelescope would receive a radiation power equal to that received from the radiation field of the cosmic source. T_A is the measured quantity directly obtained from radioastronomy. In practice one determines T_A by comparison with the known power $P_\nu^{(V)} = kT$ of a thermal "noise source", which is fed into the receiver instead of the antenna output.

Radioastronomers usually express the intensity I_ν of a cosmic source by the *brightness temperature* T_b, using the Rayleigh-Jeans radiation rule:[2]

$$I_\nu = \frac{2\nu^2}{c^2}kT_b(\nu) = 3.08 \times 10^{-28}\nu_{MHz}^2 T_b(\nu)[\text{W m}^{-2}\text{Hz}^{-1}\text{sterad}^{-1}]. \quad (4.27)$$

Since I_ν frequently depends on the frequency ν in a manner other than according to the Rayleigh-Jeans rule for a *single* temperature, T_b often has nothing to do with the effective kinetic temperature of the emitting medium and it varies with ν. If one substitutes (4.27) in (4.26) with $P_\nu = kT_A$, it follows for fixed frequency by applying the relation $\Omega_A = \lambda^2/A$ that

$$T_A = \frac{1}{\Omega_A}\int\int T_b(\vartheta, \varphi)f(\vartheta, \varphi)d\Omega \quad . \tag{4.28}$$

For antenna solid angles Ω_A small in comparison with the extent of the source one gets $T_A = T_b$. If the solid angle of the source $\Omega_Q \ll \Omega_A$, on the other hand, then $T_A = T_b \cdot \Omega_Q/\Omega_A \ll T_b$. In the general case the distribution of T_b over the source must be obtained from measurements of the distribution of T_A through inversion of the integral equation (4.28).

The *flux density* $\Phi_\nu[\text{W m}^{-2}\text{Hz}^{-1}]$ through unit surface area normal to the direction $\vartheta = 0$ (z-axis) is defined by

$$\Phi_\nu = \int\int I_\nu(\vartheta, \varphi)\cos\vartheta\, d\Omega \quad . \tag{4.29}$$

For a not too greatly extended source we can set $\cos\vartheta = 1$ if $\vartheta = 0$ corresponds

---

[2] In the mm-wave region, e.g. for the CO line at 2.6 mm mentioned in Sect. 4.1.6, the general Planck radiation formula must be used for I_ν!

to some point inside the source. If the angular resolution is high, so that Ω_A is small compared with the extent of the source, then it follows from (4.26) that $P_\nu = \frac{1}{2}AI_\nu\Omega_A = \frac{1}{2}\lambda^2 I_\nu = kT_b(\nu)$, and we obtain for the flux of the whole source

$$\Phi_\nu = \iint_{\text{source}} I_\nu d\Omega = \frac{2k}{\lambda^2} \iint_{\text{source}} T_b d\Omega \quad . \tag{4.30}$$

As a convenient flux unit one generally uses the quantity now known as 1 Jansky, abbreviated to 1 Jy:

$$1\,\text{f.u.} = 1\,\text{Jy} = 10^{-26}\text{W m}^{-2}\text{Hz}^{-1} \quad .$$

One encounters also the notations $1\,\text{mJy} = 10^{-3}\,\text{Jy}$ and $1\,\text{kJy} = 10^3\,\text{Jy}$. For example, solid angle fields of $1\,\square° \hat{=} 3 \times 10^{-4}$ sterad in the region of the southern Milky Way at the central wavelength of the 21 cm line of interstellar hydrogen lead to fluxes of the order of 10^3 Jy, which correspond to brightness temperatures T_b of about 100 K.

Results of the 21 cm Surveys: The first systematic measurements of the 21 cm line emission from the Milky Way were made with parabolic antennae with diameters of $D = 7.5\,\text{m}$ and 11 m, respectively, in the Netherlands for the northern part and in Australia for the southern part. From the rule of thumb $\alpha_{1/2} \approx 60°(\lambda/D)$ the (whole) half power beam widths of the directivity pattern were 1.7° and 1.2°, respectively – the actual values were somewhat higher. The frequency resolution of the radiospectrographs used was $\Delta\nu \approx 40\,\text{kHz}$, corresponding to a velocity resolution of $\Delta v_r = (\Delta\nu/\nu)c \approx 8\,\text{km s}^{-1}$. The line profiles obtained $T_b(\nu)$ and I_ν are generally substantially broader than the amount of the spectral resolution. The highest values of the brightness temperature T_b were about 100 K. Figure 4.16 shows a selection of profiles for the region $l = 0°$ to 180°. As the first main results one can state:

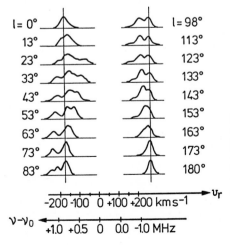

Fig. 4.16. Observed profiles of the 21 cm line of the interstellar neutral hydrogen at various galactic longitudes l on the galactic equator. Instead of the frequency ν one usually uses as abscissa the radial velocity $v_r = (\nu_0 - \nu)c/\nu_0$ with ν_0 = rest frequency of the line: $1\,\text{km s}^{-1}$ corresponds to 4.74 kHz. The two vertical lines mark the null point $\nu = \nu_0$ corresponding to $v_r = 0$ in the local standard of rest

177

1. The broad line profiles are in general displaced as a whole in comparison with the laboratory wavelengths (earth orbital motion and solar motion are eliminated!), the amount of the displacement following a double wave form of dependency on the galactic longitude (Fig. 4.16 shows only the first half of this).

2. The 21 cm radiation is strongly concentrated on the region of the galactic equator: sections perpendicular to the galactic equator often show intensity profiles with half-height widths of only 5° to 10° and therefore permit a far more accurate establishment of the base plane of a galactic coordinate system (new system, see Sect. 2.1.1) than had been possible on the basis of the apparent distribution of the stars.

A *qualitative interpretation of the observational results* for low galactic latitudes is provided by the following considerations. Just as was observed with interstellar absorption lines in stellar spectra, with the spectral and spatial resolutions given above we expect first of all a splitting of the line profiles, showing as a broadening of the phenomenon, and resulting from the cloud structure of the interstellar gas. The mean amount of splitting, however, is in general not appreciably more than $10 \, \mathrm{km \, s^{-1}}$. The great widths of the 21 cm lines must therefore have another cause: the participation of the gas in the galactic rotation with a radial gradient of orbital velocity $\Delta\Theta/\Delta R \approx 15 \, \mathrm{km \, s^{-1} \, kpc^{-1}}$ (for the vicinity of the sun). If we think of the line of sight divided into many small intervals of equal length Δr, then each of these will produce a relatively narrow line, in practice broadened only by the thermal and small scale turbulent motion of the atoms, which as a result of the differential rotation is displaced from the spectral line produced by the neighbouring interval by an amount $\Delta\lambda = \lambda\Delta v_{\mathrm{r}}/c$. Here Δv_{r} is the variation in the radial velocity of the gas (in relation to the observer) in the transition to the neighbouring interval (disregarding here the cloud structure of the gas). When integrated over the whole line of sight this gives a composite broad line.

In particular, for the directions of the centre and the anticentre $l = 0°$ and $l = 180°$, with pure circular orbit motions for all distances r, the radial velocity and hence also its variation Δv_{r} is equal to zero, so that specially narrow profiles should result. This is indeed qualitatively the case. The contributions from a fixed interval $(r, \ldots r+\Delta r)$ should show total displacements $\lambda-\lambda_0 = \lambda_0 v_{\mathrm{r}}(r)/c$ which with varying galactic longitude l follow the double wave explained in Sect. 3.4.1 (with amplitude proportional to distance r). If the gas is not too inhomogeneously distributed one can expect this pattern also for the line as a whole.

As an example let us consider the splitting of the line in the region $l = 100° \ldots 140°$. This corresponds to the behaviour shown in Fig. 4.14 of the interstellar absorption lines in the spectra of O- and B-stars of the Milky Way and so confirms the concentration of the interstellar gas in these directions on the two arms. In this comparison one naturally takes heed of the fact that the 21 cm line contributions extend over the whole line of sight, so the cross-section of the relatively broad cone included becomes very large with increasing distance.

In the past ten years a very extensive body of observational material has become available, obtained with newer radiotelescopes which are significantly larger and better. The diameters of these radio reflectors, about 65 to 100 m, give angular resolutions up to $\alpha_{1/2} \approx 10'$. At the same time the spectral resolution has been enhanced to a few kHz, corresponding to $\Delta v_r \approx 1\,\mathrm{km\,s^{-1}}$. In the presentation of results it is customary to use, instead of the wavelength or frequency, the distance from the laboratory wavelength λ_0 or frequency ν_0, expressed as the radial velocity

$$v_r' = c\frac{\lambda - \lambda_0}{\lambda_0} = c\frac{\nu_0 - \nu}{\nu_0} \quad . \tag{4.31}$$

For investigations of the galactic structure one has first of all to correct the v_R' for the effect of the motion of the earth [Sect. 2.1.3, (2.27 ff.)] and then to refer it to the local standard of rest (LSR). For this one can use the third equation (3.39) referred to galactic coordinates, with the components of the solar motion from (3.40):

$$v_r(\mathrm{LSR}) \equiv v_{rp} = v_r + U_\odot \cos l \cos b + V_\odot \sin l \cos b + W_\odot \sin b \quad .$$

Here v_r denotes the radial velocity in relation to the sun.

The publication of the data usually adopts the following form

a) Line profiles $T_b(v_r)$ for fixed positions $(l,\ b)$,
b) Contour charts showing $T_b(v_r, l)$ for a fixed b, or $T_b(v_r, b)$ for a fixed l, by lines of equal brightness temperature,
c) Distribution on the sphere: $T_b(l, b)$ for a fixed v_r (or even integrated over the whole line) is illustrated by lines of equal brightness temperature (or equal integral value).

As an example Fig. 4.17 gives a contour chart of $T_b(v_r, l)$ for $b = 0°$ and $l = 0° \dots 120°$, which will be discussed in what follows. The velocity resolution here amounts to $1.7\,\mathrm{km\,s^{-1}}$, the angular resolution $\alpha_{1/2} = 0°.6$ is adequate for a general survey. Three fundamental features of the structure claim our attention:

1) The main crest line follows the double wave of the *differential galactic rotation*. The characteristic splitting can be observed here, as for example the double structure (Perseus arm, local arm) above $l = 90°$, already explained above in connection with Fig. 4.16. The contributions also occurring at relatively high negative velocities around $l \approx 45°$ come from that part of the line of sight lying outside the circle about the galactic centre with $R = R_0$ (further in Sect. 5.4.3).

2) The numerous closed contours of small diameter must be associated with local limited regions of uniform velocity, which one may call "*clouds*" or unresolved "cloud complexes". The continuous high ridges corresponding to "arms" are evidently composed of individual clouds. With a resolution of about

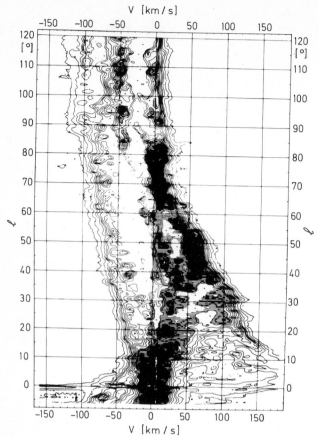

Fig. 4.17. Contour chart of the brightness temperature $T_b(v_r, l)$ of the 21 cm line for $b = 0°$. The radial velocities relate to the local standard of rest. Bandwidth of the observations $\Delta v_r = 1.7\,\mathrm{km\,s^{-1}}$, angular resolution $0°\!.6$. (After Burton, 1974)

$10'$ this fine structure stands out still more clearly down to the "cloudlets" (further in Sect. 5.4.2).

3) In the direction of the *galactic central region* the 21 cm line extends very far into the high positive and negative velocities v_r. In Fig. 4.18 we have sketched the fundamental features of a schematised contour chart extending to these high velocities. A long "beam" between $v_r = -220$ and $+200\,\mathrm{km\,s^{-1}}$, $l = 358.5°$ and $1.5°$, lies with almost complete symmetry about the central direction and $v_r = 0$. It would imply a central disc rotating with high velocity. With $R_0 = 10\,\mathrm{kpc}$ the radius of this disc is about 300 pc; observations at non-zero galactic latitudes lead to a thickness of about 200 pc. The contour extremities at $l \approx 356°$, $v_r \approx -240\,\mathrm{km\,s^{-1}}$ and $l \approx 3°$, $v_r \approx +240\,\mathrm{km\,s^{-1}}$ would first of all suggest a rapidly rotating ring between 600 and 900 pc from the centre (Ruougoor and Oort, 1960). Model calculations show, however, that these features of the chart can only be satisfactorily explained by the assump-

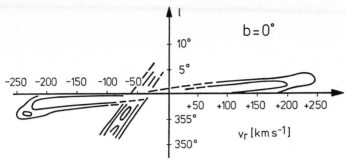

Fig. 4.18. Schematic contour chart of the brightness temperature $T_b(v_r, l)$ of the 21 cm line at $b = 0°$ for the galactic central region. Explanation in text

tion of a central rotating disc of about 1.5 kpc radius, if this disc is inclined to the galactic plane at about 22° (Burton and Liszt, 1978; Liszt and Burton, 1980). We return to this in Sect. 5.4.3.

A further structural feature of the contour chart shown in Fig. 4.18 is a chain of heights running between about $l = 354°$ (at $v_r \approx -80\,\mathrm{km\,s^{-1}}$) and $l = 3°$ (also detectable in Fig. 4.17). It passes the central direction $l = 0°$ at about $v_r = -53\,\mathrm{km\,s^{-1}}$. The source of this radiation is therefore moving radially outwards at $53\,\mathrm{km\,s^{-1}}$ and also takes part in the galactic rotation, because of the tilt of the structure (at $l = 354°$, v_r becomes about $-80\,\mathrm{km\,s^{-1}}$). The methods described in Sect. 5.4.3 at first yielded a distance from the centre for this gas arm of $R = 3\,\mathrm{kpc}$ (with $R_0 = 8.2\,\mathrm{kpc}$), so it acquired the name of the "3 kpc Arm". The continuation of the corresponding chain of heights on the contour chart up to about $l = 5°$ suggests a counterpart of this arm on the other side of the galactic centre (see also Fig. 1.12, where both arms are marked by arrows with the direction of their motion).

At galactic latitudes $|b| > 10°$ the line of sight runs for only a short distance through the relatively thin gas disc encircling the galactic centre, so only weak radiation fluxes and small velocities v_r relative to the local standard of rest would be expected. Surprisingly, the 21 cm surveys at higher latitudes in numerous regions also showed hydrogen with very high velocities $|v_r| = 30 \ldots 200\,\mathrm{km\,s^{-1}}$. It turns out that we observe here the cooler of the *high velocity clouds* discussed already in Sect. 4.1.4 (further in Sect. 5.4.5).

21 cm Line Absorption in Continuum Sources: In the continuous spectra of discrete radio sources, be they galactic diffuse nebulae (H II regions), remnants of galactic supernovae or distant extra-galactic objects (radiogalaxies, quasars), the interstellar hydrogen produces a 21 cm absorption line, which arises from the transition $f = 0 \rightarrow 1$. As in the case of interstellar absorption lines in stellar spectra, with sufficiently high spectral resolution one usually finds a splitting of the line into several narrow components, an example being shown in Fig. 4.19.

181

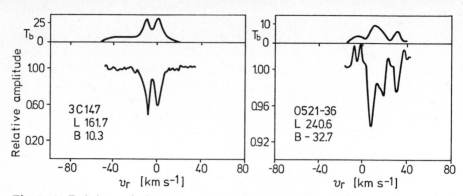

Fig. 4.19. Emission and absorption profiles of the 21 cm line. Below in each case is the interferometrically measured absorption profile in the continuous spectrum of a discrete extragalactic radio source. The upper profile shows the emission measured directly from the source. L and B denote the galactic coordinates of the direction of observation. (By permission of Hughes, Thompson, and Colvin, 1971)

The optimal representation of the observed profile by a "Gauss Analysis" yields single line widths whose interpretation as purely thermal Doppler effects: $\Delta\nu_D = (\nu/c)\sqrt{2\mathcal{R}\,T/\mu}$ with $\mathcal{R} =$ gas constant per mole and $\mu =$ molecular or atomic weight leads to kinetic temperatures T between 40 and 120 K, on average about 80 K (further in Sect. 5.4.2). For the average number of these cold interstellar clouds per unit of length along the line of sight we find a value of 3 to $4\,\mathrm{kpc}^{-1}$. The scatter of the radial velocities of the clouds (corrected for the effect of galactic rotation) turns out to be about $6\,\mathrm{km\,s}^{-1}$, so that the three-dimensional velocity dispersion (by isotropy) amounts to about $10\,\mathrm{km\,s}^{-1}$.

Zeeman Splitting of the 21 cm Line: If the emitting or absorbing hydrogen finds itself in a magnetic field, then there is, for observations in the direction of the magnetic field, a splitting of the 21 cm line into a right- and left-circularly polarised component (longitudinal Zeeman effect). For a magnetic field strength of $1\,\mu\mathrm{Gauss} = 10^{-6}\,\mathrm{Gauss} = 10^{-10}$ Tesla the separation of the components amounts to only 2.8 Hz. This amount is so small in comparison with the line width ($\gtrsim 10^4$ Hz), that it cannot be directly measured. However, using a special, differentially operating technique, which corresponds to the procedure for measuring weak magnetic fields on the sun, it is possible to detect fields of a few μGauss: one alternately uses for the observation two antennae with opposite circular polarisations and can thus directly form the difference of the two extensively overlapping line profiles of the different circular polarisations. The amplitude of the signal obtained is a measure of the amount of the Zeeman Splitting.

The greater the intensity of the line (or of the continuum in absorption observations) and the smaller the line width, the better can the amplitude be measured. Significant evidence on interstellar magnetic fields could be obtained in this way by the observation of the relatively sharp components of 21 cm absorption lines in the continuous spectra of strong radiosources. Measurements on the Taurus A source (Crab Nebula) and Cassiopeia A, which are both supernova remains, as well as Orion A (the Orion Nebula) gave field strengths (longitudinal components) between 1 and 50 μGauss. These values each relate to definite clouds which are recognisable through their radial velocities. For two clouds on the line of sight to the source Cas A relatively good accuracy was obtained. The results were $18.0 \pm 0.9 \,\mu$Gauss and $10.8 \pm 1.7 \,\mu$Gauss. Since greater field strengths are to be expected with higher density (Sect. 6.3.1), so on average the fields must be even weaker (see Sect. 4.1.8).

4.1.6 Line Emission of Interstellar CO Molecules

For the extremely low densities in the interstellar space – the 21 cm line observations give on average 1 atom cm^{-3} (see Sect. 5.4.2) – the probability of the formation of molecules outside the stronger concentrations is very low, and already existing molecules are quickly destroyed even at great distances from O- and B-stars by the unshielded cosmic radiation or by cosmic x-rays. In the widely distributed "diffuse" clouds with average visual extinctions of about $0\overset{\mathrm{m}}{.}25$, which are also sources of 21 cm line radiation, simple and particularly stable molecules can, however, be detected. Besides the optical interstellar absorption lines of CH, CH$^+$ and H$_2$ in the stellar spectra (Sect. 4.1.4), emission and/or absorption lines are observed of the hydroxyl radical OH at $\lambda = 18$ cm, the CO molecule at $\lambda = 2.6$ mm and the formaldehyde molecule H$_2$CO at $\lambda = 6$ cm – the latter only in absorption against the radio continuum of discrete sources or the 3 K background radiation (see Sect. 4.1.9).

For a systematic study of the galactic distribution of the concentrated cold gas the 2.6 mm emission line of CO (rotational transition $J = 1{\rightarrow}0$, more on interstellar molecules in Sect. 4.2.7) is particularly well suited. For C and O are, after H and He, relatively abundant elements, and the CO molecule is particularly stable. The H$_2$ molecule, still more abundant in cold dense clouds with generally high extinctions, has no radio spectrum and therefore cannot be detected. A further notable advantage of the CO molecule consists in that, besides the line of the normal C^{12}O^{16} a weaker emission is also observed with the slightly displaced wavelength of C^{13}O^{16}. If the stratum is optically thick in the normal line, and therefore permits no further statement on the thickness (Sect. 5.1.1), the optically thin case is usually still available in the line of C^{13}O^{16}, the latter being less abundant in the ratio of the isotopes C^{13}/C$^{12} \approx 1/40$ (terrestrial value $= 1/89$).

Surveys of the Milky Way in the 2.6 mm line of CO have been available since 1975. The results can be displayed in the same fashion as we have described for the 21 cm line of hydrogen, through contour charts of the brightness or antenna temperatures against the radial velocity v_r and the galactic longitude l for a fixed galactic latitude b. Figure 4.20 shows such a contour chart for $l = 352° \ldots 0° \ldots 50°$ at $b \approx 0°$. In major features there exists a similarity with the corresponding chart for neutral hydrogen (Fig. 4.17). The CO distribution, however, is much more uneven, consisting of individual concentrations

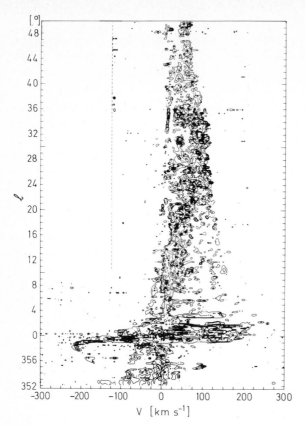

Fig. 4.20. (v_r, l) chart of the CO emission at $\lambda = 2.6\,\mathrm{mm}$ along the galactic equator ($b = 0°$) for $l = 352° \ldots 0° \ldots 50°$. (After Burton and Gordon, 1978)

with small inner velocity scatters, often of only about $2\,\mathrm{km\,s^{-1}}$. Here one can directly comprehend those dense and cold interstellar clouds, in which the hydrogen is predominantly or totally in molecular form, so that the 21 cm line cannot usually be observed. Measurements of the CO line emission as a function of the galactic latitude moreover show a smaller thickness of the molecular layer than one finds for neutral hydrogen. Most of the larger CO clouds are found in the region $l \approx 300° \ldots 0° \ldots 60°$: they must therefore lie *inside* the circular orbit of the sun with central distance $R = R_0$.

In the direction of the galactic central region around $l \approx 0°$ the contour chart shows a similar structure to that of the corresponding display of the 21 cm observations (cf. Fig. 4.18): a long "beam" extends between about $v_r = -200$ and $+200\,\mathrm{km\,s^{-1}}$ and $l = -2°$ and $+2°$. So the CO emission is here restricted fundamentally to the central, rapidly rotating (and expanding) disc already mentioned in Sect. 4.1.5. Inside this disc one finds, amongst others, two specially strong CO sources, which coincide with the radio continuum sources Sgr A and Sgr B2. The details of the contour chart, in agreement with observations of OH and H_2CO in absorption, lead to further far reaching statements on the structure and kinematics of the cold clouds in the central region, on which we shall speak briefly in Sect. 5.4.4. Also in evidence is the structural feature known as the "3 kpc Arm". Further discussion of radioline emission of interstellar origin is to be found in Sect. 4.2.3.

4.1.7 Continuous Radio Emission

Discrete Sources and Galactic Background: Continuous spectrum radiofrequency radiation extends over the whole sphere and, according to the wavelength under consideration, is more or less strongly concentrated at the galactic equator with clearly marked maxima in the direction of the centre. With increasing angular resolution of the observations, more and more discrete sources as well as the major structure are revealed. Many individual sources can be identified with optical objects (galactic nebulae, galaxies) or explained as emission from supernova remnants. As early as about 1950 the question was already raised as to what share these individual sources contributed to the total continuous radiofrequency radiation, and whether perhaps all the radiation could stem from discrete sources.

As an example, Fig. 4.21 shows a radio chart of the observed metre wave radiation with moderate angular resolution (about 2°). If one disregards the clearly identifiable discrete sources there remains a distribution with a general concentration towards the galactic equator and the galactic centre, which moreover shows a few peculiar large-scale features. The most conspicuous of these is the North Polar "Spur", an arc of radiation extending towards the north galactic pole along $l \approx 30°$. A similar structure can be detected on the south side at $l \approx 150°$: because of its location in the constellation Cetus it is known as the "Cetus Arc". Both spurs have been explained as radiation from supernova remnants. Along the galactic equator itself there are characteristic

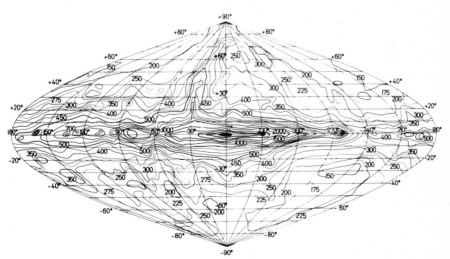

Fig. 4.21. Distribution of the galactic continuous radiofrequency radiation in the metre wave region for $\nu = 150\,\mathrm{MHz}$ ($\lambda = 2\,\mathrm{m}$) over the whole sphere in galactic coordinates. The plotted numbers without signs are brightness temperatures [K]. The chart has been drawn from three surveys, in which the angular resolution is not quite uniform and lies between about 2° and 4°. (After Landecker and Wielebinski, 1970)

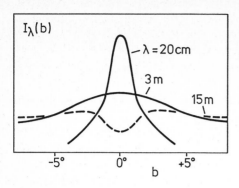

Fig. 4.22. The intensity of the continuous radiofrequency radiation as a function of the galactic latitude b for various wavelengths (schematic)

broad maxima and "steps", which can be at least partly attributed to the spiral arm structure of the galaxy (Sect. 3.3.4).

If one compares the observations of this "background radiation" for different wavelengths, one notices the following behaviour (Fig. 4.22): for wavelengths $\lambda \lesssim 30\,$cm ($\nu \gtrsim 1\,$GHz) the radiation is predominantly concentrated on the narrow galactic belt at $b = 0°$ with a half-width of only about $2°$, whereas the distribution for $\lambda \gtrsim 3\,$m ($\nu \lesssim 100\,$MHz) is essentially broader. At longer metre wavelengths, e.g. for $\lambda = 15\,$m, there appears in the middle of this belt an irregular trough. Since it is established that the H II regions, strongly concentrated at the galactic equator, radiate as discrete radio sources with continuous spectra, primarily at wavelengths $\lambda \lesssim 1\,$m (Sect. 4.2.2), it suggests an analysis of the observed distribution of the radio continuum into two components with different spectral profiles and different concentrations at the galactic equator.

For $|b| > 5°$ the observed intensity has a spectral profile which can be expressed piecewise by the formula

$$I_\nu \sim \nu^{-\alpha} \tag{4.32}$$

with positive "*spectral index*" $\alpha \approx 0.4 \ldots 0.9$ for $\nu \approx 200\,$MHz $\ldots 3\,$GHz. With the assumptions that here only the component with the broad distribution is effective, and that this has the same spectral profile for $|b| < 5°$ as for $|b| > 5°$, then the separation can be carried out. A schematic presentation of the spectral profiles of both components is given in Fig. 4.23. The spectrum of the compo-

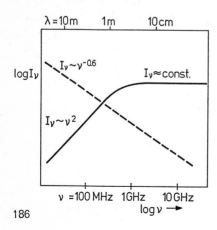

Fig. 4.23. Spectral profiles for thermal and nonthermal cosmic radiofrequency radiation (schematic)

186

nent strongly concentrated at the galactic equator is typical of H II regions and in the region $\lambda > 1$ m follows the Rayleigh-Jeans law for the longwave radiation of a black body: $I_\nu \sim \nu^2$. This component can be explained as *thermal radiation* (Sect. 5.1.3). The component with the spectrum of the form (4.32), where $\alpha > 0$, demands a completely different interpretation and is therefore usually called non-thermal radiation. Spectra of this sort are well known from supernova remnants.

Observations in the decimetre and centimetre wave region with high angular resolutions of a few arc minutes provide direct evidence of the non-thermal radiation component even at low galactic latitudes. In fact they show that the relatively thin "thermal disc" essentially consists of many discrete sources and a weak diffuse thermal component, after the removal of which there remains a *non-thermal background*, whose spectrum can be represented, for example near $\nu = 1$ GHz, by (4.32) with $\alpha \approx 0.6$. The thermal radiation comes from individual H II regions and from an extended, thin, ionised gas. The non-thermal background radiation has its origin probably in a relatively thick disc ("electron halo").

The intensity of the non-thermal background decreases steadily with increasing frequency over the whole range measurable from the earth's surface ($\nu \gtrsim 10$ MHz, $\lambda \lesssim 30$ m) – as already said, roughly proportional to $\nu^{-\alpha}$. As extraterrestrial observations with the radio-astronomical satellite RAE-1 (Radio Astronomy Explorer) between 6 and 0.4 MHz have shown, a maximum is observed at about 3 MHz, after which this radiation falls off towards low frequencies. This phenomenon may possibly be explained in a similar way to the already mentioned trough in the non-thermal galactic belt near $b = 0°$ for long metre waves (Fig. 4.22), namely, as continuous absorption in the interstellar plasma (further in Sect. 5.1.3).

Polarisation of the Galactic Background Continuum: We explain in Sect. 5.1.3 that the non-thermal radiation with a spectrum of the form (4.32), as shown by supernova remnants and the galactic background, can be readily interpreted as *synchrotron radiation* of relativistic electrons in weak interstellar magnetic fields. Synchrotron radiation is linearly polarised, the direction of polarisation (electric vector) being perpendicular to the magnetic field. For supernova remnants a high degree of polarisation is actually observed. Evidence of polarisation of the galactic background continuum was successfully obtained in 1962 after earlier fruitless attempts. An example of the distribution of the measured polarisation directions and amounts is shown in Fig. 4.24. The polarisation of the galactic background radiation is only slight and moreover shows irregular variations from place to place.

By comparison of measurements at different wavelengths it can be shown that the reason for this is to be sought in a depolarisation of the radiation through Faraday rotation on its path through the interstellar medium. This effect is discussed in the following section. Results of the measurements of polarised radiation from discrete sources show that at $\lambda = 1$ m a rotation of

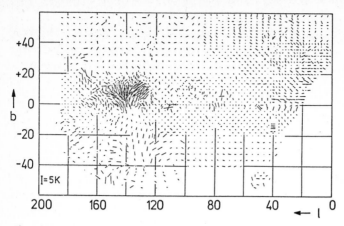

Fig. 4.24. Distribution of the directions and amounts of polarisation of the continuous galactic background radiation at $\lambda = 75$ cm in the region of the northern Milky Way. (After Berkhuijsen and Brouw, 1963)

1 radian of the plane of polarisation is caused even by a path length of 100 pc. The irregular spatial structure of the interstellar medium and the breadth of the beam of the radiotelescope have the effect that many different polarisation directions contribute simultaneously. Only polarised radiation produced in a distance range up to about 100 pc should show a predominantly uniform and only slightly rotated polarisation direction within the beam width. In other words, the information on the interstellar magnetic field contained in the observed polarisation of the galactic background radiation relates in practice only to the local region. Since Faraday rotation also occurs in the ionosphere (about $10°$ for $\lambda = 20$ cm, but more for longer waves, proportional to λ^2), the direct observational results in general still require a corresponding correction.

4.1.8 Faraday Rotation in the Interstellar Magnetic Field and Dispersion of the Radiation of Discrete Radio Sources

Faraday Rotation, Measure of Rotation and Dispersion: The radio frequency radiation of discrete galactic, as of extra-galactic, sources which emit a non-thermal continuum, is as a rule in part linearly polarised. Interpretation as synchrotron radiation (Sect. 5.1.3) provides evidence in the first place on the magnetic fields of the source themselves.

If a linearly polarised electromagnetic wave is propagated in a plasma consisting of positive ions and electrons under the influence of a magnetic field, then the polarisation plane undergoes a rotation of the amount (in radians)

$$\chi = 8.1 \times 10^5 \lambda^2 \int N_e B_\parallel dr \quad . \tag{4.33}$$

Here the wavelength λ is measured in m and the pathlength r in pc, N_e denotes the electron density [cm^{-3}] and $B_\parallel$ is the longitudinal (lying parallel to the line of sight) component of the (microscopic) magnetic field [Gauss][3]. In its passage through the interstellar plasma the radiation will undergo a rotation of its plane of polarisation through this Faraday effect. The signal issuing from the source can thus be considered as a probe for the investigation of the interstellar medium.

Besides the Faraday rotation there is also a dispersion: the velocity of propagation of the signal depends on the frequency. This effect is not noticeable with a constantly radiating source, but it is directly in evidence with pulsars, since a frequency dependence affects the arrival time of the pulses.

The establishment of the Faraday rotation requires observations of the state of polarisation of the incoming radiation at different wavelengths λ. An example is given in Fig. 4.25. The rate of change of the direction of polarisation with λ^2 must be constant and will therefore be called the *"rotation measure"*:

$$RM = \partial\chi/\partial\lambda^2 = 8.1 \times 10^5 \int N_e B_\parallel dr \, [\text{rad m}^{-2}] \quad . \tag{4.34}$$

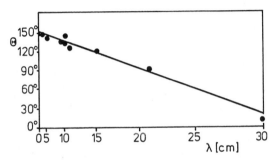

Fig. 4.25. Polarisation direction of the discrete nonthermal radio source Taurus A (Crab Nebula) as a function of the square of the wavelength λ. The position angle Θ of polarisation plotted as ordinate varies as a result of the Faraday rotation proportional to λ^2 (abscissa) arising in the interstellar medium. The true polarisation direction (150°) can be found by extrapolation to $\lambda = 0$. (After Kraus, Radio Astronomy, 2nd edition, 1986)

Since χ was measured in radians and λ in m, RM here has the dimensions [rad m^{-2}]. The sign of RM is positive if the magnetic field is directed towards the observer. The path length is from now on measured in pc, the magnetic field in Gauss.

The dispersion in the interstellar plasma causes a difference $t_1 - t_2$ between the arrival times of well-defined radiation pulses at two different frequencies ν_1 and ν_2, [MHz], given by[4]

---

[3] We use the Gaussian cgs system. Then in a vacuum the magnetic field strength $H = B$, the magnetic induction (flux density strength) measured in Gauss.

[4] For constant electron density N_e the travel time of a radiation pulse from a pulsar located at the distance r is given by $t = r/v_g$, where $v_g(\nu)$ denotes the group velocity which in general is connected with the phase velocity $v(\nu)$ by $1/v_g = d(\nu/v)/d\nu$. The phase velocity $v = c/n$ follows from the refractive index of the interstellar electron gas $n(\nu) = 1 - \varepsilon$ with $\varepsilon = (e^2/2\pi m_e)N_e\nu^{-2}$. Using $\varepsilon \ll 1$ one obtains $1/v_g \approx (1/c)[1 + (e^2/2\pi m_e)N_e\nu^{-2}]$.

$$t_1 - t_2 = 4.1 \times 10^3 DM \left(\frac{1}{\nu_1^2} - \frac{1}{\nu_2^2} \right) \quad \text{with} \tag{4.35}$$

$$DM = \int N_e dr = \overline{N}_e \cdot r [\text{cm}^{-3} \text{pc}] \quad . \tag{4.36}$$

The quantity DM is known as the *"dispersion measure"*, and it is customary also to put the path length (distance) in pc. DM can be determined from (4.35) with observations of pulsars at as many different frequencies as possible.

Observational Results: A representative selection of the numerical values obtained for DM is given in Table 4.7. The question whether the dispersion takes place in the immediate neighbourhood of the pulsar or is produced in the general interstellar medium can be settled by a glance at the dependence of the observed dispersion measures DM on the galactic latitude presented in Table 4.6. The strong increase in DM with proximity to the galactic equator shows that the dispersion effect arises quite predominantly in the interstellar space. If we assume just for illustrative purposes that the half-thickness of the effective layer is 100 pc, then for these numbers the average electron density $\overline{N}_e = DM/r$ takes values between 0.03 and 0.08 cm^{-3}.

Table 4.6. Dispersion measure of the interstellar medium from observations of 60 pulsars for various intervals of galactic latitude b. (After Pottasch, 1974)

$|b|$	<2°	2° ...5°	5° ...10°	10° ...30°	30° ...90°
DM	142	60	59	37	13

The rotation measure has been determined for numerous extra-galactic and galactic sources, including the pulsars. The separation of the purely interstellar portion from the effect originating in the region of the source, particularly for extra-galactic sources, e.g. for even very extended radio galaxies, is not possible in the individual case with satisfactory confidence. Since one finds on average an increase of the measured amounts with decreasing galactic latitude of the source under investigation, a considerable interstellar contribution must be included. A selection of the results for pulsars is given in Table 4.7.

If one considers the distribution of the total available material on RM values with regard to sign, then there is a stronger preponderance of each particular sign in large areas of the sky, for example negative values between $l \approx 30°$ and $180°$ for latitudes between about $+10°$ and $-60°$. This is shown, in particular, by the rotation amounts measured for pulsars, which − after removal of the ionospheric effect − are certainly of only interstellar origin. This is again clear evidence of the existence of a locally homogeneous component of the interstellar magnetic field.

Table 4.7. Rotation and dispersion measures for ten pulsars and the mean values of the longitudinal component of the magnetic field obtained from them. The field strengths are given in μGauss $= 10^{-6}$ Gauss $= 10^{-10}$ Tesla. (From Verschuur and Kellermann, 1974)

PSR	l	b	DM [cm^{-3} pc]	RM [rad m^{-2}]	$\overline{B}_\parallel$ [μGauss]
1642 −03	14°	+26°	35.7	+16.5	+0.58
1818 −04	26°	+5°	84.4	+70.5	+1.00
1929 +10	47°	−4°	3.1	−8.6	−3.3
2016 +28	68°	−4°	14.1	−34.6	−3.0
2111 +46	89°	−1°	141.4	−223.7	−1.95
0329 +54	145°	−1°	26.7	−63.7	−2.93
0525 +21	184°	−7°	50.8	−39.6	−0.96
0834 +06	220°	+26°	12.9	+24.5	+2.3
1133 +16	242°	+69°	4.8	+3.9	+0.99
0833 −45	264°	−3°	69.2	+33.6	+0.59

Results for the Magnetic Field: The pulsar observations are of special interest on this account, since once the dispersion measure has been determined they give the possibility also of obtaining the magnetic field via the rotation measure. From RM and DM we can obtain the mean value of the longitudinal component of the interstellar magnetic field along the line of sight:

$$\overline{B}_\parallel = \frac{\int N_e B_\parallel dr}{\int N_e dr} = \frac{1}{8.1 \times 10^5} \frac{RM}{DM} \quad . \tag{4.37}$$

Typical results for $\overline{B}_\parallel$ are included in Table 4.7. They show the same order of magnitude for the field strength as the values derived from the Zeeman Splitting for the 21 cm line. A positive sign indicates that the field is directed towards the observer. The notation PSR 1642-03, for example, means that the pulsar is at right ascension $16^h 42^m$ and declination $-3°$.

Conspicuous in Fig. 4.24 are the relatively large polarisations, directed preponderantly perpendicular to the galactic plane, at $l \approx 140°$, $b \approx 0° \ldots +10°$. The same result is found also for shorter waves. The Faraday rotation on the path of the radiation to the observer must here be practically negligible: we receive the true direction of polarisation of the emission, which is oriented in each case perpendicular to the magnetic field at the place of its origin (Sect. 5.1.3). One is led to the conclusion that the local interstellar magnetic field is perpendicular to the line of sight for $l \approx 140°$ ($B_\parallel = 0$!) and parallel to the galactic equator, so that it is aligned in the direction $l \approx 50°$, $b \approx 0°$ − in agreement with the conclusion from the optical results already mentioned in Sect. 4.1.2. That also allows us to understand specially large polarisation amounts in the direction $l \approx 140°$.

4.1.9 Interstellar Radiation Field, X- and Gamma Rays, Cosmic Particle Radiation

The interstellar space is filled with electromagnetic radiation, which is produced only in small part by the interstellar matter itself. Between the extreme ultra-violet and the far infrared the radiation field is dominated by the galactic stars. In the x-ray and gamma-ray regions the radiation comes on the one hand from galactic and extra-galactic objects (supernova remnants, stellar x-ray sources, globular clusters; active galaxies) which are usually already known optically, and on the other hand there is a diffuse component, whose contribution is of the same order of magnitude as that of all the discrete sources together. A considerable contribution to the radiation density is finally supplied by an isotropic background radiation first detected in 1965, predominantly limited to the microwave region (mm and cm waves), with the continuous spectrum of a black body at a temperature $T \approx 3\,\mathrm{K}$ ("3 K radiation"). In contrast to the ultraviolet radiation and x-rays, the interactions of these low-energy photons with the interstellar matter are of course very weak.

In addition to these electromagnetic radiations, the interstellar space is flooded with the high-energy particles of cosmic primary radiation. Their energy density is of the same order of magnitude as that of the ultraviolet radiation and x-rays. Their interactions with the interstellar matter and the interstellar magnetic field produce a considerable portion of the above mentioned diffuse electromagnetic radiation of the x-ray, gamma and radio regions. In addition they produce a weak ionisation of the generally distributed gas.

The Radiation Field of the Galactic Stars: An individual star with radius R_*, which at a distance r subtends a solid angle $\Omega_* = \pi R_*^2/r^2$, produces there a distance dependent monochromatic radiation flux $\Phi_\nu(r)$ [$\mathrm{Wm^{-2}Hz^{-1}}$]. The mean intensity of the stellar disc $\Phi_\nu(r)/\Omega_* = \overline{I}_\nu^*$ [$\mathrm{Wm^{-2}Hz^{-1}sterad^{-1}}$] is on the other hand independent of distance – when $r = R_*$ it follows that $\overline{I}_\nu^* = \Phi_\nu^*/\pi$, where Φ_ν^* denotes the flux at the stellar surface, and we have $\Phi_\nu(r) = \Phi_\nu^*(R_*^2/r^2)$! The radiation density on the other hand decreases strongly with distance r, because it is given by integration of the intensity over the very small solid angular region Ω_* and division by the velocity of light.

$$U_\nu = \frac{1}{c}\int\int \overline{I}_\nu^* d\omega = \frac{1}{c}\overline{I}_\nu^* \Omega_* \quad . \tag{4.38}$$

One frequently expresses U_ν by the intensity of an equivalent isotropic radiation field I_ν with the property $U_\nu = (4\pi/c)I_\nu$, for which we have

$$I_\nu = W\overline{I}_\nu^* \quad \text{with} \tag{4.39}$$

$$W = \frac{\Omega_*}{4\pi} = \frac{R_*^2}{4r^2} \quad . \tag{4.40}$$

I_ν can also be regarded as the mean value of the actual intensity over *all* directions. The straightforward quantity W is known as the geometric *dilution factor*. For a B0 V-star ($R_* \approx 8\,R_\odot$) for example at a distance of $1\,\mathrm{pc}$, $W \approx 10^{-14}$.

For the determination of the total interstellar radiation field of all stars one must first of all sum the contributions of the stars (index i) for each frequency ν or wavelength λ:

$$I_\lambda = \sum_i W_i \overline{I_\lambda^*}(i) e^{-\tau_\lambda^{(i)}} + I_{\mathrm{DGL}}(\lambda) \quad .$$

Here we have already taken account of the interstellar extinction $A^{(i)} = 1.086\,\tau_\lambda^{(i)}$ and the contribution of the associated Diffuse Galactic Light $I_{\mathrm{DGL}}(\lambda)$. Naturally in practice not every star is included as an individual object in the summation, but one collects the stars of the same spectral type, and so with the same spectral energy distribution. One then requires, besides the energy distributions, the abundances of the individual spectral types. We have direct knowledge of this only for the solar neighbourhood, and accordingly the results for the interstellar radiation field discussed in what follows refer also only to this region of space.

In practice only O- and B-stars contribute significantly to the total intensity I_λ in the region $\lambda < 3000\,\text{Å}$ which cannot pass through the earth's atmosphere. Knowledge of the spectral energy distribution of these star types in the ultraviolet rests on the extra-terrestrial measurements by rockets and artificial satellites, as well as on computations of model atmospheres. In the short waves the interstellar extinction is specially high (Sect. 4.1.1) and the light scattered from solid interstellar particles in some cases reaches about 50% of the amount of the direct stellar radiation (Sect. 4.1.3). At the Lyman boundary, $\lambda = 912\,\text{Å}$, very strong absorption by neutral interstellar hydrogen sets in and so sets a lower bound to the regions of the stellar spectra available even to extra-terrestrial observations.

As already described in Sects. 3.1.1 and 4.1.3, one expresses the surface brightness I_λ first of all in the units $S_{10}(\lambda)/\Box^\circ$, corresponding to the (extra-terrestrial) radiation flux $\Phi_\lambda^{(10)}$ of an A0 V-star of magnitude 10 (per square degree). If one multiplies $I_\lambda [S_{10}(\lambda)/\Box^\circ]$ by the absolute value of $\Phi_\lambda^{(10)} [\mathrm{Wm}^{-2} \text{Å}^{-1}]$, goes from $1\,\Box^\circ$ to the full solid angle 4π, and then divides by c, one obtains the interstellar radiation density $U_\lambda [\mathrm{J\,m}^{-3}\,\text{Å}^{-1}]$. Results for $\lambda = 950$ to $8000\,\text{Å}$ are presented in Table 4.8. In the ultraviolet these figures are in some cases still subject to an uncertainty factor of about two.

In the visible and infrared regions the observed interstellar radiation can be approximated by the Planck function for temperature $T = 10\,000\,\mathrm{K}$ in association with the dilution factor $W = 10^{-14}$.

The increase of U for $\lambda < 2000\,\text{Å}$ up to a maximum at $\lambda \approx 1500\,\text{Å}$ is produced by the contribution of the hot stars with spectral types O to about B5,

Table 4.8. Energy density U_λ [10^{-18} J m^{-3} Å^{-1} = 10^{-17} erg cm^{-3} Å^{-1}] of the interstellar radiation field in the neighbourhood of the sun for wavelengths from 950 to 8000 Å. (After Witt and Johnson, 1973; Jura, 1974; Zimmermann, 1965; Henry et al., 1977)

λ [Å]	U_λ	λ [Å]	U_λ
950	5	2400	2.9
1000	7.5	2500	3.0
1100	9	3000	3.0
1200	9.4	3300	3.1
1300	9.6	4250	6.8
1400	9.7	4500	7.2
1550	9.2	5000	6.6
1700	7.1	6000	5.8
1900	3.9	7000	5.2
2000	3.1	8000	4.8

but also by an increase of the diffuse galactic light, and hence of the scattering capability of the interstellar dust particles (cf. I_* and I_{DGL} in Table 4.3). The relatively low radiation density between 2000 and 3000 Å on the other hand can be explained by the specially high interstellar extinction in this wavelength region (cf. Fig. 4.3, p. 154). Since the scattered light is particularly low here, there must be a considerable portion of genuine absorption (further in Sect. 5.3.2). For $\lambda < 950$ Å the observed profile of U_λ falls well below the radiation of a black body at a temperature $T \approx 10\,000$ K, as one would expect (see above for the absorption of the interstellar hydrogen).

The total energy density of the radiation field of the galactic stars between $\lambda = 912$ Å and the infrared has been estimated at $U \approx 5 \times 10^{-14}$ Jm^{-3}. The corresponding radiation flux through an arbitrarily oriented surface in the half-space amounts to $\Phi^+ = \pi I = \pi(c/4\pi)U \approx 4 \times 10^{-6}$Wm^{-2}. The contribution of the light of all galaxies lies at least a factor of 10 lower.

Cosmic X-Rays: Below 912 Å down to about 100 Å the energy density of the interstellar radiation field is vanishingly small. This extreme ultraviolet radiation can ionise hydrogen (ionisation energy = 13.6 eV$< h\nu$) and therefore near the galactic plane it is completely absorbed in relatively short path distances – largely even after 0.1 pc with 1 H atom per cm^3 immediately below 912 Å. Since the absorption coefficient of the hydrogen in the Lyman continuum is proportional to λ^3 (see Sect. 5.1.3), the interstellar gas becomes moderately transparent again for sufficiently short wavelengths, namely in the region of soft x-rays.

The energy of the x-ray photons is usually expressed in electron volts eV or keV:

$$E = h\nu = \frac{hc}{\lambda} = \frac{12.4}{\lambda[\text{Å}]}\text{keV} \quad . \tag{4.41}$$

Photons at $\lambda = 60$ Å therefore have energies around 0.2 keV. The x-ray region extends from about 0.2 keV up to nearly 500 keV=0.5 MeV, corresponding to $\lambda = 0.02$ Å.

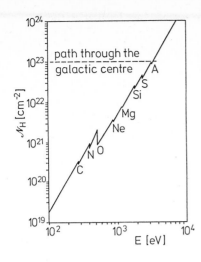

Fig. 4.26. Range of x-rays of energy E [eV] in interstellar space, assuming normal cosmic abundances of the elements, expressed as the number $\mathcal{N}_H$ of hydrogen atoms per cm^2 in the column which will attenuate the radiation by the fraction $1/e$. On the path through the galactic centre there are about 10^{23} H atoms cm^{-2}, so that transparency of the entire galaxy is to be expected for $E > 3$ keV. (After Gursky, 1973)

The transparency of the interstellar gas generally increases with increasing photon energy throughout the whole x-ray region. The photoelectric absorption by helium and the heavier atoms C, N, O, Ne, Mg, Si, S, Ar, however, always causes a dip when the binding energy of the K-shell of one of these elements is reached, after which the increase in transparency is resumed (Fig. 4.26). The evidence of these dips in the spectrum of a discrete x-ray source can therefore provide information on the abundance of the absorbing elements in the interstellar gas. For a density of one hydrogen atom per cm^3 and normal cosmic abundances of the other elements the mean free path length for 0.3 keV photons ("soft" x-rays) is only about 100 pc. Photons with 1 keV have a range of about 1 kpc. Above $E \approx 3$ keV the range increases proportional to $E^{8/3}$.

Incoming cosmic x-rays are completely absorbed by the earth's atmosphere. For the observation of the "hard" component with $E > 20$ keV, balloon ascents to heights of about 40 km indeed suffice. Measurements of the soft x-rays require heights of about 100 km, which can only be attained with rockets or artificial satellites. The first successful observations of non-solar cosmic x-rays were made in 1962 by means of Geiger counters mounted on board a rocket launched from the USA. Apart from the discovery of a discrete x-ray source, this provided evidence of an unexpectedly strong diffuse x-ray background. Our picture of the "x-ray sky" was considerably broadened above all by the systematic measurements of 1970 made by the "x-ray satellite" UHURU (Swahili word for Freedom) launched from the coast of Kenya. The third catalogue of measured discrete sources ("3 U catalogue"), published in 1974, already contained 161 discrete objects.

The energy density of the interstellar x-ray field receives a considerable contribution from an intensive *diffuse background,* which is remarkably isotropic for energies between 2 and 100 keV (variations of a few percent) and therefore undergoes no dilution effect, in contrast to the discrete sources. Because of the extensive transparency of the galactic gas layer for photons of this energy region, the isotropy can be associated with an extra-galactic origin. Discrete extra-galactic sources (quasars, radiogalaxies?) seem to provide only a portion of this radiation, of which the greater portion is probably emitted in intergalactic space. As possible explanations the following have been considered: thermal emission (Bremsstrahlung) of a very thin extra-galactic gas with temperatures around 10^7 K and inverse Compton effect between relativistic electrons and photons of the microwave background radiation (3 K radiation) discussed further below. In the region from 1 to 10 keV the intensity of this diffuse component amounts to about 10^{-11} Wm^{-2}sterad^{-1}, which gives a radiation density $U = (4\pi/c)I \approx 4 \times 10^{-19}$ J m^{-3}. Results for the spectral profile are shown in Fig. 4.27.

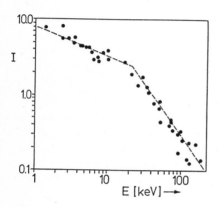

Fig. 4.27. Observed energy spectrum of the diffuse component of the cosmic x-rays in the region 1...200 keV. Ordinate: intensity I, measured in keV per cm^2 s sterad and per energy interval $\Delta E = 1$ keV. Abscissa: energy of the photons E in keV. (After Schwartz and Gursky, 1974)

For energies below 1 keV the background radiation is no longer isotropic, but shows a decrease of intensity towards the galactic equator. For the soft x-ray background observed at $E \lesssim 0.3$ keV the range is so short (see above) that an extensive galactic origin must be assumed for this diffuse radiation. This conclusion is confirmed by the evidence that the Magellanic Clouds and the Andromeda Galaxy produce no "x-ray shadow", which one would expect with extra-galactic origin of the 0.3 keV radiation. The observations can be readily explained by the assumption of a very hot thin component of the interstellar gas (temperature $\approx 10^6$ K, electron density $\approx 4 \times 10^{-3}$cm^{-3}), which fills a considerable fraction of the galactic disc between the interstellar clouds (see also 4.4.2). Such a medium is also needed for the explanation of the O VI-absorption lines mentioned in Sect. 4.1.4. The diffuse soft x-radiation can be produced by collisional excitation followed by line emission.

Cosmic Gamma Radiation: The interstellar radiation field in the gamma region $E \gtrsim 0.5$ MeV consists, like the cosmic x-rays, of contributions from discrete sources and from an unresolved "diffuse" component. The increase in range of x-rays in interstellar gas with increasing photon energy shown in Fig. 4.26 is strongly echoed here. The observations for energies below 100 GeV must nevertheless be carried out extraterrestrially, and indeed outside the radiation belt of the earth. Whilst one has been aware for some time of its possibilities, observational "Gamma Astronomy" has developed rapidly only in recent years, above all through the well-known successful measurements since 1975 in the region $E>30$ MeV with NASA's "Small Astronomy Satellite" SAS-2 and the satellite COS-B of the European Space Agency (ESA).

The discrete gamma radiation sources are in part galactic (supernova remnants, pulsars, x-ray binary stars), in part extra-galactic objects (active galaxies, quasars). For the investigation of the galactic structure the *diffuse gamma* radiation has a special significance: the measured distribution on the sphere in the energy region $\sim 0.1\ldots 5$ GeV shows a strong concentration towards the galactic plane and the galactic centre with characteristic maxima at certain galactic longitudes, in which the structure of the galactic gas disc is partially reflected (Fig. 4.28). The result is remarkably similar to the distribution of non-thermal metre wave radiation shown in Fig. 4.21 (further in Sect. 5.4.4). For the region of the galactic equator intensity values of around 10^{-4} photons with $E>100$ MeV per cm^2 s sterad are typical. The energy density of the cosmic gamma radiation with $E>10$ MeV in the solar neighbourhood amounts in order of magnitude to $U \approx 10^{-19}$ J m^{-3}.

The shape of the energy spectrum of the diffuse gamma radiation (Fig. 4.29) leads to a guiding principle for the effective production mechanisms. Discussion of the results for $b \approx 0°$ shows that in the energy region $\gtrsim 100$ MeV the main contribution arises in the interaction of the nucleon component of the cosmic radiation with the interstellar matter, from the decay of the $\pi°$-mesons so formed. A further contribution, important primarily at energies < 100 MeV, consists of the Bremsstrahlung from electrons of the cosmic radiation in their reactions with interstellar matter. Results of measurements at high galactic latitudes indicate an isotropic component with a notably steep spectral profile, which is presumably of extra-galactic origin.

3 K Background Radiation: On the longwave side of the optical region the radiation field of the galactic stars falls off steadily to the far infrared; the same is true also for the contribution of all the extra-galactic objects. Then in the millimetre waves the low energy radiation of the galactic and extra-galactic radio sources begins, increasing indeed in the metre wave region but in total making no appreciable contribution to the interstellar radiation density. Between a few tenths of a millimetre and a few decimetre wavelengths, however, there is superimposed a widespread isotropic background radiation, whose total energy density surprisingly is of the same order of magnitude as that of the radiation of all the galactic stars. This radiation, first discovered in 1965 by A.A. Penzias and R.W. Wilson, according to the latest measurements (for $\lambda < 1$ mm possible only extraterrestrially from balloons) has a continuous spectrum which corresponds to that of a black body with a temperature $T = (2.7 \pm 0.1)$ K. It is explained as a relic of the hot initial phase of the universe. For the interstellar matter the significance of this "3 K radiation" lies, amongst other things, in that it can penetrate even very dense interstellar clouds with correspondingly high

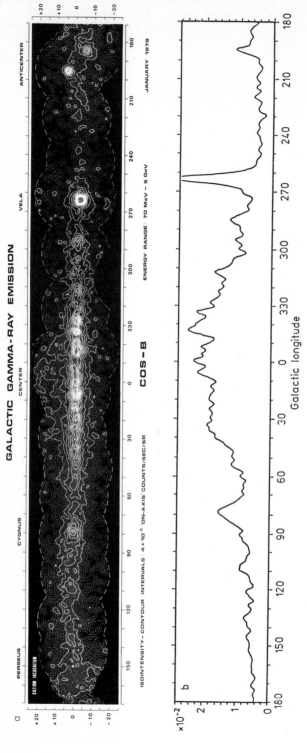

Fig. 4.28. Gamma ray chart of the Milky Way for the energy region $E = 70\,\text{MeV} \ldots 5\,\text{GeV}$ from results obtained by the European Research Satellite COS-B. Strongest individual source is the Vela Pulsar. (MPG Press picture/MPE). *Below:* Intensity profile as a function of the galactic longitude averaged over the region $-5°\ldots$ $+5°$ in galactic latitude. Unit of ordinate: counts per s and sterad. (From Mayer-Hasselwander et al., 1982)

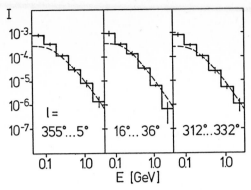

Fig.4.29. Energy spectra of the diffuse cosmic gamma radiation from three regions of the galactic equator, from measurements made by the satellite COS-B. The *broken curves* are given by the theory of the $\pi^0 - 2\gamma$ decay. Unit of ordinate: photons $cm^{-2}\,s^{-1}\,sterad^{-1}\,GeV^{-1}$. (After Paul et al., 1978)

dust extinction and prevent the temperature there sinking below about 3 K. The total radiation density, from $U = (4\pi/c)I$, with $I = \Phi^+/\pi = \sigma T^4/\pi$ (Stefan-Boltzmann Law) for $T = 2.7\,\mathrm{K}$, turns out to be $U_{3\mathrm{K}} = 4 \times 10^{-14}\,\mathrm{J\,m^{-3}}$.

Cosmic Particle Radiation: The relative abundances of the various nuclei contained in the extraterrestrial primary radiation are in large measure similar to those in the atmospheres of normal stars. Overabundant are Li, Be and B – probably through formation of these nuclei in collisions of heavy nuclei with the atoms of the interstellar gas. Protons, helium nuclei and electrons far surpass in their abundances all other particles. The kinetic energy of the observed nucleons ranges up to around $10^{20}\,\mathrm{eV}$.

Results for the extraterrestrial particle flux, measured from balloons, rockets and artificial satellites, are shown in Fig. 4.30 for protons, helium nuclei and

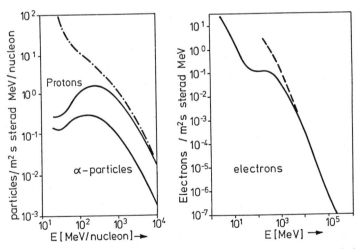

Fig.4.30. Energy spectra of protons, α-particles and electrons of the low energy cosmic radiation. The continuous curves show the observed particle fluxes. Correction for the solar wind leads to the chain-dotted profile for the protons. The profile shown as a dashed curve for the electron component comes from interpretation of the diffuse galactic radio emission as synchrotron radiation (see Sect. 4.1.7). (After Mezger, 1972)

electrons as functions of the kinetic energy E of the particles. In the region $E > 3 \times 10^9$ eV the energy spectrum can be represented in the form

$$N(E)dE = KE^{-\gamma}dE \quad . \tag{4.42}$$

For the electrons one finds the exponent $\gamma \approx 2.5$ up to the highest measured energies ($\approx 3 \times 10^{11}$ eV); for the nucleons one also finds a constant $\gamma \approx 2.5$ up to about 10^{15} eV, but $\gamma \approx 3.2$ for $E \approx 3 \times 10^{15} \ldots 10^{18}$ eV. The direction distribution of the incoming high energy particles is largely isotropic up to the highest observed energies.

The motions of these charged particles come under the influence of the interstellar magnetic field. A particle with charge $Z \cdot e$ and mass m moving *in vacuo* with velocity v perpendicular to a homogeneous magnetic field of strength B [Gauss], describes a circular orbit under the influence of the Lorentz force, in a plane perpendicular to the magnetic field. The radius of the orbit (gyration radius or Larmor radius) is given by $r_B = mcv/ZeB$. For relativistic velocities $v \approx c$ it follows that

$$r_B \approx \frac{mc^2}{ZeB} \approx 2 \times 10^{14} \frac{E}{Z \cdot B} [\mathrm{m}] \quad . \tag{4.43}$$

Here $E = mc^2 = m_0 c^2 / \sqrt{1 - (v^2/c^2)}$ denotes the kinetic energy of the particles in J. For protons with $E \approx 10\,\mathrm{GeV} = 10^{10}\,\mathrm{eV} = 1.6 \times 10^{-9}\mathrm{J}$ in an interstellar magnetic field $B = 3 \times 10^{-6}$ Gauss $= 3 \times 10^{-10}$ Tesla (cf. Sect. 4.1.8, especially Table 4.7) it follows that the gyration radius $r_B \approx 10^{11}\mathrm{m} \approx 0.7\,\mathrm{AU}$. Up to relatively high energies, then, the particles of the cosmic radiation are captured long-term in the galactic gas disc. Only the most extremely energetic particles with $E > 10^{18}\mathrm{eV}$ – they produce the big atmospheric showers – cannot be retained by the galactic magnetic field and probably have an extragalactic origin (nuclei of active galaxies, quasars).

Charged particles with energies below about $10^9\,\mathrm{eV} = 1\,\mathrm{GeV}$, on the other hand, can be swept out of the solar system again by the turbulent magnetic fields guided by the solar wind in interplanetary space, or have their particle energies changed significantly. Measurements with space probes accordingly show an increase in the flux of these "low energy" particles with increasing distance from the sun. By the profiles shown as dashed curves in Fig. 4.30 we have attempted to reconstruct the interstellar proportions. For the electron component, the dashed low energy profile has been obtained from the observations of the diffuse non-thermal radio continuum of the galactic disc, interpreted as synchrotron radiation from electrons of the cosmic radiation (cf. also Sect. 5.1.3). The total energy density of the cosmic radiation amounts from these results to $U_{CR} = 1.3 \times 10^{-13}\,\mathrm{J\,m^{-3}}$. The electrons form only about one percent of this amount.

The energy densities of the various components of the interstellar radiation field are listed for comparison in Table 4.9.

Table 4.9. Energy densities of the different radiations in the interstellar space near the sun

Component	U [J m^{-3}]
Stellar radiation	5×10^{-14}
Diffuse x-rays	4×10^{-19}
Diffuse gamma radiation	$\sim 10^{-19}$
2.7 K background radiation	4×10^{-14}
Cosmic particle radiation	1.3×10^{-13}

Since the charged high energy particles in interstellar space move on curved paths it is difficult to trace their sources. There is general agreement, however, that the origin of the cosmic radiation is closely associated with the various high energy phenomena in our galaxy (supernovae, shock fronts, pulsars) as well as with explosive events in other (active) galaxies.

4.2 Interstellar Clouds

In what follows we shall turn our attention to the observations of the definitely distinguishable objects which are recognised as sources of line and continuum radiation, or even only stand out through a weakening of the background radiation. The interpretation of these phenomena in Chaps. 5 and 6 shows that it is a question in part of the massive condensations of interstellar matter. They are first of all cold, but become heated internally after the formation of one or more hot stars. The stellar ultraviolet radiation can lead to the ionisation of the neutral hydrogen and other sorts of atom in a large region of space (H II region), which we see as a bright emission nebula. We shall consider first of all this phenomenon of hot clouds. After that the dust filled dark clouds and molecular clouds will be treated as cold aggregates of interstellar matter.

4.2.1 Diffuse Emission Nebulae: Optical Phenomena

Morphology and Classification of Bright Galactic Nebulae: The designation "bright nebula" for a non-stellar distributed optical light source − originally also including the extra-galactic objects later recognised as distant stellar systems − is to-day generally used only for the galactic phenomena. Bright nebulae with emission-line spectra are called emission nebulae or gas nebulae. If the spectrum is continuous, on the other hand, like a stellar spectrum, then one speaks of a reflection nebula (see Sect. 4.1.3). The concept of emission- or gas-nebulae however includes three fundamentally different types of phenomenon: diffuse emission nebulae or diffuse nebulae for short, planetary nebulae, and remnants of supernovae.

The most regular of these objects are the *planetary nebulae* – so-called on account of their similarity to the brightest examples of the planet discs on visual inspection in a telescope. They are approximately spherically symmetric expanding envelopes of stars. This material will indeed pass over into the interstellar medium, but can in the observed stage of evolution still be unmistakably associated with a "central star". We therefore regard the planetary nebulae as a matter of stellar physics and go into the subject no further here.

That should also be true for the comparatively rare *remnants of supernovae*, or more accurately the ejected massive shells of supernovae. Characteristic phenomena of the optical manifestation are the thin cirrus-like arcs (example: the Loop or Network Nebula in Cygnus) and in earlier stages the filamentary structures over an amorphous nebula (Crab Nebula in Taurus).

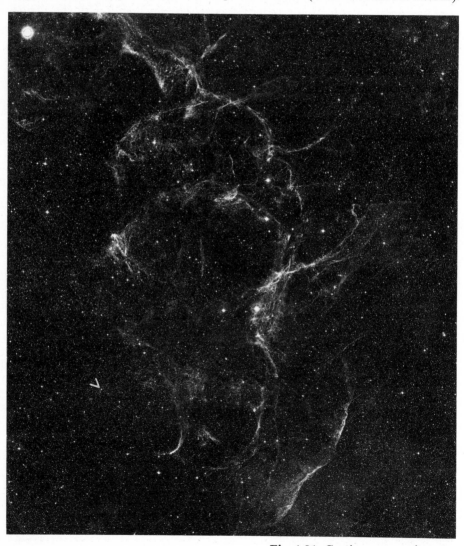

Fig. 4.31. Caption see opposite page

Fig. 4.32. Detail of the Rosette Nebula. Description of the structure in the text. (Photograph obtained with the 48-inch Schmidt Telescope on Mt. Palomar)

Assurance that in an individual case it is indeed a matter of the remnants of a supernova is usually derived, first from radio, and often also from x-ray observations. The rapidly expanding supernova shell produces a complex shockwave structure in the surrounding inhomogeneous interstellar medium. A specially impressive example is the supernova remnant in the region of the radio source Vela X (Fig. 4.31). It is embedded in the Gum Nebula, so-called after its discoverer, in the constellations Vela and Puppis of the southern Milky Way, whose angular diameter amounts to around 35°.

Fig. 4.31. The Vela Supernova Remnant. The arrow indicates the position of the Vela Pulsar PSR 0833-45, the surviving stellar object. (Photograph obtained with the Schmidt Telescope of the European Southern Observatory)

Fig. 4.33. The Lagoon Nebula M8 = NGC 6523 in the red including Hα (north is to the left). (Photograph taken with the 2.2 m Telescope at the Calar Alto Observatory)

The true *diffuse nebulae* are more or less compact bright clouds with usually an irregular structure. Examples are given in Figs. 1.4, 4.32 and 4.33 — apart from the Vela supernova remnant the Gum nebula also has essentially this character. The brightness distribution inside a diffuse nebula shows very strong variations. In most cases there is also a reflection constituent: starlight scattered from the dust component with a continuous spectrum. There are often dark channels, so-called elephants' trunks, or even larger dark regions with sharp bright edges (see Fig. 4.32); in many cases there are small dark regions, almost circular in shape, and accordingly called globules. On the sphere diffuse nebulae are more strongly concentrated towards the galactic equator than are the other types of galactic nebulae.

Catalogues and Nomenclature: The galactic nebulae which are brightest in appearance were recorded in Charles Messier's well known catalogue of 1784 – as well as star clusters and the distant stellar systems not recognised at that time as extra-galactic objects. Thus for example it allots to the Orion Nebula the designation M 42; the best known supernova remnant, the Crab Nebula, and the best known planetary nebula, the Ring Nebula in Lyra, have the designations M 1 and M 57, respectively. A far greater number of galactic nebulae is contained in the New General Catalogue (NGC) with two continuations called the Index Catalogue (IC) published by Dreyer between 1888 and 1908. In this work the three nebulae already mentioned carry the designations NGC 1976 (= M 42), NGC 1952 (= M 1) and NGC 6720 (= M 57). A comprehensive modern list was published by Lynds in 1965. This "Catalogue of Bright Nebulae" includes about 1100 objects. It is based on photographs of the "Palomar Observatory Sky Survey" of the northern sky down to declination − 33° in two spectral regions: blue (3500...5000 Å) and red (6200...6700 Å). These photographs were obtained with the powerful 48 in. (122 cm) Schmidt telescope on Mt. Palomar. Another important list is that of Sharpless (reference, for example, S 20). For the bright nebulae of the southern sky one often finds the notation RCW in association with a number, which refers to the 1960 list of Rogers, Campbell and Whiteoak. (More precise citation of these catalogues see list of References).

Exciting Stars and Dimensions: Inside a diffuse emission nebula or in its immediate neighbourhood one finds as a rule one or more bright stars with spectral types earlier than B 1, although this is a very rare class of star. Near reflection nebulae, on the other hand – defined as diffuse nebular objects with continuous spectra such as those of stars – one usually finds only "later" spectral types from B 1 onwards. The excitation of the light from diffuse emission nebulae can therefore evidently come only from a very hot star (spectral type from B 0, corresponding to $T_{eff} \gtrsim 30\,000$ K). Frequently this conclusion is stated the other way round: if the exciting or illuminating star is of a spectral type earlier than B 1, it is a question of a diffuse emission nebula. In the other case it is a reflection nebula ("Hubble's Law").

Between the apparent photographic magnitude m_{pg} of the exciting or illuminating star and the angular distance a (arc minutes) from the star to the furthest "edge" of the nebula, E. Hubble in 1922 found the following empirical relationship for 80 diffuse emission and reflection nebulae:

$$m_{pg}^* = -4.9 \log a + 11.0 \quad , \tag{4.44}$$

where the constant on the right is fixed by the particular observation (telescope aperture, exposure time, photo-material). For reflection nebulae this result is easily explained (Sect. 4.1.3). That diffuse emission nebulae, within the limits of the modest systematic precision of the old underlying observational material, follow roughly the same rule appears surprising, because this fact requires a

Table 4.10. Angular extent, distance r, greatest linear extent in Hα and exciting star (HD = Henry Draper Catalog) for a few diffuse emission nebulae

Name, Messier No. NGC No.	Apparent dimensions [arc minutes]	r [kpc]	Greatest linear extent [pc]	Exciting star Name	Spectral type
Orion Nebula M42 = NGC 1976	90 × 60	0.5	13	θ^1 Ori C	O6p
Rosette Nebula NGC 2237–46	80 × 60	1.6	37	6 stars of the NGC 2244 cluster	O5–O9 V
Trifid Nebula M20 = NGC 6514	20 × 20	2.1	12	HD 164 492	O7
Lagoon Nebula M8 = NGC 6523	45 × 30	1.4	18	stars of the NGC 6530 cluster	O5–B0
Omega Nebula M17 = NGC 6618	20 × 15	2.1	12	stars of a cluster	O4–O8 V
M16 = NGC 6611	120 × 25	2.2	77	stars of the NGC 6611 cluster	O5–B0
North America Nebula NGC 7000	120 × 30	1.2	42	Association Cyg OB2 HD 199 579	O6

quite different explanation (Sect. 5.2.1). As an empirical result the relationship can be called upon in every case for the identification of the exciting star.

Once the star has been found, then its distance r can be found in familiar fashion from the relation (2.43) if not only its apparent magnitude but also the MK spectral type (deduced from the absolute magnitude and the reddening) and a colour index (removal of the interstellar extinction!) can be determined. From the angular extent of the nebula its linear size now follows. Table 4.10 gives a selection of results. Here the angular extent, and hence the linear diameter of the nebula, refer to the emission in the Hα-line of hydrogen from the red plates of the Palomar Observatory Sky Survey.

The excitation of the Orion Nebula arises from four stars which form a trapezium: the multiple system Θ^1 Orionis (components A, B, C, D) in the centre of the nebula. Only the brightest of these stars, Θ^1 Ori C, is shown in Table 4.10, the other three are B-stars. Inside a radius of about 10 pc around the trapezium one finds other stars of early spectral type, which form a stellar association (Ic Ori); in addition this region is filled with numerous faint stars with emission lines of the nature of T Tauri stars. In the central region of the Rosette Nebula a whole star cluster of early type is contained.

Spectra of Diffuse Nebulae: Because of meagre surface brightnesses, highly resolved slit spectra can be obtained only from the brightest nebulae, even by use of the largest telescope. The best known is the spectrum of by far the brightest object, the Orion Nebula (see Fig. 1.4b). In recent times special equipment with high light-gathering power on the basis of the Fabry-Perot Interferometer has been successfully applied to the study of fainter nebulae. A list of all the observed emission lines of the Orion Nebula between 3200 and 11 000 Å is to be found in Schaifers and Voigt (1982). In the optical region from the ultraviolet to the infrared, three sorts of emission lines occur:

1) Lines which can result from the *recombination* of H^+, He^+, C^+, N^+, O^+ etc. with free electrons: Balmer series of hydrogen (up to high members!); He I spectrum, especially $\lambda\,4471$ Å and $\lambda\,5875$ Å (= D_3 line); also He II lines for many nebulae. (That all these lines must in fact result from recombination, and not from direct excitation of the initial states, is demonstrated in Sect. 5.2.1). For radio recombination lines see Sect. 4.2.3.

2) Lines which are caused by *fluorescence*: a particular emission line of the nebula supplies photons of energy $E = h\nu$, which (fortuitously) is directly equal to the excitation energy of the line under consideration. For example, the recombination line of He^+ at $\lambda\,303.780$ Å can excite the $3d^3P_2^0$-term of O^{++}, whose energy difference from the ground state corresponds to a wavelength of 303.799 Å. Subsequently several lines of the O^{++} ion in the near ultraviolet can be emitted through transitions into various lower states. This mechanism of resonance-fluorescence is particularly clearly marked in the planetary nebulae.

3) *Collision-excited lines* of ions of the elements O, N, Ne, S etc. which partly arise through transitions from low-lying metastable states to the ground state (*forbidden lines*). Well known examples are the two "green nebula lines" N_1 and N_2 at wavelengths 5007 and 4959 Å, originally ascribed to a hypothetical element "Nebulium", which were eventually identified as intercombination lines of O^{++} (transitions $2p^2\,^1D_2 \rightarrow 2p^2\,^3P_2$ and $2p^2\,^1D_2 \rightarrow 2p^2\,^3P_1$, respectively) (I.S. Bowen, 1927). These and other lines forbidden by the selection rules of the Quantum Theory often display the strongest nebular emission lines. How this is possible under the conditions of a gas nebula is explained in Sect. 5.1.2. The theory shows in particular that the excitation of the initial states of these lines follows from *electron collisions*. Below about 3500 Å allowed lines are also to be expected, which are excited by this mechanism. On the other hand many of the transitions between the low-lying fine structure levels of the various ions (e.g. inside the ^3P levels of O^{++} or of neutral O) forbidden by the rule fall in the infrared region. The observation of these infrared lines offers the advantage of extensive freedom from the interstellar extinction. For example, lines from O^{++}, S^{++}, S^{+++} and Ne^+ have been authenticated (see Sect. 4.2.4).

In 1976 emission lines of the H_2 molecule at $\lambda = 2.0$ to $2.4\ \mu m$ were also detected in the Orion Nebula. Here the forbidden (quadrupole) rotation-vibration transitions ($v = 1 \rightarrow 0$) are involved, which are probably excited in the region of the shock fronts.

The first observations of an emission nebula in the "rocket ultraviolet" were made in 1972 at the moon-base of the Apollo 16 astronauts (North America Nebula in the region 1250...1600 Å). Emission *line spectra* – down to 912 Å – were first made in more recent times with UV-spectrometers installed in artificial satellites or space probes. One of the interesting questions is that of the Lyman α-line of hydrogen at 1216 Å. A difficulty of the observation consists in that the geocorona extending out to about 100 000 km produces a dominant Lyman α background.

Besides the emission lines one usually also observes a weak *continuum*. Photometric measurements on a number of diffuse nebulae showed, in spite of strong variations in the continuum intensity I_λ^K and also in the strength of the emission lines from place to place in the nebula, a nearly constant ratio of I_λ^K for the blue spectral region to the strength of the Hβ-line: $I_\lambda^K \Delta\lambda / H\beta \approx 1.5 \times 10^{-3}$, where $\Delta\lambda = 1$ Å. This quotient turns out to be nearly the same for all nebulae investigated. Observations in the ultraviolet show in addition that $I_\lambda^K \Delta\lambda / H\beta$ increases with decreasing wavelength. I_λ^K cannot therefore be a recombination continuum, which decreases in the UV as λ^3. It suggests rather that the observed continuum is caused predominantly by scattering of starlight on the dust particles inside the nebula (cf. Sect. 4.1.3). A direct indication thereto comes from the occurrence of an absorption line in the nebular continuum at $\lambda 4686$ Å. In the spectrum of the brightest of the Trapezium stars, the O6-star Θ^1 Ori C, the He II line at $\lambda 4686$ Å is of course present in absorption! (Emission lines of the He II spectrum have not been observed in the Orion Nebula). This finding is not surprising, insofar as the structure of most diffuse nebulae already indicates the existence of large quantities of dust inside these objects.

The absolute *radiation emission* of a nebula in the light of particular emission *lines* is obtained from measurements of the (extra-terrestrial) radiation flux at the place of the earth Φ^L [W m^{-2}] in association with the distance of the nebula r. If we approximate the shape of the nebula by a sphere of equal volume with radius R_N, and denote by Φ_N^L the flux at the surface of this sphere, then the total emission of the nebula per unit time, its luminosity (in the lines under consideration), is given by

$$L_N^L = 4\pi R_N^2 \Phi_N^L = 4\pi r^2 \Phi^L \quad . \tag{4.45}$$

Introducing the angular diameter of the nebula in arc minutes a, it follows that the radiation flux at the position of the nebula Φ_N^L is given by

$$\Phi_N^L = \Phi^L \left(\frac{6.88 \times 10^3}{a} \right)^2 \quad . \tag{4.46}$$

The Hα emissions of the objects listed in Table 4.10 have an order of magnitude $\Phi^L(H\alpha) \approx 10^{-10}$ W m^{-2}, whence, taking into account the correction factors for the interstellar extinction (between 2 and 5) it follows that $\Phi_N^L(H\alpha) \approx 10^{-6}$ W m^{-2} and $L_N^L(H\alpha) \approx 10^{30}$ W. For an appreciation of this number we remind ourselves that the total luminosity of the sun is 4×10^{26} W and that of a B0-star is 2×10^{31} W. A considerable share of the (ultraviolet) radiation of the exciting star is evidently converted into Balmer emission from the nebula (more on this in Sect. 5.2.1).

Structure and Internal Motions from Observations of the Optical Nebular Lines: Monochromatic photographs by the light of particular lines show that the emitting gas inside a nebula is extremely unequally distributed. In the Orion Nebula, for example, one finds strong variations of intensity within an angular distance of a few arc seconds (Fig. 4.34). Line profiles, especially Doppler displacements, are determined not only with conventional spectrographs in association with large telescopes, but also with Fabry-Perot interferometers. In the latter case the line is first of all singled out by a preselector, (e.g. an interference filter). The interferometer then produces a picture of the whole nebula for the line with a ring pattern (Fig. 4.35). If there is, e.g. a red-shift of the line at a certain position of the nebula under consideration, then on the interferogram there is a local shrinkage of the ring there, and so a measurable displacement of a portion of the ring.

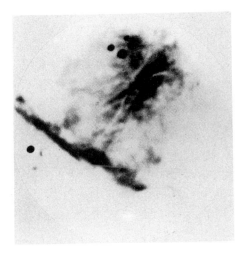

Fig. 4.34. Monochromatic photograph of M 42 in the light of the line [N II] λ 6584 Å. The Trapezium stars are to be seen near the upper edge of the picture. (From Elliot and Meaburn, 1974, by permission)

The expectation that one can directly establish a clearly marked general expansion of the entire heated gas has not proved possible to confirm directly. The internal motions are obviously more complicated. The observed structures indicate that several motion fronts or filaments have been formed, which partially overlap. Thus, for example in the Orion Nebula one finds irregular spatial variations of the radial velocity $\sqrt{(\Delta v_r)^2}$ of the order of $10\,\mathrm{km\,s^{-1}}$ from the line widths (after subtraction of the Doppler broadening for a temperature of $10^4\,\mathrm{K}$). For the strongly marked condensations, splitting of the nebular lines was also observed. Probably the line of sight here crossed two different motion fronts (more on this in Sect. 6.3.3). A surprising phenomenon are weak blue-shifted components of several emission lines of the Orion Nebula, which correspond to motions away from the nebula with velocities up to $100\,\mathrm{km\,s^{-1}}$. Possibly a part of the gas is being so strongly expelled by the radiation pressure on the dust particles or by the stellar wind of the exciting star.

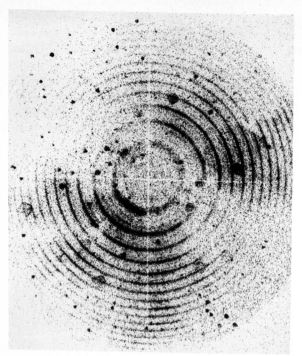

Fig. 4.35. Fabry-Perot interferogram of the diffuse emission nebula IC 1318 in the light of the Hα line of hydrogen (negative; north at top). Explanation in text. (From Hippelein)

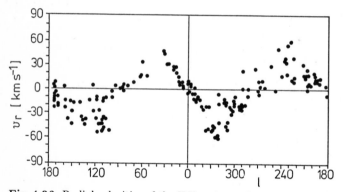

Fig. 4.36. Radial velocities of the H II regions, plotted against the galactic longitude l. The splitting of the double wave around $l = 130°$ (Perseus) and $l = 320°$ (Norma-Centaurus) can be explained by the assumption of a further spiral arm outside and inside the local arm, respectively. (After Georgelin and Georgelin, 1970)

The *radial velocities of diffuse nebulae as a whole* (mean value over each nebula), if plotted against the galactic longitude, follow the double wave of the differential galactic rotation (Fig. 4.36). In the regions $l \approx 100° \ldots 160°$ and $l \approx 300° \ldots 340°$ splits resulting from the spiral structure are clearly marked (Perseus Arm and Sagittarius Arm, respectively, cf. also Figs. 4.14, 16 and 17).

If one assumes the law of galactic rotation as given, then the "kinematic" distances of the diffuse nebulae can be derived from the observed radial velocities. This method is above all useful if the exciting stars are blotted out by masses of dust (further in Sects. 4.2.3 and 5.4.4).

4.2.2 Continuous Radio Emission from H II Regions

Thermal Radio Sources: The diffuse emission nebulae are as a rule also sources of continuous radio frequency radiation, above all in the cm region: they coincide with discrete sources of that type into which the thermal component of the galactic radio radiation mentioned in Sect. 4.1.7 can to a large degree be resolved. The crucial commonality of all these sources is a spectral profile of the radiation flux of the sort displayed in Fig. 4.37 for the radio source "Orion A" coinciding with the Orion nebula. At lower frequencies (longer waves) the observed radio continuum flux $\Phi^K(\nu) \sim \nu^2$, as for the radiation from a black body, whilst for higher frequencies (shorter waves) $\Phi^K(\nu)$ becomes nearly constant – more exactly $\Phi^K(\nu) \sim \nu^{-0.1}$. In Sect. 5.1.3 it is shown that radio radiation with these spectral properties is produced by free-free transitions of electrons in the electric fields of ions (thermal Bremsstrahlung), as occurs also in the solar corona. The site of the transition region differs somewhat from object to object. The radiation flux at the location of the nebula $\Phi_N^K(\nu)$ is obtained from $(r/R_N)^2 \Phi^K(\nu)$, and can also be written as $\Phi_N^K(\nu) = \pi I_\nu$. Comparing this, in the region of lower radio frequencies, with the isotropic intensity of a black body (Rayleigh-Jeans Law) $I_\nu = 2(\nu/c)^2 kT$, we find that the observations require in general temperatures around $T = 10^4$ K. For the actual kinetic temperature of the emitting gas one finds values of the same order of magnitude (Sect. 5.2.4). We therefore call these objects "thermal radio sources".

Not all of these radio continuum sources coincide with a visible nebula, and many have no identifiable optical counterpart. The optical manifestation

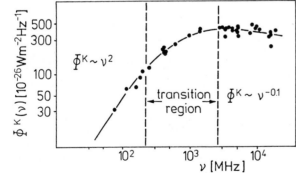

Fig. 4.37. Spectral Profile of the observed continuous radiofrequency radiation flux from the Orion Nebula $\Phi^K(\nu)$. (After Terzian and Parrish, 1970)

of the H II region[5] is of course often obscured by local or − at great distances − interstellar dust extinction. The radio phenomenon therefore usually also receives a special apellation: a rather large number of sources are known by their number in a list by G. Westerhout. The radio sources associated with the nebulae listed in Table 4.10 receive, for example, the names (in order beginning with M 42): W 10, W 16, W 28, W 29, W 38, W 37 and W 80. Besides names from lists by other authors, one often gives the values of the galactic coordinates l, b (in degrees) of the central point of the source after the letter G (= galactic source), e.g. G 209.0-19.4 for the radio source in the Orion Nebula.

The absolute values of the observed radio continuum flux $\Phi^K(\nu)$ of the objects listed in Table 4.10 are of order 10^2 Jy for the frequently used frequency $\nu = 5\,\mathrm{GHz}$, corresponding to $\lambda = 6\,\mathrm{cm}$. For example, for M 42 $\Phi^K(5\,\mathrm{GHz}) \approx 4 \times 10^2$ Jy. From (4.45) the product $r^2 \Phi^K(\nu)$ is a measure of the total emission of the source at the frequency under consideration. In Sect. 5.2.4 it is shown that this quantity characterises the degree of excitation of the H II region. In the case

$$r^2 \Phi^K(5\,\mathrm{GHz}) \geq 400\,[\mathrm{kpc}^2\,\mathrm{Jy}] \quad ,$$

hence about four times the value for the Orion Nebula, one speaks (after P.G. Mezger) of a *Giant H II Region*. Examples are W 38 (M 17) and W 37 (M 16).

Structure of the Radio Sources, Compact H II Regions: The distribution of the radio emission over the sources is usually more uniform than the structure of the visible nebula. It suggests the assumption that the latter is produced by overlying dust with strong spatial density variations, since small solid particles can attenuate only the short wave optical, not the radio frequency, radiation. Radio observations with high resolution of about $10''$ to $20''$ of course also show structures which correspond in reduced form to those of the optical picture.

In recent times numerous thermal continuum sources have been studied with radio interferometers with high angular resolution, of about $1'$ nearly down to $1''$ (California Institute of Technology and National Radio Astronomy Observatory, USA; Cambridge, England; Westerbork, Netherlands). This has shown that nearly all sources contain one or more small but intensive components, which stand out from a much weaker, extended emission distribution. One speaks of compact sources or of *compact H II regions* (Ryle and Downes, 1967; Mezger, 1967). In many cases the location of such a source is not identifiable with any visible nebular emission. The linear dimensions of compact H II regions are about 0.1 pc and extend for large objects up to about 1 pc. Examples are given in Table 4.11.

In the case of the Orion Nebula the whole central region around the Trapezium stars consists of a $2' \times 4'$ extent of this character (Fig. 1.8). Another example is the Omega Nebula M 17 = NGC 6618 with two large compact sources

---

[5] For a qualitative explanation of the concept see Chap. 1; a quantitative treatment follows in Sect. 5.2.1.

Table 4.11. Examples of compact H II regions which are contained in larger objects. The quoted positions and angular sizes of the individual sources were obtained by interferometric observations in the cm wave region

Extended object	Compact components	α_{1950}	δ_{1950}	Angular diameter [arc s]	Distance [kpc]	Linear diameter [pc]
W3 (with opt. nebula IC 1795)	A1–A5	$2^h 21^m 56\overset{s}{.}6$	$+61°52'43''$	28 … 35 ⎫	⎫	0.4 … 0.5
	C	$2^h 21^m 43\overset{s}{.}6$	$+61°52'46''$	5 ⎬ 3.1		0.071
	W3(OH), comp. A	$2^h 23^m 16\overset{s}{.}5$	$+61°38'57''$	1.5 ⎭		0.024
M42	Ori A, G 209.0–19.4	$5^h 32^m 50^s$	$- 5°25'15''$	240	0.5	0.6
M8	A1 (Ring)	$18^h 00^m 37\overset{s}{.}0$	$-24°22'50''$	15 ⎫ 1.4		0.10
	A4	$18^h 00^m 36\overset{s}{.}3$	$-24°22'52''$	46 ⎭		0.32
M17	S	$18^h 17^m 34^s$	$-16°13'24''$	220 ⎫		2.3
	N	$18^h 17^m 39^s$	$-16°10'30''$	280 ⎬ 2.1		2.9
	E	$18^h 17^m 51^s$	$-16°11'30''$	190 ⎭		2.0
W51	G 49.5–0.4d	$19^h 21^m 22\overset{s}{.}3$	$+14°25'15''$	11 ⎫ 7.3		0.4
	e	$19^h 21^m 24\overset{s}{.}4$	$+14°24'43''$	19 ⎭		0.7
W75	DR 21 A	$20^h 37^m 13\overset{s}{.}7$	$+42°08'55''$	5.7 ⎫		0.083
	B	$20^h 37^m 14\overset{s}{.}0$	$+42°09'03''$	4.0 ⎬ 3.0		0.058
	C	$20^h 37^m 14\overset{s}{.}1$	$+42°08'54''$	7.1		0.101
	D	$20^h 37^m 14\overset{s}{.}2$	$+42°09'15''$	4.2 ⎭		0.062
NGC 7538 = S 158	A1	$23^h 11^m 30\overset{s}{.}3$	$+61°12'56''$	110 ⎫		1.4
	A2	$23^h 11^m 20\overset{s}{.}8$	$+61°13'45''$	10 ⎬ 2.5		0.12
	B	$23^h 11^m 36\overset{s}{.}7$	$+61°12'00''$	12		0.15
	C	$23^h 11^m 36\overset{s}{.}7$	$+61°11'50''$	1.0 ⎭		0.013

(Fig. 4.38). The compact sources are often of still smaller angular extent and present in greater numbers. Thus the radio source W 3, discovered by older measurements at the position of the diffuse nebula IC 1795, can be resolved into 9 compact individual sources, of which 8 are located inside the main source G 133.7 + 1.2 (Fig. 4.39). With the highest resolutions attainable, inside many compact components still finer structures are detectable, for example shell or ring fragments in one of the individual sources of W 3.

At about 20′ distance from the main source of the W 3 complex, an isolated very small thermal radio continuum source was discovered, whose angular extent amounted to only about 1″, corresponding to around 2000 AU in linear measure. At the same position a strong OH molecular line "point source" had been found (OH maser, further in Sect. 4.2.8). After the discovery of more extremely small continuum sources, near diffuse nebulae or quite isolated, the name *ultracompact H II regions* was coined for them. Most of these sources coincide with molecular line point sources, which we discuss in Sect. 4.2.8.

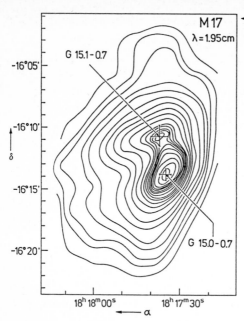

◄Fig. 4.38. Radio continuum chart of the Omega Nebula M 17 for λ = 1.95 cm. Two radio sources can be detected: M 17 N = G 15.1-0.7 and M 17 S = G 15.0-0.7. (After Schraml and Mezger, 1969)

Fig. 4.39. Contours of the individual components of the radio continuum source W 3 at 1.425 GHz on the corresponding section of the red exposure of the Palomar Observatory Sky Survey. The radio observations were obtained with the Netherlands Synthesis Radio Telescope in Westerbork. The angular resolution amounted to about 25″. (After Sullivan and Downes, 1973)

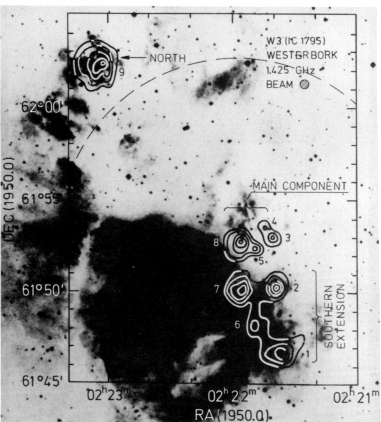

The phenomena described are today regarded as characteristic of star formation regions. For further detail see Sect. 4.2.9.

Sources in the Galactic Central Region: Because of the location of the sun amid the galactic dust layer, diffuse nebulae in the region of the galactic centre cannot be observed. The radio frequency radiation of the thermal continuum sources belonging to them, on the other hand, can propagate extensively unhindered even near the galactic plane. The first observations of galactic radio waves already showed that strong radio sources are located in the central region of our star system. The strongest of these sources, called Sagittarius A or Sgr A for short, has been selected as the null point of the galactic longitude scale − yet even the appellation G 0.0 + 0.0. Sgr A consists of four components: a thermal source "Sgr A West", a non-thermal source "Sgr A East" (supposedly a supernova remnant), a "halo" of about 6′ diameter and an extremely compact nucleus in Sgr A West. Very long (intercontinental) base-line interferometry (VLBI) at cm wavelengths for this "point source" leads one to expect a diameter $\lesssim 0''001$, corresponding to $\lesssim 10$ AU: it is probably identical with the true *nucleus of our galaxy*. The equatorial coordinates of this ultra-compact source were determined as

$$\alpha = 17^{\mathrm{h}}42^{\mathrm{m}}29\overset{\mathrm{s}}{.}29 \pm 0\overset{\mathrm{s}}{.}01 \ , \quad \delta = -28°59'18''2 \pm 0''1(1950.0)$$

and correspond to the galactic coordinates

$$l = -3'.34 \ , \quad b = -2'.75 \ \ .$$

Recent high-resolution observations ($\sim 1''$) show a spiral structure around the galactic nuclear source. Figure 4.40a gives an example.

In the vicinity of Sgr A one finds further sources, among which are several giant H II regions, embedded in an "extended H II region of lower density" (~ 10 electrons or ions per cm^3) of about 300 pc diameter and 120 pc thickness. A general view is given in the radio chart presented in Fig. 4.40b. The most prominent of these thermal sources is Sgr B, whose main component Sgr B 2 or G 0.670 −0.036 has the "excitation parameter" $r^2 \Phi^{\mathrm{K}}(5\,\mathrm{GHz}) \approx 1800\,\mathrm{kpc}^2\,\mathrm{Jy}$. Interferometer observations, moreover, show several compact sources inside Sgr B 2.

4.2.3 Radio Recombination Lines of H II Regions

Fundamentals: The recombinations of H^+, He^+ and other ions in hot interstellar clouds can produce not only emission lines of the optical region, such as perhaps the Balmer lines through transitions from states with main quantum numbers $n > 2$ into the state with $n = 2$, but also radio lines in those cases when it is a matter of transitions between the very densely clustered *high* quantum states. With hydrogen, for example, transitions from $n + 1$ to n, when n is greater than or equal to 60, lead to emission lines in the radio region. For

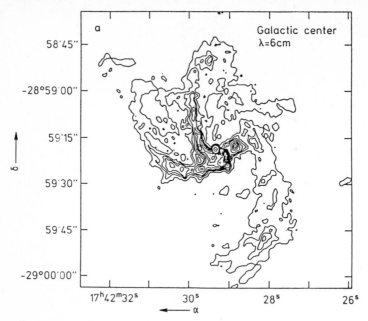

Fig. 4.40a. Contour map of the inner central region at $\lambda = 6$ cm measured with the Very Large Array of the National Radio Astronomy Observatory. Angular resolution $\sim 1''$. (By permission of Lo and Claussen, 1983)

such lines the name "radio recombination lines" has been introduced. In analogy with the differentiation between Balmer lines by the Greek letters $\alpha, \beta, \ldots$ one uses $n\alpha$ to denote the line from $n + 1$ to n, $n\beta$ the line from $n + 2$ to n, and also speaks in this connection of α-type, β-type, etc. The Hα line of the Balmer series would according to this be denoted by 2α, or more precisely with a statement of the element, by H2α. The transition from $n = 110$ to $n = 109$, corresponding to an actually observed radio recombination line, carries the indication H109α.

The frequency of a radio recombination line follows from the Rydberg formula, with $\Delta n \ll n$,

$$\nu = RcZ_{\text{eff}}^2 \left[\frac{1}{n^2} - \frac{1}{(n + \Delta n)^2} \right] \approx 2\,RcZ_{\text{eff}}^2 \frac{\Delta n}{n^3} \quad .$$

Thus the frequency of the lines decreases rapidly with increasing main quantum number n. The recombined electron in a high quantum orbit "sees" in practice only a charge $+e$, so that $Z_{\text{eff}} \approx 1$. The nuclear mass M is very much greater than the electron mass m_{e}, so the Rydberg constant $R = R_\infty[1 - (m_{\text{e}}/M)]$ varies only very slightly. Consequently the lines for different elements, but with the same n and of the same type, e.g. $n\alpha$-lines, are relatively close together. For example, the H109α and the He109α lines have the frequencies 5009 MHz and 5011 MHz. A detailed table of the frequencies of the radio recombination lines of H and He has been published by Lilley and Palmer (1968).

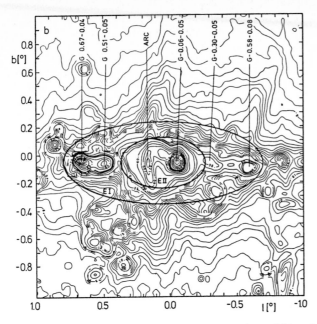

Fig. 4.40b. Radio continuum chart of the region within 1° distance of the direction towards the galactic centre for $\lambda = 6\,\mathrm{cm}$. The underlying observations were carried out with the 100 m telescope of the Max Planck Institute for Radioastronomy (Bonn); the angular resolution amounted to about $2'.6$. The curved emission region marked "arc" consists of several individual sources. The outermost of the two ellipses denotes schematically an extended ionised region (about $300\,\mathrm{pc} \times 120\,\mathrm{pc}$) of low density hydrogen ($N_e \approx 10\,\mathrm{cm}^{-3}$). Inside the smaller ellipse the electron density is on average about a factor 3 higher. (After Mezger and Pauls, 1979; by courtesy of the International Astromical Union)

Observations: The first evidence of radio recombination lines of interstellar origin was obtained in 1963. The main sources, as was to be expected, turned out to be the H II regions appearing as diffuse nebulae and (or even only) as thermal radio continuum sources. Thus, for example, the hydrogen lines H 56α (at $\nu = 35.5\,\mathrm{GHz}$), H 94α, H 109α, H 158α, H 137β, H 225γ and the helium lines He 85α, He 109α, He 134α were observed in the Orion Nebula amongst others. Like the radio continuum, this radio line radiation was not influenced by dust extinction inside the H II region or in its immediate vicinity. "Weak" radio recombination line emission was observed from extensive regions outside these relatively dense H II regions. The thin interstellar gas is likewise ionised here by hot stars (Example: Central region, see Sect. 4.2.2, last sub-section).

In complete analogy with the optical nebular lines, the profiles and Doppler displacements of the radio recombination lines provide evidence on the internal motions and on the radial velocity of an H II region as a whole. It was feared at first that the high quantum states would undergo strong displacements through collisions with charged particles and that the expected transitions either would not occur at all or would give lines with extreme Stark Effect broadening and observable only with difficulty. It turned out, however, that the line profiles are actually broadened predominantly only by thermal and turbulent macroscopic motions of the emitting atoms. With the Orion Nebula, for example, after allowance for the thermal contribution (assuming a temperature of 10^4 K), a velocity scatter between 10 and $20\,km\,s^{-1}$ was found for the inner motions, similar to the finding from the optical nebular lines (see Sect. 4.2.1).

Recombination lines with essentially smaller widths represent a characteristic feature, as found for example in the region of the W 3 complex and in the source Orion B, among them the carbon line C 157. This emission must originate in cooler regions, *"C II regions"*.

For the derivation of the distribution and motion of the H II regions in the Milky Way system it is essential to determine radial velocities of these objects as a whole. The use of radio recombination lines offers the advantage, compared with optical lines, that the interstellar extinction in the galactic dust layer plays no part. Very distant H II regions can therefore be studied. We return to this question in Sect. 5.4.4.

4.2.4 Infrared Emission from H II Regions

In the spectral region between $1\,\mu$m and about $25\,\mu$m the terrestrial atmosphere is transparent only in a number of narrow "windows" (Regions *I, J, K, ..., Q*). Likewise, if one ignores three regions of reduced atmospheric absorption at $34\,\mu$m, $350\,\mu$m, and $450\,\mu$m, capable only of restricted use (from high mountains), observations in the submillimetre region first become possible from ground based stations in the far infrared beyond $\lambda \approx 750\,\mu$m. Since the absorption occurs predominantly in the troposphere, however, observations can be successfully carried out at wavelengths between $40\,\mu$m and $750\,\mu$m from high-flying aircraft and from balloons in the stratosphere; rockets were also employed. As receivers in the near infrared, photoresistors and photodiodes were used, in the far infrared germanium bolometers, which had to be cooled with liquid helium ($-269°$C).

Infrared Emission and Radio Continuum: Infrared radiation was detected at the end of the sixties in the Orion Nebula, in the Omega Nebula (M 17) and in the Lagoon Nebula (M 8). Since then it has been shown that continuous surface IR emission is a general characteristic of H II regions. The continuous spectra of most H II regions, at wavelengths from about $5\,\mu$m up to more than $100\,\mu$m, show a strong "infrared excess" in comparison with the extrapolation of the almost constant profile of the radio radiation flux shown in Fig. 4.37. At the positions of maximal radio continuum emission in the compact H II regions one finds also, as a rule, maxima of IR emission (see Table 4.12). One finds also, however, strong IR sources outside the regions of high radio continuum emission, especially IR point sources at which no radio emission is to be seen.

As an example of the good correlation of infrared and radio continuum radiations in compact H II regions, Fig. 4.41 shows observations of M 17 at $21\,\mu$m

Table 4.12. Examples of compact H II regions which are also observed as infrared sources. The measured radiation flux at the earth at $\lambda = 20\,\mu$m and $\lambda = 6$ cm ($\nu = 5$ GHz) are denoted by $\Phi(20\,\mu$m) and $\Phi(6$ cm), respectively. L_{IR} is the total infrared luminosity of the source in units of solar luminosity $L_\odot = 4 \times 10^{26}$ W

Extended object	Compact components		$\Phi(20\,\mu$m)	$\dfrac{\Phi(20\,\mu m)}{\Phi(6\,cm)}$	L_{IR}
	Radio source	IR source	[Jy]		$[L_\odot]$
W3	A1–A5	IRS 1	2×10^3	70	3×10^5
	C	IRS 4	3×10^3	500	–
	W3(OH)	IRS 8	2×10^2	300	2×10^5
M42	G209.0–19.4	IRe 1	1.4×10^5	500	4×10^5
M8	A1–A4		1.3×10^3	30	5×10^4
M17	S	IRe 1	5×10^4	220 $\Big\}$	5×10^6
	N	IRe 2a	3×10^4	140	
W51	G49.5–0.4d	IRS 2	1.5×10^3	140 $\Big\}$	5×10^6
	e	IRS 1	–	–	
W75	DR 21 D	DR 21N	1×10^2	50	6×10^4
NGC 7538 = S 158	B	IRS 2	7×10^2	500 $\Big\}$	3×10^4
	C	IRS 1	2×10^2	1700	

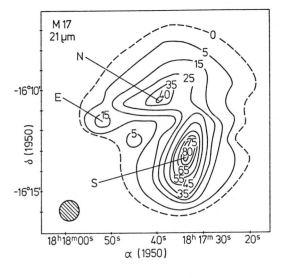

Fig. 4.41.
Contour chart of the infrared emission from the Omega Nebula M 17 at $\lambda = 21\,\mu$m. Unit of plotted values of intensity: 10^{-17} W m^{-2} Hz^{-1} sterad^{-1}. (After Lemke and Low, 1972)

for comparison with the radio map of Fig. 4.38. Similar results are available for many other H II regions. In the W 3 complex an IR source is to be found at three of the four radio continuum sources, but other IR sources are also observed. In the Orion Nebula the maximum of the radio emission falls in the region of the Trapezium stars, and one finds here also a clearly marked maximum of IR emission, often called, after its discoverers Ney and Allen, the "NA object" or also the "Trapezium Nebula".

About $1'$ northwest of the Trapezium, however, there is another strong IR source in a region that is not prominent in either the optical or the radio continuum. The position of this source is indicated in Fig. 1.8. At higher reso-

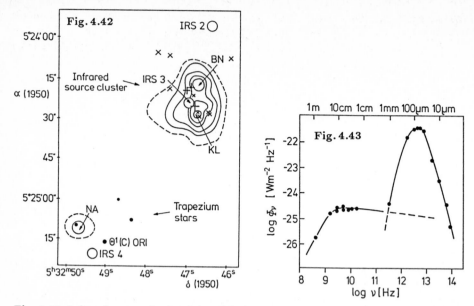

Fig. 4.42. Infrared sources in the Orion Nebula. The lines of equal intensity refer to measurements at $\lambda = 20\,\mu$m with an angular resolution of $5''$. The interval between successive full lines amounts to $3 \times 10^{-15}\,\mathrm{W\,m^{-2}\,Hz^{-1}\,sterad^{-1}}$. IRS 2, IRS 3 and IRS 4 are unresolved infrared sources which were observed at $\lambda < 20\,\mu$m. The large and small crosses give the positions of OH masers and H_2O masers, respectively, which we discuss in Sect. 4.2.8. Further explanation in text. (By permission of Wynn-Williams and Becklin, 1974)

Fig. 4.43. Energy distribution in the spectrum of the infrared source IRS 1 in the compact H II region W 3(A). The spectral profile consists of two components: the thermal Bremsstrahlung of the ionised gas and the thermal emission of the solid particles mixed in with the gas. (After Mezger and Wink, 1975)

lution it appears that this consists of two sources (Fig. 4.42). The northern one, brighter in the 3 to $10\,\mu$m region, is known as the "BN Object" after Becklin and Neugebauer, and has a diameter smaller than $2''$. The southerly one, called the "KL Object" after Kleinmann and Low, extends for about $30''$ – hence at first usually called the "Infrared Nebula" – but has a central kernel with only about $2''$ diameter. Near these objects were later discovered another two unresolved point sources (diameter less than $2''$), so that one now often speaks of the Kleinmann-Low IR Source Cluster and also includes in this the BN Point Source.

Spectra and Interpretation of IR Emission: Measurements of the IR radiation flux from H II regions at various wavelengths between $1\,\mu$m and 1 mm in broad bands lead to spectral profiles, of which Fig.4.43 gives a typical example. It can be approximately represented by the superposition of the extrapolated, almost constant, radio continuum (Fig. 4.37) on a Planck curve for a relatively low temperature. In the case of the Trapezium Nebula, for example, one finds a "colour temperature" $T \approx 100\,\mathrm{K}$, for the KL Nebula $T \approx 130\,\mathrm{K}$ (exterior) up to 200 K (in the kernel). These results suggest a simple explanation, which we can accordingly give now, namely that we are dealing here with the warm radiation from the small solid particles mixed with the gas, which are heated by the stars in the H II region.

If we assume for the sake of simplicity that the dust particles are small perfectly black spheres of radius a, then by equating the energy $4\pi a^2 \sigma T_g^4$ emitted by a grain with temperature T_g, with the energy $\pi a^2 \Phi(r) = \pi a^2 (R/r)^2 \sigma T_{\text{eff}}^4$, absorbed from the radiation field of a star (effective temperature T_{eff}, radius R) at a distance r, we obtain

$$T_g^4 = \frac{1}{4}\left(\frac{R}{r}\right)^2 T_{\text{eff}}^4 \quad.$$

At a distance $r = 0.1\,\text{pc}$ from an O5-star with $T_{\text{eff}} = 37\,500\,\text{K}$ and $R = 18\,R_\odot$ one obtains, for example, $T_g \approx 50\,\text{K}$. The absorbed radiation comes almost entirely from the UV region, the emitted radiation on the other hand lies in the far infrared. The actual particles, with $a \approx \lambda_V \approx 0.5\,\mu\text{m}$, as a rule have an absorbing and emitting capability which decreases at longer wavelengths ($\lambda > a$, comparison with an antenna!). Therefore they can radiate less effectively as black bodies and will on that account attain considerably higher temperatures still. A more precise interpretation of the observations must naturally also take into account the gradient of the temperature of the dust with r.

As energy sources for warming the dust one must above all consider the stars which also excite the luminosity of the nebula. One or more O-stars emit energy amounts of about $10^5 L_\odot$ to a few times $10^6 L_\odot$, per unit of time. In the compact regions with high dust extinction this energy will be either totally or to a large extent absorbed by the dust, and must then be radiated in the IR region. By integration of the observed radiation flux over the whole IR region, and taking into account the distances of the objects, one in fact finds IR luminosity of the compact H II regions of this order of magnitude; for example, for the "Trapezium Nebula" $L(\text{IR}) \approx 4 \times 10^5 L_\odot$ and for M 17 altogether around $5 \times 10^6 L_\odot$ (see Table 4.12).

In many H II regions the exciting stars are obscured behind dense masses of dust, e.g. in the Omega Nebula M 17 and in W 3. As in the case of the generally distributed interstellar dust, however, here also the extinction decreases with increasing wavelength. Thus, with an image converter camera for the near infrared, a star could be detected in one of the compact H II regions of the W 3 complex (named IRS 1 or W 3A) whose observed spectral energy distribution can readily be accounted for by the assumption that it is an O5-star with its radiation undergoing an extinction of 14 magnitudes in the visual ($\lambda \approx 5500\,\text{Å}$) (cf. Fig. 4.44). At the same time one can explain to the first approximation the observed behaviour, shown in Fig. 4.43, in the spectrum of these compact H II regions in the near infrared. With the assumption (established in Sect. 5.1.3) that the true flux in the region between $5\,\mu\text{m}$ and $1\,\mu\text{m}$ is proportional to $\nu^{-0.1}$ and undergoes a dust extinction with the normal wavelength dependency (see Sect. 4.1.1), one obtains the observed values if one sets $A_V \approx 14^{\text{m}}$.

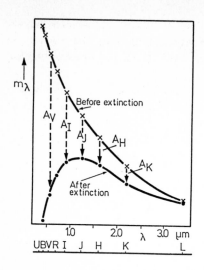

Fig. 4.44. Change in the spectral energy distribution of a star (*upper curve*) because of the wavelength-dependent interstellar extinction

Other IR sources, like the BN object or a similar IR point source discovered in M 17 (the Kleinmann-Wright Source) are very probably proto-stars, which are still embedded in a dense, dust laden shell. One therefore calls stars in this embryonic state *"cocoon stars"*. The colour temperatures of such objects reach a few 100 K. The dust shell often absorbs the optical radiation of the starlike kernel almost completely; it also produces amongst other things the characteristic 10 μm absorption (see Sect. 4.1.1, especially Fig. 4.4) in the IR spectra of these sources. The IR luminosities of the objects of this sort known to-day lie between about $3 \times 10^2\ L_\odot$ and $3 \times 10^4\ L_\odot$. In the case of the KL Nebula the observations mentioned in Sect. 4.2.7 furnish a strong argument that here it is a matter of the centre of a cooler cloud lying behind the Orion Nebula but outside the H II region (see Sect. 4.2.9).

Infrared Line Emission: Observations of compact H II regions with higher spectral resolution have made possible the detection of the infrared forbidden emission lines of various ions mentioned in Sect. 4.2.1. The lines [Ar III] $\lambda 8.99\ \mu$m, [S IV] $\lambda 10.52\ \mu$m, [Ne II] $\lambda 12.8\ \mu$m, [S III] $\lambda 18.71\ \mu$m, [O III] $\lambda 51.8\ \mu$m and $\lambda 88.35\ \mu$m may be mentioned as examples. In several H II regions, moreover, some of the recombination lines of the Brackett and the Pfund Series of hydrogen lying in the near infrared have also been detected. The discovery of emission lines of molecular hydrogen at 2.0 and 2.4 μm has already been alluded to in Sect. 4.2.1.

The Galactic Central Region: IR observations here show several specially strong compact sources, which are most intense at the locations of the radio sources Sgr A and Sgr B2. In the centre of the IR source coinciding with Sgr A West there is a small region of only about 20″ extent which produces most of the emission in the near IR in the central galactic region, and which has already been mentioned in Sect. 3.1.3 (see esp. Fig. 3.5). Noteworthy here is the observation of infrared line emission of Ne^+ at $\lambda\, 12.8\,\mu m$, which is concentrated strongly at the region of the nucleus and indicates several ionised gas clouds with velocities up to $\pm 260\,km\,s^{-1}$ which are located within about 1 pc from the ultra-compact radio source mentioned in Sect. 4.2.2.

Strong emission in the medium and far infrared is observed in the region inside about $R = 150\,pc$, and hence in a volume that is predominantly filled with ionised gas (Sect. 4.2.2). It must be here a question of thermal radiation from dust particles which are warmed primarily by the ultraviolet radiation of hot stars. Moreover, IR emission in the submillimetre region occurs in a wider region which extends over at least 6° in galactic longitude and 2° in galactic latitude around the direction of the centre. Absolute measurements lead to a total luminosity of the order of $10^{10}\,L_\odot$.

This suggests that this energy also is radiated by dust particles which have been warmed by stars of the central bulge.

4.2.5 Dark Clouds: Optical Data

The visible phenomena of cold interstellar clouds are rather inconspicuous compared with the splendid optical manifestations of a developed H II region. One observes a local increase of the interstellar extinction of the starlight, as already described in Sect. 4.1.1 and illustrated in Figs. 1.5 and 4.2, or even a local weakening of the surface brightness of a diffuse nebula, as with the well known Horsehead Nebula in Orion (Fig. 4.45). Dark clouds of small apparent diameter with distinct, often circular, boundaries are called *Globules*. At the position of a dark cloud there is often continuous scattered light (see Sect. 4.1.3), and often a clearly visible reflection nebula, to be seen.

A *catalogue* of dark clouds was first made available in 1919 by Barnard on the basis of his photographic survey of the Milky Way at the Lick Observatory. The number of objects included at first was 182, and a later edition included 349 dark clouds. Recent catalogues, some with maps, by Khavtassi (1958; 797 objects), Schoenberg (1964; 1456 objects), Lynds (1962; 1802 objects, northern hemisphere) and Feitzinger and Stüwe (1984; 820 objects, southern hemisphere) are more precisely cited in the list of references. The designation of a dark cloud often consists of the initial of one of these authors, followed by the number in the catalogue. The dark cloud shown in Fig. 4.2 south of the star ϱ Ophiuchi, for example, carries the designation B 42 after Barnard's Catalogue, whilst Lynds' Catalogue shows that as several individual structures L 1686, L 1688, L 1690, L 1692 and L 1696. The boundaries of dark clouds often vary from catalogue to catalogue.

The *distribution* on the sphere of the dark clouds included in the catalogues shows a strong concentration at the galactic equator. Only relatively few clouds have latitudes $|b| > 10°$. This agrees with the general impression gained from wide-angle photographs of the Milky Way (Fig. 1.6).

Extinction Δm and distance of a dark cloud can in several cases be derived by a simple and intuitive method, developed by M. Wolf (1923). In it one

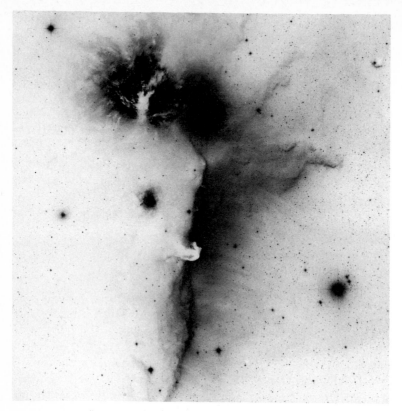

Fig. 4.45. The bright nebula IC 434 south of the belt star Zeta of Orion, on which the dark "Horsehead Nebula" is projected (on the negative). Above, left of centre, the nebula NGC 2024. © 1960 National Geographic Society-Palomar Observatory Sky Survey. Reproduced by permission of the California Institute of Technology

compares the profile of the cumulative number of stars $N(m)$ as a function of the apparent magnitude m for the dark cloud with the corresponding profile for a directly neighbouring, undisturbed comparison field. It is a basic assumption here that the star density profile $D(r)$ in the directions of the dark cloud and of the comparison field are fully identical. For the sake of simplicity we shall first of all assume the same absolute magnitude M_0 for all stars. The apparent magnitude m then represents (in the extinction-free case) a straightforward measure of the distance. We approximate the dark cloud by a thin absorbing screen at a distance r_1, at which a star appears to have magnitude $m_1 = M_0 - 5 + 5\log r_1$. The profiles of $N(m)$ or more appropriately $\log N(m)$ for the dark cloud and the comparison field must then coincide up to $m = m_1$ (Fig. 4.46). Stars fainter than m_1 however find themselves behind the screen and are attenuated by a constant amount Δm : the curve for the dark cloud is shifted to the right by the amount Δm. Because of the actual presence of strong scatter of the absolute magnitudes of the stars, the plot of the observed star numbers in such a *Wolf Diagram* does not show a sharply marked step,

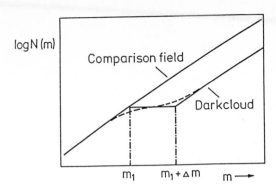

Fig. 4.46. Schematic Wolf Diagram. Explanation in text

but a smeared smooth profile like that shown in Fig. 4.46 by a broken line. A confident statement of distance is therefore rarely possible. On the other hand the extinction Δm of the cloud can be determined relatively well if the two curves well above m_1 can be determined, where they run parallel with one another, separated by a distance Δm in the direction of the abscissa.

The case of uniform luminosity of the stars can be approximated more closely by selecting for the investigation only stars of a certain MK spectral group. The amount of work, however, then becomes very much greater.

The most frequently used method to-day for deriving the extinction in the visible and also the distance of a dark cloud aims at the determination of the reddening for the greatest possible number of stars inside and in the immediate vicinity of the dark cloud. The method is simplest if the MK spectral types of the stars are known or can be determined. Thereupon the intrinsic colours and absolute magnitudes are known, and after measurement of the apparent magnitudes in at least two colours, e.g. B and V, the visual extinction values $\Delta m_V = A_V$ and the distances r of the stars are obtained in the well known manner (see Sect. 2.2.2). The extinction can now be plotted as a function of the distance. Typical results of a far-reaching investigation by Neckel and Klare (1980), whose data were based on around 12 000 stars distributed along the

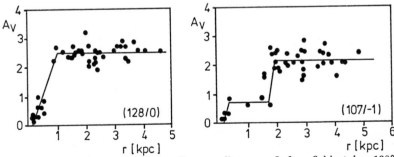

Fig. 4.47. Examples of extinction-distance diagrams. *Left*: a field at $l = 128°$, $b = 0°$, in which the line of sight encounters only one cloud, with $r<1$ kpc and $\Delta m_V \approx 2.5$ mag, up to distances of at least 4 kpc. *Right*: a field at $l = 107°$, $b = -1°$, in which direction there is a nearby cloud at $r = 200$ pc with $\Delta m_V \approx 0.8$ mag, and a more distant cloud at $r \approx 1.8$ kpc with $\Delta m_V \approx 2.2-0.8 = 1.4$ mag. The scatter in the A_V-values arises in considerable measure from a variation in the extinction from place to place, and hence from star to star, in the region of the dark cloud. (After Neckel and Klare, 1980)

Table 4.13. Examples of dark clouds and globules. Galactic coordinates l and b, distance r, extinction in V-region Δm_V, approximate value D for the linear diameter of the object (from angular size and distance). In most cases Δm_V and r are based on star counts, partly also on reddening; the visual extinctions of the small globules represent a lower limit

Constellation	l	b	r [pc]	Δm_V [mag]	D [pc]
Aquila	42°	0°	100...300	3	~10
Cygnus: E,S and W from North America Nebula	80°	0°	600	3	–
Large globule B 361 = L 970 in Cygnus	89.3°	– 0.6°	600	2	1.8
Complex in Taurus	170°	−15°	50, 175, 600	3, 0.8, 1	~10
Individual cloud L 1534	174.1°	−14.0°	50	6.5	1
Large globule B 227 = L 1570 in Orion	190.6°	– 0.8°	400	3	0.6
Small globules in the Rosette Nebula (Fig. 4.32)	206.4°	– 1.9°	1000	~8	0.01...0.04
"Coalsack"	302°	0°	170	1...3	~10
Cloud south of ϱ Oph	353°	+17°	200	2...8	3

whole Milky Way band, are shown in Fig. 4.47. Here Δm_V and r_1 can be directly inferred. Table 4.13 gives a selection of results for individual objects which were obtained by this method and/or by star-counts.

The systematic increase of the interstellar extinction with increasing distance has the effect that at larger distances an increasing number of stars with higher extinction values remain unconsidered, because as a result of these high extinctions they already fall below the threshold of the observational method. This selective effect means, for example, that for a threshold of $m_V = 10^m$, beyond which usually no spectral types are available, unbiased results can be expected only up to $r \approx 1500\,\mathrm{pc}$. One can attain fainter threshholds by photometric spectral classification on the basis of a multicolour photometry. This makes possible the measurement of the magnitudes U, B and V as well as the Hβ index − a measure for the strength of the Hβ line − the determination of not only the intrinsic colour, and hence the reddening, but also the absolute magnitude of a star.

4.2.6 Dark Clouds of High Extinction and IR Sources

Observations in the visible region with a magnitude threshhold at $m_V \approx 10$, like those which form the basis for the diagrams displayed in Fig. 4.47, lead almost exclusively to extinction values $A_V < 5$. This is not surprising, since even an intrinsically bright star with $M_V = -5$ at a distance $r = 1\,\mathrm{kpc}$ with $A_V = 5$ will only have an apparent magnitude $m_V = 10$. There may accordingly be clouds with substantially higher extinction values than $A_V = 5$, which this method would not cover. That this is indeed the case, observations in the infrared have shown. Here the dust extinction is considerably less − at $\lambda =$

2.2 μm, for example, it amounts to only about 10% of the visual value – so that dark clouds are still penetrable and hence evidence can be obtained on their extinction values. An example of this has already been discussed in Sect. 4.2.4 in conjunction with Fig. 4.44. Photographs with the already mentioned image-converter camera at $\lambda \approx 1 \mu$m showed in another example, the dense dark cloud in the region of the Omega Nebula M 17, numerous strongly reddened stars with extinction values in the visual of about 10 to 12 magnitudes. They probably form a star cluster, whose earliest types are the exciting stars of the large H II region.

Also in the typical dark cloud complexes which are not associated with large H II regions – though possibly with reflection nebulae – infrared observations have led to the discovery of whole star clusters whose visible radiation undergoes an extinction $A_V \gtrsim 10$. As a result of the high extinction these extremely reddened stars are often quite undetectable in the visible. Likewise studies conducted on the dark clouds around the star ϱ Ophiuchi (Fig. 4.2) in the near infrared led to the discovery of several dozen point sources, at whose locations the red exposures of the Palomar Sky Survey (threshhold $m_R = 20$) in many cases showed no visible object. From the observed colours and the assumption that the wavelength dependence of the dust extinction is similar to that in regions of lower interstellar extinction, the individual stars gave values for A_V beyond 20^m. Taking as basis the distance of the Ophiuchus dark cloud $r_1 = 170$ pc, found by the methods described above, one finds for a considerable portion of the objects the absolute magnitudes of A- and B-stars. Early B-stars should, if they are inside the cloud, produce small H II regions. Radio observations have actually led to the discovery of compact H II regions.

The "anomalous" recombination lines C 167α (1400 MHz) and C 110α (4875 MHz) are also observed. They presumably come from regions of ionised carbon (C II regions) in the vicinity of these B-stars. Surface IR emission from the dust of the dark clouds is again explained by their warming by the embedded stars.

Observations of this kind on dense dark clouds are of special interest, since in all likelihood they give insight into the star forming region in an early stage of development. The stellar objects in such clouds must in part be very young, protostars also being found there.

4.2.7 Molecular Clouds

As already mentioned in Chap. 1, the interstellar clouds have in recent times become frequently studied objects, because of the occurrence of molecules in them. Through the molecular lines observed predominantly in the centimetre and millimetre regions, radioastronomy has here developed a new and fruitful field.

Molecules Observed, Transitions: The fundamental features of the description of the energy states and spectra of molecules are presented in Appendix

B. The lines in the microwave region are caused mostly by *transitions between the rotational states* of the molecules. For the majority of the diatomic and the polyatomic linear molecules one can in the first approximation start from the model of a rigid rotator and one obtains as a result equidistant spectral lines. The more precise prediction of line frequencies, however, must take into account, amongst other things, the effect of the centripetal force on the mutual separation distances of the individual atoms, from which small systematic frequency shifts arise.

Lines can also occur in the microwave region through transitions between sub-levels, into which the individual rotational states split if special coupling relations exist. Important examples are the Λ-*doubling* (see Appendix B) and the *hyperfine structure splitting*. Figure 4.48 explains this for the case of the OH molecule. The lines at $\lambda = 18\,\text{cm}$ arise from the transitions between the two levels, into which the ground state $^2\Pi_{3/2}$, $J = 3/2$ is split through Λ-doubling. The hyperfine structure causes a further splitting of each of these individual levels, so that altogether four lines are possible. Hyperfine structure quantum numbers F and frequencies ν for these lines are contained in Table 4.14. Specially noteworthy is the fact that these transitions – in contrast to the hyperfine structure transition of the 21 cm line – are not forbidden. The transition probabilities are about four orders of magnitude higher than for the 21 cm line.

A further possibility for the production of molecular lines in the microwave region is the *Inversion Doubling* (see Appendix B). The best known example is the NH_3 molecule, for which the transitions between the inversion levels for $(J, K) = (1,1), (2,2), (3,3)\ldots$ are very strong. The frequencies of the lines produced lie around 24 GHz, corresponding to $\lambda \approx 1.3\,\text{cm}$.

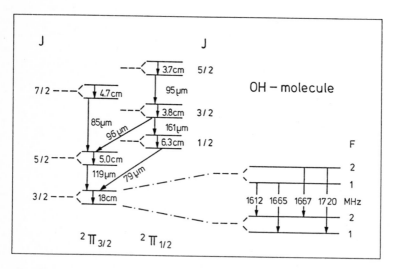

Fig. 4.48. Energy level diagram of the OH molecule. The lowest Λ-doublet terms are shown on the *left*; on the *right* is the hyperfine structure splitting of the ground state, which leads to the creation of four lines instead of a doublet

Table 4.14. Interstellar molecules detected in the microwave region represented by a selection of observed lines. The observed transitions of isotopic variants of these molecules are not included, with the exception of $C^{13}O^{16}$. Unless stated otherwise, the electronic ground state is $^1\Sigma$ and the vibration state $v = 0$; F denotes the total angular momentum quantum number including spin. The quoted frequency ν refers to a molecule at rest relative to the observer. (After Winnewisser, Churchwell and Walmsley (1979))

Molecule		Transition	ν[MHz]	Observed in source				
Symbol	Name			Sgr B2	Ori A	W 51	W 3	others
OH	Hydroxyl radical	$^2\Pi_{3/2}; J = \tfrac{3}{2}, F = 1-2$	1612.231	•	•	•	•	•
		$1-1$	1665.402	•	•	•	•	•
		$2-2$	1667.359	•	•	•	•	•
		$2-1$	1720.530	•	•	•	•	•
SiO	Silicon monoxide	$v = 0; J = 2-1$	86846.891	•				•
		$v = 1; J = 1-0$	43122.027	•				
		$2-1$	86243.350	•				
SiS	Silicon monosulphide	$J = 6-5$	108924.267	•				•
NO	Nitric oxide	$^2\Pi^{+}_{1/2}; J = \tfrac{3}{2}-\tfrac{1}{2}, F = \tfrac{5}{2}-\tfrac{3}{2}$	150176.54	•				
		$^2\Pi^{-}_{1/2}; J = \tfrac{3}{2}-\tfrac{1}{2}, F = \tfrac{5}{2}-\tfrac{3}{2}$	150546.50	•				
NS	Nitric monosulphide	$^2\Pi_{1/2}; J = \tfrac{5}{2}-\tfrac{3}{2}, F = \tfrac{7}{2}-\tfrac{5}{2}$	115153.835	•				
		$\tfrac{5}{2}-\tfrac{3}{2}$	115156.799	•				
SO	Sulphur monoxide	$^3\Sigma^{-}; J, N = 2,2-1,1$	86093.95	•	•			•
		$3,2-2,1$	99299.87	•	•	•		
		$2,3-1,2$	109252.10	•				
		$4,3-3,2$	138178.60	•	•			•
CO	Carbon monoxide	$J = 1-0$	115271.204	•	•	•	•	•
		$2-1$	230537.974	•	•			
		$3-2$	345795.900	•				
$C^{13}O^{16}$		$J = 1-0$	110201.370	•	•	•	•	•
		$2-1$	220398.714	•	•			
CN	Cyanogen radical	$^2\Sigma^{+}; N = 1-0$ $J = \tfrac{3}{2}-\tfrac{1}{2}, F = \tfrac{5}{2}-\tfrac{3}{2}$	113490.982	•		•		•
CH	Methyladyne radical	$^2\Pi_{1/2}, J = \tfrac{1}{2}, F = 0-1$	3263.794	•	•	•	•	•
CS	Carbon monosulphide	$J = 1-0$	48991.000	•	•	•	•	•
		$2-1$	97981.007	•	•	•	•	•
		$3-2$	146969.039	•		•	•	•

229

Table 4.14 (continued)

Molecule		Transition	ν[MHz]	Observed in source				
Symbol	Name			Sgr B2	Ori A	W 51	W 3	others
H_2O	Water	$6_{16}-5_{23}$	22235.080	·		·	·	·
		$3_{12}-2_{20}$	183310.091	·				
H_2S	Hydrogen sulphide	$1_{11}-1_{10}$	168762.762	·		·	·	·
SO_2	Sulphur dioxide	$8_{17}-8_{08}$	83688.074	·	·			
HNO	Nitrosyl hydride	$1_{01}-0_{00}$	81477.49	·				
NH_2^+	Diazenylium	$J=1-0, F_1=2-1$	93173.75	·	·	·	·	·
OCS	Carbonyl sulphide	$J=6-5$	72976.784	·				
C_2H	Ethynyl radical	$^2\Sigma, N=1-0, J=\frac{3}{2}-\frac{1}{2}$						
		$F=2-1$	87317.05	·	·	·	·	·
		$1-0$	87328.70	·	·	·	·	·
NH_3	Ammonia	$J,K=1,1; F=2-2$	23694.506	·	·	·	·	·
		$J,K=2,2; F=3-3$	23722.634	·	·	·	·	·
HCN	Hydrogen cyanide	$F=1-0, F=2-1$	88631.847	·	·	·	·	·
HNC	Iso-hydrogen cyanide	$J=1-0$	90663.592	·	·	·	·	
HCO	Formyl radical	$1_{01}-0_{00}; J=\frac{3}{2}-\frac{1}{2}, F=2-1$	86670.65				·	·
HCO^+		$J=1-0$	89188.545	·	·	·	·	·
H_2CO	Formaldehyde	$1_{10}-1_{11}$	4829.660	·	·	·		·
		$2_{11}-2_{12}$	14488.474	·	·	·	·	·
		$2_{12}-1_{11}$	140839.502	·	·	·	·	
C_3N	Cyanoethynyl radical	$^2\Sigma; J=9-8$	$\begin{cases}89045.7\\89064.4\end{cases}$					·
HNCO	Isocyanic acid	$1_{01}-0_{00}, F=2-1$	21981.471	·				
H_2CS	Thioformaldehyde	$3_{12}-2_{11}$	104616.977	·	·	·		
C_4H	Butadiynyl radical	$N=9-8$	$\begin{cases}85633.9\\85672.4\end{cases}$					·
H_2CNH	Methylenimine	$1_{10}-1_{11}; F=2-2$	5283.786	·				
NH_2CN	Cyanamide	$5_{14}-4_{13}$	100629.500	·				
HCOOH	Methanoic acid	$1_{10}-1_{11}$	1638.805	·				
		$2_{11}-2_{12}$	4916.312	·				
HCCCN	Cyanoacetylene	$J=1-0, F=2-1$	9098.332	·				·
		$J=5-4$	45490.307	·	·	·		
H_2C_2O	Ketene	$5_{14}-4_{13}$	101981.387	·	·			

Table 4.14 (continued)

Molecule		Transition	ν[MHz]	Observed in source				
Symbol	Name			Sgr B2	Ori A	W 51	W 3	others
CH$_3$OH	Methyl alcohol	$2_0 - 1_0$	$\begin{cases} 96\,741.42 \\ 96\,744.58 \\ 96\,755.51 \end{cases}$	. . .	. . .	. .	. .	. .
CH$_3$CN	Methyl cyanide	$J, K = 6,0 - 5,0$ $6,1 - 5,1$	110\,383.494 110\,381.362	. .				
NH$_2$CHO	Formamide	$1_{10} - 1_{11}; F = 2 - 2$	1\,539.851	.				
CH$_3$NH$_2$	Methylamine	$2_{02} - 1_{01}; F\ 3 - 2$	8\,777.38	.	.			
CH$_3$CHO	Acetaldehyde	$1_{10} - 1_{11}$	1\,065.075	.				
CH$_3$CCH	Methylacetylene	$J, K = 5,0 - 4,0$	85\,457.29	.	.			
CH$_2$CHCN	Vinylcyanide	$10_{0,10} - 9_{0,9}$	94\,276.638	.				
HC$_5$N	Cyandiacetylene	$J = 1 - 0$	2\,662.662	.				.
HC$_7$N	Cyantriacetylene	$J = 8 - 7$	9\,024.014					.
CH$_3$CH$_2$CN	Ethylcyanide	$10_{0,10} - 9_{0,9}$	88\,323.72			.		
HCOOCH$_3$	Methyl formate	$1_{10} - 1_{11}$	$\begin{cases} 1\,610.249 \\ 1\,610.906 \end{cases}$			. .		
CH$_3$OCH$_3$	Dimethylether	$2_{11} - 2_{02}$	$\begin{cases} 31\,105.26 \\ 31\,106.20 \\ 31\,107.12 \end{cases}$	. . .	. . .			
CH$_3$CH$_2$OH	Ethyl alcohol	$6_{06} - 5_{15}$ $4_{14} - 3_{03}$	85\,265.46 90\,117.51	. .	. .			
HC$_9$N	Cyanotetracetylene	$J = 18 - 17$ $25 - 24$	10\,458.634 14\,525.862					. .

The molecules detected in interstellar space through observations of lines in the centimetre and millimetre regions are listed in Table 4.14, where at least one of the observed transitions is given for each molecule. A few of these molecules, including CO, OH, CS, NH$_3$ and H$_2$CO, can be identified in many different directions, predominantly in the region of the Milky Way near the H II regions and in the regions of high visual extinction. The complicated organic molecules, however, were found only in a relatively few sources. The greatest profusion of molecular types is shown by the molecular clouds observed in the region of the great H II region Sgr B2; a few of the complex molecules were also found in Sgr A and in the region of the Orion Nebula.

Line Emission from Clouds: The systematic surveys of the emission of the widespread CO molecule, which is also observed in diffusion clouds of low extinction Δm_V, have already been reported in Sect. 4.1.6. In dense clouds with $\Delta m_V > 5^m$ other molecules such as OH, CS, NH_3 and H_2CO are also found in emission. The maximal numerical densities in such condensations are about 10^4 H_2 molecules per cm^3, as shown in Sect. 5.2.4. The following comments apply mainly to the relatively abundant CO.

In dense *dark clouds* the observed CO line profiles are frequently flattened in the middle, sometimes there are even central depressions (Fig. 4.49). This indicates that the emitting gas is already optically thick for the wavelength of the line, and therefore is largely opaque. This is confirmed by the lines of $C^{13}O^{16}$, which are also observed: these are each about a third as strong as the lines of the normal $C^{12}O^{16}$, whereas the isotope ratio C^{12}/C^{13} in interstellar gas (determined from the molecules $H_2C^{12}O$ and $H_2C^{13}O$ emitting in an optically thin layer) is about 40 (terrestrial value: 89). The $C^{12}O^{16}$ lines must therefore be about 40 times as strong as those of $C^{13}O^{16}$ in the optically thin case! In the middle of the $C^{12}O^{16}$ line the intensity should reach that of the radiation of a black body with the temperature of the gas of the thick cloud. The brightness temperature T_b measured there (minus the 2.7 K for the cosmic background radiation) is then approximately equal to the kinetic temperature of the gas T. By this method T is found to have the extremely low values 6...20 K, which justifies one in speaking of the dense clark clouds as cold clouds.

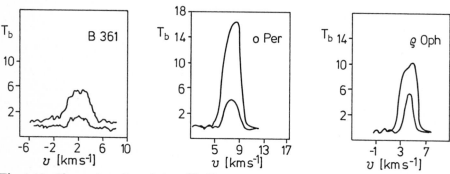

Fig. 4.49. Observed profiles of the $C^{12}O^{16}$ line $J = 1 \rightarrow 0$ and the corresponding line of $C^{13}O^{16}$ (the lower profile in each case) for the great globule B 361, a large dark cloud near the star o Persei and the dark cloud south of ϱ Ophiuchi. Ordinate: brightness temperature in Kelvin; abscissa: velocity relative to the local standard of rest in $km\,s^{-1}$. (From measurements by Milman et al., 1975)

The half-height widths of the CO line profiles are mostly between 1 and $5\,km\,s^{-1}$. A temperature of 20 K, on the other hand, corresponds with a thermal Doppler broadening of only about $0.1\,km\,s^{-1}$ [see (5.203)]. There must therefore be internal macroscopic motions amounting to several $km\,s^{-1}$. These could, amongst other possibilities, be caused by contraction of the whole cloud as a result of the excess of gravitation compared with the gas pressure. We return to this question in Sect. 5.2.4.

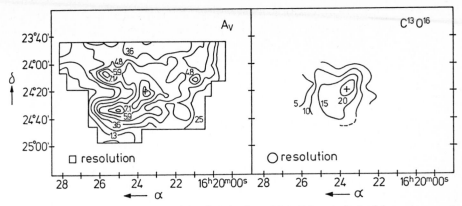

Fig. 4.50. Contours of the visual extinction A_V from 1.3 to 7.1 mag, derived from star counts, and the $C^{13}O^{16}$ emission ($J = 1 \rightarrow 0$) for the ϱ Ophiuchi dark cloud integrated over the whole line. Coordinates for 1950.0. (By permission of Myers et al., 1978)

The intensity distribution of the CO emission over a dark cloud usually shows a correlation with the distribution of the visual extinction. This is demonstrated in Fig. 4.50 for the case of the ϱ Oph cloud. Moreover one often observes strengthened molecular line radiation at IR sources.

The most intense molecular line sources with the greatest profusion of kinds of molecules are observed in the *vicinity of H II regions* (see Table 4.14). With amounts over $10\,\mathrm{km\,s}^{-1}$, the line widths here are often still greater than in dark clouds: in the extreme case Sgr B2 around $50\,\mathrm{km\,s}^{-1}$. In addition it is characteristic that the kinetic temperatures − derived from the optically thick CO lines − are distinctly higher, namely between 20 and about 100 K. An especially thoroughly studied example is the Orion Nebula. The maximum of the emission here is centred on the KL-object (see Fig. 4.42). The extent of the emission region, however, is not the same for all sorts of molecule: the one with the greatest diameter, of about $1°$, is CO; for HCN the extent amounts to about $6'$, for CS and H_2CO (emission at $\lambda = 2.1\,\mathrm{mm}$) only around $3'$. The range of intensity peaks for all these molecules shows a north-south oriented ridge (Fig. 4.51). Moreover "peaks" and "valleys" occur, especially numerous in the distribution of $C^{13}O^{16}$. The line widths correspond to velocity dispersions of around $4\,\mathrm{km\,s}^{-1}$. At the positions of strong infrared emission the kinetic temperature reaches values up to about 80 K.

The large molecular clouds associated with the strong radio continuum sources Sgr A and Sgr B2 (Sect. 4.2.2) of the galactic centre show a richness of different sorts of molecules, which so far is found in no other sources. Sgr B2 probably coincides with the most massive cloud in the entire galaxy (see Table 5.9, p. 301). Two examples of observational results for this object are shown in Fig. 4.52. The maximum of the thermal continuum radiation of the H II region nearly coincides with the centre of the molecular line emission and absorption. With a distance of 10 kpc the linear diameter of the molecular cloud works

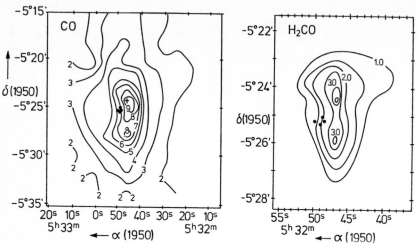

Fig. 4.51. Contour charts for the $C^{12}O^{16}$ line ($J = 1 \to 0$) and the 140 GHz line of the H_2CO molecule in the region of the Orion Nebula. The principal maximum coincides in each case with the KL infrared source. The position of the Trapezium stars is shown for comparison. (After Liszt et al., 1974 [CO] and Thaddeus et al., 1971 [H_2CO]; by permission)

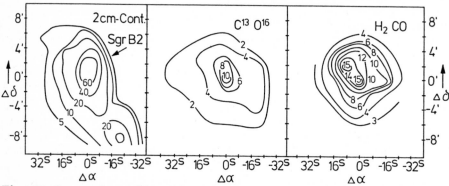

Fig.4.52. Contour charts of the molecular cloud associated with Sgr B2. *Left*: antenna temperature T_A of the thermal continuum radiation at $\lambda = 2$ cm (maximal value $= 100$ corresponds to $T_A = 3.6$ K). *Centre*: column density of the $C^{13}O^{16}$ molecules (unit 6×10^{16} cm^{-2}). *Right*: column density of the H_2CO molecules (in arbitrary units) derived from observations of the 2 cm H_2CO line in absorption against the thermal continuum. The origin was chosen in each case at the position of an OH point source at $\alpha(1950) = 17^h 44^m 11^s$, $\delta(1950) = -28°22'30''$. (After Scoville, Solomon, and Penzias, 1975)

out at around 40 pc. Outside the nucleus of the cloud the kinetic temperature amounts on average to about 20 K.

4.2.8 Maser Point Sources

In molecular clouds, often close by compact H II regions and/or IR sources, strong emission from OH at $\lambda = 18$ cm and also from H_2O at $\lambda = 1.35$ cm

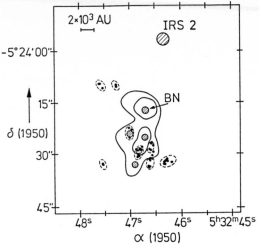

Fig. 4.53. H_2O maser sources in the Kleinmann-Low Nebula (points) from VLB Interferometer observations at 22 GHz (1.35 cm). "Activity centres" are enclosed with *dashed lines*. Contours of the infrared emission at 21 μm are shown for comparison; hatched circles mark the positions of compact infrared sources (cf. also Fig. 4.42). (After Genzel et al., 1978)

is observed in many cases from regions of very small extent. Interferometric measurements, in part very long baseline interferometry (VLBI), show for the KL- and BN-infrared sources, for example, that often a whole group of individual sources with angular diameters from 0.1″ down to about 0.0003″ fill an "activity centre" of about one to two arc seconds extent (Fig. 4.53). From the observed radiation fluxes $\Phi_\nu = I_\nu \Omega$, with Ω = solid angle of the individual source, one finds extraordinarily high intensities I_ν. Equation (4.27) gives brightness temperatures $T_b \approx 10^{10} \ldots 10^{15}$ K! On the other hand the widths of the line components originating from the individual sources are relatively small, corresponding to a velocity dispersion of 1 km s^{-1} and less. The interpretation as a pure thermal Doppler broadening leads to kinetic temperatures less than 100 K. The individual sources have different radial velocities; they extend over a range about 10 to 20 km s^{-1} wide. Moreover the radiation is circularly polarised − especially strongly for the OH sources.

These objects unquestionably cannot be thermal sources. The low line widths in conjunction with the extremely high intensities admit only the intensification mechanism of stimulated emission. We have here before us, therefore, a "natural" maser: the occupation of the two energy levels, between which the observed transition takes place, must be inverted by a certain process, i.e. the upper level is strongly over-populated, so that at each absorption transition (from lower to upper) a profusion of emission transitions (from upper to lower) follows (Sects. 5.1.2, 5.2.4).

A common spectral property of the OH maser sources found in molecular clouds consists in that one of the two inner of the four 18 cm lines ($\Delta F = 0$) at 1665.40 and 1667.36 MHz occurs as the strongest emission − these lines also have the greatest transition probabilities. The two outer lines (1612 and 1720 MHz) are considerably weaker; they are usually called satellite lines. These sources are called Type I OH masers. In another type of OH maser source (Type II) either only the satellite line at 1720 MHz is observed in emission (Type IIa) or the 1612 MHz line is the strongest (Type IIb). These maser sources are found in long-period variables (Mira stars) and late supergiants with infrared excess, i.e. with dust shells. Well known examples are NML Cyg and VY CMa.

For the 1.35 cm line of H_2O it is a matter of a pure rotation transition. The hyperfine structure splitting is here so small that the components overlap. The H_2O maser emission is not always associated with the occurrence of OH masers. The H_2O masers are smaller than the OH sources, also the spatial and velocity structures obtained interferometrically are in general different. H_2O masers show very great variations in their intensity (usually on a time scale of days to months); with OH masers they are less strong. To-day CS and H_2CO maser sources are also known.

For a few of the interstellar masers observed with specially high angular resolution, Table 4.15 lists the measured maximal radiation fluxes, the linear total diameter of the region enclosing in each case several activity centres, and the linear dimensions of the individual sources (rounded values). The positions of the activity centres and individual sources in the Orion Nebula can be ascertained from Figs. 4.53 and 4.42. The OH maser source W 3 (OH) found in W 3 practically coincides with an ultracompact H II region, whose diameter is of order 10^3 AU. This continuum source is optically thick in the cm wavelengths and was discovered at $\lambda = 2$ cm. The H_2O maser in W 3 lies in the neighbourhood of the source IRS 5 observed in the long-wave infrared.

Table 4.15. Observational results for a few regions of interstellar maser emission which were studied with very high angular resolution. (After Burke, 1975)

Source (molecule)	Distance [kpc]	Maximal flux Φ_ν [Jy]	Total diameter [AU]	Largest individual source [AU]	Number of individual sources
Orion A (H_2O)	0.5	1.3×10^4	10^4	1	30
W3 (OH)	3.1	3×10^3	10^4	10	7
W49 (H_2O)	14	4×10^4	10^4	<25	9
M17 (H_2O)	2.1	9×10^2	10^3	<4	4

Repeated measurements with the VLBI technique have recently made possible the determination of the relative proper motions of H_2O masers. For the individual sources of the Orion molecular cloud values of a few thousandths of an arc second per year were found. This is interpreted to mean that the gas of several individual sources is moving outwards from a general centre with velocities in each case of about $20 \, km \, s^{-1}$ (Downes et al., 1980, Genzel et al., 1981).

Summarising, one can say that the linear dimensions of the individual maser sources, of from 1 to 10 AU, are comparable with the dimensions encountered in the solar system. The measured radiation fluxes, in combination with the distances, imply (under the assumption of isotropic emission) luminosities in the observed lines of about 10^{20} up to 10^{26} W, and hence up to the luminosity of the sun in the whole spectrum. The observations suggest that maser sources occur in regions of formation of massive stars. They are probably ejected as localised condensations from the envelopes of such stars.

4.2.9 Interstellar Clouds and Star Formation

It has already been indicated repeatedly in the foregoing sections that the hot H II regions and the cold clouds of the interstellar medium do not occur in a mutually unrelated manner. On the one hand one finds them in close spatial proximity and there is often mutual interpenetration, on the other hand a cold cloud can be transformed into a bright emission nebula under the influence of a star newly emerging in it. These connections will be further elucidated in the following discussion.

The various features observed in the region of the Orion Nebula, which are included in Fig. 4.54, lead to a model, sketched in Fig. 4.55, that takes as the starting element a vast, massive molecular cloud with intermixed dust. This cold cloud is detected by the line emissions from CO and other molecules, as well as by strong extinction in the visual, and is known as OMC 1 (Orion Molecular Cloud 1). In its border regions facing towards the terrestrial observer, and thus appearing to us as projected on the body of the cloud lying behind it, there

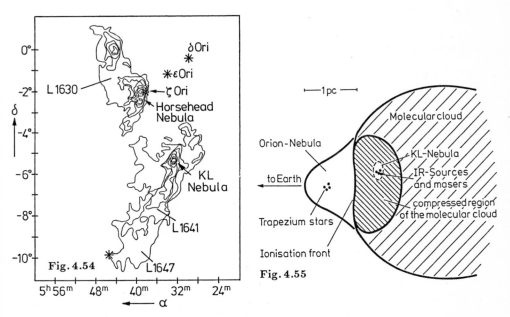

Fig. 4.54. Contour chart for the CO line emission of the molecular cloud complex in Orion with indications of the positions of dark clouds and the Kleinmann-Low infrared nebula. (After Kutner, Tucker, Chin, and Thaddeus, 1977)

Fig. 4.55. Model of the Orion Nebula after a suggestion of Zuckerman (1973). Explanation in text

came into existence massive stars of high surface temperature (Trapezium), whose ultraviolet radiation created an H II region which is still today relatively compact. Its heating up is accompanied by a local expansion which is colliding with the side of the cold cloud facing us. (The stellar wind emanating from the young stars may also play a part here). The slender carbon recombination lines observed presumably stem from the transition region between the H II region and the molecular cloud. The intensity maxima of the molecular emissions come from the compressed regions of the cold cloud in the background, where other stars also are in the process of being formed and are found as IR sources in the neighbourhood of the KL-object. It may also be a matter of existing protostars still in the contraction phase or of stars with dense dust envelopes, in any case at a significantly earlier stage of development than that of the Trapezium stars. The encircling dust transforms their radiation into the infrared and is probably also partly responsible for the occurrence of the maser sources.

Nowadays a whole series of other star formation regions is known, in which a similar scenario is encountered in co-existence with various interstellar phenomena. Particularly thoroughly investigated regions are the W 3/W 4/W 5

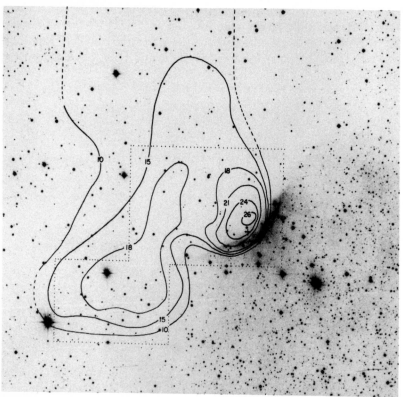

Fig. 4.56. The H II region S 140 (negative) with adjacent CO emission, infrared source (+) and dark cloud, which in the region of the molecular cloud considerably obscures the stellar background. (Blair et al., 1978; photograph reproduced from the Palomar Sky Survey © 1960 National Geographic Society, California Institute of Technology)

complex and M 17, where the H II regions and a multitude of young stars occur at one end of a giant molecular cloud, 100 pc in extent. In Fig. 4.56 yet a further example is illustrated. Here also we observe the H II region not in front of, but beside the molecular cloud, whose outline is clearly recognisable by the contours of the CO emission. Near the edge of the cloud, close to the highest CO radiation temperatures, there is an IR source similar to the BN-object in the Orion Nebula. The considerably reduced stellar background (dark cloud L 1204) in the region of the molecular emission demonstrates here in a particularly impressive way the close coupling of interstellar dust and molecules.

The fact that the young stars are often found at the edge of the molecular clouds suggests the supposition that the star formation is set off by an impact from the outside, then penetrates from the edge into the interior of the cloud and there exhausts itself. In Sect. 6.3.4 we shall return to this picture of the "burning cigar" with young stars and H II regions as the ash.

However, this is not necessarily always so: in the case of S 106, an object reently investigated thoroughly at the Max-Planck-Institut für Astronomie in Heidelberg using the telescopes on the Calar Alto, Spain, the star formation appears to emanate from the centre of a molecular cloud. This bipolar nebula is however also of interest for other reasons. It has attracted attention as an intense IR source located in the middle of a compact H II region, surrounded by an extensive molecular cloud with H_2O masers. Everything accordingly indicates that S 106 is a young object.

By the use of image converter techniques on photographs in the near infrared from the Calar Alto, the star exciting the H II region could be identified (Fig. 4.57). It is of spectral type O9 or B0. A dust layer enveloping it causes an extinction of 21 magnitudes in the visual, and first becomes penetrable at

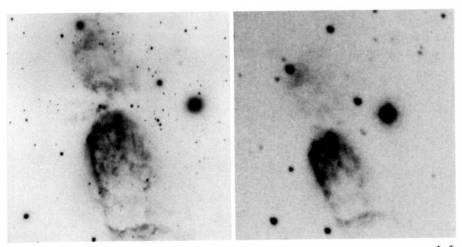

Fig. 4.57. Bipolar nebula S 106 from photographs with an image converter camera. *Left*: wavelength around 1 μm. *Right*: 0.7 μm. At the longer wavelength the central exciting star between the two luminous gas clouds can be clearly discerned. Its rapidly decreasing brightness towards the visual is due to the high extinction of its enveloping dust disc. (Photographs taken at the Calar Alto Observatory)

wavelengths $\gtrsim 1\,\mu$m. The associated heating causes longwave infrared emission from the dust.

S 106, however, evidently does not have a spherically symmetric structure. The dust appears to be concentrated essentially on a relatively thin disc which stands out clearly as a gap between the two luminous gas clouds, and which we see nearly edge-on. In the polar directions, however, it is presumably clear, or considerably more transparent, than in the equatorial plane, so in these directions the ultraviolet stellar radiation can emerge and excite the nebular luminosity. Spectroscopic observations have shown that the gas of these clouds is moving apart with velocities of up to $70\,\mathrm{km\,s}^{-1}$ in opposite directions, apparently propelled in this way by the central star.

The anisotropic structure of S 106 points to the influence of rotation, as it has been possible recently to confirm by measurements with the Coudé spectrograph of the 2.2 m telescope at the Calar Alto. The rotation axis corresponds roughly with the line of symmetry connecting the two gas clouds and the star, and penetrating the dust disc perpendicularly. This is nearly the same direction as that observed for the rotation axis of the extended molecular cloud enveloping S 106 (Fig. 4.58). The two maxima of the molecular emission, between which the luminous nebula is located, suggests an annular structure, whose principal plane almost coincides with the orientation of the dust disc in the nebula. This accordingly indicates a structural integrity of the young objects and their mother-cloud. This contradicts however, among other things, the interpretation that we see the emission nebula rather as a projection in the centre of the cloud. The maser source at one of the maxima of the molecular radiation is presumably an indicator that further star formation activities may be expected there.

Perhaps these structural features of S 106 are typical of an early phase in the evolution of rotating stars.

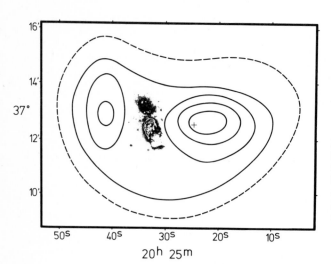

Fig. 4.58. The bipolar nebula S 106 and the molecular cloud surrounding it, depicted by lines of equal NH_3 emission (from Little et al., 1979). The cross marks the position of a temporary H_2O maser

5. Physics of the Interstellar Matter

5.1 Radiation in the Interstellar Gas

5.1.1 Radiation Transport

Radiation Transport Equation: Let us follow the propagation of radiation of frequency ν along the line of sight towards the observer, in the region of an interstellar cloud (Fig. 5.1). Let ε_ν denote the *emission coefficient* of the gas, so defined that $\varepsilon_\nu d\omega$ represents the radiation energy emitted per unit of volume, time and frequency in the solid angle element $d\omega$ (oriented towards the observer). Moreover let I_ν be the *intensity* of the radiation, so that $I_\nu d\omega$ gives the energy, per unit of time and frequency, flowing across unit area perpendicular to the line of sight within the solid angle $d\omega$. On the element of the path between s and $s + ds$ the intensity undergoes an increase of $\varepsilon_\nu d\omega\, ds$. On the other hand it undergoes, in general, attenuation proportional to I_ν itself and the length of the path element ds, and which one writes as $\kappa_\nu I_\nu d\omega\, ds$. The proportionality factor κ_ν, which in general depends on the frequency, is the *absorption coefficient* referred to unit length [cm^{-1}].

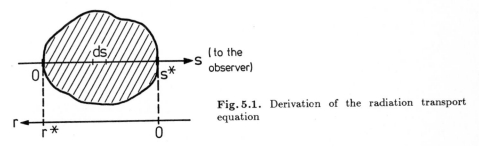

Fig. 5.1. Derivation of the radiation transport equation

The balance of emission and absorption $dI_\nu d\omega \equiv \varepsilon_\nu d\omega\, ds - \kappa_\nu I_\nu d\omega\, ds$ leads directly to the radiation transport equation

$$\frac{dI_\nu}{ds} = -\kappa_\nu I_\nu + \varepsilon_\nu \quad . \tag{5.1}$$

In the case of pure absorption $\varepsilon_\nu \equiv 0$ and an incident intensity $I_{\nu,0}$ at $s = 0$, integration leads simply to the absorption law

$$I_\nu(s^*) = I_{\nu,0} \exp\left\{ -\int_0^{s^*} \kappa_\nu(s)ds \right\} \quad .$$

The value of the integral of the absorption coefficient appearing in the exponential function is called the optical thickness or optical depth of the cloud.

Let us now transform to distance *from* the observer, measured into space by the coordinate r, with the property $dr = -ds$, and introduce as new variable the corresponding *optical depth* τ_ν by $d\tau_\nu = \kappa_\nu dr = -\kappa_\nu ds$ (cf. Fig. 5.1)

$$\tau_\nu = \int_0^r \kappa_\nu dr = -\int_{s^*}^s \kappa_\nu ds$$

($\tau_\nu = 0$ for $r = 0$ or $s = s^*$). Then it follows that

$$\frac{dI_\nu}{d\tau_\nu} = I_\nu - \frac{\varepsilon_\nu}{\kappa_\nu} \quad . \tag{5.2}$$

In order to determine I_ν it is therefore necessary to know the quotient called the *source function*

$$\frac{\varepsilon_\nu}{\kappa_\nu} = S_\nu \quad . \tag{5.3}$$

Under the conditions of interstellar space S_ν frequently depends on the intensity I_ν. Equation (5.2) can however be formally integrated without difficulty. For that purpose we multiply (5.2) by $\exp(-\tau_\nu)$; then using (5.3) we can write for the result

$$\frac{d[I_\nu \exp(-\tau_\nu)]}{d\tau_\nu} = -S_\nu\, e^{-\tau_\nu} \quad .$$

Integration leads to

$$I_\nu(\tau_\nu)e^{-\tau_\nu} = -\int^{\tau_\nu} S_\nu(\tau_\nu')e^{-\tau_\nu'} d\tau_\nu' \quad .$$

Writing down this equation once for $\tau_\nu = 0$ and then for the optical thickness of the whole cloud $\tau_\nu = \tau_\nu^* = \tau_\nu(r^*)$ and taking the difference of the two equations we arrive at

$$I_\nu(\tau_\nu = 0) - I_\nu(\tau_\nu^*)e^{-\tau_\nu^*} = -\int_{\tau_\nu^*}^0 S_\nu(\tau_\nu')e^{-\tau_\nu'} d\tau_\nu' \quad .$$

If we denote the background intensity $I_\nu(\tau_\nu^*) = I_\nu(s = 0) = I_\nu(r = r^*)$ by $I_{\nu,0}$, then the result for the observed intensity can be written as

$$I_\nu = I_\nu(\tau_\nu = 0) = \int_0^{\tau_\nu^*} S_\nu(\tau_\nu)e^{-\tau_\nu} d\tau_\nu + I_{\nu,0}e^{-\tau_\nu^*} \quad . \tag{5.4}$$

This result can clearly also be directly derived as the integral of the contributions of the different path elements $\varepsilon_\nu(r)\exp[-\tau_\nu(r)]dr$, taking account of the fact that $dr = d\tau_\nu/\kappa_\nu$ and $\varepsilon_\nu/\kappa_\nu = S_\nu$. If S_ν also depends on I_ν, then (5.4) represents an integral equation for the intensity.

For a first interpretation of the observed intensity it is often sufficient to assume a spacewise constant source function S_ν in the whole emitting region. Then (5.4) leads to

$$I_\nu = S_\nu[1 - e^{-\tau_\nu^*}] + I_{\nu,0}e^{-\tau_\nu^*} \tag{5.5}$$

with the limiting cases

$$I_\nu = \begin{cases} \tau_\nu^*(S_\nu - I_{\nu,0}) + I_{\nu,0} & \text{for optically thin layer} \quad \tau_\nu^* \ll 1 \\ S_\nu & \text{for optically thick layer} \quad \tau_\nu^* \gg 1 \end{cases} \tag{5.6}$$

(in the first case we have used $e^{-x} \approx 1 - x$ for $x \ll 1$).

"Local Thermodynamic Equilibrium" (LTE): If the source function is given by the Planck Function for the local temperature $T = T(\tau_\nu)$,

$$S_\nu = B_\nu(T) = \frac{2h\nu^3}{c^2} \frac{1}{\exp(h\nu/kT) - 1} \quad, \tag{5.7}$$

one has an important limiting case denoted in what follows, as is customary, by the abbreviation LTE. Insofar as thermodynamic equilibrium is assumed to be not fully established, as the temperature is not required to be constant, I_ν can deviate from $B_\nu(T)$ by an arbitrary amount.

In the radiofrequency region, when $\lambda > 1$ cm and $T > 10$ K, $h\nu/kT = 1.44/\lambda\,[\text{cm}]\,T[\text{K}] \ll 1$, so that one may use the Rayleigh-Jeans approximation

$$B_\nu(T) = \frac{2\nu^2}{c^2}kT \quad. \tag{5.8}$$

Radioastronomers also formally express the observed intensity I_ν in these relations by a *brightness temperature* $T_b(\nu)$ [see Sect. 4.1.5, (4.27)] and accordingly write, for a spacewise constant temperature,

$$T_b = T(1 - e^{-\tau_\nu^*}) + T_{b,0}e^{-\tau_\nu^*} \quad, \tag{5.9}$$

where $T_{b,0}$ corresponds to the background intensity $I_{\nu,0}$. For an optically thick medium ($\tau_\nu^* \gg 1$) one therefore has $T_b = T$, whilst in the optically thin case ($\tau_\nu^* \ll 1$) it follows that $T_b = \tau_\nu^* T + T_{b,0}$.

The optical thicknesses are determined by the absorption coefficient, which depends primarily upon how many atoms or molecules contained in a unit volume are in just that energy state which can absorb radiation of the frequency under consideration. If one wishes to draw conclusions on the density of the relevant atomic or molecular species from the observed intensity I_ν or brightness temperature T_b, one therefore needs to know the excitation and ionisation conditions, i.e. the relative populations of the different energy states, of these gas particles. In strict thermodynamic equilibrium these population ratios can be obtained from the formulae of Boltzmann and Saha (Appendix D) as a function of temperature and electron density. If these two parameters are known, then one can, for example in the case T = const, solve (5.5) or (5.9) for τ_ν^* and hence for the density multiplied by the geometric layer thickness – except when $\tau_\nu^* \gg 1$, in which case I_ν is determined by T alone.

Under *interstellar conditions* there must be doubts from the start about the usefulness of the assumption of thermodynamic equilibrium. This arises if the distribution of the gas particles over the different possible energy states is caused by collisional processes. Then *one* temperature characterises the kinetic energies of the particles, the population ratios *and* the radiation field. With the extremely low densities in the interstellar gas, however, collisions between the particles are relatively uncommon and the radiation field frequently does not correspond to the kinetic temperature of the gas. For example, the observed interstellar energy density of starlight (see Sect. 4.1.9, Table 4.9) would require a temperature of only about 3 K in the case of LTE, whereas the photon energies correspond to the high temperatures of stellar atmospheres! Accordingly, the theories of excitation and ionisation, as well as the calculation of the source function $S_\nu = \varepsilon_\nu/\kappa_\nu$, must be extended to the general case of interstellar conditions. For this it is necessary to give detailed consideration to all significant processes of excitation, de-excitation, ionisation and recombination. The applicability of the considerably simpler LTE assumption can subsequently be examined for the individual case. For example, LTE can be assumed if only certain energy levels of relevance to the radiation under consideration are populated according to the Boltzmann formula. Strong deviations from LTE arise predominantly with the line radiation, which we therefore treat in what follows without any supposition of LTE.

5.1.2 Line Emission and Absorption Under Interstellar Conditions

Emission and Absorption Coefficients: For the following discussion we resolve ε_ν and κ_ν into the two parts, line radiation (upper index L) and continuous radiation (upper index C): $\varepsilon_\nu = \varepsilon_\nu^L + \varepsilon_\nu^C$, $\kappa_\nu = \kappa_\nu^L + \kappa_\nu^C$. The total spontaneous line emission of the atom or molecule of a certain element in the transition from the upper state n to the lower state m, per cm^3, s and sterad is given by

$$\int_{\text{line}} \varepsilon_\nu^L d\nu = \frac{1}{4\pi} h\nu_{nm} A_{nm} N_n \quad . \tag{5.10}$$

Here ν_{nm} is the frequency of the line, A_{nm} the Einstein *transition probability* and N_n the number density of atoms or molecules in the starting state n. The stimulated emission is proportional to I_ν and is therefore included as a negative term in the usual way with the absorption.

The energy absorbed per second in the inverse transition from m to n in $1\,\mathrm{cm}^3$ of matter amounts on the other hand per sterad to

$$\int_{\mathrm{line}} \kappa_\nu^{\mathrm{L}} I_\nu d\nu = \frac{1}{4\pi} h\nu_{nm} N_m B_{mn} U(\nu_{nm}) - \frac{1}{4\pi} h\nu_{nm} N_n B_{nm} U(\nu_{nm}) \quad .$$

Here $B_{mn}U(\nu_{nm})$ denotes the probability of a transition $m{\to}n$ by absorption in the radiation density $U(\nu_{nm})$; the second term represents the contribution (treated as negative absorption) of the stimulated emission.[1] The radiation flowing in the unit solid angle, $U_\nu/4\pi$ is equal to $I_\nu dt$, where dt is the time the radiation needs to pass through the unit volume, i.e. to cover the unit length. Therefore $dt = 1/c$, where c is the velocity of light, and one obtains

$$U_\nu = \frac{4\pi}{c} I_\nu \quad .$$

If the intensity I_ν inside the line may be regarded as constant, it further follows from this relation that

$$\int_{\mathrm{line}} \kappa_\nu^{\mathrm{L}} d\nu = \frac{1}{c} h\nu_{nm}(N_m B_{mn} - N_n B_{nm}) \quad . \tag{5.11}$$

The coefficients $\varepsilon_\nu^{\mathrm{L}}$ and κ_ν^{L} themselves are each given by multiplication of their integrals (5.10) and (5.11) by the normalised "line profiles" $\psi(\nu)$, assumed equal for emission and absorption:

$$\varepsilon_\nu^{\mathrm{L}} = \frac{h\nu_{nm}}{4\pi} \psi(\nu) A_{nm} N_n \quad , \tag{5.12}$$

$$\kappa_\nu^{\mathrm{K}} = \frac{h\nu_{nm}}{c} \psi(\nu)(B_{mn} N_m - B_{nm} N_n) \quad .$$

A simple relation between the three Einstein transition coefficients A_{nm}, B_{nm} and B_{mn} can be derived by consideration of strict thermodynamic equilibrium. In this case (5.10) and the corresponding integral over $\kappa_\nu^{\mathrm{L}} I_\nu$ must be equal and we get

$$A_{nm} N_n = N_m B_{mn} U(\nu_{nm}) - N_n B_{nm} U(\nu_{nm}) \quad .$$

Here we express the ratio N_n/N_m by the Boltzmann Formula [Appendix D]

$$N_n/N_m = (g_n/g_m)\exp\{-(h\nu_{nm}/kT)\} \quad , \tag{5.13}$$

---

[1] Note that other definitions of B_{nm} and B_{mn} are also used in the literature.

where g_n, g_m are the statistical weights of the lower and upper states, and replace $U(\nu)$ by $(4\pi/c)I_\nu$ with $I_\nu = B_\nu(T)$. If we now take the limiting case $h\nu_{nm}/kT \ll 1$, with $B_\nu(T)$ from (5.8), we arrive at the conclusion that the following relations must be universally valid:

$$A_{nm} = \frac{8\pi h\nu^3}{c^3} B_{nm} \ , \qquad g_m B_{mn} = g_n B_{nm} \quad . \tag{5.14}$$

With the relation on the right-hand side of (5.14) the expression for κ_ν^L can be written

$$\kappa_\nu^L = \frac{h\nu_{nm}}{c} \psi(\nu) B_{mn} N_m \left(1 - \frac{g_m}{g_n} \frac{N_n}{N_m} \right) \quad \text{or} \tag{5.15a}$$

$$\kappa_\nu^L = \frac{h\nu_{nm}}{c} \psi(\nu) B_{nm} \Delta N \ \text{ with } \ \Delta N = \frac{g_n}{g_m} N_m - N_n \quad . \tag{5.15b}$$

The expression (5.15b) is frequently applied if the stimulated emission has a strong influence.

Having regard to (5.14) one obtains from (5.12) and (5.15a) the following expression for the source function of the line radiation:

$$S_\nu^L = \frac{\varepsilon_\nu^L}{\kappa_\nu^L} = \frac{2h\nu^3}{c^2} \left[\frac{g_n}{g_m} \frac{N_m}{N_n} - 1 \right]^{-1} \quad . \tag{5.16}$$

From (5.13) there follows in particular the inequality $(N_n/N_m)(g_m/g_n) \leq 1$. We show below that $(N_n/N_m)(g_m/g_n)$ in the interstellar gas is usually considerably smaller than in thermal equilibrium from (5.13), and so the higher states are underpopulated. The observational results mentioned in Sect. 4.2.8, on the other hand, leave no doubt that in interstellar condensations the converse case, the overpopulation of higher states or *"population inversion"*, with $(N_n/N_m)(g_m/g_n) > 1$ or $\Delta N < 0$, also occurs. The stimulated emission then predominates over the absorption, κ_ν^L and S_ν^L become negative, and as a result the incident radiation is reinforced: we have maser or laser operation.

The B_{nm} and B_{mn} can be obtained from A_{nm} by (5.14), and these coefficients themselves can be obtained by quantum mechanical calculation of the corresponding dipole matrix elements:

$$A_{nm} = \frac{64\pi^4 \nu^3}{3hc^3} P_{nm}^2$$

with P_{nm} being the value of the dipole matrix element for the transition $n \to m$. In the simple case of the important radio molecular lines which arise in pure rotational transitions in the ground vibrational state $v = 0$ of diatomic or linear molecules, we have

$$P_{J,J+1}^2 = \mu^2 \frac{J+1}{2J+1} \ \text{ and } \ \nu = 2B(J+1) \quad ,$$

where J denotes the rotational quantum number of the lower level, μ the permanent *dipole moment* and B the rotational constant of the molecule. For the CO molecule, for example, $\mu = 0.112$ Debye (1 Debye $= 10^{-18}$ electrostatic units), which for the 2.6 mm line ($J = 1{\to}0$) with $B = 57.6$ GHz gives $A_{21} = 6 \times 10^{-8}\,\mathrm{s}^{-1}$.

Statistical Equilibrium: The ratios of the state populations N_m/N_n necessary for the calculation of the source function S_ν^L and the line absorption coefficients κ_ν^L or the optical depth τ_ν^L will now be obtained free from the assumption of thermodynamic equilibrium. In this general case of non-LTE one simply assumes that no changes occur with time in the population numbers, so that a steady state exists. Let us consider a certain quantum state which, in the interest of simplicity, is characterised by *one* quantum number n; then the number of all transitions into this state n in unit time must be equal to the number of all removals from n to other − lower or higher − states m. Let R_{nm} and R_{mn} denote the numbers of transitions per unit of time and volume from n to m and from m to n, respectively, caused by radiation processes, and C_{nm}, C_{mn} the corresponding transition rates by collision processes. Then obviously for every state n there is a balance equation

$$-\frac{dN_n}{dt} = N_n \sum_m (R_{nm} + C_{nm}) - \sum_m N_m(R_{mn} + C_{mn}) = 0 \quad , \tag{5.17}$$

where the first sum represents the removals and the second sum the acquisitions. To obtain the population densities N_n or the ratios N_n/N_1 this system of so-called *statistical equations* is to be solved. The symbol m can also be spread over continuously distributed states, whereby direct ionisation from n and direct recombination into n are included. A similar system of equations exists for the different stages of ionisation, whose population in a steady state is achieved by equality of the numbers of ionisation and recombination processes (Sect. 5.2.1).

We shall consider the detailed formulation of the statistical equations for the case when only bound states are to be considered, and besides the radiation processes only inelastic collisions occur with a certain kind of particle whose number density is N'.

The *transition rates* in the statistical equations are given as follows:

Departures from n to higher states m :

Photoexcitations (absorptions) $R_{nm} = B_{nm}U_\nu$

Collisional excitations $\qquad C_{nm} = N'Q_{nm}$

$\left.\begin{array}{c}\\ \\ \\ \end{array}\right\}\; m > n$.

Here Q_{nm} denotes the *rate coefficient* = number of transitions $n{\to}m$ per unit of volume and time caused by collisions at a number density of the colliding particles $N' = 1$; U_ν is the radiation density.

Departures from n to lower states m :

Photo de-excitations (sponta- $\quad R_{nm} = A_{nm} + B_{nm}U_\nu$
neous and stimulated emission)

$\left.\phantom{\begin{matrix}a\\a\\a\\a\end{matrix}}\right\}$ $m<n$.

Collisional de-excitations $\qquad C_{nm} = N'Q_{nm}$

Entries to n from higher states m :

Photo de-excitations (sponta- $\quad R_{mn} = A_{mn} + B_{mn}U_\nu$
neous and stimulated emission)

$\left.\phantom{\begin{matrix}a\\a\\a\\a\end{matrix}}\right\}$ $m>n$.

Collisional de-excitations $\qquad C_{mn} = N'Q_{mn}$

Entries to n from lower states m :

Photoexcitations (absorptions) $R_{mn} = B_{mn}U_\nu$

$\left.\phantom{\begin{matrix}a\\a\end{matrix}}\right\}$ $m<n$.

Collisional excitations $\qquad C_{mn} = N'Q_{mn}$

By equating the sums of the removals and the entries, and introducing the radiation density $(4\pi/c)I_\nu$, we now obtain the statistical equations (5.17) in the form

$$N_n\left\{\sum_m B_{nm}\frac{4\pi}{c}I_\nu + \sum_{m<n} A_{nm} + N'\sum_m Q_{nm}\right\}$$

$$= \sum_m N_m B_{mn}\frac{4\pi}{c}I_\nu + \sum_{m>n} N_m A_{mn} + N'\sum_m N_m Q_{mn} \quad . \qquad (5.18)$$

For each n there exists such an equation. Because of the occurrence of the intensity of the radiation field I_ν the whole system must in general be solved simultaneously with the radiation transport equation, which in the case under consideration, from (5.1) with (5.12) and (5.15a), becomes

$$\frac{dI_\nu}{ds} = -\frac{h\nu_{nm}}{c}B_{mn}N_m\left(1 - \frac{g_m}{g_n}\frac{N_n}{N_m}\right)\psi(\nu)I_\nu$$

$$+ \frac{h\nu_{nm}}{4\pi}A_{nm}N_n\psi(\nu) \quad , \qquad (5.19)$$

where $\psi(\nu)$ again denotes the line profile of the emission and absorption coefficents.

Collisional Processes and Kinetic Temperature: We have already indicated the way to calculate the Einstein coefficients B_{nm}, B_{mn} and A_{nm}. The derivation of the collisional rate coefficients Q_{nm} and Q_{mn} however still needs explaining. The collision interaction of two gas particles depends on the individual profiles of the electric potential and on the relative velocity. The effective

248

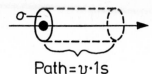 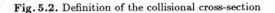

Fig. 5.2. Definition of the collisional cross-section

Path $= v \cdot 1\,\mathrm{s}$

particle cross-sections correspond roughly to the particle radii, for which the potential and kinetic energies are of equal size. The atoms or molecules under consideration and the particles occurring as collision partners may have the velocities v_A and v_C, so that the relative velocities are given by $v = v_A - v_C$. We introduce as "collisional cross-section" $\sigma(v)$ of the atom (molecule) in relation to its collision partner a surface lying perpendicular to v, which by the motion of the particles sweeps out in $1\,\mathrm{s}$ the volume $v\sigma(v)$ (Fig. 5.2), so that

$$N'v\sigma(v) = C(v) \tag{5.20}$$

just indicates the number of collision partners encountered, and hence the number of collisions which an arbitrarily selected atom undergoes on average in $1\,\mathrm{s}$. Here N' again denotes the number density of collision partners. If it is a matter of *inelastic collisions*, which lead to an excitation $(m \to n)$ of the atom (molecule), we write more exactly for the collisional cross-section $\sigma_{mn}(v)$ or $\sigma_{nm}(v)$. The coefficients Q_{mn}, Q_{nm} introduced above are given by averaging over all pertinent relative velocities: $Q_{mn} = v \cdot \sigma_{mn}(v)$ and $Q_{nm} = v \cdot \sigma_{nm}(v)$. One also needs to know the *velocity distributions* of the two gas particles. If only elastic collisions were to occur, then these would necessarily in interstellar space be Gaussian random distributions of the velocity components, and hence Maxwell distributions of the velocity magnitudes. If, however, the atoms, ions or molecules are excited by inelastic collisions, for example with relatively high energy electrons, with consequent radiation, then the excitation energy is removed from the velocity field and the high energy "tail" of the velocity distribution is reduced in comparison with the Maxwell distribution. More exact study leads to the following results. Collisions of H or He atoms with each other or between their ions and free electrons with energies up to about $10\,\mathrm{eV}$ are in the main predominantly elastic. The mainly inelastic collisions with heavier atoms or with molecules are very uncommon on account of the meagre abundance of these gas particles in comparison with H or He atoms, and therefore of minor significance to the energy exchange. Not only in the regions of predominantly neutral hydrogen (H I regions), but also in regions of high kinetic temperature, in which the hydrogen is almost completely ionised (H II regions), the relative deviations of the velocity distribution of the electrons from the Maxwell distribution are expected to be less than 10^{-5}. For neutral H atoms the corresponding deviations are less than $1\,\%$. Accordingly there should in general be no differences worth mentioning between the kinetic temperatures of the different types of gas particle inside a volume of a few free path lengths diameter. A thorough discussion is given by Spitzer (1978, Chap. 2).

The statement that the velocity distribution of the particles in the interstellar gas should be approximated well by the Maxwell distribution is of great significance for studies of the physical state of the interstellar medium.

249

If the two particle types involved in a collision interaction have Maxwellian velocity distributions of the temperature T, then this is also true of the *distribution of the relative velocities v*:

$$\varphi(v;T) = \frac{1}{\sigma^3(2\pi)^{3/2}}\exp(-v^2/2\sigma^2)dv_x\,dv_y\,dv_z$$

$$= \frac{4\pi}{\sigma^3(2\pi)^{3/2}}v^2\exp(-v^2/2\sigma^2)dv \tag{5.21}$$

with $v^2 = v_x^2 + v_y^2 + v_z^2$ and

$$\sigma^2 = \overline{v_x^2} = \overline{v_y^2} = \overline{v_z^2} = \left(\frac{1}{m_A} + \frac{1}{m_C}\right)kT = \left(\frac{1}{A} + \frac{1}{A_C}\right)\mathcal{R}T \tag{5.22}$$

(Introduction of polar coordinates: $dv_x\,dv_y\,dv_z = v^2\sin\vartheta\,dv\,d\vartheta\,d\varphi$ and integration over ϑ and φ). Here m_A and m_C denote the masses, $A = m_A/m_H$ and $A_C = m_C/m_H$ the atomic weights of the collision partners (index C for the "collision particle"); k is the Boltzmann Constant, $\mathcal{R} = 8.314 \times 10^7$ erg K^{-1} mol^{-1} the Gas Constant. The averaging of $v \cdot \sigma_{nm}(v)$ over all v gives the collisional rate for de-excitations $n \rightarrow m$ with $m < n$

$$Q_{nm} = \overline{v \cdot \sigma_{nm}(v)} = \int_0^\infty v \cdot \sigma_{nm}(v)\varphi(v;T)dv \quad . \tag{5.23}$$

Between the coefficients of the collisional rates for excitations and de-excitations Q_{mn} and Q_{nm} there exists the general relation

$$g_m Q_{mn} = g_n Q_{nm}\exp(-\Delta E/kT)$$

$$= g_n Q_{nm}10^{-(5040\Delta E[\mathrm{eV}]/T)} \tag{5.24}$$

with $\Delta E = E_n - E_m > 0$. This result follows by appealing to the strict thermodynamical balance, which contains in particular the detailed balance $N_n N' Q_{nm} = N_m N' Q_{mn}$, where N_m/N_n is given by (5.13).

In the *case of electron collisions* it is customary to write the result of the integration (5.23) in the form ($m_C = m_e =$ electron mass):

$$Q_{nm} = \frac{h^2}{(2\pi m_e)^{3/2}(kT)^{1/2}}\frac{\Omega_{nm}}{g_n}$$

$$= 8.63 \times 10^{-6}\frac{\Omega_{nm}}{T^{1/2}g_n}[\mathrm{cm}^3\,\mathrm{s}^{-1}] \quad . \tag{5.25}$$

The collision efficiency Ω_{nm} introduced here has values mostly of order unity for the transitions, important in the interstellar gas, between the lower excited states and the ground state of the ions of C, N, O, Si, Ne etc.

Of the heavy particles occurring as collision partners, *hydrogen atoms are* first to be considered because of their great cosmic abundance. If the hydrogen is predominantly neutral (H I regions), then the H atoms undergo for example the following processes as collision partners: (1) excitation of hyperfine structure levels in the ground state of the neutral H atom (change of electron spin), (2) excitation of lowlying fine structure levels of ions such as C^+, O^+, O^{++} etc. and (3) excitation of rotational states of the hydrogen molecule.

In order to assess the *order of magnitude of the collisional rates* for the interaction between H atoms we shall set

$$Q_{nm} = \overline{v \cdot \sigma_{nm}(v)} \approx \overline{v} \cdot \overline{\sigma}_{nm} , \qquad \text{where} \tag{5.26}$$

$$\overline{v} = \int_0^\infty v \varphi(v; T) dv = 2 \sqrt{\left(\frac{1}{m_A} + \frac{1}{m_C} \right) \frac{2kT}{\pi}}$$

$$= 1.46 \times 10^4 \sqrt{\left(\frac{1}{A} + \frac{1}{A_C} \right) T} \tag{5.27}$$

with $A = A_C = 1$. When $T = 100\,\text{K}$ it follows for example that $\overline{v} \approx 10^5\,\text{cm}\,\text{s}^{-1}$. With the radius of the H atom being around 10^{-8} cm we obtain $\overline{\sigma} \approx 10^{-16}\,\text{cm}^2$ and $Q_{nm} \approx 10^{-11}\,\text{cm}^3\,\text{s}^{-1}$. For the overall local density of neutral hydrogen in the interstellar space $N_H \approx 1\,\text{cm}^{-3}$ (see Sect. 5.4.2) it follows that on average each atom has only one collision about every $10^{11}\,\text{s} \approx 3000\,\text{years}$. We cannot here go into a more precise calculation of $\sigma_{nm}(v)$, and hence of Q_{nm}. Typical results are collected in Table 5.1. We refer the reader to the literature listed in the appendix – values of Ω_{nm} are also given by Spitzer (1978, p. 74).

Excitation Under Interstellar Conditions: Let us discuss the population ratios arising under the combined effect of radiative and collisional processes in the interstellar gas by means of idealised examples, which nevertheless provide fundamental insight. It will be assumed that ionisation and recombination processes occur very much more rarely than transitions between bound states. This assumption is justified in Set. 5.2.1. The statistical equations can then be formulated with consideration only of transitions between bound states.

1) Example: The atom or molecule has only two energy states, the ground state $m = 1$ and an excited state $n = 2$. The transition $2 \rightarrow 1$ corresponds to a *line in the optical region*, so that we can take the radiation field as diluted black body radiation at temperature T_* and describe it as

$$I_\nu = W \cdot B_\nu(T_*) \tag{5.28}$$

W denotes the "dilution factor" introduced in Sect. 4.1.9. Let the collision partners of the atoms or molecules under consideration be electrons with number density $N' = N_e$. The statistical equations (5.18) then reduce to one equation:

$$N_1 \left(B_{12} \frac{4\pi}{c} I_\nu + N_e Q_{12} \right) = N_2 \left(B_{21} \frac{4\pi}{c} I_\nu + A_{21} + N_e Q_{21} \right) ,$$

from which it follows for the required ratio of population numbers that

$$\frac{N_2}{N_1} = \frac{B_{12}\frac{4\pi}{c}I_\nu + N_e Q_{12}}{B_{21}\frac{4\pi}{c}I_\nu + A_{21} + N_e Q_{21}} \cdot \tag{5.29}$$

If one now expresses B_{12} and B_{21} in terms of A_{21}, and Q_{12} in terms of Q_{21} by means of (5.14) and (5.24), and further substitutes for I_ν (5.28) the Planck Function (5.7) for the temperature T_*, then it follows that

$$\frac{N_2}{N_1} = \frac{g_2}{g_1}\frac{N_e(Q_{21}/A_{21})\exp(-h\nu/kT) + W[\exp(h\nu/kT_*) - 1]^{-1}}{1 + N_e(Q_{21}/A_{21}) + W[\exp(h\nu/kT_*) - 1]^{-1}} \cdot \tag{5.30}$$

In the interstellar space, far from stars, $W \ll 1$. The radiation field in the visible region can be described approximately by $W = 10^{-15}$ and $T_* = 10^4$ K. In diffuse nebulae also in general $W \approx 10^{-14}$ (cf. Sect. 4.1.9). The numerical values of Q_{21}, A_{21} and N_e discussed subsequently prove that $W \ll N_e(Q_{21}/A_{21})$. The terms in (5.30) involving the factor W can therefore be ignored for photon energies of only a few eV, i.e. excitations by radiation play no role in the case considered here. One has simply

$$\frac{N_2}{N_1} = \frac{g_2}{g_1}\frac{\exp(-h\nu/kT)}{1 + (A_{21}/N_e Q_{21})} \cdot \tag{5.31}$$

For LTE in particular the population numbers N_1^* and N_2^* assume the ratio (Boltzmann Formula)

$$\frac{N_2^*}{N_1^*} = \frac{g_2}{g_1}\exp(-h\nu/kT) \quad .$$

The deviations from this are frequently expressed by the coefficients $b_n = N_n/N_n^*$ (for arbitrary n), for which, taking account of (5.31),

$$\frac{b_2}{b_1} = \frac{1}{1 + (A_{21}/N_e Q_{21})} \quad .$$

Only for $A_{21}/N_e Q_{21} \ll 1$, so that the collisional processes are completely dominant, does the case of thermodynamic equilibrium $b_2/b_1 = 1$ occur.

With Ω_{nm}/g_n of order unity and kinetic temperatures of the colliding particles $T \approx 10^2 \ldots 10^4$ K, (5.25) leads to values for Q_{21} between about 10^{-7} and 10^{-6}. For the upper state, from which *allowed transitions* lead to the lower state, one can set $A_{21} \approx 10^8$ s^{-1}. From this it follows that $A_{21}/N_e Q_{21} \approx 10^{14}/N_e$ to $10^{15}/N_e$. For the electron densities $N_e \lesssim 10^{-1}$ cm^{-3} existing at great distances from hot stars (H I regions) as also under the conditions in the diffuse gas nebulae with values up to $N_e \approx 10^4$ cm^{-3}, it follows that $b_2/b_1 \ll 1$. In contrast to the case of LTE practically all the atoms or molecules are in the ground state: the rare collisional excitation 1→2 is always followed at once by a radiative transition 2→1, so that averaged over time scarcely an atom (molecule) exists in the upper state.

The situation is different with the *forbidden transitions* from the so-called metastable energy levels into lower states. As an example we select the green nebular lines of the O III spectrum $N_1 \lambda 5007$ Å and $N_2 \lambda 4959$ Å, which arise from forbidden transitions between lowlying fine structure levels: $2p^2\,^1D_2 - 2p^2\,^3P_2$ and $2p^2\,^1D_2 - 2p^2\,^3P_1$ with the transition probabilities $A_{21} = 2.1 \times 10^{-2}\,\mathrm{s}^{-1}$ and $7.1 \times 10^{-3}\,\mathrm{s}^{-1}$ (magnetic dipole radiation). With these figures for typical gas nebulae, in which one can assume a kinetic temperature $T \approx 10^4\,\mathrm{K}$ and hence $Q_{21} \approx 10^{-7}$, and $N_e \approx 10^3$ to $10^4\,\mathrm{cm}^{-3}$, it follows that $A_{21}/N_e Q_{21} \approx 10^5/N_e \approx 10$ to 10^2. Here the population of the upper state reaches 1 to 10 % of its value for thermodynamic equilibrium. In the extremely low densities of the interstellar gas the lowlying metastable levels with (relatively) long lifetime are therefore very strongly favoured in comparison with levels of normal lifetime: in our example in the ratio of about $10^{10}{:}1$. For the line emission $\varepsilon_\nu^L \sim N_2 A_{21}$. The product $N_2 A_{21}$ is now of about the same order of magnitude for the forbidden lines as for the allowed lines, so in this case the forbidden lines become as strong as, or even stronger than, the allowed ones.

2) Example: Let us consider again the atom or molecule with only two energy states, the transition $2 \to 1$ corresponding however to a *radioline*. For the radiation field in the radiofrequency region, the dilution factor as a rule lies close to unity, so that I_ν is now characterised by a brightness temperature T_b with the property $I_\nu = 2(\nu^2/c^2)kT_b$ (Rayleigh-Jeans law for $h\nu/kT_b \ll 1$). Now, when we again substitute throughout for N_e the density of the collision partners N', instead of (5.30) we get

$$\frac{N_2}{N_1} = \frac{g_2}{g_1} \frac{\exp(-h\nu/kT) + \frac{A_{21}}{N'Q_{21}}\frac{kT_b}{h\nu}}{1 + \frac{A_{21}}{N'Q_{21}} + \frac{A_{21}}{N'Q_{21}}\frac{kT_b}{h\nu}} \ . \tag{5.32}$$

When $A_{21}/N'Q_{21} \ll 1$ this expression transforms into the Boltzmann Formula. One often however formally sets

$$\frac{N_2}{N_1} = \frac{g_2}{g_1}\exp(-h\nu/kT_{\mathrm{ex}}) \tag{5.33}$$

with the *excitation temperature* T_{ex}, which in general differs from the kinetic temperature T. Since in the radio region not only kT but also kT_{ex} are large compared with $h\nu$, and therefore $\exp(-h\nu/kT) \approx 1 - h\nu/kT$ for both temperatures, it follows from the right-hand sides of (5.32) and (5.33) that

$$T_{\mathrm{ex}} = \frac{T + xT_b}{1 + x} \quad \text{with} \tag{5.34}$$

$$x = \frac{A_{21}}{N'Q_{21}}\frac{kT}{h\nu} \ .$$

Here x represents the ratio between the mean lifetimes of the starting states of the emission for the effectiveness separately of the collisional processes $t_{col} = (N'Q_{21})^{-1}$ [s] and of the radiation processes t_r, given by

$$t_r = [B_{21}(4\pi/c)I_\nu + A_{21}]^{-1} = \{A_{21}/[1 - \exp(-h\nu/kT)]\}^{-1}$$
$$= \{A_{21}kT/h\nu\}^{-1}[s] \quad,$$

so that we can also write

$$T_{ex} = \frac{t_s T + t_{col}T_b}{t_r + t_{col}} \quad. \tag{5.35}$$

If, for example, the radiative processes permit a considerably longer lifetime than the collisional processes: $t_r \gg t_{col}$, then the collisional processes are dominant and we have $T_{ex} = T =$ kinetic temperature.

Let us now discuss the *applications* to three actual cases. First of all population of the starting state of the *21 cm emission line* from neutral hydrogen (cf. Sect. 4.1.5). In the regions from which this radiation comes the ionisation is very low (H I regions), so that free electrons for excitation are excluded. As collision partners, only the relatively numerous neutral H atoms themselves can play a role. We therefore set $N' = N_H =$ number density of the (neutral) hydrogen atoms. We take the collisional rate coefficient Q_{21} from Table 5.1 at the kinetic temperature: for $T = 10^2$ K (see Sect. 5.4.2) $Q_{21} \approx 1 \times 10^{-10}$ cm^3 s^{-1}. On the other hand, for the excitation energy of the 21 cm line $\Delta E = h\nu = 5.9 \times 10^{-6}$ eV and for the transition probability $A_{21} = 2.87 \times 10^{-15}$ s^{-1}. By using $kT/h\nu = T/11\,600\,\Delta E$ [eV] we obtain as the condition for equality of the excitation temperature T_{ex} and the kinetic temperature T:

$$x \approx 4 \times 10^{-4}\frac{T}{N_H} \ll 1 \quad \text{or} \quad N_H \gg 4 \times 10^{-4}T \quad, \tag{5.36}$$

if the comparatively low temperature of the cosmic background radiation $T_b = 2.7$ K may be taken for the radiation field.

Table 5.1. Rate coefficients for de-exciting collisions by H atoms and H$_2$ molecules. For comparison the last column gives the values of $\bar{v}\cdot\bar{\sigma}$ for elastic collisions between H atoms, in which the total kinetic energy of the collision partners is the same before and after the collision. [After Spitzer (1978, 1968)]

T [K]	Q_{nm} [cm^3 s^{-1}]				$\bar{v}\cdot\bar{\sigma}$ [cm^3 s^{-1}]
	H–H (Hyperfine structure levels $n=1$)	H–C$^+$ ($^2P_{3/2}$-level)	H–H$_2$ ($J=2$ of H$_2$)	H$_2$–CO ($J=1$ of CO)	H–H
10	2.3×10^{-12}	7.7×10^{-10}	9.6×10^{-13}	1.8×10^{-12}	9.5×10^{-11}
100	9.5×10^{-11}	8.0×10^{-10}	3.0×10^{-12}	3.7×10^{-12}	3.2×10^{-10}
1000	2.5×10^{-10}	9.7×10^{-10}	4.2×10^{-11}		8.0×10^{-10}

In the cold interstellar clouds, for which $T \lesssim 10^2$ K and $N_H > 0.1$ cm^{-3}, this condition is fulfilled. For the thin "warm" gas component, however (cf. Sect. 5.4.2), one has to reckon with an underpopulation of the upper state in comparison with a Boltzmann distribution for $T_{ex} = T$, so far as the excitation is caused purely by collisions.

In the case of the *CO emission line* at $\lambda = 2.6$ mm ($J = 1 \rightarrow 0$) one can start with the fact that the collisional excitations are caused primarily by H$_2$ molecules, since the hydrogen at the place of origin of this radiation must be primarily in molecular form. The temperatures here are between about 10 and 30 K, so that from Table 5.1 we must expect a collisional de-excitation rate $Q_{21} \approx 1 \times 10^{-11}$ cm^3 s^{-1}; the probability of spontaneous radiation transitions amounts to $A_{21} = 6 \times 10^{-8}$ s^{-1}. With $h\nu = \Delta E = 4.8 \times 10^{-4}$ eV it then follows that $x \approx 1 \times 10^3 (T/N')$. Only in very thick clouds with $N' = N(H_2) \gtrsim 1 \times 10^4$ cm^{-3} and $T \approx 10$ K can one expect LTE with $T_{ex} = T$. At lower densities, on the other hand, radiative excitation becomes important – at very low temperatures by photons of the cosmic background radiation with $T_b = 2.7$ K.

As the final example let us deduce the population of the starting state for the *optical interstellar absorption line of the CN molecule* listed third in Table 4.4 of Sect. 4.1.4. This line arises from an electronic transition from the first excited rotational level $J = 1$ of the ground state. Here the probability of radiative transitions $J = 1 \rightarrow 0$ is relatively high, with $A_{21} \approx 10^{-5}$ s^{-1} in the denotation previously used, and the excitation energy ΔE of the level $J = 1$ corresponds – as for the CO line – to a frequency falling in the principal region of the cosmic background radiation ($\Delta E \approx kT_b$ with $T_b = 2.7$ K). Accordingly one finds that $x \gg 1$ or $t_r \ll t_{col}$ and so from (5.34) or (5.35) $T_{ex} = T_b = 2.7$ K. The significance of the collisions declines markedly in favour of excitation by the 3 K radiation field. The interstellar CN molecules are therefore sensitive indicators of the cosmic background radiation.

Recombination Lines: Here the population of the starting state n arises either from direct recombination of the ions in this state or as the result of recombinations in higher states followed by radiative transitions to n, and hence only through processes leading "downwards". For very large n, and so for the radio recombination lines, excitation from lower states (by collisions) can become important. For the sake of simplicity we formulate the statistical equations here only for the case of the optical lines of hydrogen, in which the terms for stimulated emission and collisional processes do not appear. In Sect. 5.2.1 it is shown that in an H II region ionisations are very much more uncommon than these transitions (ionisation rate $\approx 10^{-8}$ s^{-1} compared with $A_{nm} \approx 10^4 \ldots 10^8$ s^{-1}). After the recombination in a certain bound state the further downward leading processes follow very rapidly, and thereafter the atom spends a relatively long time in the ground state before eventually another ionisation occurs. We can accordingly omit the ionisation processes from the statistical equations for the bound states and obtain ($n = 1, 2, \ldots$)

$$N_n \sum_{m=1}^{n-1} A_{nm} = \sum_{m=n+1}^{\infty} N_m A_{mn} + N_p N_e \alpha_{0,n}(T_e) \quad . \tag{5.37}$$

On the left stand the removals from n into the lower levels, on the right the entries into n from the higher levels and from direct recombinations in n. N_p and N_e denote the number densities of the protons and electrons, $\alpha_{0,n}(T_e)$ is the recombination coefficient ($=$ probability of a recombination per s into the state n with $N_p = N_e = 1 \, \mathrm{cm}^{-3}$; further discussion in Sect. 5.2.1) with $T_e =$ kinetic temperature of the electron gas ("*electron temperature*"). It is further assumed that the gas nebula under consideration is optically thin in all the recombination lines, so that no radiative excitation processes (from $m = 1, 2, \ldots, n-1$ to n) are to be taken into account.

In thermodynamic equilibrium the Saha Equation and the Boltzmann Formula are valid, which in the case of hydrogen ($T = T_e$) lead to

$$\frac{N_p N_e}{N_1} = \left(\frac{2\pi m_e k T_e}{h^2} \right)^{3/2} \exp(-\chi_0 / k T_e) \tag{5.38}$$

[see Appendix D for $i = 0$ with $N_0 \approx N_{0,1} \equiv N_1$ in the denotation used here and $2u_1/u_0 = 1$] and

$$\frac{N_n}{N_1} = n^2 \exp(-\chi_{0,n} / k T_e) \tag{5.39}$$

[see Appendix D with $g_{0,n} = 2n^2$]. χ_0 and $\chi_{0,n}$ denote the ionisation energy and excitation energy of the state n of the neutral atom. If one again introduces the coefficients $b_n = N_n / N_n^*$, where N_n^* are the population numbers calculated from (5.39) for LTE, then after elimination of N_1 with the help of (5.39), (5.38) is transformed into the important formula

$$N_n = b_n n^2 \left(\frac{h^2}{2\pi m_e k T_e} \right)^{3/2} \exp[-(\chi_0 - \chi_{0,n})/k T_e] N_p N_e \quad . \tag{5.40}$$

If one substitutes this expression for N_n and N_m in (5.37) we obtain a system of equations for the b_n coefficients:

$$\frac{\alpha_{0,n}}{n^2} \left(\frac{2\pi m_e k T_e}{h^2} \right)^{3/2} \exp[-(\chi_0 - \chi_{0,n})/k T_e] + \sum_{m=n+1}^{\infty} b_m A_{mn}$$

$$= b_n \sum_{m=1}^{n-1} A_{nm}.$$

As we see, in the case under consideration the b_n do not depend on the density, but only on T_e. We cannot here go into the methods of solution of this system of equations. Numerical results for $n = 2$ up to 7 at $T_e = 10^4$ K are given in Table 5.2 (Case A).

Table 5.2. Values of the b_n coefficients for the population of the energy states of the hydrogen in the case (A) of a gas nebula, optically thin in the lines, with $T_e = 10\,000$ K. Values for the case (B) of a nebula, optically thick in the Lyman lines $(1{\rightarrow}n)$, are given in the last line. [From Burgess (1958)]

n	2	3	4	5	6	7
$b_n(A)$	0.0038	0.036	0.087	0.136	0.176	0.027
$b_n(B)$	–	0.108	0.183	0.245	0.290	0.321

For large n, such as occur with the radio recombination lines, collision processes play an essential role, because atoms in very high quantum states have collisional cross-sections higher by many powers of ten than in the lowest states with small orbital radii of the electrons. Accordingly the b_n then first become dependent also on N_e, but they approximate to the value $b_n = 1$. In addition the stimulated emission from the continuous radiation field of the gas can also no longer be neglected. Figure 5.3 shows results for $n>20$ at $T_e = 10\,000$ K.

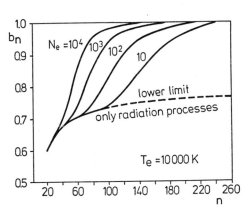

Fig. 5.3. Dependence of the b_n coefficients on the principal quantum number n for the electron temperature $T_e = 10\,000$ K. Parameter of the family of curves is the electron density N_e [cm^{-3}]. [After Sejnowski and Hjellming (1969)]

Line Intensities as a Function of Density and Temperature: The strengths of the lines to be expected cannot be determined from a discussion of the population ratios alone. In general one has to solve the radiation transport equation (5.2) with the source function given by (5.16), simultaneously with the statistical equations (5.17) or (5.18). We first of all consider two cases in which radiation excitation and stimulated emission play no essential role, so that the statistical equations are decoupled from the radiation transport equation: collision-excited lines of heavier ions and optical recombination lines in an optically thin gas nebula. Afterwards we shall go on to discuss the more general case of radio line emission by molecules, where also the radiation processes, and in particular the emission stimulated by the radiation field, must be included.

The intensity of an emission line I_ν^L is given in general as the difference between the intensity I_ν *observed* in the line and the (continuous) background intensity $I_{\nu,0}$, which one finds by observations in a frequency lying directly next to the line. In what follows we take as basis the model of an emitting cloud already considered in Sect. 5.1.1. Then I_ν is given by (5.4), where the source function in the line S_ν^L can depend on $I_{\nu,0}$ (radiation excitation, stimulated emission!).

For the collision excited optical emission lines and for the optical recombination lines of gas nebulae $I_{\nu,0} \ll I_\nu$ [see discussion after (5.30)]. In the optically thin case $\tau_\nu^L \ll 1$, since $S_\nu^L d\tau_\nu^L = \kappa_\nu^L S_\nu^L dr = \varepsilon_\nu^L dr$ one then finds from (5.4) simply, using (5.12),

$$I_\nu^L = I_\nu - I_{\nu,0} = \int_0^l \varepsilon_\nu^L\, dr = \frac{h\nu_{nm}}{4\pi}\psi(\nu)A_{nm}\int_0^l N_n\, dr \quad . \tag{5.41}$$

For the sake of simplicity we shall in the following assume constant density and temperature. Then the ratio of the strengths of two emission lines with the same lower level m and the same profile shape $\psi(\nu)$ is according to (5.12) given by $I_{nm}^L/I_{n'm}^L = (\nu_{nm}/\nu_{n'm})(A_{nm}/A_{n'm})(N_n/N_{n'})$. In the approximation of the two-level atom, for *forbidden emission lines*, which are excited by electron collisions (density and temperature of the electrons N_e and T_e), it then follows from (5.31) that

$$\frac{I_{nm}^L}{I_{n'm}^L} = \frac{\nu_{nm}A_{nm}g_n}{\nu_{n'm}A_{n'm}g_{n'}}\left[\frac{1+(A_{n'm}/N_e Q_{n'm})}{1+(A_{nm}/N_e Q_{nm})}\right]$$

$$\times \exp\left\{-\frac{h\nu_{nm}-h\nu_{n'm}}{kT_e}\right\} \quad . \tag{5.42}$$

If the two upper levels n and n' lie close together one may set the exponential factor practically equal to unity and one can determine the *electron density* N_e from the observed ratio $I_{nm}/I_{n'm}$. The best known examples of application are the two lines [O II] $\lambda 3728.9\,\text{Å}$ and $\lambda 3726.2\,\text{Å}$ with the upper levels $^2D_{5/2}$ and $^2D_{3/2}$, respectively, and the lower level $^4S_{3/2}$ (ground state). The situation with the line pair [S II] $\lambda 6716.4\,\text{Å}$ and $\lambda 6730.8\,\text{Å}$ is completely analogous. The calculated dependence of the intensity ratios for these two line-pairs on the product $N_e\sqrt{10^4/T_e}$ is shown in Fig. 5.4. The dependence on $\sqrt{T_e}$ arises in the coefficients of the collision rates [see (5.25)]. For the approximate value $T_e = 10^4\,\text{K}$ the abscissa gives the electron density directly. With the result for N_e a more exact value can be obtained for the *electron temperature* T_e from the intensity ratio $I_{nm}^L/I_{n'm'}^L$ of two lines at significantly different frequencies ν_{nm} and $\nu_{n'm'}$. An example of this is the ratio of the [O III] lines $\text{I}(\lambda 4959 + \lambda 5007)/\text{I}(\lambda 4363)$, which for $T_e = 7000\,\text{K}$ to $14\,000\,\text{K}$ decreases from 10^3 to 10^2.

The total intensity of an *optical recombination line* is obtained from (5.41) with (5.12) and (5.40) by integration over the whole line

$$I_{nm}^L = \frac{h\nu_{nm}}{4\pi}A_{nm}\int_0^l N_n\, dr = F_{n,m}(T_e)\int_0^l N_p N_e\, dr \quad . \tag{5.43}$$

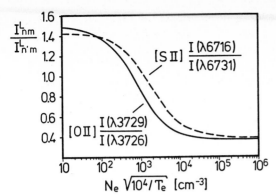

$\dfrac{I^L_{nm}}{I^L_{n'm}}$

1.6
1.4
1.2
1.0
0.8
0.6
0.4

[S II] $\dfrac{I(\lambda 6716)}{I(\lambda 6731)}$

[O II] $\dfrac{I(\lambda 3729)}{I(\lambda 3726)}$

10 10^2 10^3 10^4 10^5 10^6

$N_e \sqrt{10^4/T_e}$ [cm^{-3}]

Fig. 5.4. Calculated dependence of the line ratios $I(\lambda 3729)/I(\lambda 3726)$ and $I(\lambda 6716)/I(\lambda 6731)$ in H II regions on the product $N_e\sqrt{10^4/T_e}$, where T_e is expressed in [K]. [After Osterbrock (1974)]

For hydrogen-like atoms A_{nm}, and hence the function $F_{n,m}(T_e)$, can be represented by a closed expression with a quantum mechanical correction factor lying close to unity, the so-called Gaunt Factor $g_{n,m}$. We have from (5.14)

$$A_{nm} \sim \nu^3_{nm}(g_m/g_n)B_{mn} \quad , \qquad \text{where}$$

$$B_{mn} \sim g_m^{-1}(m^{-2} - n^{-2})^{-3}m^{-3}n^{-3}\nu^{-1}_{nm}g_{n,m} = g_m^{-1}R^3 Z^6 \nu^{-4}_{nm}m^{-3}n^{-3}g_{n,m}$$

and

$$g_n = 2n^2 \quad , \quad g_m = 2m^2 \quad .$$

The right-hand expression follows by use of the Rydberg formula (see Sect. 4.2.3). Since

$$\chi_0 - \chi_{0,n} = -E_n = RhZ^2/n^2 \quad ,$$

where R is the Rydberg Constant in Hz, and $N_p = N_e - N(\text{He}^+)$, we obtain for the whole line the expression

$$I^L_{nm} = 34.24\frac{g_{n,m}}{n^3 m^3}\frac{b_n}{T_e^{3/2}}\exp\left\{\frac{158\,000}{n^2 T_e}\right\}\frac{E}{1 + [N(\text{He}^+)/N_p]}$$

$$[\text{erg cm}^{-2}\,\text{s}^{-1}\,\text{sterad}^{-1}] \quad . \tag{5.44}$$

Here E denotes the so-called *emission measure*

$$E = \int_0^l N_e^2 dr \quad , \tag{5.45}$$

which is here expressed in pc cm^{-6}.

For the radio recombination lines with $n\gtrsim 100$, already $b_n \approx 1$ for moderate densities (cf. Fig. 5.3) and one is approaching the situation $S^L_\nu = B_\nu(T_e)$, which is valid in LTE, and so $\varepsilon^L_\nu = \kappa^L_\nu B_\nu(T_e)$. In contrast to the optical case the superimposed continuous radiation of the gas is sufficiently strong in relation

259

to the line radiation to cause a noticeable stimulated emission. This not only introduces additional terms into the statistical equations (5.37), but also causes the factor

$$1 - \frac{g_m}{g_n} \frac{N_n}{N_m} = 1 - \frac{b_n}{b_m} \exp(-h\nu/kT_e) \quad , \tag{5.46}$$

occurring in the line absorption coefficient (5.15a) to be different from unity, where now $h\nu/kT_e \ll 1$. With deviations from LTE b_n/b_m becomes greater than unity when $n > m$ (cf. Fig. 5.3), so that the factor (5.46) can even become negative. Then a strengthening of the continuum intensity occurs in the region of the line and the line intensity lies above the LTE value (maser). From measurements of the absolute intensities of several suitably chosen radio recombination lines T_e and E can be derived, and, since $b_n = b_n(T_e, N_e)$, even N_e provided n is not too large.

In the case of *radiolines of molecules* the intensity of the background $I_{\nu,0}$ is often comparable with I_ν itself. We limit ourselves to the optically thin case $\tau_\nu^L \ll 1$ and again assume spacewise constant density, temperature and source function. Then from the upper equation (5.6) we have

$$I_\nu^L = I_\nu - I_{\nu,0} = (S_\nu^L - I_{\nu,0})\tau_\nu^L = (S_\nu^L - I_{\nu,0})\kappa_\nu^L l \tag{5.47}$$

l being the length of the line of sight inside the cloud. We satisfy ourselves with discussion of the case of weak lines: $I_\nu - I_{\nu,0} \ll I_{\nu,0}$. Then we may replace the intensity ocurring in the statistical equations by $I_{\nu,0}$ and instead of (5.30) we get — in the approximation of the two-level system —

$$\frac{N_2}{N_1} = \frac{g_2}{g_1} \frac{N'(Q_{21}/A_{21})\exp(-h\nu/kT) + I_{\nu,0}(c^2/2h\nu^3)}{1 + N'(Q_{21}/A_{21}) + I_{\nu,0}(c^2/2h\nu^3)} \quad , \tag{5.48}$$

where we have again denoted the number density of collision partners generally by N'. The source function follows from the substitution of (5.48) in (5.16). As has been shown by Kegel (1976), the result can be written in the form

$$S_\nu^L = \frac{\beta B_\nu(T)}{1+\beta} + \frac{I_{\nu,0}}{1+\beta} \quad \text{with} \tag{5.49}$$

$$\beta = \frac{N'Q_{21}}{A_{21}}[1 - \exp(-h\nu/kT)] \quad . \tag{5.50}$$

Here the numerator of the first term in (5.49) has been recast using the identity $e^{-u} = (1 - e^{-u})/(e^u - 1)$, $(u = h\nu/kT)$. The first term in (5.49) represents the part of the source function which is determined only by the collisional processes, and is accordingly independent of the radiation field; when $\beta \gg 1$ it becomes $B_\nu(T)$. In analogous fashion, from (5.48) and (5.15a) we obtain the line absorption coefficient

$$\kappa_\nu^L = \frac{h\nu}{c} B_{12} \psi(\nu) N_1 \frac{1+\beta}{1+N'(Q_{21}/A_{21}) + I_{\nu,0}(c^2/2h\nu^3)} \quad . \tag{5.51}$$

Finally, substituting (5.49) and (5.51) in (5.47) and integrating over the line,

$$I_{21}^L = \frac{h\nu}{c} N_1 l \frac{B_{12}\beta[B_\nu(T) - I_{\nu,0}]}{1+N'(Q_{21}/A_{21}) + I_{\nu,0}(c^2/2h\nu^3)} \quad . \tag{5.52}$$

For a finite line strength to result, the kinetic temperature T must obviously be greater than the brightness temperature of the background defined by $I_{\nu,0} = 2(\nu/c)^2 k T_b$. This is also clearly understandable, since $T \to T_b = 2.7\,\mathrm{K}$ indicates transition into the case of thermodynamic equilibrium at this temperature, at which there is no line. The transition $T_{ex} \to T_b$ for $A_{21}/N'Q_{21} \gg 1$ or $\gg h\nu/kT$ following from (5.34) at first produced the paradoxical conclusion that with increasing Einstein Coefficients A_{21} the radio molecular lines must grow weaker, and from the existence of the lines it was deduced that $T_{ex} > T_b$ and $A_{21}/N'Q_{21} \ll 1$, so that collisional processes must dominate in every case. Thereupon lower limits were estimated for the density N', and so in the case of the CO line for the density of the molecular hydrogen.

With the help of (5.52), however, one can show (for the optically thin case) that the line strength does not decrease with increasing A_{21}: if the radiative processes dominate, so that $N'Q_{21}/A_{21} \ll 1$ and thereby also $\beta \ll 1$, then it follows that $I_{21}^L \sim N_1 N' l$, and the observed intensity becomes proportional to the square of the density. At the same time it is independent of A_{21}, since $B_{12} \sim B_{21} \sim A_{21}$ and so A_{21} drops out of the product βB_{12}. If on the other hand the population ratios are determined by collisional processes, then $N'Q_{21}/A_{21} \gg 1$, and so $\beta \gg 1$, whence $I_{21}^L \sim A_{21} N_1 l$. This is the case, as we have already seen, for the 21 cm line and, for sufficiently high densities, also for the CO line $\lambda = 2.6\,\mathrm{mm}$. (Use of the total density $N = N_1 + N_2$ instead of the density of the molecules in the ground state N_1 brings no change in these results). Because of the appearance of the temperature as second unknown parameter, direct conclusions on the density or the column density are in general only possible if the intensities of several lines of the same atom or molecule have been observed. For this it is necessary, of course, to extend the non-LTE theory to atoms or molecules with more than two energy states.

If one moves to arbitrary optical thickness τ_ν^L, then the radiation transport equation can be decoupled from the statistical equations only when $N'Q_{21}/A_{21} \gg 1$, and hence $T_{ex} = T$. Then (5.5) (LTE) is valid and for $T = $ constant we have

$$I_\nu^L = I_\nu - I_{\nu,0} = [B_\nu(T) - I_{\nu,0}](1 - e^{-\tau_\nu^L}) \quad . \tag{5.52a}$$

If T is known, then the optical thickness of the molecular cloud $\tau_\nu^L \approx \kappa_\nu^L l$ can be obtained by measuring I_ν^L, and from (5.51) with $\kappa_\nu^L \sim N_1 \beta \sim N_1 N'$, also a conclusion on the density. We cannot here go into approximate solutions when

$N'Q_{21}/A_{21} \lesssim 1$, but we quote in Sect. 5.2.4 selected numerical results relating to this case.

5.1.3 Continuous Emission and Absorption

In the interstellar gas the following radiation processes with continuous spectrum play a significant role. (1) Free-free transitions of thermal electrons in ion fields produce the emission or absorption of radiofrequency radiation in the H II regions, and also in regions of very high temperature optical and x-ray emission (thermal bremsstrahlung) (see Sect. 4.1.4 O VI lines; Sect. 4.1.9). (2) Bound-free transitions by absorption of ultraviolet stellar radiation lead to complete or partial ionisation of the gas particles; free-bound transitions contribute to the optical continua of the H II regions, but are usually unimportant at radiofrequencies. (3) Non-thermal electrons of the cosmic radiation, which move with relativistic velocities in interstellar magnetic fields, cause synchrotron radiation which can occur in practically all frequency regions.

Free-Free Transitions of Thermal Electrons: In passing a positive ion an electron according to the classical theory emits a radiation pulse, resulting in a constant Fourier spectrum for low frequencies. Averaging over all incident electron velocities under the assumption of a Maxwell distribution, however, causes at high frequencies − corresponding to high velocities − an exponential decrease. For hydrogen-like ions of charge Ze the exact result for the emission coefficient is as follows:

$$\varepsilon_\nu^C(\text{f-f}) = \frac{8}{3}\left(\frac{2\pi}{3}\right)^{1/2}\frac{Z^2 e^2}{m_e^{3/2} c^3 (kT)^{1/2}} g_{\text{ff}} N_e N_i \exp(-h\nu/kT)$$

$$\approx 5.44 \times 10^{-39}\frac{Z^2 g_{\text{ff}}}{T^{1/2}} N_e N_i \exp(-h\nu/kT)[\text{erg cm}^{-3}\,\text{s}^{-1}\,\text{sterad}^{-1}\,\text{Hz}^{-1}]$$

$$(5.53)$$

Here m_e denotes the mass of the electron, c the velocity of light, N_e and N_i the number densities of the electrons and the ions [cm^{-3}] and g_{ff} the "Gaunt factor", a quantum mechanical correction factor.

The exponential factor has the effect that, in the temperatures of typical H II regions $T \approx 10^4$, emission occurs essentially only in the radiofrequency region $(h\nu/kT \ll 1)$ and possibly in the infrared. For the Gaunt factor then the following expression, valid when $\nu \gg \nu_p = (e^2 N_e/\pi m_e)^{1/2}$ $(=\text{plasma frequency})$, has been calculated:

$$g_{\text{ff}} = \frac{\sqrt{3}}{\pi}\left\{\ln\frac{(2kT)^{3/2}}{\pi Z e^2 m_e^{1/2}\nu} - \frac{5\gamma}{2}\right\} = \frac{\sqrt{3}}{\pi}\left\{\ln\frac{T^{3/2}}{Z\nu} + 17.7\right\}$$

$$(5.54)$$

where $\gamma = $ Euler's Constant $= 0.577$. It therefore has only a weak dependence on T and ν, which in the relevant radio region can be approximated by

$$g_{ff} \sim T^{0.15} \nu^{-0.1} \quad . \tag{5.55}$$

The total free-free emission per cm^3 is obtained by integration of (5.53) over all frequencies

$$4\pi\varepsilon_{ff} = 4\pi \int_0^\infty \varepsilon_\nu^C (\text{f-f}) d\nu$$

$$= 1.426 \times 10^{-27} Z^2 T^{1/2} N_e N_i \bar{g}_{ff} \quad [\text{erg cm}^{-3} \text{s}^{-1}] \quad . \tag{5.56}$$

The Gaunt factor $\bar{g}_{ff}$ averaged over all frequencies, for $T/Z^2 = 10^4 \ldots 10^6$, lies between 1.25 and 1.45.

In passing a positive ion an electron can also acquire energy from the radiation field and thereby attain a higher kinetic energy. The free-free absorption coefficient $\kappa^C(\text{f-f})$ can be calculated from $\varepsilon^C(\text{f-f})$ by consideration of thermodynamic equilibrium, because in this case Kirchhoff's Law is valid: $\varepsilon_\nu = \kappa_\nu B_\nu(T)$. Using the Rayleigh-Jeans approximation for $B_\nu(T)$ in the radio region, we obtain, with $\varepsilon_\nu^C(\text{f-f})$ from (5.53)

$$\kappa_\nu^C (\text{f-f}) = \frac{\varepsilon_\nu^C (\text{f-f})}{2(\nu/c)^2 kT} = \frac{4}{3} \left(\frac{2\pi}{3}\right)^{1/2} \frac{Z^2 e^6}{m_e^{3/2} c (kT)^{3/2}} \frac{1}{\nu^2} \frac{N_e N_i}{g_{ff}} \quad . \tag{5.57}$$

The optical thickness formed from this can often be approximated by the expression

$$\tau_\nu^C (\text{f-f}) = 8.24 \times 10^{-2} Z^2 T^{-1.35} \nu^{-2.1} \int N_e N_i \, dr \quad . \tag{5.58}$$

Here ν is in GHz and the path length r is expressed in pc.

Intensity of the Thermal Radio Continuum: We restrict ourselves to the two limiting cases of small and large optical thickness [Sect. 5.1.1, (5.6)] and set $I_{\nu,0} = 0$. If a Maxwellian distribution of the velocities of the free electrons for the electron temperature T_e pertains, we can set $S_\nu = B_\nu(T_e) = 2(\nu^2/c^2)kT_e$. In the longwave region only the f-f transitions are important for the optical thickness, so remembering that $N_i \sim N_e$ (in the case of pure hydrogen $N_i = N_e$) and using (5.58) we obtain

$$I_\nu = \begin{cases} \tau_\nu^C B_\nu(T_e) \sim \nu^{-0.1} T_e^{-0.35} E & \text{for} \quad \tau_\nu^C \ll 1 \\ B_\nu(T_e) \sim \nu^2 T_e & \text{for} \quad \tau_\nu^C \gg 1 \end{cases} \quad . \tag{5.59}$$

Here E is the emission measure defined by (5.45). The observed profile of the radio continuum of H II regions, described in Sect. 4.2.2 (see Fig. 4.37), can be interpreted then as the result of the transition from the optically thick to the optically thin case with increasing frequency: $\tau_\nu^C \sim \nu^{-2.1}$. If one determines T_e from absolute values of $I_\nu \sim \nu^2 T_e$ at small ν, then the observed intensity in the region $I_\nu \sim \nu^{-0.1} \approx$ const. gives the emission measure, from which at the (approximately) known radial extent l of the H II region a mean value of the electron density can be obtained, or more accurately the root mean square value $(\overline{N_e^2})^{1/2}$. We present an overall summary and discussion of the results in Sect. 5.2.4.

Bound-Free and Free-Bound Transitions: We restrict ourselves to hydrogen and hydrogen-like atoms. We characterise the free state by a continuously variable quantum number $\overline{\kappa}$, defined by means of the kinetic energy of the free electron $E_{\overline{\kappa}} = (m_e/2)v^2 = RhZ^2/\overline{\kappa}^2 > 0$ analogously to the energy of the bound state. The bound state n has the negative energy $E_n = -RhZ^2/n^2$. ($R = 2\pi^2 e^4 m_e/h^3 =$ Rydberg Constant in frequency measure, $Z =$ nuclear charge number). For the absorption coefficient for the bound-free transition $n \to \overline{\kappa}$ we have

$$\kappa_\nu^C(\text{b-f}) = \frac{64\pi^4 m_e\, e^{10} Z^4}{3\sqrt{3}ch^6 n^5 \nu^3} g_{\mathrm{nf}} N_{0,n} \approx 3 \times 10^{29} \frac{Z^4}{n^5 \nu^3} g_{\mathrm{nf}} N_{0,n} \quad . \tag{5.60}$$

The Gaunt Factor g_{nf} lies near to unity in the optical region and depends only weakly on ν: for $\lambda = 912\,\text{Å}$ to $50\,\text{Å}$, $g_{1\mathrm{f}}$ for hydrogen moves between 0.8 and 1.0.

For the emission at the recombination of free electrons with velocities $v \ldots v + dv$ into the bound state n, corresponding to (5.12) one can write

$$\varepsilon_v^C(\overline{\kappa} \to n)dv = (h\nu/4\gamma)(v)dv\, Q_{\overline{\kappa}n}(v)N_1 \quad \text{with}$$

$$N_e(v)dv = N_e\varphi(v; T)dv \quad \text{and} \quad Q_{\overline{\kappa}n}(v) = v\sigma_{\overline{\kappa}n}(v) \quad .$$

Here N_1 denotes the number density of the ions, $\varphi(v; T)$ is the Maxwell distribution (because of $m_e \ll$ ion mass, the velocity is practically equal to the absolute velocity) and $\sigma_{\overline{\kappa}n}(v)$ is the *effective cross-section for the recombination $\overline{\kappa} \to n$*. We now form $\varphi(v; T)$ from (5.21) with (5.22), and in the exponential factor we express v by means of ν and also dv in terms of $d\nu$ with the help of the relations

$$h\nu = \tfrac{1}{2}m_e v^2 + \overline{\chi}_n \;, \qquad h\, d\nu = m_e v\, dv \;,$$

where $\overline{\chi}_n = \chi_0 - \chi_{0,n} = -E_n =$ difference between ionisation energy and excitation energy (from the ground state $n = 1$) of the neutral atom. Then it follows that

$$\varepsilon_v^C(\text{f-b}) = \frac{m_e^{1/2} h^2 \nu}{(2\pi kT)^{3/2}} v^2 \sigma_{\overline{\kappa}n}(v) N_e N_1 \exp[-(h\nu - \overline{\chi}_n)/kT] \quad . \tag{5.61}$$

By consideration of the case of detailed thermodynamic equilibrium with the help of the Kirchhoff Law $\varepsilon_\nu^C = \kappa_\nu^C B_\nu(T)$ an important relation between

$v^2 \sigma_{\overline{\kappa}n}(v)$ and κ_ν^C can now be obtained. For this we assume that all ions are in the ground state: $N_1 = N_{1,1}$ and express $N_{1,1}$ by means of the Saha and Boltzmann formulae [Appendix D] in terms of $N_{0,n}$ as well as N_e and T. With $B_\nu(T)$ from (5.7) it then follows ("Milne relation") that

$$
v^2 \sigma_{\overline{\kappa}n}(v) = \left(\frac{h\nu}{m_e c} \right)^2 \frac{g_{0,n}}{g_{1,1}} \frac{\kappa_\nu^C(\text{b-f})}{N_{0,n}} \quad . \tag{5.62}
$$

The expression (5.60) produces

$$
v^2 \sigma_{\overline{\kappa}n}(v) = \frac{128 \pi^4 e^{10} Z^4}{3\sqrt{3} m_e c^3 h^4} \frac{1}{\nu} \frac{g_{nf}}{n^3} \quad . \tag{5.63}
$$

Now the emission coefficient for free-bound transitions (recombinations) can be calculated from (5.61), which can be of importance for the optical continuum of the gaseous nebulae (Balmer and Paschen continuum: $n = 2$ and $n = 3$ with longwave limits at $\lambda = 3647\,\text{Å}$ and $8206\,\text{Å}$, respectively).

Here let us mention a completely different possibility of continuous emission from H II regions: the state $n = 2$ of the H atom is split into the fine structure levels $2\,^2S_{1/2}$ and $2\,^2P_{1/2,\,3/2}$, of which the first is metastable. In the transition from this into the ground state the emitted radiation energy can consist of two photons: $h\nu' + h\nu'' = h\nu_{12}(\text{Ly}\alpha) = 10.2\,\text{eV}$, where the distribution of the different frequencies and wavelengths is continuous and symmetric about $\lambda = 2431\,\text{Å}$. At low electron densities ($N_e \ll 10^4\,\text{cm}^{-3}$) this *two-photon emission* becomes noticeable in comparison with the H I continuum. See, for example, Osterbrock (1974, p. 73).

Synchrotron Radiation: The non-thermal galactic background radiation in the radiofrequency region described in Sect. 4.1.7 can be readily interpreted in the following way. If an individual particle with charge Ze and mass m moves in a vacuum with velocity v and energy $E = m_0 c^2 / \sqrt{1 - (v^2/c^2)}$ perpendicular to a homogeneous magnetic field of strength B, then it must describe a circular path whose radius is given by (4.43) p. 200. A charge accelerated in this way radiates electromagnetic waves (magnetobremsstrahlung). The magnetobremsstrahlung of relativistic electrons, known as synchrotron radiation, is of particular interest in astrophysics because they are contained in cosmic radiation (cf. Sect. 4.1.9).

For velocities $v \ll c$ the radiation of a charge in the magnetic field is produced in a wide region of the angle θ between the radiation direction and the velocity vector (cyclotron radiation). For electrons with relativistic velocities this is still true only in relation to a coordinate system moving with the electron. In the system of the stationary observer the result is that the radiation is directed only in a very narrow cone of angle $\overline{\theta} = m_0 c^2 / E = \sqrt{1 - (v^2/c^2)}$ (Fig. 5.5). In each case, however, the radiation is linearly polarised (electric vector perpendicular to the magnetic field), if the observer is in the plane of the electron orbit.

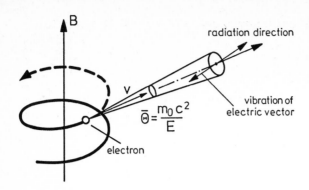

Fig. 5.5. Synchrotron radiation of an electron with velocity $v \simeq c$ in a magnetic field B. Explanation in text

The radiation cone of an individual circling electron impinges on the observer for only a very short time, thus giving him only a very short pulse of radiation, resulting in a relatively high frequency spectrum. The energy radiated in all directions per second and frequency interval $\Delta\nu = 1\,\mathrm{Hz}$ is given by exact analysis as

$$P(\nu; E) = \frac{\sqrt{3}e^2 B}{m_e c^2} F\left(\frac{\nu}{\nu_c}\right) \tag{5.64}$$

with the "critical frequency"

$$\nu_c = \frac{3eB}{4\pi m_e c}\left(\frac{E}{m_e c^2}\right)^2 \approx 16.0 \times BE^2 [\mathrm{MHz}] \quad . \tag{5.65}$$

Here B is in Gauss and the energy in the right-hand expression of (5.65) is in MeV. The function $F(x)$ climbs for small x from $F(0) = 0$ to a maximum value at $x = 0.29$ and then falls off like $\sqrt{x}\,e^{-x}$ approximately. A table for $F(x)$ is given by Ginzburg and Syrovatski (1965; p. 312). The maximum of $P(\nu; E)$ therefore occurs at $\nu_m = 0.29\nu_c$. A selection of (rounded) standard values for ν_m and the corresponding wavelengths λ_m for an interstellar magnetic field of 3×10^{-6} Gauss is given in Table 5.3.

If there are on average in a volume of $1\,\mathrm{cm}^3$ $N_e(E)dE$ electrons with energies in the interval $E \ldots E + dE$, then the continuous emission coefficient is given by the integration

$$\varepsilon_\nu^C(\mathrm{syn}) = \frac{1}{4\pi}\int_0^\infty P(\nu; E)N_e(E)dE \quad . \tag{5.66}$$

Table 5.3. Position of the synchrotron radiation maximum of an individual electron as a function of the electron energy E in a magnetic field $B = 3 \times 10^{-6}$ Gauss

E [MeV]	10^3	10^4	10^5	10^6	10^7
ν_m [MHz]	14	1.4×10^3	1.4×10^5	1.4×10^7	1.4×10^9
λ_m		20 m	20 cm	2 mm	20 μm	2000 Å

In astrophysical applications one usually takes for $N_e(E)$ the power law

$$N_e(E) = KE^{-\gamma}[\text{cm}^{-3}\,\text{erg}^{-1}] \tag{5.67}$$

(for a certain energy region $E_1 \leq E \leq E_2$), which is suggested by observational results for the particles of the cosmic radiation, and which allows the integration to be carried out in closed form. If we insert the numerical values of the physical constants occurring, the result directly follows:

$$\varepsilon_\nu^C(\text{syn}) = 1.35 \times 10^{-22} a(\gamma) K B^{(\gamma+1)/2} \left(\frac{6.26 \times 10^{18}}{\nu[\text{Hz}]}\right)^{(\gamma-1)/2}$$

$$[\text{erg cm}^{-3}\,\text{s}^{-1}\text{sterad}^{-1}\,\text{Hz}^{-1}] \quad . \tag{5.68}$$

The dimension of K is $\text{erg}^{\gamma-1}\,\text{cm}^{-3}$, if E is given in ergs. The factor $a(\gamma) \approx 0.1$ varies only weakly with γ. We have assumed at first that all the electrons move perpendicularly to the magnetic field. A more realistic assumption is an isotropic distribution of the electron motions and, if radiation over a very widely extended region of interstellar space is involved, perhaps also a random distribution of magnetic field directions. For the last case the factor $a(\gamma)$ is given in Table 5.4 for a small selection of values.

Table 5.4. Numerical values of the factor $a(\gamma)$ in (5.68). [After Ginzburg and Syrovatskii (1965)]

γ	1	2	3	4	5
$a(\gamma)$	0.283	0.103	0.0742	0.0725	0.0922

In the result (5.68) it is above all noteworthy that in the optically thin case the spectral intensity distribution of the synchrotron radiation follows a power law similar to the energy spectrum of the electrons:

$$I_\nu = \int_0^l \varepsilon_\nu^C(\text{syn})dr \sim \nu^{-\alpha} \qquad \text{or} \tag{5.69a}$$

$$T_b = \frac{c^2}{2\nu^2 k} I_\nu \sim \nu^{-\beta} \tag{5.69b}$$

with the spectral indices

$$\alpha = \frac{\gamma - 1}{2} \qquad \beta = \frac{\gamma + 3}{2} \quad . \tag{5.70}$$

The observed values given in Sect. 3.17 for the spectral index α for the radiation of the "non-thermal galactic disc" correspond, according to (5.70), to exponents of the electron energy spectrum $\gamma \approx 1.8 \ldots 2.8$, where the smaller values, found at lower frequencies, relate to energies $E<5\,\mathrm{GeV}$. A straightforward comparison with direct extra-terrestrial observations of $N_e(e)$ must be restricted to the region $E\gtrsim 5\,\mathrm{GeV}$ because of the influence of the solar wind on the propagation of "low energy" charged particles in the interplanetary space (cf. Sect. 4.1.9). The direct measurements here give $\gamma \approx 2.5$, in very satisfactory agreement with the radio astronomical result (cf. also Sect. 4.1.9, Fig. 4.30). From the observed absolute values of the radio continuum intensity $I_\nu = \varepsilon_\nu^C l$, and an assumption on the effective geometric layer thickness l, one can even determine the constant K of (5.68), introduced in (5.67). Hence, with an estimate of the magnetic field strength, the total energy density of the electrons of the cosmic radiation can be calculated. If on the other hand one assumes the relatively uncertain value of the total energy density of the electrons from the direct measurements, then the strength of the magnetic field can be estimated. By this method one finds $B \approx 10^{-5}$ Gauss.

For the degree of polarisation P defined by (4.19) one finds with a homogeneous magnetic field and isotropic distribution of electron motions

$$P = (\gamma + 1)/(\gamma + \tfrac{7}{3}) \quad . \tag{5.71}$$

With the values given above for γ one finds that $P \approx 0.7$. With a completely random orientation of the magnetic fields it naturally follows that over a long path length $P = 0$.

The relations (5.68) to (5.71) are strictly valid only for synchrotron radiation in a vacuum. In some cases one must take account of the fact that emission and propagation of the radiation take place in a plasma. Since the refraction index here is smaller than unity, the phase velocity becomes larger than c ("Razin-Tsytovich" effect), so that a relativistic electron cannot stay in phase with the waves it has created. The result of this is a strong *attenuation* of the synchrotron radiation for frequencies

$$\nu < \nu_T = 2\nu_p^2/3\nu_H \approx 20 N_e/B\,[\mathrm{Hz}] \quad .$$

Here $\nu_p = \sqrt{e^2 N_e/\pi m_e}$ is the plasma frequency and $\nu_H = v/2\pi r_H = eB/2\pi m_e c$ the gyration frequency of the relativistic electrons ($v \approx c$). Under the conditions in the interstellar space outside H II regions $N_e<1\,\mathrm{cm}^{-3}$ and with $B = 3 \times 10^{-6}$ Gauss one obtains for the "cut-off frequency" $\nu_T \approx 7\,\mathrm{MHz}$. In addition to this, the propagation of magnetobremsstrahlung can also be attenuated by thermal free-free absorption. If the emitting region contains sufficiently many relativistic electrons, even self-absorption by these same electrons in the magnetic field can become important (magnetic attenuation of the radiation). One can ascertain the absorption coefficients for *synchrotron self-absorption* in principle by a relation corresponding to (5.11) in Sect. 5.1.2, in which the Einstein Coefficients B_{nm} and B_{mn} are expressed by the known relations (5.14) with $A_{nm} = P(\nu; E)/h\nu$ and $P(\nu; E)$ from (5.64). For isotropic distribution of electron motions, where all electrons however see the same magnetic field $B_\perp$, we get the result

$$\kappa_\nu^C(\text{syn}) = 0.019 b(\gamma)(3.5 \times 10^9)^\gamma K B_\perp^{(\gamma+2)/2} \nu^{-(\gamma+4)/2}[\text{cm}^{-1}] \quad . \qquad (5.72)$$

When $\gamma = 1\ldots5$, $b(\gamma)$ lies between 0.65 and 0.36. With the approximate value $K \approx 3 \times 10^{-17} \text{erg}^{\gamma-1} \text{cm}^{-3}$ obtained in the manner described above, for $\gamma \approx 2.5$ and $B = 3 \times 10^{-6}$ Gauss, using (5.72) itself for the low frequency $\nu = 1\,\text{MHz}$ and the path length $l = 10\,\text{kpc}$, one obtains $\tau_\nu^C(\text{syn}) = \kappa_\nu^C(\text{syn})l \ll 1$.

5.2 State of the Interstellar Gas

5.2.1 Ionisation

Ionisation Equilibrium: Let $N_{i,n}$ denote the number density of the i-fold ionised atoms of a certain element which are in the excitation state n, so that $i = 0, 1, 2, \ldots$ correspond to the neutral atom, single ion, double ion, etc. The number density of all i-fold ionised atoms is then given by $N_i = \Sigma N_{i,n}$. Suppose that in an ionisation or recombination process always only one electron is lost or gained, so that there are no transitions $i \rightarrow i-2$ or $i+2 \rightarrow i$; then in a stationary state there exists, in analogy with (5.17) in Sect. 5.1.2, the condition of equilibrium of the numbers of ionisations and recombinations in the form

$$N_i(R_{i,i+1} + C_{i,i+1}) = N_{i+1}(R_{i+1,i} + C_{i+1,i}) \quad . \qquad (5.73)$$

Here $R_{i,i+1}$ denote the photo-ionisation rates, $C_{i,i+1}$ the collisional ionisation rates and $R_{i+1,i}$ and $C_{i+1,i}$ the recombination rates of the $(i+1)$-fold ions with free electrons and radiation away of a photon (photorecombination) or with the release of free nascent energy to an additional collision partner (three-body recombination). In the low densities of the interstellar gas, $C_{i+1,i}$ (proportional to $N_e N'$, and so to the square of the density) can usually be neglected, in contrast to $R_{i+1,i} \sim N_e$.

Ionisation of Hydrogen: We can write (5.73) as

$$N_0(R_{0,1} + C_{0,1}) = N_1 R_{1,0} \quad . \qquad (5.74)$$

In Sect. 5.1.2 it was shown that under interstellar conditions practically all the atoms and ions of an element are in the ground state $n = 1$: $N_i \approx N_{i,1}$. For the calculation of the *photo-ionisation rate* one accordingly has to integrate over $\nu \geq \nu_0$ the number $4\pi\kappa_\nu^C(1 \rightarrow \text{f})I_\nu d\nu/h\nu$ of quanta absorbed in $1\,\text{cm}^3$ and $1\,\text{s}$ from all directions per frequency interval $d\nu$:

$$N_0 R_{0,1} = N_{0,1} R_{0,1} = 4\pi \int_{\nu_0}^{\infty} \frac{\kappa_\nu^C(1 \rightarrow \text{f})I_\nu}{h\nu} d\nu \qquad (5.75)$$

$\kappa_\nu^C(1{\rightarrow}\mathrm{f})$ is given by (5.60), so that $N_{0,1}$ cancels; ν_0 is fixed by the ionisation energy $\chi_0 = h\nu_0$ and corresponds to the wavelength of the Lyman limit $\lambda_0 = 912\,\text{Å}$.

Relatively large values for $R_{0,1}$ are to be expected in the *neighbourhood of hot stars*, in H II regions. If we take as an estimate $I_\nu = W\overline{I_\nu^*}$ with $W \approx 10^{-14}$ [see Sect. 4.1.9, (4.40)] and $\overline{I_\nu^*} \approx B_\nu(T_\mathrm{b})$ with $T_\mathrm{b} \approx T_\mathrm{eff} \approx 40\,000\,\text{K}$ (about an O5 star), then (5.75) leads to a photo-ionisation rate of the order of $R_{0,1} \approx 10^{-8}\,\text{s}^{-1}$. This figure is remarkably small, even in comparison with the A_{nm} for forbidden transitions. In other words, the waiting time of an H atom in the neutral state for the next ionisation amounts to about 10^8 s and is accordingly appreciably longer than the lifetime of the metastable bound states.

At great distances from hot stars the interstellar radiation field is so weak that other sources of a weak ionisation must also be considered. For example, the hydrogen here can be ionised by the *diffuse component of the cosmic x-ray emission* (4.1.9) to a small extent. Since the continuous absorption coefficient (5.60) decreases rapidly with increasing frequency ν and energy $E = h\nu$ of the ionising radiation, x-rays are in practice ineffective when $E > 1\,\text{keV}$. For the important region of soft x-ray emission: $E = 0.2\ldots1\,\text{keV}$ the observations are relatively uncertain. Ionisation rates are estimated to be of the order of $10^{-17}\,\text{s}^{-1}$.

In addition to this photo-ionisation, *collision ionisation by particles of the cosmic radiation* must also be discussed. Analogously with (5.20), we write

$$C_{0,1} = N'Q_{0,1} = N'\overline{v\sigma_{\mathrm{CR}}(v)} \quad . \tag{5.76}$$

Here N' denotes the number density of the high energy particles. For collisional ionisation of hydrogen (ionisation energy χ_0) by particles with charge Ze and velocity v the effective cross-section at energies over $0.3\,\text{MeV}$ is given by Bethe (1933)

$$\sigma_{\mathrm{CR}}(v) = \frac{2\pi e^4 Z^2}{m_e \chi_0 v^2} \times 0.285 \times \left\{ \ln \frac{2m_e v^2}{[1 - v^2/c^2]\chi_0} + 3.04 - \frac{v^2}{c^2} \right\} \quad . \tag{5.77}$$

As we can see, $\sigma_{\mathrm{CR}}(v) \sim v^{-2}$. It therefore turns out that only relatively low energy particles (E less than about $100\,\text{MeV}$), whose interstellar flux must be obtained by extrapolation (Sect. 4.1.9), can contribute to the ionisation. From the observed fluxes values for $C_{0,1}(\mathrm{H})$ of the order of $10^{-17}\,\text{s}^{-1}$ were estimated.

In the region of the hot component of the interstellar gas (Sect. 4.1.4: O VI lines, Sect. 4.1.9: soft x-rays) with temperature considerably above $10\,000\,\text{K}$, *collision ionisations by thermal electrons* become important. For electron energies of about $10^2 \ldots 10^3\,\text{eV}$ the collisional cross-section σ of hydrogen amounts to a few $10^{-17}\,\text{cm}^2$, so that from (5.27) $Q_{0,1} \approx \overline{v}\sigma$ takes values of the order of $10^{-11}\sqrt{T}\,\text{s}^{-1}$.

The rate of *recombinations* with photon emission is generally put in the form

$$R_{i+1,i} = N_e \sum_n \alpha_{i,n} = N_e \alpha_i \quad . \tag{5.78}$$

The $\alpha_{i,n}$ is the recombination coefficient for the transition from the $(i+1)$-fold ion to an i-fold ion in the nth excitation state by electron capture; α_i is the total recombination coefficient. We have

$$\alpha_{i,n} = \overline{v\sigma_{\overline{\kappa}n}(v)} = \int_0^\infty v\sigma_{\overline{\kappa}n}(v)\varphi(v;T)dv \quad , \tag{5.79}$$

where $\sigma_{\overline{\kappa}n}(v)$ denotes the effective cross-section, already derived in Sect. 5.1.3, for the recombination of a free electron with velocity v — corresponding to the continuous quantum number $\overline{\kappa}$ — into the state n.

Let us now reconsider the case $i = 0$. For hydrogen and hydrogen-like atoms $\sigma_{\overline{\kappa}n}(v)$ is given by (5.63). If we there replace the factor $h\nu$ in the denominator by $(m_e/2)v^2 + (\overline{\kappa}_0 - \overline{\kappa}_{0,n})$ and choose for $\varphi(v;T)$ in (5.79) the Maxwell distribution for the kinetic temperature of the electrons, then we get

$$\alpha_{0,n} = \frac{2^9\pi^5 e^{10}Z^4}{m_e^2 c^3 h^3}\left(\frac{m_e}{6\pi kT}\right)^{3/2}\exp(\overline{\chi}_n/kT)F\left(\frac{\overline{\chi}_n}{kT}\right)\frac{g_{nf}}{n^3} \tag{5.80}$$

with

$$F(x) = -\mathrm{Ei}(-x) = \int_x^\infty \frac{e^{-t}dt}{t}$$

(exponential integral) and $\overline{\chi}_n = \chi_0 - \chi_{0,n} =$ ionisation energy out of the state n.

For the important "partial" recombination coefficients

$$\alpha_0^{(j)} = \sum_{n=j}^\infty \alpha_{0,n} \quad \text{with} \quad \alpha_0^{(1)} = \alpha_0 \; , \quad \alpha_0^{(2)} = \alpha_0 - \alpha_{0,1} \tag{5.81}$$

we obtain the following expression

$$\alpha_0^{(j)} = \frac{2.06 \times 10^{-11}Z^2}{T^{1/2}}\Phi_j\left(\frac{\overline{\chi}_1}{kT}\right)[\mathrm{cm}^3\,\mathrm{s}^{-1}] \quad . \tag{5.82}$$

For higher atoms, here considered as hydrogen-like, this is only a crude approximation, in particular for the lower states. Z is set equal to $1, 2, \ldots$, according to whether it is a matter of the recombination of simple ions, second degree ions, and so on. Approximate values are given in Table 5.5 for the function Φ_1 and Φ_2, which vary only slowly. For hydrogen at $T = 10^4$ K the values to be used in what follows are $\alpha_0^{(1)} = \alpha_0 \approx 4 \times 10^{-13}\,\mathrm{cm}^3\,\mathrm{s}^{-1}$ and $\alpha_0^{(2)} = \alpha_0 - \alpha_{0,1} \approx 2.6 \times 10^{-13}\,\mathrm{cm}^3\,\mathrm{s}^{-1}$.

In the region of the hot component of the interstellar gas the condition for the ionisation equilibrium (5.74) for the hydrogen is

$$\frac{N_1}{N_0} = \frac{C_{0,1}}{R_{1,0}} = \frac{Q_{0,1}}{\alpha_0^{(1)}} \quad . \tag{5.83}$$

Table 5.5. Numerical values of the two functions Φ_1 and Φ_2 in (5.82) for the recombination coefficients $\alpha_0^{(j)}$ as functions of T/Z^2. The functions Ψ_j are explained in Sect. 5.2.3. [After Spitzer (1978)]

$T\,[\mathrm{K}]/Z^2$	30	125	1×10^3	4×10^3	8×10^3	3×10^4	1×10^6
Φ_1	4.7	4.0	3.0	2.4	2.1	1.5	0.36
Φ_2	3.9	3.3	2.3	1.6	1.3	0.8	–
Ψ_1	4.2	3.6	2.6	2.0	1.7	1.1	–
Ψ_2	3.5	2.8	1.8	1.2	0.92	0.50	–

For $T = 10^6\,\mathrm{K}$ with $Q_{0,1} \approx 10^{-11}\sqrt{T} = 10^{-8}$ (see above) and $\alpha_0^{(1)} \approx 10^{-14}$ we have practically complete ionisation with $N_1/N_0 \approx 10^6$!

H II Regions: Let us now turn to ionisation equilibrium of the hydrogen in the neighbourhood of a hot star. The sum of the number densities of the neutral hydrogen atoms N_0 and the ions (protons) N_1 is constant: $N_0 + N_1 = N_\mathrm{H}$. Let the relative abundance of the hydrogen be so great that practically all free electrons are produced by its ionisation. The electron density N_e will then be equal to the density of the protons N_1. From the estimated values for $R_{0,1}$ and $C_{0,1}$, collision ionisations can be neglected in comparison with photon ionisations. Then the statistical equations (5.74) become

$$N_0 R_{0,1} = N_1 R_{1,0} = N_1 N_\mathrm{e}\alpha_0 \quad . \tag{5.84}$$

With $R_{0,1} \approx 10^{-8}$ for a very early spectral type, and the value $\alpha_0 \approx 4 \times 10^{-13}$ valid for $T \approx 10^4\,\mathrm{K}$, we obtain $N_1/N_0 \approx 2.5 \times 10^4/N_\mathrm{e}$. For electron densities $N_\mathrm{e} \approx 10^2 \dots 10^3\,\mathrm{cm}^{-3}$, such as exist in typical diffuse nebulae, a high ionisation is accordingly produced: $N_1 \gg N_0$ since $R_{1,0} \ll R_{0,1}$. In words: few neutral atoms, but a high ionisation probability correspond in the equilibrium to many ions with low recombination probability.

If one sets $N_1 = N_\mathrm{e}$ (pure hydrogen nebula), then the foregoing estimate of N_1/N_0 gives for the number of atoms in the ground state $N_{0,1} \approx N_0 \approx N_\mathrm{e}^2/2.5 \times 10^4$, and so for example for $N_\mathrm{e} = 10^2$ to 10^3 the value of $N_{0,1} \approx 0.4 \dots 4 \times 10^{-3}$. Thereupon one obtains for the absorption coefficients $\kappa_{\nu 0}^\mathrm{C}$ at the Lyman limit from (5.60) (with $N_n \equiv N_{0,n}$ at $n = 1$) values between 0.4×10^{-17} and 0.4×10^{-19}. The optical thickness $\tau_{\nu 0} = 1$ is therefore already reached at path lengths of 2.5×10^{17} to 2.5×10^{19} cm, corresponding to about 0.1 to 10 pc. We therefore restrict ourselves in what follows to homogeneous and static gas clouds, which are optically thick in the Lyman continuum. The problem however then presents itself of radiation transport through the cloud.

It should be noticed in particular that in recombinations which lead directly to the ground state $n = 1$ a diffuse radiation field $I_\nu^{(\mathrm{d})}$ with $\nu \leq \nu_0$ is caused, which also contributes to the ionisation. In the expression (5.75) for $R_{0,1}$ one has therefore to set $I_\nu = I_\nu^{(\mathrm{s})} + I_\nu^{(\mathrm{d})}$, where $I_\nu^{(\mathrm{s})}$ is the intensity coming directly from the star. We can however evidently omit $I_\nu^{(d)}$ if in the right-hand side of (5.84) we take account only of recombinations into states $n > 1$, and so replace α_0 by $\alpha_0^{(2)} = \alpha_0 - \alpha_{0,1}$. If we introduce the degree of ionisation x by $x = N_1/(N_0 + N_1) = N_1/N_H$, then it follows that $N_e = N_1 = xN_H$ and also $N_0 = (1 - x)N_H$, and we obtain from (5.84) first of all the relation

$$(1 - x)N_H R_{0,1} = x^2 N_H^2 \alpha_0 \tag{5.85}$$

hence

$$(1 - x)N_H 4\pi \int_{\nu_0}^{\infty} \frac{\kappa_\nu^C I_\nu^{(s)}}{h\nu} d\nu = x^2 N_H^2 \alpha_0^{(2)} \quad . \tag{5.86}$$

In the spherically symmetric case to be considered here the left hand side of the radiation transport equation (5.1) is to be replaced by

$$\frac{dI_\nu}{ds} = \frac{\partial I_\nu}{\partial r}\frac{dr}{ds} + \frac{\partial I_\nu}{\partial \vartheta}\frac{d\vartheta}{ds} \quad .$$

Here r denotes the distance from the centre of the star and ϑ the angle from the radial direction. From Fig. 5.6a one gets

$$dr = ds \cos \vartheta \ , \quad r\, d\vartheta = -ds \sin \vartheta \quad .$$

Thereby one obtains

$$\frac{\partial I_\nu}{\partial r} \cos \vartheta - \frac{\partial I_\nu}{\partial \vartheta}\frac{\sin \vartheta}{r} = -\kappa_\nu I_\nu + \varepsilon_\nu \quad . \tag{5.87}$$

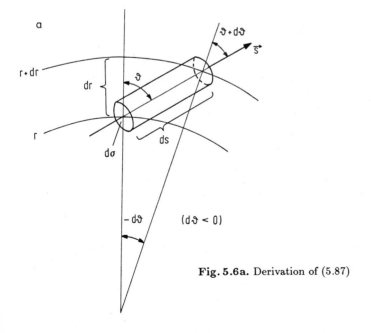

Fig. 5.6a. Derivation of (5.87)

273

By integration over all directions (solid angle element $d\omega = \sin \vartheta \, d\vartheta \, d\varphi$) we accordingly obtain

$$\frac{d\Phi_\nu}{dr} + \frac{2}{r}\Phi_\nu \equiv \frac{1}{r^2}\frac{d}{dr}(r^2\Phi_\nu) = -4\pi\kappa_\nu I_\nu + 4\pi\varepsilon_\nu \quad . \tag{5.88}$$

Here Φ_ν denotes the radiation flux, defined by

$$\Phi_\nu = \iint I_\nu(\vartheta)\cos\vartheta \, d\omega \, [\text{erg cm}^{-2}\,\text{s}^{-1}\,\text{Hz}^{-1}] \quad . \tag{5.89}$$

The factor 4π in front of $\kappa_\nu I_\nu$ on the right-hand side of (5.88) arises from the fact that the intensity of an isotropic radiation field I_ν is defined by (cf. Sect. 4.1.9)

$$\iint \overline{I}_\nu^* \, d\omega = \overline{I}_\nu^* \Omega_* = 4\pi I_\nu \quad ,$$

where I_ν^* denotes the intensity emitted from the star with radius R_*, which occupies only the small solid angle given by $\Omega_* = \pi R_*^2/r^2$. Because of the continuous absorption of the stellar radiation on the way from the surface of the star to the field point r, however, $\overline{I}_\nu^*$ has to be replaced by $\overline{I}_\nu^* \exp(-\tau_\nu)$, where the optical thickness is determined by

$$d\tau_\nu = \kappa_\nu^C \, dr = \kappa_{\nu,\text{atom}}^C N_0 \, dr = \kappa_{\nu,\text{atom}}^C(1-x)N_{\text{H}}\, dr \quad . \tag{5.90}$$

Since $\kappa_\nu^C \sim \nu^{-3}$ for $\nu<\nu_0$ the shortwave radiation is attenuated least, so that there must be a shift of the intensity distribution I_ν over ν towards the shorter wavelengths.

To the direct stellar radiation must also be added the diffuse UV-radiation field with intensity $I_\nu^{(d)}$ and the radiation flux $\Phi_\nu^{(d)}$, which is emitted by those electrons which recombine directly into the ground state $n = 1$, since this radiation with $\lambda<912$ Å also contributes to the ionisation. Finally it must be said that $\varepsilon_\nu = \varepsilon_\nu^{(d)}$ represents just the emission power of this diffuse UV-radiation. If we divide the whole equation by $h\nu$ and integrate over ν from the Lyman limit ν_0 to ∞ and multiply by $4\pi r^2$, we obtain

$$\frac{d}{dr}\left[4\pi r^2 \int_{\nu_0}^{\infty}(\Phi_\nu^{(s)} + \Phi^{(d)})_\nu \frac{d\nu}{h\nu}\right]$$

$$= -4\pi r^2\left[4\pi \int_{\nu_0}^{\infty}\frac{\kappa_\nu(I_\nu^{(s)} + I_\nu^{(d)})}{h\nu}d\nu - 4\pi \int_{\nu_0}^{\infty}\frac{\varepsilon_\nu^{(d)}}{h\nu}d\nu\right] \quad \text{with} \tag{5.91}$$

$$\Phi_\nu^{(s)} = \overline{I}_\nu^*\,\text{e}^{-\tau_\nu}\,\Omega_* \quad \text{and} \quad I_\nu^{(s)} = \overline{I}_\nu^*\,\text{e}^{-\tau_\nu}\,\Omega_*/4\pi = \overline{I}_\nu^*\,\text{e}^{-\tau_\nu}\,W \quad . \tag{5.92}$$

The first term in the square bracket on the right-hand side represents $N_0 R_{0,1}$

from (5.75), and can be replaced by $x^2 N_{\mathrm{H}}^2 \alpha_0$ from (5.85). The second term gives the number of electron captures into the ground state $n = 1$, and is therefore given by $x^2 N_{\mathrm{H}}^2 \alpha_{0,1}$. The expression in the square bracket on the left-hand side of (5.91) just gives the total number of photons in the Lyman continuum with $h\nu > h\nu_0$, flowing per second through the spherical surface with radius r. In what follows we call this quantity $L_c(r)$ and can write (5.91) more briefly as

$$\frac{dL_c(r)}{dr} = -4\pi r^2 x^2 N_{\mathrm{H}}^2 \alpha_0^{(2)} \quad . \tag{5.93}$$

This equation states clearly that at each distance r from the centre of the star the decrease in the total number of all outward flowing photons per unit path length is equal to the total number of all recombinations occurring there into excited states $n > 1$, because these recombinations produce photons which can cause no ionisation of the hydrogen and therefore escape unhindered — in contrast to the photons caused by recombinations into the ground state.

The behaviour of the degree of ionisation $x = x(r)$ can be obtained from (5.86) and (5.90), given the intensity distribution in the stellar spectrum $\overline{I}_\nu^*$ and the total density N_{H} with $I_\nu^{(s)} = \overline{I}_\nu^* \exp(-\tau_\nu) R_*^2 / 4r^2$. This problem was first treated by B. Strömgren (1939). It turned out that as r increases, $x(r)$ is at first practically constant and equal to unity, and then within a relatively short distance it sinks to zero (see Fig. 5.6b). One can accordingly define a radius r_{s}, the so-called *Strömgren Radius*, within which the hydrogen is almost completely ionised and so forms an H II region, whilst beyond r_{s} the hydrogen remains neutral. Here r_{s} naturally depends on $\overline{I}_\nu^*$ and N_{H}.

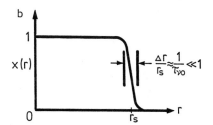

Fig. 5.6b. Definition of the Strömgren radius r_{s} by the radial profile of the degree of ionisation $x(r)$ in an H II region

This result can to a certain extent be clearly understood: let us follow, in the H II region, a radius vector from the star, going from the inside to the outside. First of all $N_1 \gg N_0$ and at the Lyman limit the optical thickness $\tau_{\nu 0}$ is much less than unity. When $\tau_{\nu 0} \approx 0.1$ is reached, the radiation field is appreciably reduced by absorption. This has the consequence that from here onwards there is less ionisation and so N_0 increases at the expense of N_1. The absorption coefficient, being proportional to the density of neutral hydrogen present, also increases; this causes a further reduction in the radiation field, and so on. After a relatively short further path length, a high optical thickness and a high share of neutral atoms must therefore result. The Strömgren radius r_{s} indicates in practice the distance at which the UV-photon flux $L_c(r)$ has fallen to zero, $L_c(r_s) = 0$. If we set the degree of ionisation $x = 1$ for $r \leq r_s$ in (5.93), then integration from $r = R_*$ to $r = r_s$ gives

$$L_c(r_s) - L_c(R_*) = -L_c(R_*) = -4\pi N_H^2 \alpha_0^{(2)} \tfrac{1}{3} r_s^3$$

(since $R_* \ll r_s$). Since $\Phi_\nu^{(d)}(R_*) \ll \Phi_\nu^{(s)}(R_*) = \pi \overline{I}_\nu^*$, it follows that

$$r_s^3 = 3R_*^2 \int_{\nu_0}^{\infty} \frac{\pi \overline{I}_\nu^*}{h\nu} d\nu \Big/ N_H^2 \alpha_0^{(2)} \quad . \tag{5.94}$$

The integral represents the number of Lyman continuum quanta N_L which issue from the surface of the star per cm^2 in 1 second. With the value quoted above $\alpha_0^{(2)} = \alpha_0 - \alpha_{0,1} = 2.6 \times 10^{-13} \,\mathrm{cm}^3\,\mathrm{s}^{-1}$, valid for $T = 10^4\,\mathrm{K}$, we obtain

$$r_s = 1.23 \times 10^{-7} \left(\frac{R_*}{R_\odot}\right)^{2/3} N_L^{1/3} N_H^{-2/3} [\mathrm{pc}] \quad . \tag{5.95}$$

The product $u = r_s N_H^{2/3} \approx r_s N_e^{2/3}$ $[\mathrm{pc\,cm}^{-2}]$, whose cube is essentially equal to the *total* flux of the Lyman continuum quanta $L_c(R_*)$, is often known in the literature as the *excitation parameter*. To calculate N_L one uses model stellar atmospheres, to which can be assigned specified MK-spectral types and stellar radii. Results for $L_c(R_*) = 4\pi R_*^2 N_L$ and r_s are contained in Table 5.6. T_L is the temperature of a black body which emits just N_L quanta per cm^2 per s in the region $\lambda < \lambda_0 = 912\,\text{Å}$.

At sufficiently low densities very extensive H II regions can evidently occur. Observations of "diffuse" or "weak" emission from optical (Hα) and radio recombination lines (Sect. 4.2.3) have shown that such "low density H II regions" actually exist.

Table 5.6. Strömgren radii r_s of H II regions round Main Sequence stars of various spectral types ($Sp.$). M_V is the visual absolute magnitude of the star, $L_c(R_*) = 4\pi R_*^2 N_L$ is the total number of photons with $\lambda \leq 912\,\text{Å}$ emitted from the star per second. For T_L see text. The values of r_s are valid for hydrogen density $N_H = 1\,\mathrm{cm}^{-3}$ and are to be divided by $N_H^{2/3}$ for other densities. [After Osterbrock (1974)]

Sp	M_V	T_L [K]	$L_c(R_*)$ [photons s^{-1}]	r_s [pc]
O5	−5.6	48 000	4.7×10^{49}	108
O6	−5.5	40 000	1.7×10^{49}	74
O7	−5.4	35 000	6.9×10^{48}	56
O8	−5.2	33 000	4.0×10^{48}	51
O9	−4.8	32 000	1.7×10^{48}	34
O9.5	−4.6	31 000	6.9×10^{47}	29
B0	−4.4	30 000	4.7×10^{47}	23
B0.5	−4.2	26 200	6.8×10^{46}	12

Ionisation of Helium, after hydrogen the most abundant element $- N(\text{He})/N(\text{H}) \approx 0.1$ (Sect. 5.2.5) $-$ requires photon energies $h\nu > 24.6\,\text{eV}$, corresponding to wavelengths $\lambda < 504\,\text{Å}$ (single ionisation), and is therefore to be expected to an appreciable extent only for exciting stars with spectral types earlier than about O 8. For later types the number of photons with $h\nu > 24.6\,\text{eV}$ is very much smaller than the number of photons leading to the ionisation of hydrogen, so that helium is ionised only to a distance $r \ll r_\text{s}$ (H II region). On the other hand, for the earliest types the He^+ region cannot be further extended than the H^+ (H II) region: for stellar photons which can ionise helium can also ionise hydrogen. Ionisation of He^+, for which $h\nu > 54.4\,\text{eV}$ is needed, is practically absent from normal H II regions.

Values of the radius r_s of the H II region calculated for a pure hydrogen nebula are only slightly changed by taking account of the helium, so long as the electron density N_e remains below $10^3\,\text{cm}^{-3}$. For a thorough discussion, see Osterbrock (1974, p. 21).

Extinction by dust particles, which are present in considerable quantity in the denser H II regions, can appreciably modify the results of the foregoing discussion. The Strömgren radius r_s is decreased, and since one can count on an approximately constant dust/gas ratio, the reduction of r_s increases with the density. The most important parameter for determining the reduction factor r'_s/r_s is the optical thickness $\tau_{\nu\text{d}}$ of the dust for the path length r_s immediately below the Lyman limit $\lambda = 912\,\text{Å}$. The connection with the corresponding visual extinction A_V of the starlight inside the H II region is given approximately by $\tau_{\nu\text{d}} = 4\,A_\text{V}$. We restrict ourselves here to presenting a few numerical results for r'_s/r_s as a function of A_V in Table 5.7.

Table 5.7. Reduction factors of the Strömgren radii r'_s/r_s by dust particles inside the H II regions as a function of the internal visual extinction A_V. [After Spitzer (1978)]

A_V [mag]	0.1	0.5	1.0	2.0	5.0	10
r'_s/r_s		0.91	0.70	0.56	0.42	0.25	0.15

In conclusion it must be pointed out that the simple homogeneous and static model of an H II region considered here represents a strong idealisation of the true conditions. We have ignored dynamic processes, such as the expansion associated with the heating of a gas, the propagation of ionisation fronts and the creation of shockwaves by the motion of the gas. We shall come back to this in Sect. 6.3.3.

H I Region: We must assume that practically all O- and B-stars are surrounded by more or less extensive H II regions, from which no radiation with $\lambda < 912\,\text{Å}$ can emerge. In the other regions of the interstellar space there is only stellar radiation with $\lambda > 912\,\text{Å}$, corresponding to $E = h\nu < 13.6\,\text{eV}$, available.

The energy of these longer wave photons is only sufficient for the ionisation of higher elements, such as C, Si, Fe etc. Weak ionisation of the hydrogen is caused, however, by the cosmic radiation and diffuse cosmic x-rays. The resulting *degree of ionisation* can be determined from the statistical equations (5.74). We can assume that $N_1 \ll N_0 = N(\text{H I})$. In (5.78) for $R_{1,0}$ we take the electron density to be

$$N_e = N_1 + x_0 N(\text{H I}) \quad . \tag{5.96a}$$

where x_0 represents the relative portion of the electrons coming from other sources. Then (5.74) leads to a quadratic equation for $N_1/N(\text{H I})$, whose solution gives

$$\frac{N_1}{N(\text{H I})} = \frac{x_0}{2} \left\{ \sqrt{1 + \frac{4(R_{0,1} + C_{0,1})}{x_0^2 N(\text{H I}) \alpha_0^{(2)}}} - 1 \right\} \quad . \tag{5.96b}$$

To take account of the fact that all Lyman quanta emitted are immediately absorbed again, we have replaced α_0 by $\alpha_0^{(2)}$. With normal element abundances, the minimal value for x_0 that can be considered to result in complete single ionisation of C, Si, and Fe is $x_0 \approx 5 \times 10^{-4}$. A further contribution to the electron density can come from the photo effect on dust particles described in Sect. 5.2.3.

Taking the estimates $R_{0,1} + C_{0,1} = 1 \times 10^{-17}\,\text{s}^{-1}$ already referred to, and $x_0 = 5 \times 10^{-4}$, we calculate for a diffuse cold cloud with $T = 100\,\text{K}$ and $N(\text{H I}) = 20\,\text{cm}^{-3}$, and also for a "warm" H I region with $T = 4000\,\text{K}$ and $N(\text{H I}) = 1\,\text{cm}^{-3}$ – using Table 5.5 on p. 272 to obtain $\alpha_0^{(2)}$ from (5.82) – the numerical values $N_1/N(\text{H I}) \approx 1 \times 10^{-4}$ and 5×10^{-3}, respectively.

The *electron densities* are given in these examples by $N_e \approx 1 \times 10^{-2}\,\text{cm}^{-3}$. Ionisation by low energy cosmic radiation and perhaps diffuse cosmic x-rays plays only a subordinate role in the case of the cold cloud, but in the case of the warm H I gas it dominates the assumed portion $x_0 N(\text{H I})$. Values of the order of $N_e = 10^{-2}\,\text{cm}^{-3}$ were derived (see below) from observations of interstellar absorption lines in stellar spectra for clouds of low density $N(\text{H I}) \approx 10 \ldots 100\,\text{cm}^{-3}$. For a thin warm H I medium with $N(\text{H I}) \approx 0.1 \ldots 0.01\,\text{cm}^{-3}$ and $T = 4000\,\text{K}$, (5.96b) becomes $N_1/N(\text{H I}) \approx 5 \times 10^{-3}/\sqrt{N(\text{H I})}$, so that relative electron densities can reach a few per cent. The average values ($\overline{N}_e \approx 5 \times 10^{-2}\,\text{cm}^{-3}$) derived from pulsar observations (Sect. 4.1.8) should not be used directly for comparison, since they are based to a considerable extent on contributions from extensive H II regions of low density, from which the observed diffuse emission of the recombination lines appears to come (Sect. 4.2.3).

The low energy particles of cosmic radiation and the photons of the diffuse cosmic x-rays cannot penetrate far into *dense clouds*. For particles with energy of 1 MeV the product of the mean free path length l and the number density of the hydrogen nuclei N_H is of the order of $l N_\text{H} \approx 10^2\,\text{pc cm}^{-3}$. Similar values are obtained for x-ray photons in the energy region of a few 10^2 eV. For $N_H \gtrsim 10^3$ cm^{-3} this gives $l \lesssim 0.1$ pc. A further important screening effect is produced by the embedded dust particles. Low energy cosmic radiation and x-ray photons can therefore cause neither ionisation of atoms nor ionisation or dissociation of molecules already formed in the interior of cold clouds with number densities of gas particles above about 10^3 cm^{-3}. The penetration depth increases rapidly, however, with increasing energy. For particle energies of 5 MeV the product lN_H is already of the order of 10^3 pc cm^{-3}. The high energy particles of the cosmic radiation will therefore produce a very weak ionisation of atoms and molecules in the thick clouds. According to recent investigations, relative electron densities $N_e/N(H_2) \approx 10^{-8} \ldots 10^{-7}$ are to be expected.

Ionisation of the Higher Elements: Let us consider first of all the examples Na and Ca with the ionisation energies $\chi_0 = 5.1$ and 6.1 eV, respectively, in diffuse H I clouds and in typical H II regions. In both regions the ionisation equilibrium is determined primarily by photo ionisation and photo recombinations. We therefore write the statistical equations (5.73) in the form

$$\frac{N_{i+1}}{N_i} N_e = \frac{R_{i,i+1}}{\alpha_i} \quad , \tag{5.97}$$

where α_i denotes the total recombination coefficient of the $(i+1)$-fold ions. To calculate $R_{i,i+1}$ from (5.75) we use the interstellar radiation field (Sect. 4.1.9) for the H I clouds, and the radiation field of the exciting star for the H II regions.

For *H I clouds of low density* with $T \approx 10^2$ K one finds as a rough approximation that $R_{0,1}/\alpha_0 \approx 1$ for Na and 10^2 for Ca. Accordingly, with $N_e \approx 10^{-2}$ it follows that $N_1/N_0 = N(Na^+)/N(Na^\circ) \approx 10^2$ and $N(Ca^+)/N(Ca^\circ) \approx 10^4$. Sodium is therefore present almost entirely in the form of Na$^+$ ($\chi_1 = 47.3$ eV, $N(Na^{++})/N(Na^+) \ll 1$). Calcium occurs almost entirely as Ca$^+$ and Ca^{++} ($\chi_1 = 11.9$ eV), Ca^{++} and Ca$^+$ have abundances of the same order). In *H II regions* with $T \approx 10^4$ K one obtains values higher by about one or two powers of ten: $R_{0,1}/\alpha_0 \approx 10$ for Na and 10^4 for Ca, but electron densities are considerably higher. With the typical values $N_e \approx 10^2 \ldots 10^3$ cm^{-3} found for diffuse nebulae, one obtains, for example, $N(Na^+)/N(Na^\circ) \approx 10^{-2} \ldots 10^{-1}$ and $N(Ca^+)/N(Ca^\circ) \approx 10 \ldots 10^2$. For both elements there is a lower ionisation level than in the H I region. The reason is the much higher electron density, which leads to a correspondingly higher recombination rate.

A discussion of carbon with $\chi_0 = 11.3$ eV leads to similar results: even the carbon is predominantly ionised in *H I regions*. In contrast, N and O, with ionisation energies scarcely higher than that of hydrogen, are neutral. A weak ionisation of oxygen, which of course amounts to only a few per cent of that of hydrogen, can be caused by the charge-exchange reaction $O^+ + H \rightarrow O + H^+$.

In sufficiently dense clouds ionisations of atoms and recombinations of ions are also possible through collisions with molecules. A notable example is the chemical exchange reaction $O^+ + H_2 \rightarrow OH^+ + H$, after the completion of which the energy of a captured electron can be applied to the dissociation of the OH$^+$ ion just created: $OH^+ + e \rightarrow O + H$ ("dissociative recombination").

From observed intensity ratios of the interstellar absorption lines of different ionisation levels, e.g. Na I and Na II, the electron densities in H I clouds can be derived from (5.97). For diffuse clouds with H I densities between 10^2 and 10^3 cm^{-3} the values obtained in this way are $N_e \approx 10^{-2} \dots 10^{-1}$ cm^{-3}, corresponding to $N_e/N(\text{H I}) \approx 10^{-4}$.

In the *hot component* of the interstellar gas with temperatures between about 4×10^4 K and 10^6 K (Sects. 4.1.4, 9) the higher elements – as we have already explained for hydrogen – are ionised predominantly by collisions with the abundant high velocity thermal electrons. The ratios of the individual ionisation levels, analogously to (5.83), are independent of the electron density. On the recombination side, account must be taken of "dielectronic recombinations", in which the energy of the captured electron is first of all used for the excitation of the outer bound electron. Numerical results for the fractions $N_i/\Sigma N_i$ are available for various elements of which interstellar absorption lines of higher ions have been observed (Sect. 4.1.4). Assuming normal elemental abundances the ratios of the line strengths could be calculated as functions of the temperature. This type of interpretation of the strengths of the O VI lines in relation to the lines of N V, which were also observed, indicated $T \approx 6 \times 10^5$ K and number densities of hydrogen nuclei $N_{\text{H}} \approx 10^{-4}$ cm^{-3}.

5.2.2 Formation and Dissociation of Interstellar Molecules

In the cool regions of the interstellar gas with kinetic temperatures $T < 100$ K, given a long-term state of thermodynamic equilibrium, most of the abundant elements would exist in molecular form or be contained in chemical compounds with other elements. The observations show, however, that this is the case only in the thickest cold clouds, whilst the very widespread low density cool gas, including many small diffuse clouds, consists predominantly of atoms and ions. The interstellar conditions here evidently fall far short of the state of thermodynamic equilibrium, and the interpretation of the observations requires a special discussion of the statistical balance between the continuous formation and destruction of molecules.

The *essential formation processes* of interstellar molecules are:

1. Direct joining of atoms in the gas phase with output of the binding energy in the form of a photon, for example $C + H \rightarrow CH + h\nu$.
2. Catalytic molecule formation on the surface of interstellar dust particles: atoms getting stuck to the surface of a particle one after another combine into a molecule, whose binding energy is given off as warmth to the solid particle. The molecule created vaporises off into space.
3. Chemical exchange reactions between already formed molecules, either with each other or with atoms or ions. Examples are the formation of CH_2^+ and of H_2:

$$CH^+ + H_2 \rightarrow CH_2^+ + H \quad ; \quad CH^+ + H \rightarrow C^+ + H_2 \quad .$$

Important processes which can lead to the *destruction* of interstellar molecules are:

1. Photodissociation by the interstellar radiation field.
2. Dissociative recombination. Examples:

$$OH^+ + e \rightarrow O + H \; ; \quad CH^+ + e \rightarrow C + H \quad .$$

3. Chemical exchange reactions. Examples of the dissociation of H_2 and of CH^+ are given above.

Molecular Hydrogen: Let us start from the position that hydrogen atoms in H I regions of the interstellar space exist in the ground state $1\,^2S$. Two H atoms in close proximity can then, according to whether the two electron spins are parallel or anti-parallel, be in one of the two electronic states $^3\Sigma_u^+$ or $^1\Sigma_g^+$ (see Appendix B). To achieve the stable state $^1\Sigma_g^+$ directly, a rotation-vibration transition is necessary with emission of a radiation quantum. These transitions are forbidden, however, for electric dipole radiation with a homopolar bound H_2 molecule: The molecule formed from nuclei of the same kind possesses no electric dipole moment — the centres of mass and charge are not separated by an external field — so no rotation-vibration radiation can be emitted or absorbed. The transitions associated with electric quadrupole and magnetic dipole radiation have only extremely small transition probabilities. Direct formation of H_2 molecules in the ground state $^1\Sigma_g^+$ with emission of the binding energy is therefore not possible in practice. On the other hand if there is originally the repulsion bound state $^3\Sigma_u^+$, then a transition into the stable state $^1\Sigma_g^+$ with radiation emission is not allowed, because it involves an intercombination. The molecule formation would therefore only be readily possible if one of the two atoms was originally in a state with the principal quantum number $n > 2$, since the resulting higher electronic states, e.g. $^1\Sigma_u^+$, have allowed transitions into the ground state $^1\Sigma_g^+$. Occupation of the higher states, however, in the low densities and temperatures of the general extended interstellar gas far from hot stars, is extremely improbable.

Since the exchange reactions also — because of the great abundance of the hydrogen compared with the other gas particles — can play no great role for the formation of H_2, one reaches the conclusion that the hydrogen molecules observed in the interstellar space cannot arise in the pure gas phase. The combination of two H atoms after their adsorption by the surface of an interstellar dust particle turns out to be the only important process for the formation of H_2. In the potential field of the crystal lattice of such a solid particle this is easily possible. The binding energy (4.5 eV) released in the formation of the H_2 molecule on the crystal lattice is sufficient to allow the molecule formed to evaporate into space. The dust particles therefore act as catalysts in the molecule formation.

An H_2 molecule, once formed, can be destroyed by the interstellar radiation field if the density is not too high. Photo-ionisation and direct photodissociation ($h\nu > 15.42$ and $14.67\,\mathrm{eV}$, respectively, see Appendix B, Fig. B.1) are not to be expected in H I regions (otherwise the H atoms would not be neutral there), so that one would at first expect there a relatively high abundance of the H_2 molecule, even at low densities $N_H \approx 1\,\mathrm{cm}^{-3}$. It must however be taken into account that an indirect photo-dissociation of the H_2 molecule is possible also by stellar radiation with $\lambda > 912\,\text{Å}$. After absorption in one of the Lyman Bands lines ($X^1\Sigma_g^1 - B^1\Sigma_u^+$) at $\lambda \leq 1108\,\text{Å}$ (see Appendix B, Fig. B.1) the molecule will revert to the electronic ground state $^1\Sigma_g^+$ with emission of radiation. This transition can lead to the lowest vibrational state $v = 0$, but also first into a higher level $v > 0$ — this indirect excitation is also known as "photopumping". If the end level has $v > 14$, then dissociation occurs (dissociation energy $= 4.476\,\mathrm{eV}$).

In order to estimate the relative number density of the hydrogen molecules continuously present by means of an *example*, we start from the statistical equilibrium conditions:

$$N(\mathrm{H\,I})R_{0,1} = N(H_2)R_{1,0} \quad , \tag{5.98}$$

where we have denoted the formation rate by $R_{0,1}$ and the dissociation rate by $R_{1,0}$.

We first of all calculate $R_{0,1}$. The number of collisions $[\mathrm{s}^{-1}]$ between an H atom with velocity v and dust particles with the cross-section σ_d $[\mathrm{cm}^2]$ and the number density N_d $[\mathrm{cm}^{-3}]$ is given by $v\sigma_d N_d$. If a fraction γ of these collisions lead to the formation of an H_2 molecule (after collisions with two H atoms), then it follows that

$$R_{0,1} = \tfrac{1}{2}\gamma v\sigma_d N_d \quad . \tag{5.99}$$

For a typical small diffuse H I cloud with $A_V = 0.3\,\mathrm{mag}$ and a diameter of $5\,\mathrm{pc}$ (4.14), with $Q_{ext} \approx 1$, leads to the value $\sigma_d N_d \approx 2 \times 10^{-20}\,\mathrm{cm}^{-1}$. With $v = \bar{v}$ from (5.27) for $T = 100\,\mathrm{K}$ and $A_d \gg A = 1$ and also, tentatively, $\gamma = 0.3$, the molecule formation rate $R_{0,1} \approx 4 \times 10^{-16}$.

For the total dissociation rate we write

$$R_{1,0} = \beta k \quad ,$$

where β denotes the number of transitions per second from the electronic ground state into the upper electronic state and k is the fraction of transitions from the upper electronic state into vibrational levels $v > 14$ of the lower state (dissociation). For the band spectrum of the H_2 molecule with $\lambda > 912\,\text{Å}$, β is calculated from the line absorption coefficient per neutral H atom $\kappa_\nu^L / H(\mathrm{H\,I})$ and the intensity I_ν of the interstellar radiation field, using the expression analogous to (5.75)

$$\beta = 4\pi \sum \int \frac{\kappa_\nu^L}{N(\mathrm{H\,I})} \frac{I_\nu}{h\nu} d\nu \tag{5.100}$$

(summation over all lines). In the interior of the cloud I_ν is to be replaced by $I_\nu e^{-\tau_\nu}$, where τ_ν denotes the optical depth, reckoned from the outside to the inside, of each point under consideration. For $\tau_\nu = 0$, with the interstellar radiation field in the solar neighbourhood (Sect. 4.1.9), $\beta \approx 5 \times 10^{-10}\,\mathrm{s}^{-1}$; for $\tau(V) \approx A_V = 0.3$, in the region of the Lyman bands, corresponding to about $\tau_\nu = 1$ (cf. Table 4.1, p. 156), the value is lowered to $\beta = 2 \times 10^{-10}\,\mathrm{s}^{-1}$. The factor k was calculated at 0.1 – detailed discussion in Spitzer (1978, pp. 124f). It then follows that $R_{1,0} \approx 2 \times 10^{-11}\,\mathrm{s}^{-1}$.

In our example of a diffuse low density cloud the relative abundance of the H_2 molecule is given by (5.98) as

$$\frac{N(H_2)}{N(H\,I)} = \frac{R_{0,1}}{R_{1,0}} \approx 2 \times 10^{-5} \quad .$$

This result is in qualitative agreement with the analyses of the weak interstellar H_2 lines observed in nearby stars (Sects. 4.1.4, 5.2.5), which are produced in regions with densities $N(H\,I) < 10\,\mathrm{cm}^{-3}$.

With increasing dust and gas densities the formation rate $R_{0,1} \sim N_d \sim N_H$ (the number density of all hydrogen nuclei) becomes greater, but above all with increasing extent of the cloud (column density!) the radiation field gets weaker, so the dissociation rate $R_{1,0}$ must decrease by a factor of around $\exp(-\tau_\nu) = \exp(-3A_V)$. Because of the high extinction by the dust in the ultraviolet, the interstellar radiation field can already penetrate only the outer layers of clouds with visual extinctions A_V of a few magnitudes. In the inner region – for densities $\gtrsim 10^3\,\mathrm{cm}^{-3}$ – the hydrogen must then be primarily in molecular form. The observations of the 21 cm line emission from the directions of the dark cloud complexes in Taurus and in Ophiuchus actually show no increased radiation, but in places rather a deficit, in spite of the considerably higher gas densities.

CO, OH and H_2O in "Diffuse" Clouds: The molecules detected in the interstellar space can be dissociated by the UV portion of the interstellar radiation field. A rough measure for the stability against photo-dissociation, and hence for the *lifetime* of the molecule, is the thermal dissociation energy or binding energy. More exactly, the lifetime depends on three factors: the absorption cross-section, the probability of a dissociation per absorbed photon, and the radiation field. A few results of calculations are presented in Table 5.8. CO has the highest binding energy of all interstellar molecules and thereby achieves the longest lifetime, whilst the other molecules are already dissociated in considerably shorter times. This fact explains qualitatively why the CO molecule is the most widely distributed. These figures moreover indicate that the empirically substantiated long-lived existence of CO and a few other molecules in the relatively transparent diffuse clouds can be achieved only by correspondingly effective formation processes.

Since a discussion of the various formation and dissociation processes for the numerous kinds of molecule detected in the interstellar space would inflate

Table 5.8. Lifetimes of molecules in the unattenuated interstellar radiation field. [After Mitchell et al. (1978)]

Process	Binding energy [eV]	Lifetime [years]
$CO + h\nu \rightarrow C + O$	11.1	7×10^3
$CS + h\nu \rightarrow C + S$	8.0	3×10^3
$CN + h\nu \rightarrow C + N$	7.7	7×10^2
$HCN + h\nu \rightarrow CN + H$	5.8	3×10^2
$H_2O + h\nu \rightarrow OH + H$	5.2	1×10^2
$NH_3 + h\nu \rightarrow NH_2 + H$	4.3	7×10^1
$H_2CO + h\nu \rightarrow H + HCO$	3.6	5×10^1

the compass of this book, we restrict ourselves to a few examples. Most of these molecules can be formed by reactions in the gas phase; for a few molecules, however, the formation on the surface of the dust particles, so essential for H_2, must also be of significance. For direct formations an essential role is played by positive ions, which above all originate from the influence of the cosmic radiation (H^+, He^+!) and subsequent charge exchange reactions. The reason for this is that reactions between molecules and positive ions at the low temperatures of the interstellar medium occur just as frequently as at higher temperatures – in contrast to the lower rates of reactions with neutral particles.

Hence, after the formation of O^+ by the exchange reaction $O + H^+ \rightleftharpoons O^+ + H$, first of all OH^+ can be formed:

$$O^+ + H_2 \rightarrow OH^+ + H \tag{5.101}$$

and thereafter also H_2O^+ by

$$OH^+ + H_2 \rightarrow H_2O^+ + H \quad . \tag{5.102}$$

Eventually OH and H_2O are formed by recombination in relative abundance.

CO can result from the destruction of OH by C^+ (formed by the interstellar radiation field):

$$C^+ + OH \rightarrow CO + H^+ \quad . \tag{5.103}$$

A further possibility is the reaction

$$CH^+ + O \rightarrow CO + H^+ \quad , \tag{5.104}$$

since CH^+ is observed in considerable abundance – the analysis of its interstellar absorption lines in the visible leads in typical cases to $N(CH^+)/N(H\,I) \approx 10^{-7}$.

The destruction of CO is caused chiefly by photo-dissociation. The reaction rates of the various processes available till now are in part still subject to uncertainties. Without going into details of the calculation of the equilibrium, we give

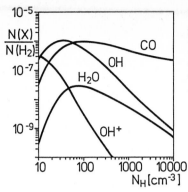

Fig. 5.7. Calculated relative abundances $N(X)/N(H_2)$ of the molecules X = OH, OH$^+$, H$_2$O, and CO in chemical equilibrium for interstellar clouds of low density and kinetic temperature $T = 100$ K at optical depth $\tau(V) = 0.2$. These were based on normal elemental abundances and an ionisation rate of the hydrogen by the cosmic radiation $R_{0,1} = 10^{-17}$ s^{-1}. [After Barsuhn and Walmsley (1977)]

in Fig. 5.7 as examples a few results for the relative abundances of OH, H$_2$O and CO compared with H$_2$ in clouds with densities up to $N_H = 10^4$ cm^{-3}. It may be seen that, for typical diffuse clouds with $N_H \approx 10^2 \dots 10^3$ cm^{-3}, the values of $N(CO)/N(H_2)$ are barely 10^{-6}, which with $N(H_2)/N_H \approx 0.1 \dots 1$ lead to values for $N(CO)/N_H \approx 10^{-7} \dots 10^{-6}$. Values of this order of magnitude were confirmed in a few cases from the observed intensities of the absorption lines of CO in the ultraviolet (ε Per, 139 Tau, 10 Lac, α Cam, ξ Per, ζ Oph; see also Sect. 5.2.5).

Molecules in "Dense" Clouds: The most important difference from the "diffuse" clouds is the screening of the interstellar ultraviolet radiation between 912 and about 2000 Å by the dust particles. Dissociation takes place correspondingly seldom and the gas is therefore almost completely molecular so far as it is not picked up by solid particles. The second possibility allows it to be expected that the elemental abundances in the gas of dense clouds sometimes lie considerably below the normal "cosmic" values (Sect. 5.2.5). The question must even be raised, why all elements above helium are not completely deposited on solid particles. An explanation for the appearance of large amounts of CO and N$_2$ would follow from the assumption that these molecules are created in an earlier stage of the cloud and thereby a considerable portion of the existing C, O and N was used up. Because of their high vapour pressure, CO and N$_2$ would remain in the gas phase if they were once present in the first place. Starting from this, ion-molecule reactions could result in the formation of molecules such as HCN, NH$_3$ and H$_2$CO. The ions needed for this (H$_2^+$, He$^+$) could be formed by the high energy cosmic radiation (Sect. 5.2.1: H I regions).

5.2.3 Heat Balance and Kinetic Temperature of the Gas

Heat Equilibrium: The most important parameter characterising the thermodynamic state of the interstellar gas is the kinetic temperature T. The particular value of T pertaining obviously depends on the heat supply and the heat losses of the gas. The total gain and the total loss of thermal energy, per

cm^3 and per second, are usually denoted by Γ and Λ, respectively. Γ and Λ in general depend on T and on the density of the gas particles N. For example, the cooling of the gas by emission of line radiation, arising from transitions from collisionally excited fine structure levels into the lowest state, must in general become weaker with decreasing temperature, and hence lower energy of the colliding particles. If a stationary state has become established, then there is heat equilibrium

$$\Gamma = \Lambda \ .$$

With the values of density N and of pressure $P = NkT$ given, this condition determines the equilibrium temperature T_E, which can be calculated from a knowledge of the functions $\Gamma(N, T)$ and $\Lambda(N, T)$.

To assess the effectiveness of various heating and cooling mechanisms it is often useful to estimate the times necessary for the establishment of equilibrium or more generally for a significant change of temperature. A rough measure for this is the ratio between the total internal energy of the gas per cm^3 $U = (3/2)NkT$ and the contribution of the net cooling rate per cm^3 per s $\Gamma - \Lambda$ (<0 for $T>T_E$), this ratio being known as the *cooling time*:

$$t_C = \frac{(3/2)NkT}{|\Gamma - \Lambda|}[s] \ . \tag{5.105}$$

We have defined t_C by the relation $1/t_C = (-dU/dt)/U$, hence immediately $U \sim \exp(-t/t_C)$. At constant density, however, $U \sim T$, so $T \sim \exp(-t/t_C)$.

In general the behaviour of temperature with time outside the equilibrium state can be obtained from the energy law: the net heat gain per gramme of gas $dq/dt = (\Gamma - \Lambda)/\varrho$ is used for the raising of the internal energy per gramme $u = U/\varrho$ and the work performed by the gas:

$$\frac{dq}{dt} = \frac{du}{dt} + P\frac{dV}{dt} = \frac{\Gamma - \Lambda}{\varrho} \ ,$$

where $V = 1/\varrho$ for 1 g of gas, so that $dV = -d\varrho/\varrho^2$. We also have $P = NkT$ and $d\varrho/\varrho = dN/N$, N being the particle density. For a monatomic gas $\varrho u = (3/2)NkT$, so, using the relation $\varrho\, du = d(\varrho u) - u\, d\varrho$, we transform the above equation into

$$\frac{d}{dt}\left(\frac{3}{2}NkT\right) - \frac{5}{2}kT\frac{dN}{dt} = \frac{3}{2}kN\frac{dT}{dt} - kT\frac{dN}{dt} = \Gamma - \Lambda \ . \tag{5.106}$$

For a given density this is a differential equation for $T(t)$. For constant pressure P_0, for example, replacing N by P_0/kT and assuming that $\Gamma - \Lambda<0$, we find an exponential rate of fall of temperature with $t_K = (5/2)P_0/|\Gamma - \Lambda| = (5/2)NkT/|\Gamma - \Lambda|$. The difference compared with (5.105) in the coefficient is due to the negative work performed by the gas.

Heating of the Gas: For the heat supply Γ the following processes may contribute:

1) *Photo-ionisation* of neutral atoms produce free electrons, which transfer part of their kinetic energy to the other gas particles by collisions. The heating of the H II regions is caused solely by this process, in which the electrons are produced predominantly by ionisation of the hydrogen. The energy of an electron liberated by a photon of frequency ν is equal to the excess of the energy $h\nu$ taken from the incident radiation field I_ν over the ionisation energy of the neutral atom $\chi_0 = h\nu_0$. If the atoms are in their ground state, the mean energy of the liberated electrons is given by averaging, using as weighting factors the numbers of absorbed photons $\kappa_\nu^C I_\nu / h\nu$ in the various frequencies of the Lyman continuum per unit of wavelength:

$$\overline{E}_e = \frac{\int\limits_{\nu_0}^{\infty} h(\nu - \nu_0)\kappa_\nu^C I_\nu\, d\nu / h\nu}{\kappa_\nu^C I_\nu\, d\nu / h\nu} \quad . \tag{5.107}$$

The "initial temperature" T_0, defined by $\overline{E}_e = (3/2)kT_0$, of the electrons created is obtained by taking a diluted Planck radiation $I_\nu = W B_\nu(T_*)$ from the exciting star with the surface temperature T_*, approximately $T_0 \approx (2/3)T_*$ (for $h\nu_0 > kT_*$)[2]. The stronger attenuation of the stellar radiation at lower frequencies ($\kappa_\nu^C \sim \nu^{-3}$) has the effect that $\overline{E}_e$ corresponds to a higher frequency with increasing distance from the star and hence increases accordingly.

In the ionisation equilibrium each photo-ionisation must be followed by a recombination, in which the kinetic energy of a free electron $m_e v^2/2$ is again removed. It is expedient to take account of this loss of energy immediately, since it must happen in every case: the number of photo-ionisations is equal to the number of all recombinations into the various bound states $N_e N_1 \alpha_0 = N_e N_1 \Sigma v \sigma_{\overline{\kappa}} n(v)$ see (5.78 and 79). The net contribution to the heat supply is given by multiplying this quantity by the difference $\overline{E}_e - \frac{1}{2}m_e v^2$ after averaging over all relative velocities v (Maxwell distribution). We write this in the form ("Net heating rate")

$$\Gamma_{\mathrm{Ph}} = N_e N_1 \left\{ \alpha_0 \overline{E}_e - \frac{1}{2}m_e \sum_n \overline{v^3 \sigma_{\overline{\kappa}} n(v)} \right\} \quad . \tag{5.108}$$

2) *Collisional ionisation* of neutral hydrogen atoms in the H I regions *by particles of the cosmic radiation.* Let $\overline{\Delta E}$ denote the average energy gain of the gas per free electron created, then the heat supply can be represented by

$$\Gamma_{\mathrm{CR}} = N(\mathrm{H\,I})C_{0,1}\overline{\Delta E} \quad . \tag{5.109}$$

---

[2] One sets $\kappa_\nu^C \sim \nu^{-3}$ and uses Wien's approximation $B_\nu(T_*) \sim \nu^3 \exp(-h\nu/kT_*)$, valid when $h\nu/kT_* \gg 1$. One introduces the new integration variable $x = h\nu/kT_*$, and with $x_0 = h\nu_0/kT_* \gg 1$ applying the corresponding series development of the exponential integral, it follows that $\overline{E}_e \approx kT_*/[1 - (1/x_0)] \approx kT_*$ (A.S. Eddington, 1926).

Here $N(\mathrm{H\,I})$ denotes the number density of the neutral H atoms and $C_{0,1}$ the ionisation rate of the H atoms. A value for $\overline{\Delta E}$ is obtained from the following argument. In an ionising collision a photon of the important lower MeV region loses only a very small portion of its energy: the liberated electron receives about 20 to 40 eV. In pure hydrogen these primary electrons produce further ionisations and excitations with subsequent radiation emission until their energy has fallen below the excitation energy of the second quantum state, viz. 10.2 eV. The remaining energy is lost in elastic collisions. More detailed calculations, with $N(\mathrm{H\,I}) > 1\,\mathrm{cm}^{-3}$, lead to a value of the net heat gain per ionised H atom of $\overline{\Delta E} \approx 3\,\mathrm{eV}$.

3) *Photo effect on the surfaces of the interstellar dust particles.* Photons of the stellar radiation can liberate electrons which transmit a part of their kinetic energy to the ambient gas and thereby raise its thermal energy. This mechanism of photoelectric heating seems to be vitally important for the heat equilibrium of the H I regions, although here only radiation with $\lambda > 912$ Å is available.

The number of photons with frequency ν which impinge on the surface of a dust particle, per unit area from a given direction per unit solid angle, is given by $I_\nu/h\nu$. The total effective particle surface area in a volume of $1\,\mathrm{cm}^3$, for radiation from a given direction, is equal to $N_\mathrm{d}\sigma_\mathrm{d}Q_{\mathrm{abs}}(\nu)$, where N_d is the number density, σ_d the geometric cross-section and $Q_{\mathrm{abs}}(\nu)$ the efficiency factor of the dust particles (Sect. 5.3.1). Not every impinging photon liberates an electron. With an electron yield y_e (the fraction of electrons actually liberated) the number of free electrons produced by radiation of frequency ν from all directions per second is therefore equal to $4\pi(I_\nu/h\nu)N_\mathrm{d}\sigma_\mathrm{d}Q_{\mathrm{abs}}(\nu)y_\mathrm{e}(\nu)$.

One can approximately set $y_\mathrm{e} = 0$ for $\lambda > 1100$ Å, $y_\mathrm{e} = \overline{y}_\mathrm{e} =$ constant for $\lambda = 912 \ldots 1100$ Å. If $\overline{E}_\mathrm{e}$ denotes the mean kinetic energy of a photo electron which is liberated by radiation of this wavelength region, then the heating rate by photo effect is

$$\Gamma_{\mathrm{PE}} = 4\pi \int \frac{I_\nu}{h\nu}\,d\nu\,e^{-\overline{\tau}}N_\mathrm{d}\sigma_\mathrm{d}\overline{Q}_{\mathrm{abs}}\overline{y}_\mathrm{e}\overline{E}_\mathrm{e} \quad, \tag{5.110}$$

where the integral extends from $\lambda = 912$ Å to $\lambda = 1100$ Å. the factor $e^{-\overline{\tau}}$ is needed if the volume under consideration lies in the interior of the cloud at an optical depth $\overline{\tau}$ (for $\lambda \approx 1000$ Å). The kinetic energy of photo electrons $E_\mathrm{e}(\nu)$ can be measured in the laboratory. Since the material of the interstellar particles is not known with certainty the available results give only a rough estimate. As a plausible mean value for small particles $\overline{E}_\mathrm{e} = 5\,\mathrm{eV}$ has been proposed. By means of the relations (4.14), p. 158, and (5.195), p. 345, with $E_{\mathrm{B-V}} = A_\mathrm{V}/3$ and $Q_{\mathrm{ext}}(V) \approx 0.5$ and also $\mathcal{N}_\mathrm{H} = N_\mathrm{H}r$, we can express $N_\mathrm{d}\sigma_\mathrm{d}$ in terms of N_H :

$$N_\mathrm{d}\sigma_\mathrm{d} \approx 1 \times 10^{-21} N_\mathrm{H} \quad. \tag{5.111}$$

If one takes for I_ν the interstellar radiation field of the solar neighbourhood

(Sect. 4.1.9), then it follows from (5.110) [after Spitzer (1978; p. 146)] that

$$\Gamma_{PE} \approx 2 \times 10^{-25} N_H e^{-\bar{\tau}} \bar{y}_e [\text{erg cm}^{-3} \text{s}^{-1}] \quad . \tag{5.112}$$

For ultraviolet radiation with $\lambda > 1100 \text{Å}$ the electron yield $\bar{y}_e$ lies around 0.1. For very small dust particles, with radii less than 100Å, $\bar{y}_e$ can reach values close to unity. Only a portion of the dust can therefore contribute significantly to the photo electric heating.

4) *H_2 molecules formed on the surfaces of dust particles and evaporated into space* can contribute to the heating in the H I regions in a manner similar to the photo electrons, since these molecules receive part of their liberated binding energy (4.48 eV) as kinetic energy — the other part is used in overcoming the binding of the resulting molecule to the solid particle and perhaps also to the excitation of the molecule. In the dissociation of H_2 the resulting H atoms bring a contribution to the kinetic energy of comparable magnitude, which also raises the thermal energy of the gas. Estimates of the thermal energy actually realised for each H_2 molecule created lead one to expect approximately a tenth of the binding energy, and so about 0.5 eV. For the formation rate given in (5.99), $v = \bar{v}$ from (5.27) at $T = 100 \text{K}$ and $A_C \gg A = 1$, $\gamma = 0.3$ and $N_d \sigma_d$ from (5.111), we find that

$$\Gamma_{H_2} \approx 2 \times 10^{-29} N_H N(\text{H I}) [\text{erg cm}^{-3} \text{s}^{-1}] \quad . \tag{5.113}$$

5) *Thermalisation of macroscopic motions*: The observed variations of the velocity of the interstellar gas about the systematic motion of the galactic rotation are usually greater than the speed of sound (Sect. 6.3.1, Table 6.5), so that the formation of *shock waves* is to be expected, in which energy of translation is dissipated into thermal energy. In the transformation of the kinetic energy of the macroscopic motion of an H atom $\frac{1}{2} m_H v^2$ into thermal energy $(3/2)kT$, for example in the case $v = 8 \text{km s}^{-1}$, the resulting temperature T is 2500 K! We discuss this possibility at the conclusion of this section.

Cooling of the Gas: Loss of heat can arise from the following processes:

1) *Collisional excitation with subsequent emission of radiation*. Inelastic collisions between the gas particles lead to the excitation of certain atoms and ions ($n \rightarrow m$). If the atom or ion returns to its starting state, with radiation of the energy difference $E_m - E_n = h\nu$ acquired from its "collision partner", then this quantity of energy must be booked as a loss, if the medium is optically thin at the frequency ν. In the calculation of the total energy loss, however, one has to deduct the energy given back to the gas particle in de-exciting collisions. Since we can assume that in interstellar space all atoms and ions are in their ground states, we get, by application of the collisional rates introduced and explained in Sect. 5.1.2,

$$\Lambda_{\rm col} = N' \sum_m (E_m - E_1)(N_{i,1}Q_{1m} - N_{i,m}Q_{m1}) \quad . \tag{5.114}$$

N' denotes the number density of the collision particles. Inelastic collisions between electrons and ions are the most important cooling mechanism in both H I and H II regions. If the density is sufficiently high, so that many molecules are present, then a corresponding role is played by excitations of rotational states of these molecules by collisions with H_2. In this case it must be taken into account if appropriate, that the emitted radiation can no longer directly escape.

2) *Free-free emission of electrons* in ionised regions. For this we have

$$\Lambda_{\rm ff} = 4\pi\varepsilon_{\rm ff} \tag{5.115}$$

given by (5.56).

3) *Collisions of gas particles with solid particles.* In sufficiently dense clouds with high dust content, atoms and molecules can transfer a part of their thermal energy to the dust particles and thereby warm them. The solid particles on the other hand give off energy by radiation in the far infrared.

4) *Heat conduction* can influence the thermal equilibrium for temperatures above a few 10^4 K. For the hot component of the interstellar gas this kind of energy transfer in the boundary regions to the cool clouds is important.

Heat Equilibrium in H II Regions: The dominating energy source is the photo-ionisation of the hydrogen by the radiation of a hot star. Assuming diluted black body radiation $I_\nu = WB_\nu(T_*)$ at the colour temperature T_* of the star in the UV as well as continuous absorption in the Lyman continuum $\kappa_\nu^C(\text{b-f}) \sim \nu^{-3}$, one can first of all calculate $\overline{E}_e$ as a function of T_*. The result can be written in the form $\overline{E}_e = kT_*\psi_0$, where $\psi_0 = 0.98\ldots0.87$ for $T = 4000\ldots30\,000$ K. In (5.108) $N_1 = N_{\rm p}$ is now the number density of the protons and $\alpha_0 = \alpha_0^{(1)}$ is the total recombination coefficient of the hydrogen from (5.82). After carrying out the weighted averaging in the negative term of (5.108) with the Maxwell distribution (5.21) and inserting the numerical values of the constants, we obtain [after Spitzer (1968; p. 131)]

$$\Gamma_{\rm Ph} = \frac{2.9 \times 10^{-27} N_e N_{\rm p}}{T^{1/2}} \{T_*\psi_0\Phi_1(T) - T\Phi_1(T)\} \quad . \tag{5.116}$$

Numerical values of the functions $\Phi_1(T)$ and $\Psi_1(T)$ are given in Sect. 5.2.1, Table 5.5. Ψ_1 differs only by a constant factor from $\sum(v^3\sigma_{\overline{\kappa}_n})$, the summation being carried out from $n = 1$ to ∞.

The result (5.116) is strictly speaking valid only in the neighbourhood of the star. In the rest of the H II region direct recombinations into the ground state $n = 1$ produce an additional diffuse radiation field, which is to be taken into account in (5.107). If one only wants information on the average conditions in the whole H II region, then $\overline{E}_e$ may be taken as the average energy gain for each photon emitted from the star $(h\nu > h\nu_0)$ – since every one of these photons is absorbed somewhere or other in the H II region! In other words, κ_ν^C is to be omitted from (5.107). If one now sets $\overline{E}_e = kT_*\overline{\psi}$, then for $T_* = 4000\ldots30\,000\,\mathrm{K}$ the values $\overline{\psi} = 1.05\ldots1.34$ are obtained. As in the derivation of (5.86), the influence of the diffuse radiation field can now be simply taken into account by beginning the summations in the expressions for Φ_1 and Ψ_1 with $n = 2$. The functions obtained are denoted by Φ_2 and Ψ_2 (Table 5.5). As an *example* Fig. 5.8 portrays the resulting profile for the heating rate $\Gamma_{\mathrm{Ph}}/N_e N_p$ as a function of T, when $T_* = 32\,000\,\mathrm{K}$. In the case of pure hydrogen this energy source would be opposed only by the energy loss through free-free emission $\Lambda_{\mathrm{ff}}/N_e N_p$, whose profile is also shown in Fig. 5.8. As may be seen, the heat equilibrium $\Gamma_{\mathrm{Ph}} = \Lambda_{\mathrm{ff}}$ is then reached at the relatively high temperature $T_E \approx 27\,000\,\mathrm{K}$.

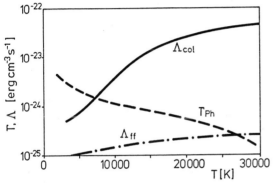

Fig. 5.8. Heating and cooling rates for an H II region with $N_e = N_p = 1$, caused by a star with surface temperature $T_* = 32\,000\,\mathrm{K}$, as a function of the gas temperature T. For other values of N_e and N_p ($\lesssim 10^2\,\mathrm{cm}^{-3}$) the ordinate scale denotes $\Gamma/N_e N_p$ and $\Lambda/N_e N_p$ with dimension $\mathrm{erg\,cm}^3\,\mathrm{s}^{-1}$. Explanation in text. [After Spitzer (1968)]

In actual H II regions the loss of kinetic energy by electron collision excitation of ions of the elements N, O, Ne is the decisive effect: it is large compared with Λ_{ff} and causes cooling down to temperatures below $10\,000\,\mathrm{K}$ (Sect. 5.2.4). On the one hand low-lying metastable fine-structure levels inside the ground term ($\Delta E \lesssim 0.1\,\mathrm{eV}$) are excited; on the other hand, by means of the electrons in the "tail" of the Maxwell distribution, terms 2 to 5 eV above the ground term are excited, for which the metastable starting state of the green nebular lines of the O^{++} ion ($^1D_2$) is an example [see Sect. 5.1.2, after (5.31)]. The energy so removed from the electron gas is radiated (Sect. 3.2.1) in the form of forbidden lines in the visible and in the infrared spectral regions and is lost by the H II region.

For low densities $N_p \approx N_e \lesssim 10^2 \, \text{cm}^{-3}$ practically every collisional excitation is followed by the emission of a photon and de-exciting collisions can be neglected. Since $N_{i,1} = N_p$ and $N' = N_e$ it follows from (5.114) that $\Lambda_{\text{col}} \sim N_e N_p$. The calculated loss rate $\Lambda_{\text{col}}/N_e N_p$ for this case under the assumption of normal abundances of the elements O, N and Ne under consideration is plotted in Fig. 5.8. As the equilibrium temperature one then obtains $T_E \approx 7500 \, \text{K}$, in good agreement with values derived from observations of typical H II regions (Sect. 5.2.4). For considerably higher densities the de-exciting collisions become significant, whereby the right-hand expression in parentheses in (5.114) becomes smaller, and accordingly Λ_{col} grows more slowly than in proportion to N_p and hence T_E becomes somewhat larger.

For a rough estimate of the *cooling time* t_C *of an H II region* we enquire into the duration of cooling from $T = 10\,000 \, \text{K}$ to $T_E = 7500 \, \text{K}$. For $T = 10\,000 \, \text{K}$ one reads from Fig. 5.8 an energy loss rate $|\Gamma_{\text{Ph}} - \Lambda_{\text{col}}| \approx 3 \times 10^{-24} \, N_p^2$. Then (5.105) gives

$$ t_C \approx \frac{2 \times 10^4}{N_p} [\text{years}] \quad . \tag{5.117} $$

In normal H II regions with $N_p \approx 10^2 \ldots 10^3 \, \text{cm}^{-3}$ heat equilibrium is therefore very rapidly established. The recombinations of the ionised hydrogen on the other hand [for $T \approx 10^4 \, \text{K}$ with $\alpha_0 = \alpha_0^{(2)}$ from (5.82)] require a time of the order of

$$ t_R = \frac{1}{R_{1,0}} = \frac{1}{N_e \alpha_0} \approx \frac{10^5}{N_e} \approx \frac{10^5}{N_p} [\text{years}] \quad , \tag{5.118} $$

which is considerably longer than t_C. After the "extinction" of the ionising star, the static model H II region being considered cools down before the recombination: it remains a "fossil" cool H II region.

Heat Equilibrium in the H I Region: The decisive difference compared with the H II regions is a very much smaller value of the energy supply Γ, whilst the energy loss function $\Lambda_{\text{col}}(T)$ has roughly the same profile. Therefore a relatively low equilibrium temperature results. Contributions to Γ_{Ph} are, first of all, the ionisation of elements with $\chi_0 < 13.6 \, \text{eV}$ by radiation with $\lambda > 912 \, \text{Å}$. One obtains about $\overline{E}_e = 2 \, \text{eV} \approx 3 \times 10^{-12} \, \text{erg}$, and at temperatures $T \approx 10^2 \, \text{K}$ and $N_H < 10^2 \, \text{cm}^{-3}$ this value is large compared with $\frac{1}{2} m_e v^2 \approx kT$, so that in (5.108) the second term can be omitted, giving: $\Gamma_{\text{Ph}}(\lambda > 912 \, \text{Å}) \approx 4 \times 10^{-4} \, N_e N_H \overline{\alpha}_0 \overline{E}_e$. Here the coefficient takes account of the low (normal) abundance of the elements under consideration (C, Si, Fe, etc.) compared with hydrogen, and $\overline{\alpha}_0$ denotes a mean approximate value of $\alpha_0 = \alpha_0^{(2)}$ from (5.82) with $Z = 1$ for these elements — for C^+, the most important ion, the shell with $n = 1$ is occupied, so that only recombination lines for $n > 2$ occur. One obtains

$$\Lambda_{\mathrm{Ph}}(\lambda>912) \approx 2 \times 10^{-26}\, T^{-1/2}\Phi_2(T)x N_{\mathrm{H}}^2 \ [\mathrm{erg\,cm^{-3}\,s^{-1}}] \quad , \qquad (5.119)$$

where the relative electron density $x = N_{\mathrm{e}}/N_{\mathrm{H}}$ is introduced.

Further contributions to Γ_{Ph} are supplied by ionisations, primarily of hydrogen, by the diffuse cosmic x-rays. A heat source of the same order of magnitude is the collisional ionisation of interstellar atoms by cosmic rays. For the estimate of the order of magnitude it is sufficient to calculate these contributions from (5.109) just for hydrogen, with $\overline{\Delta E} \approx 3\,\mathrm{eV} \approx 5 \times 10^{-12}$ erg. With $C_{0,1} = 10^{-17}\,\mathrm{s^{-1}}$ it follows that

$$\Gamma_{\mathrm{CR}} \approx 5 \times 10^{-29} N_{\mathrm{H}} [\mathrm{erg\,cm^{-3}\,s^{-1}}] \quad . \qquad (5.120)$$

The heating rate through the photo effect on dust particles lies about an order of magnitude higher. With the choice of $\overline{y}_{\mathrm{e}} = 0.3$ and $\overline{\tau} = 0.3.$, corresponding to $\tau(V) \approx 0.1$, (5.112) gives

$$\Gamma_{\mathrm{PE}} \approx 4 \times 10^{-26} N_{\mathrm{H}} \quad . \qquad (5.121)$$

The rate of heating by H_2 molecule formation Γ_{H_2} first reaches the same order of magnitude as Γ_{PE} when $N_{\mathrm{H}} \approx 10^3\,\mathrm{cm^{-3}}$.

The cooling rate in H I clouds of low density consists solely of Λ_{col}. Since the effective cross-sections for collisional excitation of neutral atoms are smaller than for ions, the predominantly neutral atoms O, N and Ne make lower contributions to Λ_{col}. On the other hand, the relative density of singly ionized carbon – the most abundant of the elements with ionisation energy $<13.6\,\mathrm{eV}$ – is practically the same as in H II regions (Sect. 5.2.1), so that excitations of C^+, besides Si^+, Fe^+ and a few others by collisions with neutral H atoms and electrons may be assumed. It turns out that, with temperatures below 100 K, the collisional excitation of C^+ is dominant, and for $T>100\,\mathrm{K}$ Si^+ and Fe^+ also become effective; for sufficiently high densities (and low temperatures) the excitation of H_2 molecules becomes important. Since Λ_{col} is directly proportional to the element abundances, depletions such as are found for C in diffuse clouds (Sect. 5.2.5) cause a corresponding fall in the cooling rate.

Examples of results for the cooling function $\Lambda_{\mathrm{col}}(T)/N_{\mathrm{H}}^2$ for various values of the relative electron density $x = N_{\mathrm{e}}/N_{\mathrm{H}}$ in H I regions are shown in Fig. 5.9. For $x<10^{-3}$ only collisions with neutral H atoms still play a role. In addition are shown: the heating function $\Gamma_{\mathrm{Ph}}/N_{\mathrm{H}}^2$ by stellar radiation with $\lambda>912\,\text{Å}$ for $x = 10^{-2}$ and the constant values for the heating rate by the cosmic rays $\Gamma_{\mathrm{CR}}/N_{\mathrm{H}}^2 = 5 \times 10^{-29}/N_{\mathrm{H}}$ and by the photo effect $\Gamma_{\mathrm{PE}}/N_{\mathrm{H}}^2 = 4 \times 10^{-26}/N_{\mathrm{H}}$, for $N_{\mathrm{H}} = 1\,\mathrm{cm^{-3}}$, and $N_{\mathrm{H}} = 4$ and $30\,\mathrm{cm^{-3}}$.

As may be seen, the heating from the stellar radiation alone would lead to a very low equilibrium temperature, namely $T_{\mathrm{E}} \approx 16\,\mathrm{K}$. Because $\Gamma_{\mathrm{CR}}/N_{\mathrm{H}}^2 {\sim} 1/N_{\mathrm{H}}$ the heating by cosmic rays can give higher temperatures only for very low densities. Although the energy gain $\overline{\Delta E}$ for each electron created is expected to be higher when $N_{\mathrm{H}} {\lesssim} 0.1\,\mathrm{cm^{-3}}$, nevertheless it would be scarcely possible to

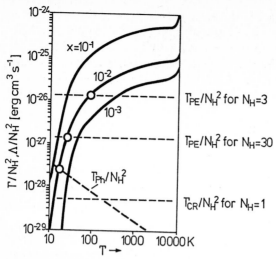

Fig. 5.9. Cooling rates $\Lambda/N_{\rm H}^2$ for the H I region as a function of the gas temperature T for three different values of the relative electron density $x = N_e/N_{\rm H}$ (*continuous curves*). For comparison are shown also as *dashed curves* the heating rate by "long wave" stellar radiation $\Gamma_{\rm Ph}/N_{\rm H}^2$, by cosmic radiation $\Gamma_{\rm CR}/N_{\rm H}^2$ and by photo effect on solid particles $\Gamma_{\rm PE}/N_{\rm H}^2$, explained in text

explain the temperatures of the observed warm H I gas, of a few 10^3 K (see Sect. 5.4.2) in this way. Since the discovery of a widely distributed hot "coronal" gas component it appears to be very probable that the warm gas forms low density envelopes around the cold clouds surrounded by the hot gas. These transition layers can be heated by photons of the soft x-rays which originate in the hot gas. The x-rays ionize only a small part of the hydrogen atoms.

The heating rate by the photo effect on dust particles in typical small diffuse clouds with $N_{\rm H} \approx 30\,{\rm cm}^{-3}$ is an order of magnitude higher than $\Gamma_{\rm Ph}/N_{\rm H}^2$. Because of the steep climb of the cooling curve one indeed obtains for $x = 10^{-2}$ still a low equilibrium temperature, $T_{\rm E} \approx 24$ K. A depletion of C by about a factor of 0.1, which is compatible with observations, would however displace all cooling curves correspondingly downwards. The resulting temperature then corresponds to the intersection of the cooling curve with $\Gamma_{\rm PE}/N_{\rm H}^2$ for $N_{\rm H} = 3\,{\rm cm}^{-3}$ and lies at $T_{\rm E} \approx 80$ K, in better agreement with the values derived from observations (Sects. 5.4.2). At low densities the heating by the photo effect is considerably less effective − amongst other things because of the higher positive charge on the dust particles.

The *cooling time of the warm component of the H I gas* is given by (5.105). For example, with $T = 1000$ K as the starting temperature, for which the middle curve of Fig. 5.9 approximately gives $|\Gamma - \Lambda| \approx 5 \times 10^{-26}\,N_{\rm H}^2$, the result is

$$t_{\rm C} \approx \frac{10^5}{N_{\rm H}}\text{years} \quad . \tag{5.122}$$

For the same density this value is higher than that for the H II regions.

Equation of State and Two-Phase Model of the H I Gas: The static theory of the heating of the interstellar gas presented offers the possibility, assuming a relatively high heating rate of the order of $\Gamma \approx 10^{-26}\,\mathrm{erg\,cm^{-3}\,s^{-1}}$, of a simple explanation of the simultaneous occurrence of cold H I clouds and a "warm" H I gas (see Sect. 5.4.2). By elimination of the electron density from the condition of heat equilibrium

$$\Gamma(N_\mathrm{H}, N_\mathrm{e}, T_\mathrm{E}) = \Lambda(N_\mathrm{H}, N_\mathrm{e}, T_\mathrm{E})$$

using (5.96), the equilibrium temperature T_E can be obtained as a function of N_H or of the total number density of the atoms $N \approx N_\mathrm{H} + N_\mathrm{He} \approx 1.1\,N_\mathrm{H}$. The result is presented qualitatively in Fig. 5.10a. The most notable feature here is that for densities $N \gtrsim 1\,\mathrm{cm^{-3}}$ the kinetic temperatures are $T_\mathrm{E} \lesssim 10^2\,\mathrm{K}$, whilst for densities below about $0.3\,\mathrm{cm^{-3}}$ relatively high equilibrium values $T_\mathrm{E} \approx 10^4$ result. The rapid fall of the kinetic temperature with increasing $N > 0.3\,\mathrm{cm^{-3}}$ is a consequence of the increasing effectiveness of the cooling mechanism, there about proportional to N^2, whilst $\Gamma \sim N$.

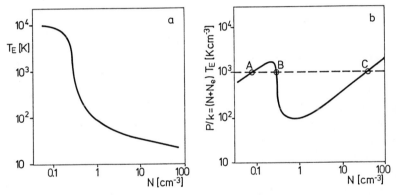

Fig. 5.10. (a) Equilibrium temperature T_E of the H I gas as a function of the total number density of the atoms N. (b) Gas pressure $P/k = (N + N_\mathrm{e})T_\mathrm{E}$ in units of $\mathrm{K\,cm^{-3}}$ for the case of heat equilibrium, as a function of N. Explanation in text

If one now considers the profile of the corresponding total gas pressure $P = (N + N_\mathrm{e})kT_\mathrm{E} \approx NkT_\mathrm{E}$ with variable N, plotted in Fig. 5.10b, the following consideration is suggested. We start from a thin gas with $N \lesssim 0.1\,\mathrm{cm^{-3}}$, which according to Fig. 5.10a is in heat equilibrium at $T_\mathrm{E} \approx 10^4\,\mathrm{K}$. A sufficiently slow compression of the gas to higher densities, in which the heat equilibrium is continually re-established, leads after the crossing of $N \approx 0.3\,\mathrm{cm^{-3}}$ in spite of increasing density to a rapid fall of pressure, because now the temperature is falling very steeply. When $N > 1\,\mathrm{cm^{-3}}$, T_E at last begins to sink so slowly with increasing N that P begins to climb again. With densities in the region of $0.3\,\mathrm{cm^{-3}}$ and temperature of a few 10^2 to a few $10^3\,\mathrm{K}$ any compression of the gas would therefore lead to cooling and condensation.

Accordingly one can expect two stable phases of the interstellar gas: with $N < 0.3\,\mathrm{cm}^{-3}$ and $T \approx 10^4$ K, and on the other hand with $N > 1\,\mathrm{cm}^{-3}$ and $T \lesssim 10^2$ K. Between these two lies a transition region with falling pressure in spite of increasing density, that is to say, in which the gas is thermally unstable. If one identifies the warm phase with a thin intercloud medium, that is in hydrostatic equilibrium with the clouds (points A and C in Fig. 5.10b), then the picture of typical H I clouds in the spiral arm regions is simply explained as the consequence of the passage of a compression wave through the warm intercloud gas, as is postulated in the density wave theory of the spiral structure (Sect. 6.2 and 6.3.3).

If the pressure temporarily rises and then falls back again, in a profile $P(t)$, then the accompanying temperature profile $T(t)$ for quasi-steady changes can be obtained by integration of (5.106) with $N = P/kT$. Numerical calculations with starting densities $N \approx 0.1\,\mathrm{cm}^{-3}$ give cooling times $t_C \approx 10^6$ years, in qualitative agreement with our estimate (5.122). The formation of interstellar clouds through thermal instability would therefore require only about 10^6 years.

Non-Steady Theory of the Heating of Interstellar Gas, Hot Component: The steady-state theory of the observed temperatures in the regions with predominantly neutral hydrogen ignores the occurrence of relatively short-lived, but very productive energy sources. Observations of components of the interstellar absorption lines in stellar spectra (Sect. 4.1.4) and the 21 cm line (Sect. 4.1.5) with velocities much greater than $20\,\mathrm{km\,s}^{-1}$ relative to the galactic rotation motion, and therefore far above the normal peculiar velocities, suggest that the interstellar gas is occasionally traversed by density waves with propagation velocities of this magnitude. Whilst the thermal expansion of H II regions over rather large distances can scarcely attain such high velocities, it appears to be possible with supernova explosions (Sect. 6.3.3) and perhaps also with strong stellar winds from massive early star types. In a few cases a far extended system of shock fronts can actually be observed, probably caused by several supernovae (see Fig. 4.31). In these high velocity shock fronts the transformation of mechanical into thermal energy is extraordinarily large, so that temperatures of order 10^6 K must occur (Sect. 6.3.3), particularly in the thin gas of the region lying immediately behind the shock front. The observations of locally intensified diffuse soft x-rays (Sect. 4.1.9) as well as interstellar absorption lines of highly ionised atoms (O VI lines: Sect. 4.1.4) can thus be readily explained, and it offers a heating mechanism which is variable with time.

We are therefore led to the following alternative non-steady model representation of the thermodynamic state of the interstellar gas outside the dense clouds. At intervals of time which are longer than the recombination time (5.118), local heating by the explosion wave of a supernova takes place – with a velocity of $20\,\mathrm{km\,s}^{-1}$ the shock front already covers a distance of 1 pc in 5×10^4 years! Thereafter the ionised hot gas slowly cools off and eventually recombines. For very high temperatures $T \gtrsim 3 \times 10^7$ K this cooling is primarily the result of bremsstrahlung $\Lambda \approx \Lambda_{\text{ff}}$ (5.56). In the temperature range $T \approx 10^5 \ldots 3 \times 10^7$ K the collisional excitation of ions dominates, with subsequent line emission in the UV and x-ray regions, and we have approximately (after Raymond, Cox and Smith)

$$\frac{\Lambda_{\text{col}}}{N N_{\text{e}}} \approx 1.6 \times 10^{-22} \left(\frac{T}{10^6} \right)^{-0.6} [\text{erg cm}^3 \, \text{s}^{-1}] \quad . \tag{5.123}$$

At $T = 1 \times 10^4$ K the Λ_{col} falls off unsteadily with lower temperatures, to the values two or three orders of magnitude smaller which underlie (5.117) (suppression of collisional excitation of H). The cooling through the different temperature ranges therefore occurs at different speeds. The cooling time from an arbitrary starting temperature T_1 down to the considerably lower temperature T_2,

$$t_{\text{C}} \approx \frac{(3/2)Nk(T_1 - T_2)}{\Lambda(T)} \approx \frac{(3/2)NkT_1}{\Lambda(T)} \tag{5.124}$$

amounts to about $10^6/N_{\text{e}}$ years from $T_1 \approx 10^7$ K to $T_2 \approx 10^6$ K, only about $4 \times 10^3/N_{\text{e}}$ years from 10^6 K to 10^5 K, and in the last phase from 10^4 K to 10^2 K a time of the order of a few $10^5/N_{\text{e}}$ years.

In a sufficiently large region of space, therefore, one expects to find side by side interstellar gas of all possible temperatures, with abundances roughly proportional to the respective relevant cooling times. Hot ionised gas with $T \approx 10^6$ K and also partially recombined "warm" or already cold gas with $T \approx 10^2 \ldots 10^4$ K should be relatively widespread, which in broad measure corresponds to the observational results. That gas in the temperature range $2 \times 10^4 \ldots 10^5$ K is also present appears from the evidence of absorption lines, e.g. of Si III, Si IV, C III, C IV and N II in the ultraviolet (among others, from the observations with the Euro-American satellite "International Ultraviolet Explorer", IUE). The quantitative discussion requires a detailed model of the structure of the interstellar medium. In Sect. 6.3.2 a short description will be given of an attempt at a global model, which is based on the scenario of heating by explosion waves from supernovae.

5.2.4 State of Molecular Clouds and H II Regions

The internal structure of the dense cold clouds, in particular the behaviour of the various physical state variables, is so far known only in broad outline. The situation is similar with regard to the compact H II regions, which can always be associated with certain molecular clouds, and even the structure of normal "developed" H II regions is still not fully explained in detail (see also Sect. 6.3.3). We therefore restrict ourselves here primarily to the portrayal of these important interstellar objects by mean parameters of state.

Parameters of State of Molecular Clouds: In the derivation of *"column densities"* of the CO molecule

$$\mathcal{N}(CO) = \int N(CO)dr = \overline{N}(CO)l$$

from the intensities of the 2.6 mm line, it was frequently assumed for the density of the molecular hydrogen that $N(H_2) \gtrsim 5 \times 10^4$ cm^{-3}, since specially simple relationships (LTE) were then obtained in the discussions in Sect. 5.1.2, in connection with (5.35) and (5.52). Whether the H_2 density in the molecular clouds fulfills this condition and how large it actually is, can only be answered by an interpretation of observations of molecular lines which does not assume LTE from the outset, but proceeds from a simultaneous solution of the statistical equations and the radiation transport equation (Sect. 5.1.2). In this way one obtains answers on the column densities of the molecules observed in the microwave region *and* on the densities of the H_2 molecules present as collision partners.

The kinetic temperatures T needed for this can be set approximately equal to the maximal values of the brightness temperatures T_b^* in the profiles of the 2.6 mm line of CO ($C^{12}O^{16}$), $J = 1{\to}0$: in Sect. 4.2.8 two arguments have already been put forward that the typical molecular clouds are optically thick in this line. This is confirmed, for example, by the observation that the CO lines $J = 2{\to}1$ and $J = 1{\to}0$ as a rule have roughly equal intensities up to near the edge of a cloud, whereas in the optically thin case the intensity of the 2→1 line should be several times that of the 1→0 line − the ratio of the two depends somewhat on the temperature.

The radiation transport problem has a new dimension through the observed fact that the widths of the molecular lines of the individual clouds are always several times − sometimes more than ten times − the thermal Doppler widths (Sect. 4.2.8): hence the kinematics of the molecular clouds comes into play. For the *explanation of the line broadening* the following possibilities have been discussed:

1) systematic contracting motion of the whole cloud with a radial velocity of the form

$$V(r) = ar^\alpha \quad ,$$

where r denotes the distance from the centre of the assumed spherical cloud;

2) rotation of the whole cloud;

3) random motions in the cloud which can be treated formally as microturbulence: Gaussian profile of the absorption and emission coefficients with the velocity dispersion

$$\sigma_v = \sqrt{\frac{RT}{\mu} + \xi_t^2}$$

($\mathcal{R}$ = gas constant, μ = molecular weight and ξ_t = "turbulent velocity");
and

4) composition of the cloud out of small condensations with individual motions, i.e. a complex "clumpy" structure.

In a gravitationally unstable molecular cloud even with originally homogeneous structure the two types of motion (1) and (3) are indeed to be expected: large, inwardly directed velocities as a result of the collapse, and flows of turbulent nature as the velocities occurring, especially in the interior, surpass the shell velocity (formation and dissipation of shock waves, etc.). In cases (1) and (2) one has in principle the same situation as for the 21 cm line radiation of the differentially rotating galactic disc (Sect. 4.1.5): a certain narrow part of the line profile is produced by a closely bounded part of the cloud. In case (3), on the other hand, the intensity at a particular part of the profile is contributed by a relatively large region along the line of sight. In all cases the line broadening has the effect that the optical depth is considerably reduced. Accordingly with most of the observed lines even the centre of the molecular cloud can be observed, and it is understandable that often no self absorption is observed in the optically thick CO lines. Even in the CO lines, one can "see" relatively deep into the clouds, so that the contour maps of the brightness temperature $T_b(CO)$ reflect to the first approximation the radial variation of the kinetic temperature in the cloud. Concerning case (4): the existence of small scale structure in some molecular clouds has been shown recently by very high resolution observations.

As an example of the derivation of the densities in molecular clouds we select a study of the lines of carbon monosulphide CS. This molecule has a considerably larger dipole moment than CO ($\mu = 2.0$ compared with $\mu = 0.1$) and hence correspondingly large spontaneous emission probabilities A_{nm} for the three observed transitions $J = 3{\to}2$, $J = 2{\to}1$ and $J = 1{\to}0$. The absolute line strength then depends especially sensitively on density variations, as emerged from the discussion in connection with (5.52) for the simple two-level system: if the collision de-excitations dominate, $I_{nm}^L \sim A_{nm} N_{nm} l$, whereas if radiative de-excitations prevail, I_{nm}^L is proportional to the square of the density. For the excitation of the starting levels, collisions with H_2 molecules and — in proportion — with helium come principally into consideration. Charged particles have indeed very much larger effective cross-sections than these neutral gas particles; however, in the molecular clouds the electrons, which arise for discussion in the first place because of their high velocities, are not sufficiently abundant (see Sect. 5.2.1) to make a noticeable contribution. For the transition of CS, $J = 1{\to}0$, we have $A_{10} \approx 2 \times 10^{-6}\,\mathrm{s}^{-1}$ and (for $T \approx 30\,\mathrm{K}$) Q_{10} (CS-H_2) $\approx 4 \times 10^{-11}\,\mathrm{cm}^3\,\mathrm{s}^{-1}$, so that the condition $N' Q_{10}/A_{10} \gg 1$ requires high densities $N' = N(H_2) \gg 2 \times 10^5\,\mathrm{cm}^{-3}$. In order to be able to include also regions with densities of the order of $10^4\,\mathrm{cm}^{-3}$, it is necessary, in the calculation of the population ratios, not to set the excitation temperature equal from the outset to the kinetic temperature [see Sect. 5.1.2, (5.32 ff.)]. The results discussed in what follows are based on solutions of the statistical equations for the lowest ten energy levels of the CS molecule.

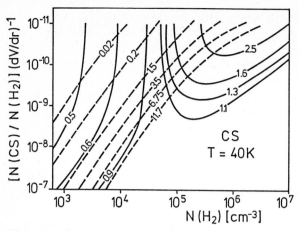

Fig. 5.11. Contours of equal antenna temperature T_A^* of the CS line $J = 2{\to}1$ (*dashed*) and equal line ratio $T_A^*(J = 2{\to}1)/T_A^*(J = 1{\to}0)$ (*continuous*) in a diagram with abscissa $N(H_2)\,[cm^{-3}]$ and ordinate $[N(CS)/N(H_2)](dV/dr)^{-1}$ $[(km\,s^{-1}\,pc^{-1})^{-1}]$ for a contracting model cloud with kinetic temperature $T = 40$ K. Further explanation in text. [By permission of Linke and Goldsmith (1980)]

Figure 5.11 shows as an example results of model calculations for the case of pure contraction motion with $V(r){\sim}r$, hence $\alpha = 1$, for constant kinetic temperature $T = 40$ K. The maximal antenna temperature T_A^* in the line CS, $J = 2{\to}1$, (for a half-power beam width of $2'$, see Sect. 4.1.5) and the line ratio $T_A^*(J = 2{\to}1)/T_A^*(J = 1{\to}0)$ were calculated for given values of the H_2 density and the quantity $[N(CS)/N(H_2)](dV/dr)^{-1}$. Diagrams of the sort shown have, of course, been constructed for other values of the kinetic temperature T. After determination of T from CO observations, the hydrogen density $N(H_2)$ and the parameter $[N(CS)/N(H_2)](dV/dr)^{-1}$ can now be read off approximately as the coordinates of the intersection of the curves corresponding to the two observed quantities $T_A^*(J = 2{\to}1)$ and $T_A^*(J = 2{\to}1)/T_A^*(J = 1{\to}0)$. For example, the value $T = 40$ K used in Fig. 5.11 was found for the molecular cloud associated with the H II region W 51, and the measured values for $T_A^*(J = 2{\to}1)$ and the line ratio are in this case 3.5 K and 1.60, respectively. From this we obtain $N(H_2) = 1 \times 10^5\,cm^{-3}$ and $[N(CS)/N(H_2)](dV/dr)^{-1} = 1 \times 10^{-10}$ $(km\,s^{-1}\,pc^{-1})^{-1}$. From the line widths values of $dV/dr \approx 2\ldots6\,km\,s^{-1}\,pc^{-1}$ were estimated generally, so it follows that the relative CS abundance $N(CS)/N(H_2) \approx (2\ldots6) \times 10^{-10}$.

The observed ratio $T_A^*(J = 2{\to}1)/T_A^*(J = 1{\to}0)$ remains practically constant over the whole line profile. Within the framework of the contraction model this means − since $[N(CS)/N(H_2)](dV/dr)^{-1}$ is always found to be greater than 1×10^{-9} $(km\,s^{-1}\,pc^{-1})^{-1}$ − that the density $N(H_2)$ remains about constant throughout the whole of the observed region of the cloud. In a collapsing cloud, however, one expects a significant increase in density towards the centre. This was noticed in particular with infrared observations of the thermally radiating dust in large molecular clouds; also various observations of other molecules indicate density gradients. On the other hand calculations for pure microturbulence with assumed density gradients give a self-reversal of the CO lines, which is not in general observed. As a way out, therefore, combined models were proposed, in which as well as large scale radial motions of the sort assumed above, there were also, principally in the central region, small scale turbulent motions with $\xi_t \lesssim 10\,\mathrm{km\,s^{-1}}$ associated with radial gradients of density and of kinetic temperature. The observations available up to now certainly do not permit the derivation of an *unambiguous* model, characterised by profiles $V(r), \xi_t(r), T(r)$ and $N_{H_2}(r)$.

On certain mean values of these parameters, however, relatively reliable statements can be made. For the central part of typical molecular clouds the observations of the CS lines lead to *densities* in the region of $N(H_2) \approx 2 \times 10^4 \ldots 2 \times 10^5\,\mathrm{cm^{-3}}$ and *kinetic temperatures* $T \approx 15 \ldots 60\,\mathrm{K}$; for objects which were already known as dark clouds, however, the lower limit of densities found is about $10^3\,\mathrm{cm^{-3}}$ and the temperatures are only about $10\,\mathrm{K}$. Individual results for a selection of molecular clouds are presented in Table 5.9. Here one can broadly distinguish two different kinds of molecular clouds: (1) cold clouds with $T \approx 10\,\mathrm{K}$ and $\Delta m_V \approx 2^m \ldots 10^m$ and (2) centrally warmed clouds with $T \gtrsim 30\,\mathrm{K}$ and $\Delta m_V > 30^m$. Clouds of the second kind, which are generally denser and as a rule associated with strong infrared sources and H II regions, are found to have a lower CO/H_2 ratio than the typical cold dark clouds, namely about

Table 5.9. Parameters of state of a few molecular clouds. As a rule the densities fall off steeply towards the outside, so the diameters given only represent approximate values. The visual extinction Δm_V for Ori A to Sgr B2 was estimated from the CO column density: $\Delta m_V \approx 5 \times 10^{-17}\, N(CO)$. The number density of the hydrogen molecules $N(H_2)$ and the kinetic temperature T relate to the inner region, from which the line emission of the CS molecule comes. T_g is the temperature of the dust grains estimated from the observed thermal emission, and $\mathcal{M}$ denotes the total mass of the molecular hydrogen

Object		Diameter [pc]	Δm_V [mag]	$N(H_2)$ [cm⁻³]	T [K]	T_g [K]	$\mathcal{M}$ [$\mathcal{M}_\odot$]
Dark clouds							
ϱ Oph		3	6	2×10^4	30	20	2×10^3
L 134		1.2	6	1×10^4	10	10	3×10^2
Globules							
B 227		0.6	3	6×10^4	10	(20)	10
B 335		0.5	10	2×10^4	10	(20)	20
Molecular clouds associated with H II regions							
Ori A (OMC 1)	"Nucleus":	0.5	} 280	{ 2×10^5	80	>100	5×10^2
	"Shell":	5		{ 10^3	20	80	10^4
M17 S		5	50	5×10^4	50	75	10^5
W 3 ("Nucleus")		1	250	1×10^5	30	80	10^4
Sgr B 2	"Nucleus":	5	} 1000	{ 5×10^4	100	>100	5×10^5
	"Shell":	40		{ 1×10^4	20	40	3×10^6

2×10^{-5} compared with the value 8×10^{-5} given in Sect. 5.4.4. "Warm" clouds are probably in a later stage of gravitational contraction, in which with higher densities a larger portion of the carbon is bound in the solid particles. In the case of the dark cloud at ϱ Ophiuchi there is warming by embedded stars with spectral types later than B 2, which were detected in the infrared.

Interpretation of the Maser Point Sources: The small line widths, as well as the small apparent diameters of the observed sources (Sect. 4.2.9) may be explained in the following manner. Let us suppose that the absorption coefficient for the line under consideration is negative in the emitting region, considered to be spherical and homogeneous [see Sect. 5.1.2 after (5.16)]. In this case $-\tau_\nu^* \gg 1$, $S_\nu < 0$ and the expression (5.5) in Sect. 5.1.1 for the intensity on a certain line of sight becomes

$$I_\nu = (-S_\nu + I_{\nu,0})\exp(-\tau_\nu^*) = (|S_\nu| + I_{\nu,0})\exp(|\tau_\nu^*|) \quad . \tag{5.125}$$

The exponential factor describes the reinforcement of the spontaneous emission and of the background radiation. Assuming for the line profile of the absorption coefficient just the thermal Doppler effect, with the Doppler width $\Delta\nu_D$, then when the ratio of the line intensity at $\nu = \nu_1$ to that at the line centre is $I_{\nu 1}/I_{\nu 0} = e^{-1}$ the difference of the optical depths is given by

$$|\tau_{\nu 0}^*| - |\tau_{\nu 1}^*| = |\tau_{\nu 0}^*| - |\tau_{\nu 0}^*|\exp\left\{-\left(\frac{\nu_1 - \nu_0}{\Delta\nu_D}\right)^2\right\} = 1 \quad .$$

For sufficiently large values of $|\tau_{\nu 0}^*|$ the line width $\nu_1 - \nu_0$ therefore becomes small compared with the thermal Doppler width. Then it follows that

$$\frac{\nu_1 - \nu_0}{\Delta\nu_D} = \frac{1}{\sqrt{|\tau_{\nu 0}^*|}} \quad . \tag{5.126}$$

This is indeed easily understood: the exponential connection between I_ν and $|\tau_\nu^*|$ has the effect that the reinforcement of the line centre is considerably enhanced and a narrow line is the result. The interpretation of the observations by (5.126) would require optical thickness $|\tau_\nu^*| \approx 30$.

In a similar way, at a fixed wavelength, for example $\nu = \nu_0$, the decrease of the optical thickness of the emission region from the middle to the edge leads to a decrease in the apparent diameter of the source. Whereas the object in the case $\tau_\nu^* \gg 1$ would appear as a uniformly luminous disc, now, with $\tau_\nu^* \ll -1$, there is a sharp maximum around the central line of sight. From (5.125) it also follows eventually that temporal variations of the optical thickness must lead to intensity fluctuations of exaggerated magnitude.

The cause of the *population inversion* assumed here in the interstellar maser sources can so far be understood only in broad outline. The molecules are first of all put into a higher state (represented in Fig. 5.12 by Level 3) by a "pumping process". From here preferred spontaneous transitions – mostly lying in the infrared – can take place into the upper state of the maser line (Level 2),

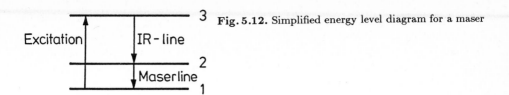

Fig. 5.12. Simplified energy level diagram for a maser

but not into the lower (Level 1): $A_{32} \gg A_{31}$. If also $A_{32} \gg A_{21}$ and C_{21}, and the depopulation of the lower state (1) is sufficiently strong, one has a continual overpopulation of the upper state (2). There are three kinds of pumping process:

1) Photon pumping: higher states are excited by absorption of photons (1→3 in Fig. 5.12).

2) Collisional pumping: higher states are excited by collisions (1→3 in Fig. 5.12).

3) Chemical pumping: molecules are created preferentially in certain excited states, from which transitions can then take place into the lower lying starting state of a maser line.

A difficulty with the assumption of photon pumping is that the radiation suitable for this coming from the outside is always still more strongly absorbed in the region of the maser source than the frequency of the maser line. In attempts to interpret the OH maser, for example, all three pumping mechanisms have been invoked. The observed intensity ratios of the four OH maser lines could be made understandable by calculations for the case of collisional excitation (by H_2) taking into account radiation transport in the infrared lines.

State Parameters of HII Regions: We limit ourselves primarily to statements of "broad analysis" on the basis of the radio observations unaffected by dust extinction. Thereby the dense compact HII regions can also be included.

The *electron temperature* T_e can be obtained by three different methods which have already been mentioned: (1) from the absolute measured thermal radiocontinuum at low frequencies from Sect. 5.1.3 (5.59) for $\tau_\nu^C \gg 1$; (2) the observed intensity ratio of two collisionally excited optical emission lines with sufficiently widely separated starting levels from (5.42) and (3) the observed intensities of radio recombination lines [see Sect. 5.1.2 (5.44 ff.)]. The last named procedure, with a suitable choice of the lines, can lead in a simple way to a good approximation for T_e. One forms the ratio of the radio recombination line relative to the neighbouring continuum for the case of small optical thicknesses $\tau_\nu^L \ll 1$ and $\tau_\nu^C \ll 1$ (high frequencies) with the measured brightness temperatures in the line and the continuum $T_L = \tau_\nu^L T_e$ and $T_C = \tau_\nu^C T_e$. From Sect. 5.1.2 (5.15a) and (5.40) with $N_e = N_p + N(He^+)$ and the definition (5.45), taking into account the stimulated emission ($h\nu_{nm}/kT_e \ll 1$), it follows that

$$\tau_\nu^L = \text{const } b_m T_e^{-5/2} \exp\left\{-\frac{\chi_0 - \chi_{0,m}}{kT_e}\right\}$$

$$\times \frac{E}{1 + [N(\text{He}^+)/N_p]}\psi(\nu) \quad , \tag{5.127}$$

whilst τ_ν^C is given by (5.58). The exponential factor is approximately equal to unity. Expressing the integral of the brightness temperature $T_L(\nu)$ over the line by the product $T_L^*\Delta\nu_L$, where T_L^* is the maximal value of $T_L(\nu)$ and $\Delta\nu_L$ the line width, then for hydrogen lines in the LTE approximation ($b_m \approx 1$) we have

$$\frac{T_L^*\Delta\nu_L}{T_C} \approx 2.3 \times 10^4 \nu^{2.1} \frac{T_e^{-1.15}}{1 + [N(\text{He}^+)/N_p]} \quad . \tag{5.128}$$

Here $\Delta\nu_L$ is expressed in kHz and ν in GHz. Equation (5.128) allows T_e to be determined as a function of N_e or E. Model calculations for lines at 5 GHz (H 109α, H 137β), taking deviations from LTE and the effect of collisional broadening into account show that the result for H II regions with emission measures $E < 10^6$ pc cm^{-6} obtained from (5.128) represents a good approximation to the mean electron temperature. Typical results for T_e are contained in Table 5.10.

One way to obtain the local *electron density* N_e starts from (5.42) and has already been explained there: one calculates as a function of N_e the ratio of

Table 5.10. Mean electron temperature T_e, emission measure E, root mean square electron density $\left(\overline{N_e^2}\right)^{1/2}$, excitation parameter u, total number of Lyman continuum photons per s L_c and total mass of ionised hydrogen $\mathcal{M}(\text{H II})$ for a selection of H II regions

Object	T_e [K]	E [pc cm^{-6}]	$\left(\overline{N_e^2}\right)^{1/2}$ [cm^{-3}]	u [pc cm^{-2}]	L_c [s^{-1}]	$\mathcal{M}(\text{H II})$ [$\mathcal{M}_\odot$]
W 3, A1-A5	8400	2×10^7	6×10^3	83	4×10^{49}	10
W 3(OH)	10000	1×10^9	2×10^5	54	3×10^{48}	0.1
M 42	8200	6×10^6	5×10^3	55	7×10^{48}	10
NGC 2237-46	8000	3×10^4	20	80	2×10^{49}	1×10^4
M 20	8000	5×10^4	1×10^2	50	5×10^{48}	2×10^2
M 8	8000	4×10^5	6×10^2	64	1×10^{49}	2×10^2
M 17, main source (S)	7700	5×10^6	2×10^3	170	2×10^{50}	10^2
M 16	8000	4×10^5	2×10^2	120	7×10^{49}	7×10^2
W 51, main source	7300	5×10^7	8×10^2	190	3×10^{50}	10^2
W 75, DR 21 A	8400	5×10^7	2×10^4	36	2×10^{48}	0.2
B		5×10^7	3×10^4	27	7×10^{47}	0.1
C		9×10^7	3×10^4	49	4×10^{48}	0.4
D		4×10^7	3×10^4	27	7×10^{47}	0.1
NGC 7538 A1	7900	8×10^5	1×10^3	60	8×10^{48}	33
A2		2×10^6	4×10^3	14	1×10^{47}	0.1
B		7×10^6	6×10^3	26	7×10^{47}	0.3
C		1×10^7	1×10^5	12	7×10^{46}	0.002
NGC 7000	7000	4×10^3	10	100	4×10^{49}	2×10^4

two collisonally excited optical emission lines with closely neighbouring upper levels, and from the resulting relationship one can obtain a value for N_e from any observed line ratio. Other methods find the *emission measure E* defined by (5.45) and thence obtain the mean square of the electron density $\overline{N_e^2} = E/l$. The first of these methods applies the upper equation (5.59), with τ_ν^C from (5.58), to measurements of the radio continuum at high frequencies. The second applies (5.44) to observations of recombination lines, especially radio lines of this sort because of the interstellar extinction in the visible. Here it is assumed that at least approximate values for T_e are known.

Because of the different ways of taking an average, complete agreement between the N_e obtained from (5.42) and the values of $\sqrt{\overline{N_e^2}} = \sqrt{E/l}$ obtained from the two other methods, is to be expected only in the ideal case of a homogeneous H II region. The reality is evidently very different from this, since the first method gives actual values of N_e which are usually several times larger than the result for $\sqrt{\overline{N_e^2}}$. It suggests that the electron densities N_e determined from line ratios are primarily related to *inhomogeneities* with higher densities. Then $\sqrt{\overline{N_e^2}}$ would necessarily always be smaller than this N_e: if, on a line of sight whose total length inside the H II region is l, there are n isolated condensations with equal thickness a and equal electron density N_e (histogram profile), then it follows that

$$\overline{N_e^2} = \frac{1}{l}\int_0^l N_e^2(r)\,dr = \frac{n}{l}N_e^2 a < N_e^2 \tag{5.129}$$

when $na < l$. Values of na/l of order 10^{-1} seem not implausible.

The connection between the observed *radiation flux in the radio continuum* $\Phi^C(\nu)$ and the emission measure E is given by (5.59) with (5.58) in the case $\tau_\nu^C \ll 1$, which is valid for sufficiently high temperatures, as

$$\Phi^C(\nu) = \int I_\nu d\Omega = \overline{I}_\nu \Omega = \text{const}.T_e^{-0.35}\nu^{-0.1}E\Omega \quad . \tag{5.130}$$

If the solid angle Ω subtended by the source is known, then E can be calculated. If one knows the Strömgren radius r_s from the angular diameter and the distance of the H II region, then from $E = \overline{N_e^2}l$ with $l = 2r_s$, one obtains the mean square value of the electron density $\overline{N_e^2}$ and moreover the *excitation parameter* $u = r_s(\overline{N_e^2})^{1/3}$. From (5.94) this quantity determines the total flux of the *Lyman continuum quanta* ionising the hydrogen:

$$L_c = \frac{4\pi}{3}u^3\alpha_0^{(2)} \approx 5.05 \times 10^{46}T_e^{-0.8}u^3 \text{ [photons s}^{-1}] \quad . \tag{5.131}$$

Here the expression (5.82) for $\alpha_0^{(2)}$ has been approximated by $4.1 \times 10^{-10}T_e^{-0.8}$; u is in pc cm^{-2}. The excitation parameter can be expressed directly in terms of the observed radiation flux $\Phi^C(\nu)$: let r be the distance of the H II region, assumed spherical, from the observer, then $\Omega = \pi r_s^2/r^2$ and one has from (5.130)

$$\Phi^C(\nu) \sim E\Omega = \overline{N_e^2} 2r_s \pi r_s^2/r^2 \sim r_s^3 \overline{N_e^2}/r^2 = u^3/r^2 \quad .$$

From (5.131) one obtains the important relation [after Mezger (1972)]

$$L_c \approx 4.8 \times 10^{48} T_e^{-0.45} \nu^{0.1} \Phi^C(\nu) r^2 \,[\text{photons s}^{-1}] \quad , \tag{5.132}$$

where ν is in GHz, $\Phi^C(\nu)$ in Jy and r in kpc. If the ionisation is caused by a single star, then its spectral type can be assessed with the help of Table 5.6 from the value obtained for L_c. The result for L_c can often only be explained by the assumption of several exciting stars.

The *total mass of ionised hydrogen* in a spherical H II region with uniform number density of protons N_p amounts to

$$\mathcal{M}(\text{H II}) = \frac{4\pi}{3} r_s^3 N_p m_p \quad . \tag{5.133}$$

If the helium is only singly ionised, then $N_p \approx N_e - N(\text{He}^+)$ with $N(\text{He}^+) \approx 0.1 N_p$.

The difference between N_p and N_e is unimportant, however, compared with the uncertainty of the volume, for which, knowing $\Omega = \pi r_s^2/r^2$, we can write $(4/3) r_s r^2 \Omega$. If r_s was determined from the angular extent of the H II region, then the distance is involved to the third power. For non-uniform density the mean value $\overline{N_e}$ should be used. If instead one uses $\sqrt{\overline{N_e^2}}$, then in general the mass obtained is too great[3] since $\overline{N_e^2} \geq (\overline{N_e})^2$.

Values of the parameters of state derived in the way described for a selection of normal and compact H II regions are presented in Table 5.10 (for observational data on these objects, see Tables 4.10, 4.11 and 4.12).

Between the mean densities and the linear diameters of the H II regions so far known there is a statistical connection: the smaller the diameter, the greater is the mean density. Starting from this one can set up a broad *classification scheme for the H II regions*, which begins with the most compact objects and ends with the most extended objects, as proposed in Table 5.11. It is probable that this scheme at the same time represents a sequence of stages of development. At the beginning we have therefore put the presumed forerunner of a compact H II region: a newly formed massive star, still enveloped in a shell which is opaque to Lyman photons because of the high dust content. Infrared observations show that objects of this sort usually occur in groups in a molecular cloud. The large "developed" H II regions with complex structures

---

3

$$\overline{N_e^2} = \overline{(\overline{N_e} + \Delta N_e)^2} = (\overline{N_e})^2 + 2\overline{N_e (\Delta N_e)} + \overline{(\Delta N_e)^2} = (\overline{N_e})^2 + \overline{(\Delta N_e)^2} \geq (\overline{N_e})^2$$

since $\overline{(\Delta N_e)} = 0$ and $\overline{(\Delta N_e)^2} \geq 0$. In particular for the case of randomly distributed condensations this follows from (5.129) in connection with the similarly formed linear mean value $\overline{N_e} = (n/l) N_e a$: one obtains $\overline{N_e^2}/(\overline{N_e})^2 = l/na \geq 1$.

Table 5.11. Broad classification scheme for the H II regions, after a suggestion by Churchwell (1975). Explanation in text

Name of the object observed	Observed characteristics	Diameter [pc]	N_e [cm^{-3}]	Stage of development	Examples
Compact infra-red source at $\approx 20\ \mu m$	Continuous emission at $\simeq 20\ \mu m$, but not at $2\ \mu m$; no radio continuum; optically not detectable	$\lesssim 0.05$	$> 5 \times 10^3$	Pre-Main Sequence O-star in a shell, opaque to optical radiation, including UV ("cocoon star")	IRS 5 in W 3
Compact H II region	Thermal radiocontinuum; infrared continuum; usually not detectable in the visible	$0.05\ldots 1$	$\gtrsim 10^3$	Earliest phase of development of an H II region	Components in W 75, DR 21 and in NGC 7538 (see Table 4.10)
Complex H II region of inter-mediate density	Continuous and line radiation in the optical, infrared and radio-frequency regions	$1\ldots 100$	$10\ldots 10^3$	Expanding H II region	M 42, M 17
Large diffuse H II region		> 100	$\lesssim 10$	Very far expanded H II region	NGC 7000 (North America Nebula)

are therefore as a rule not the result of the formation of an individual massive star, but emerge from such a *group*.

5.2.5 Elemental Abundances in the Interstellar Gas

It is very probable that the relative abundances of the chemical elements seen in the interstellar matter as a whole are the same as in the stars of early spectral type, which were formed from this medium less than 10^7 years ago. The interactions of the various components: atoms, molecules, solid particles with one another may, however, lead to differences of the elemental abundances in the gas compared with the values found for the young stellar Population I. If one wishes to recognise the main effective processes and judge their consequences for the development of the interstellar medium, then one requires first of all quantitative statements on the actual chemical composition of the gas in the various regions. The empirical basis for this is provided by observations of absorption or emission lines. From the line strengths, however, one obtains immediately at best the column densities of the atoms or ions in the starting states of the individual lines. Transition to the total value for all atoms or ions of an element requires knowledge of the excitation and ionisation ratios. The problem posed is therefore inseparably bound up with the determination of the temperature and the degree of ionisation or the electron density for each of the regions included.

For the diffuse clouds observations of interstellar absorption lines in stellar spectra form the primary basis for the analysis. We therefore mention first of all the general method for obtaining the column densities of the atoms, ions or molecules which are in the lower level of the transition causing the line. For the derivation of relative elemental abundances the results on excitation and ionisation ratios already obtained in Sect. 5.1.2 and Sect. 5.2.1 can in part be used. After a discussion of the results, including the abundance of the molecular hydrogen, we shall apply ourselves to the analysis of H II regions.

Method of Analysis of Interstellar Absorption Lines: In the pure absorption case considered here the *equivalent width* is given as a measure of the line strength by

$$A_\lambda = \int\limits_{\text{line}} \frac{I_C - I_\lambda}{I_C} d\lambda = \int\limits_{\text{line}} \left(1 - e^{-\tau_\lambda^L}\right) d\lambda \quad . \tag{5.134}$$

Here I_λ and I_C denote the intensities in the line at wavelength λ and in the neighbouring continuum of the stellar spectrum. One writes A_λ if one refers to a spectrum over λ (and not ν); A_λ has the dimension of a wavelength. The second form of the integrand follows since $I_\lambda = I_C \exp(-\tau_\lambda^L)$, where τ_λ^L is the optical depth, reckoned from the observer to the star, and formed from the line absorption coefficient κ_λ^L. For the visual and UV regions considered here, the stimulated emission can be neglected in (5.15a) from Sect. 5.1.2 (factor $1 - \exp(-h\nu/kT_{ex}) \approx 1$ because $h\nu/kT_{ex} \gg 1$), and after transforming from ν to λ one obtains

$$\tau_\lambda^L = \frac{h\nu_{nm}}{c^2}\lambda_0^2\psi(\lambda)B_{mn}\mathcal{N}_m = \frac{\pi e^2}{m_e c^2}\lambda_0^2\psi(\lambda)f_{mn}\mathcal{N}_m \qquad (5.135)$$

with the "column density"

$$\mathcal{N}_m = \int N_m\, dr \qquad (5.136)$$

and the "oscillator strength" for absorption

$$f_{mn} = \frac{m_e h\nu_{nm}}{\pi e^2}B_{mn} = \frac{m_e c}{8\pi^2\nu_{nm}^2}A_{nm}\frac{g_n}{g_m} \qquad . \qquad (5.137)$$

Here λ_0 is the wavelength of the centre of the line, $\psi(\lambda)$ denotes the line profile normalised to unity. In the reduction of B_{mn} to A_{nm} (5.14) has been applied.

The connection between the observed equivalent width A_λ and the required column density $\mathcal{N}_m$ of the atoms, ions or molecules in the starting state of the interstellar line – as a rule the ground state (resonance lines) – has the normalised line profile $\psi(\lambda)$ as an important contributory factor. At the low densities in the interstellar gas $\psi(\lambda)$ for an individual cloud is given by a convolution of the pure profile for "radiation damping"

$$\psi_N(\lambda) = (\Delta\lambda_N/\pi)[(\Delta\lambda)^2 + (\Delta\lambda_N/2)^2]^{-1}$$

and for Doppler broadening (corresponding to a Maxwellian velocity distribution)

$$\psi_D(\lambda) = (\Delta\lambda_D\sqrt{\pi})^{-1}\exp[-(\Delta\lambda/\Delta\lambda_D)^2]$$

with $\Delta\lambda = \lambda - \lambda_0$. The relevant parameters are the natural line widths $\Delta\lambda_N = (\lambda_0^2/2\pi c)\gamma_{rad}$ with $\gamma_{rad} =$ damping constant for radiation damping and the Doppler width

$$\Delta\lambda_D = \frac{\lambda_0}{c}\sqrt{2\overline{(\Delta v_r)^2}} = \frac{\lambda_0}{c}b \qquad . \qquad (5.138)$$

Here $b/\sqrt{2} = \sqrt{\overline{(\Delta v_r)^2}} = \sqrt{\overline{(v_r - \bar{v}_r)^2}}$ is the dispersion of the velocity components along the line of sight of the absorbing gas particles, which is associated with thermal and turbulent (macroscopic) motions inside the cloud. For the latter a Gauss distribution is also assumed as a formal approximation. The relation between A_λ and $\mathcal{N}_m$ can be put in the form of a family of curves with the parameter $\Delta\lambda_N/\Delta\lambda_D$. These are the so-called *curves of growth*: in the present case they determine the damping profile solely by means of the constant of radiation damping γ_{rad}, which is easily calculated for each line, so that only $\Delta\lambda_D$ or the velocity-spread parameter b enters as the unknown quantity. For this one usually plots $\log(A_\lambda/\lambda_0)$ against $\log(\mathcal{N}_m f_{mn}\lambda_0)$. Such theoretical curves of growth are shown in Fig. 5.14 for a few lines (with various γ_{rad}) for the parameter value $b = 6.5\,\text{km}\,\text{s}^{-1}$.

Three regions of the curves of growth can be distinguished:

1) *Weak lines* with $\tau_\lambda^L \ll 1$ even in the centre of the line. Then it follows that $1 - \exp(-\tau_\lambda^L) \approx \tau_\lambda^L$ and one immediately obtains from (5.134) and (5.135)

$$\frac{A_\lambda}{\lambda_0} = \frac{\pi e^2}{m_e c^2}\mathcal{N}_m f_{mn}\lambda_0 \quad . \tag{5.139}$$

The curves begin with a linear climb, which is independent of b.

2) *Moderately strong lines* with $\tau_\lambda^L \gtrsim 1$ in the centre of the line, but $\tau_\lambda^L \ll 1$ at $\Delta\lambda \gtrsim 3\Delta\lambda_D$, i.e. no noticeable damping "wings". Here we can set $\psi(\lambda) = \psi_D(\lambda)$ and obtain

$$\frac{A_\lambda}{\lambda_0} = \frac{\Delta\lambda_D}{\lambda_0}2F(\tau_0^L) = \frac{2b}{c}F(\tau_0^L) \qquad \text{with} \tag{5.140}$$

$$F(\tau_0^L) = \int\limits_0^\infty [1 - \exp(-\tau_0^L e^{-x^2})]dx \tag{5.141}$$

and the optical depth for the line centre

$$\tau_0^L = \frac{\pi e^2}{m_e c^2}\lambda_0^2\psi(\lambda_0)\mathcal{N}_m f_{mn} = \frac{\sqrt{\pi}e^2}{m_e c}\frac{1}{b}\mathcal{N}_m f_{mn}\lambda_0 \quad . \tag{5.142}$$

The factor 2 appears in (5.140) because the integral with respect to $x = \Delta\lambda/\Delta\lambda_D$ extends from $-\infty$ to $+\infty$. For large τ_0^L, $F(\tau_0^L) \approx \sqrt{\ln \tau_0^L}$, and so the result is a very flat profile of the curves. The reason for this tendency to "saturation" is simply that in the region of the line centre I_λ approximates to the value zero, so that with increasing $\mathcal{N}_m$ no further contribution to A_λ can come from there. Numerical values of the function $F(\tau_0^L)$ are contained in Table 5.12.

3) *Very strong lines.* If $\mathcal{N}_m f_{mn}\lambda_0$ is very large, the "wings" of the damping distribution cause a renewed growth of the equivalent width. In this case $I_\lambda = 0$ for $|\lambda - \lambda_0| = |\Delta\lambda| \lesssim \Delta\lambda_N$ and it affects the profile of $\psi(\lambda)$ only in the region $|\Delta\lambda| \gg \Delta\lambda_N$, in which $\psi(\lambda) \approx \psi_N(\lambda) \approx (\Delta\lambda_N/\pi)/(\lambda - \lambda_0)^2$. The integration over λ can then be carried out in closed form with the result

310

Table 5.12. Numerical values for the theoretical curves of growth with pure Doppler broadening

τ_0^L	0.10	0.40	1.0	3.0	10	100	1000	10 000
$\log \tau_0^L$	-1.00	-0.40	0.00	0.48	1.00	2.00	3.00	4.00
$\log F(\tau_0^L)$	-1.07	-0.510	-0.192	0.075	0.220	0.354	0.436	0.494

$$\frac{A_\lambda}{\lambda_0} = \frac{\lambda_0}{c} \sqrt{\frac{e^2}{m_e c}} \sqrt{\mathcal{N}_m f_{mn} \gamma_{\text{rad}}} \tag{5.143}$$

("square root" section). For each line γ_{rad} is an individual constant. If m is the lowest level, then

$$\gamma_{\text{rad}} = \gamma_n + \gamma_m \approx \gamma_n = \sum_m A_{nm} \quad .$$

With the help of (5.137), γ_{rad} can be expressed in terms of the oscillator strengths f_{mn}.

Very strong interstellar lines occur in the UV. The best known example is the Lyman α-line of hydrogen $\lambda 1216$ Å. Many lines, including those of the visual region, however, fall in the middle, "flat" region of the curves of growth. Obtaining $\mathcal{N}_m$ from A_λ then runs up against the following difficulties: (1) the parameter b cannot be taken as known, primarily because it depends not only on the kinetic temperature of the gas, but also on the as yet unknown dispersion of the turbulent velocities; (2) if the line consists of *unresolved* components with different radial velocities (clouds!), there is in general a different behaviour of the curves of growth from those for the single cloud assumed above. For example, if two unresolved components are equally strong and not overlapping, then the linear portion of the curve of growth extends twice as far, and the flat portion accordingly lies considerably higher (Fig. 5.13). Not until the damping

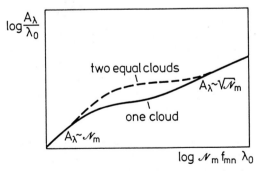

Fig. 5.13. Curves of growth for a single cloud with the column density $\mathcal{N}_m$ (*continuous curve*) and for two clouds equal to one another, which together also have a column density $\mathcal{N}_m$, and whose line profiles do not overlap owing to different radial velocities, but which are not resolved by the spectrograph and so appear as one line (*dashed*). Explanation in text

portion does the curve approach the profile for the single cloud because of the overlapping of the components now occurring in any case.

A way out is offered by the *construction of an "empirical curve of growth"*. So long as the damping plays no role, the various lines of all elements for a certain line of sight, or even for several different directions with similar ratios, should lie on the same curve of growth. If one has several lines with the same starting level m (lines of a multiplet), then the profiles of the curves can be found, to within a shift of the abscissa, by plotting $\log(A_\lambda/\lambda_0)$ against $\log(f_{mn}\lambda_0)$ from the observations themselves. The transition to the $\log(\mathcal{N}_m f_{mn} \lambda_0)$ scale is achieved by displacement in the abscissal direction until a fit is obtained with the theoretically known linear growth portion, which is independent of b (5.139). The comparison with theoretical curves in the middle region provides an effective value of b or $\Delta\lambda_{\rm D}$. An example of this is shown in Fig. 5.14.

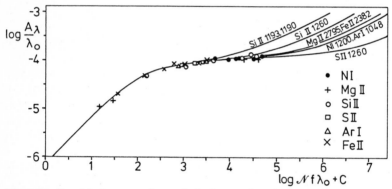

Fig.5.14. Empirical curves of growth for interstellar absorption lines of the dominant ionisation stages of a few elements in the spectrum of the star ζ Ophiuchi. The family of curves was calculated for $b = 6.5\,{\rm km\,s}^{-1}$ with the damping constants $\gamma_{\rm rad}$ for each of the plotted lines. [By permission of Morton (1975)]

For the components of the doublets D_1/D_2 at $\lambda\,5890/96\,\text{Å}$ of NaI, and H/K at $\lambda\,3934/69\,\text{Å}$ of CaII, the following, so-called *Doublet Ratio Method* is often applied. In both cases the ratio of the f-values of the components $f^{(1)} : f^{(2)} = 1 : 2$, so that for the "doublet ratio", we have, from (5.140) with $\lambda_0^{(1)} \approx \lambda_0^{(2)}$,

$$\frac{A_\lambda^{(2)}}{A_\lambda^{(1)}} = \frac{F(2\tau_0^{(1)})}{F(\tau_0^{(1)})} \tag{5.144}$$

where $\tau_0^{(1)}$ denotes the optical depth at the centre of the line (5.142) for the component with the smaller f-value. If one plots the doublet ratio (5.144) against $\tau_0^{(1)}$, then a family of curves with parameter b is obtained, which for $\tau_0^{(1)} \to 0$ take the value 2, and in the saturation region ($\tau_0^{(1)} \gg 1$, $\gamma_{\rm rad} = 0$) tend to the value unity.

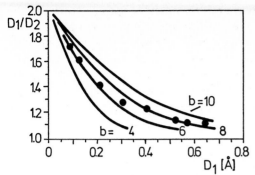

Fig.5.15. Calculated doublet ratio curves for the equivalent widths of the interstellar Na D-lines for the parameter values $b = 4, 6, 8$ and $10\,\mathrm{km\,s^{-1}}$, and observed values for stars of the local arm. [After Münch (1968)]

For the abscissa one chooses not $\tau_0^{(1)}$, but $A_\lambda^{(1)}$ calculated for that purpose. As an example, Fig. 5.15 shows the resulting group of curves for the D-lines. The plotted points are observed values for a few stars of the northern Milky Way. As one may see, the values of the parameter lie between 6 and $8\,\mathrm{km\,s^{-1}}$. The purely thermal motions of the Na atoms at $T = 100\,\mathrm{K}$ would amount only to $b = \sqrt{2\mathcal{R}T/\mu} \approx 0.3\,\mathrm{km\,s^{-1}}$!

Abundances of the Elements in H I Clouds, Share of H_2: Most of the neutral atoms or ions detected by interstellar absorption lines are in their ground states [see discussion in Sect. 5.1.2 after (5.31)]. The column densities $\mathcal{N}_m$ accordingly already represent practically all the atoms of each ionisation stage under consideration. The transition to the total column density of the element, summed over all the ionisation stages, requires a discussion such as is carried out in Sect. 5.2.1 in connection with (5.97). Frequently one ionisation stage is dominant. In the case of sodium it is the single ion Na^+, further examples being Mg^+, Si^+, S^+, Fe^+, whilst N and O are practically neutral. If one rests the analysis on the lines of one such dominant ionisation stage, then the influence of the uncertainties of temperature and electron density on the result for the sum of all ionisation stages can be kept small. These are predominantly moderately strong lines in the ultraviolet (cf. Tables 4.4 and 4.5), for which nevertheless a modification of the curves of growth due to the cloud structure must be expected.

The relations for atomic hydrogen are simple: the very strong Ly α-line already falls in the square-root section of the growth curve, so this is determined purely by a sufficiently exact knowledge of the damping constant γ_{rad}. Because of the great line widths the components arising from individual clouds overlap extensively. Practically all the neutral H atoms are in the ground state $m = 1$, the starting state for the line.

The column density of the hydrogen in molecular form is obtained by analysis of the Lyman bands lying below $1110\,\text{Å}$. It can be assumed here that practically all the H_2 molecules are in the lowest vibration state $v = 0$ of the electronic ground state $^1\Sigma_u^+$. The total column density $\mathcal{N}(H_2)$ then follows by summation of the column densities $\mathcal{N}(J)$ of the H_2 molecules in the rotational levels $J = 0, 1, \ldots$, derived from the strengths of the individual lines.

313

The available results for the elemental abundances relative to hydrogen $\mathcal{N}(\mathrm{X})/\mathcal{N}(\mathrm{H})$ show, for a considerable fraction of the 25 or so elements included, more or less marked deficits compared with the values found for the atmospheres of the stars of Population I. If certain kinds of atom are bound in molecules or solid particles, these anomalies should occur principally in regions of higher densities. Using the reddening as the always available orientation measure for the density of the interstellar medium on the path of the radiation from the star to the observer (Sect. 4.1.1), one can actually show that the depletions are primarily in the regions with higher interstellar extinction $E_{\mathrm{B-V}}>0.3$, and therefore occur in the clouds. As an example, results are presented in Fig. 5.16 for the particularly thoroughly studied star ζ Ophiuchi (O 9V, $r = 200\,\mathrm{pc}$, $E_{\mathrm{B-V}} = 0.32$).

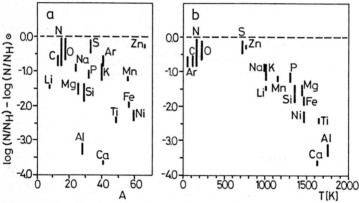

Fig. 5.16a,b. Elemental abundances in the interstellar H I gas on the path to the star ζ Ophiuchi relative to the values for the solar atmosphere, plotted (**a**) against the atomic weight A and (**b**) against the condensation temperature T_{c} of the element concerned. (After Spitzer and Jenkins, with permission, from the Annual Review of Astronomy and Astrophysics, © 1975 by Annual Reviews Inc.)

The *depletions* are particularly marked, at about 10^{-3}, for Ca, whilst for Si, Fe, Ti one finds about 10^{-2} and for C, N, O about 1/5. For lines of sight with $E_{\mathrm{B-V}}<0.05$, on the other hand, elemental abundances are close to normal. This seems also to be true for interstellar gas of high peculiar velocity ($>20\,\mathrm{km\,s^{-1}}$) which produces absorption lines in spectra of stars with $E_{\mathrm{B-V}}>0.05$.

Since the formation of polyatomic molecules or condensation into solid particles is scarcely possible under the conditions of the diffuse interstellar clouds included here, this suggests that free atoms or ions are assimilated by already existing dust particles. Here atoms with smaller ionisation energy χ_0 should as a rule be capable of being bound more easily. The elements with smaller χ_0 do actually show more marked depletions – from N, O and C χ_0 systematically decreases through Si, Fe, Ti down to Ca. Li, Na and K with χ_0-values similar to that of Ca, but relatively small depletions, do not obey this rule. These, however, are the only elements detected in the interstellar medium which can form a chemically saturated molecule with an H atom. If this happens on the surface of a dust particle, then such particle is immediately expelled and is soon dissociated by radiation in the interstellar space.

The fact that only small depletions are found in clouds with high velocity could be attributed to the partial destruction of the dust particles by shock waves (see McCray, 1979).

The results for the *abundance of hydrogen molecules* are often expressed by the ratio of the H atoms bound in H_2 to the free H atoms:

$$f = \frac{2\mathcal{N}(H_2)}{\mathcal{N}(H)} = \frac{2\overline{N}(H_2)}{\overline{N}(H)} \quad . \tag{5.145}$$

For stars with $E_{B-V} < 0.1$ one finds generally only very small values $f < 10^{-4}$, usually between 10^{-5} and 10^{-6}. On the other hand, for $E_{B-V} \approx 0.1\ldots0.4$ (stars with higher E_{B-V} have not so far been studied because of their low apparent magnitude) values of 5% to 70% molecular hydrogen are found. Already in the typical dark clouds with visual extinction values $A_V = 3E_{B-V}$ of a few magnitudes, the hydrogen must be overwhelmingly in the form of H_2, in agreement with the considerations in Sect. 5.2.2.

Abundances of the Elements in H II Regions: The starting point consists of observations of emission lines: for hydrogen and helium optical and radio recombination lines, for the higher elements the collisionally excited optical lines. In the case of small optical thickness, which is always the case for the optical emission lines, the line strengths can be expressed in the form

$$I^L = \int N_i N_e F(T_e) dr \tag{5.146}$$

where N_i denotes the number density of the ions (or atoms) responsible for the emission. The transition to N_i or $\mathcal{N}_i$, the corresponding column density, requires the assumption of a model for the H II region. In other words, we need the shape of the distribution of T_e, N_e and N_i, and use the observed intensity ratios only for determining the relative element abundances which are constant over the whole H II region. Such a model can be constructed for the radiation characteristics of the ionising star from the equations already given for the ionisation equilibrium [Sect. 5.2.1, (5.84 ff.)], the radiation transport [Sect. 5.2.1, (5.87)] and the heat equilibrium (Sect. 5.2.3). Incompleteness and inaccuracies of the observational data, and the difficulties of taking account in the model of the random inhomogeneities, make this expenditure of effort appear sensible only for a few objects, such as perhaps the Orion Nebula.

Hydrogen and helium are represented by their optical recombination lines, in particular $H\beta$, He I λ 5876 and He II λ 4686 Å. The ratio $N(He^+)/N(H)$ can also be determined from radio recombination lines. For the other elements only the collisionally excited lines are at our disposal.

With the simplest assumptions: uniform temperature and uniform density, (5.52), with N_i and N_e instead of N_1 and N', gives the following expression for the collisionally excited optical lines in the approximation of the two-level atom $[I_{\nu,0} = 0, h\nu/kT \gg 1, \beta = N_e Q_{21}/A_{21} \ll 1$, Wien's approximation for $B_\nu(T)$; B_{12} expressed in terms of A_{21} by (5.14)]:

$$I^L = \frac{h\nu}{4\pi} N_i N_e l Q_{21}(T_e) \frac{g_2}{g_1} \exp\left\{-\frac{h\nu}{kT_e}\right\} \ . \tag{5.147}$$

For the optical recombination lines it also follows from (5.43), p. 258, that

$$I^L = N_i N_e l F(T_e) \tag{5.148}$$

with $N_i = N_p = N(H)$ for hydrogen lines as well as $N_i = N(He^+)$ and $N_i = N(He^{++})$ for lines of the neutral or singly ionised helium. In the formation of the line ratios N_e drops out and only the dependence on T_e still remains.

The available *results* for the elemental abundances in H II regions extend essentially over H, He, N, O, Ne, S and Fe. Here the ratio $N(He)/N(H)$ can be obtained for a larger number of objects, but abundances of higher elements only for a small number of diffuse nebulae, including in the first place the Orion Nebula M 42. Results of an investigation of four H II regions are contained in Table 5.13. The accuracy of these values should lie at $\Delta \log N \approx \pm 0.3$. Compared with the atmospheres of the sun and other stars of Population I there seem to be no significant discrepancies for the elements considered here.

Table 5.13. Logarithms of the relative elemental abundances in four H II regions, with numbers of atoms normalized to $\log N(H) = 12.0$. The corresponding values for the solar atmosphere are also shown for comparison. [After Hawley (1978)]

Element	M 42	M 8	M 16	M 20	Sun
H	12.0	12.0	12.0	12.0	12.0
He	11.0	11.0	11.1	11.1	10.8
N	7.4	7.8	7.7	7.5	7.9
O	8.6	8.5	8.7	8.4	8.9
Ne	8.0	7.6	8.2	8.1	8.0
S	7.2	7.2	7.2	6.9	7.2

Indirect arguments lead one to expect a slow systematic increase of the abundances of the higher elements with decreasing distance of the H II region from the galactic centre. The application of (5.128) to numerous H II regions indicates that T_e becomes smaller with decreasing distance from the galactic centre: on average from about 8000 K in the solar neighbourhood $R \approx R_0$ to about 6000 K at $R \approx 5$ kpc. Because the cooling of the H II regions is due to the excitation of ions of the heavier elements, in particular of oxygen, one concludes from this a corresponding increase in the abundances by about a factor of 2, as R decreases from R_0 to 5 kpc. Analyses of the optical emission lines from H II regions with differing galactocentric distances have confirmed this directly for the abundancy ratios O/H and N/H.

5.3 The Interstellar Dust Grains

5.3.1 Optics of Small Solid Particles

Mie's Theory: The most important basis for the quantitative interpretation of the observed phenomena of the interstellar dust, and hence for making pronouncements on the size, shape, composition, mass and number density of the dust grains, is the theory of the scattering of light by small solid particles. It deals with the diffraction of electromagnetic waves by a small solid body through the solution of Maxwell's equations subject to the boundary conditions pertaining at its surface. For homogeneous spheres this problem was solved exactly by Gustav Mie in 1908. Assuming that the incident wave is plane and linearly polarised (electric and magnetic field strengths E_0, H_0, intensity I_0), then Mie's theory derives the components of the scattered wave ($E_\perp$, $H_\perp$ and $E_\parallel$, $H_\parallel$, intensities $I_\perp$, $I_\parallel$) vibrating perpendicular and parallel to the plane of vision (see Fig. 5.17), as a function of the angle of scatter Θ, of the *complex refractive index of the grain material* $\tilde{n} = n - ik$ and of the parameter

$$\alpha = \frac{2\pi a}{\lambda} \tag{5.149}$$

with $a = $ *grain radius* — α is accordingly the ratio of the grain circumference to the wavelength. All the components are obtained in the form of series expansions.

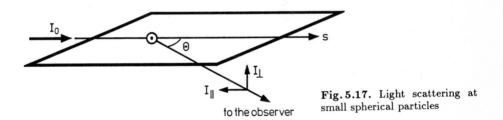

Fig. 5.17. Light scattering at small spherical particles

By averaging over all possible polarisation directions of the incident wave, one can with this solution derive in particular the normalised angular distribution of the intensity of scattered natural light, the *scattering function* or *phase function* $p(\Theta)$, in this case axially symmetric. The scattered radiation is in general partly polarised. On the other hand, calculation of the Poynting Vector and integration over all directions gives the amounts of the radiation energy scattered and absorbed by the grains. Hence one can obtain the ratio, of particular importance in astrophysics, of these amounts of energy to the incident energy per unit area, the *"extinction cross-section"* of the particle C_{ext}. We have $C_{\text{ext}} = C_{\text{sca}} + C_{\text{abs}}$, where C_{sca} — in the generally accepted notation — relates to the fraction of scattered light and C_{abs} to the fraction of absorbed light. If the imaginary part of the refractive index vanishes, then $C_{abs} = 0$. We also introduce the *efficiency factor* Q_{ext} by

$$C_{ext} = \pi a^2 Q_{ext} \tag{5.150}$$

and set

$$Q_{ext} = Q_{sca} + Q_{abs} \quad . \tag{5.151}$$

With a number density of uniform dust grains $N_d[\text{cm}^{-3}]$ the relative change of radiation intensity in a path element ds as a result of the presence of $N_d ds$ grains per cm^2 is given by $dI/I = -\pi a^2 Q_{ext} N_d ds$. The "absorption" coefficient per unit length, defined in the usual way (see Sect. 5.1.1), is accordingly given by

$$\kappa_\lambda = \pi a^2 Q_{ext} N_d \quad . \tag{5.152}$$

The ratio of the energy scattered in all directions by a volume element to the energy taken from the radiation field is called the *albedo* of the grains and is given by

$$\gamma = \frac{Q_{sca}}{Q_{ext}} = \frac{Q_{sca}}{Q_{sca} + Q_{abs}} \quad . \tag{5.153}$$

The case $\gamma = 1$ corresponds to perfectly reflecting grains, whilst when $\gamma = 0$ there is no scattered light at all.

The degree to which the scattered light is directed forward can evidently be determined in a manner analogous to radiation flux density, with a $p(\Theta)$-weighted mean value of $\cos \Theta$ over all solid angles:

$$g = \overline{(\cos \Theta)} = \int \int p(\Theta) \cos \Theta \, d\omega \quad . \tag{5.154}$$

One calls g the *asymmetry parameter* or also the phase factor. With completely forward scattering $\cos \Theta$ is in the narrow cone around $\theta = 0$ equal to unity and hence $g = 1$; for completely isotropic scattering $p(\Theta) = 1/4\pi$, so $g = 0$, since the integrals over the front and rear hemispheres are equal, but of opposite sign.

The momentum flux transferred to a grain is equal to Φ/c, where Φ denotes the effective radiation energy flux. Φ consists of the energy $I_0 \pi a^2 Q_{abs}$ absorbed by the grain in unit time and the scattered energy $I_0 \pi a^2 Q_{sca}$, reduced by the fraction of the energy flux continuing in the original direction $I_0 \pi a^2 Q_{sca}\overline{(\cos \Theta)}$. To calculate the radiation pressure on the grain one therefore introduces an efficiency factor with the definition

$$Q_{pr} = Q_{abs} + Q_{sca} - Q_{sca}\overline{(\cos \Theta)} = Q_{ext} - Q_{sca}\overline{(\cos \Theta)} \tag{5.155}$$

(pr stands for pressure). The force exerted on the grain is accordingly given by

$$K_{\text{pr}} = \frac{I_0}{c} \pi a^2 Q_{\text{pr}} \quad . \tag{5.156}$$

With regard to the analytical representations for the phase functions and the efficiency factors, reference should be made to the literature listed in the Appendix. We restrict ourselves here to the qualitative explanation of the results and accordingly quote typical examples of the quantitative numerical evaluations. For a fixed refractive index the phase function and the efficiency factors depend in characteristic fashion on the ratio α (5.149). In the limiting case of very small grains $\alpha \ll 1$ the scatter is directed in equal strength forwards and backwards (equal maxima at $\Theta = 0°$ and $180°$, minimum at $\Theta = 90°$, together with axial symmetry, so $g = 0$). For the calculation of Q_{ext} and Q_{sca} it is then sufficient to take the first term of each of the corresponding series expansions (for $\tilde{n}$ complex or real, respectively):

$$Q_{\text{ext}} = -4\alpha \, \text{Im}\left\{\frac{\tilde{n}^2 - 1}{\tilde{n}^2 + 2}\right\} \text{ or } \frac{8}{3}\alpha^4\left(\frac{n^2 - 1}{n^2 + 2}\right)^2 \quad ,$$

$$Q_{\text{sca}} = \frac{8}{3}\alpha^4 \, \text{Re}\left\{\frac{\tilde{n}^2 - 1}{\tilde{n}^2 + 2}\right\}^2 \quad , \tag{5.157}$$

where Im and Re mean that the imaginary or real part, respectively, is to be taken. This is the case of Rayleigh scattering: $Q_{\text{sca}} \sim \lambda^{-4}$.

For larger α, say $\alpha > 0.1$, a continually increasing fraction of the radiation is scattered forwards with increasing particle radius ("Mie effect"). The scattered light shows a polarisation dependent on the direction, which for $\alpha < 1$ is nearly complete at $\Theta = 90°$ — the electric vector vibrates predominantly at right-angles to the plane of vision. With $\alpha \gtrsim 1$ the polarisation, which is then in general weaker, has its maximum at other angles. An example of the dependence of the degree of forward scatter $g = \overline{(\cos \Theta)}$ on α is shown in Fig. 5.18. The underlying refractive index $\tilde{n} = n = 1.33$ corresponds to that of pure water ice. The efficiency factor Q_{ext} follows the behaviour shown in Fig. 5.19. For very large α, Q_{ext} tends to the limiting value 2 [cf. the comment in Sect. 4.1.1 in connection with (4.13)]. For very small α one has in this case $Q_{\text{ext}} = Q_{\text{sca}} \sim \alpha^4$

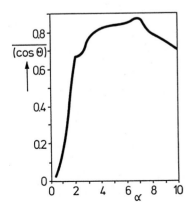

Fig. 5.18. Asymmetry parameter $g = \overline{(\cos \Theta)}$, defined by (5.154), as a function of $\alpha = 2\pi a/\lambda$ for spherical solid particles with real refractive index $n = 1.33$. [After Wickramasinghe, Kahn and Mezger (1972)]

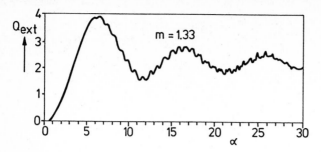

Fig. 5.19. Efficiency factor Q_{ext} for homogeneous spheres with real refractive index $n = 1.33$ as a function of $\alpha = 2\pi a/\lambda$. [After Van de Hulst (1957)]

and hence $Q_{\text{abs}} = 0$. It is above all typical, even for refractive indices found with a finite imaginary part, for there to be an approximately linear increase of $Q_{\text{ext}}(\alpha)$ inside a certain region near $\alpha \approx 1$, which in this example extends up to $\alpha \approx 4$, and is steeper for metallic particles. This means that small spheres with radii of the order of the wavelength of the incident light can cause a reddening, such as one sees for the interstellar extinction in the visible, namely approximately

$$A_\lambda \sim Q_{\text{ext}} \sim \lambda^{-1} \quad .$$

Scattering by Nonspherical Particles: Natural light scattered by small spheres is unpolarised at $\Theta = 0$ because, in contrast to $\Theta \neq 0$, no plane containing the line of sight to the observer is unique. The interpretation of the interstellar polarisation of starlight therefore requires a fraction of uniformly oriented *anisotropically* scattering particles, which cause different amounts of extinction for different vibration directions of the incident light. One thinks perhaps of axisymmetric elongated grains. For the sake of simplicity we consider here only the case when the light impinges perpendicularly to the long axis (axis of symmetry) of the grain. Then one has to differentiate between the extinction cross-sections $C_\parallel$ and $C_\perp$ for the electric vector vibrating parallel and perpendicular to this axis, respectively. The corresponding efficiency factors $Q_\parallel$ and $Q_\perp$ are obtained by dividing by the grain's geometric cross-section.

Analytic expressions for the calculation of $Q_\parallel$ and $Q_\perp$ are known only for a few special cases. Of these, the long cylinder (length $l \gg$ radius a) and the axisymmetric ellipsoid with semiaxis $\ll \lambda/2\pi$ find application. An example of the results available is shown in Fig. 5.20a. To obtain the efficiency factors for other grain shapes, including the elongated or flattened ellipsoids of rotation with semiaxes $\gtrsim \lambda/2\pi$, use is made of the fact that all scattering characteristics always depend on the ratio of grain size to wavelength: for substantially longer wavelengths the same scattering characteristics will be shown by correspondingly enlarged bodies of the same material and shape. By irradiating such bodies of convenient size with microwaves, and measuring the scattered radiation, it is therefore possible to determine empirically the required phase

320

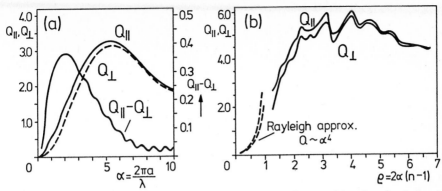

Fig. 5.20. (a) Efficiency factors for long cylinders of radius a with refractive index $\tilde{n} = 1.33 - 0.05\,\mathrm{i}$. $Q_{\parallel}$ and $Q_{\perp}$ refer to cylinder axes parallel and perpendicular, respectively, to the electric vector of the incident wave, which propagates perpendicularly to the cylinder axes. $Q_{\parallel} - Q_{\perp}$ determines the degree of linear polarisation. **(b)** Efficiency factors $Q_{\parallel}$ and $Q_{\perp}$ determined by the microwave analogy method for ellipsoids of rotation with ratio of rotation axis: small axis = 2:1 and refractive index $\tilde{n} = 1.33 - 0.05\,\mathrm{i}$. [After Greenberg (1968)]

functions and efficiency factors. An example of the results obtained by J.M. Greenberg and fellow workers by this method of microwave analogy is shown in Fig. 5.20b. One finds quite generally that the difference between $Q_{\parallel}$ and $Q_{\perp}$, and with it the degree of polarisation, is greatest between $\varrho = 2\alpha(n-1) = 0$ and the value of ϱ for the first maximum. This is true also for flattened particles.

For the optical thicknesses $\tau_{\parallel}$ and $\tau_{\perp}$ formed with $Q_{\parallel}$ and $Q_{\perp}$, one has the corresponding intensities

$$I_{\parallel} = \tfrac{1}{2} I_0\, e^{-\tau_{\parallel}} \text{ and } I_{\perp} = \tfrac{1}{2} I_0\, e^{-\tau_{\perp}} \tag{5.158}$$

with the property $I = I_{\parallel} + I_{\perp} =$ intensity of the attenuated radiation. If $Q_{\parallel} > Q_{\perp}$, as one finds in the important region before the first maximum of the Q-curves, then it follows that $I_{\parallel} < I_{\perp}$. This means that incident natural light is polarised perpendicularly to the major particle axis.

The degree of polarisation, defined in Sect. 4.1.2 (4.22), can be expressed in magnitudes using (5.158):

$$\Delta m_{\mathrm{p}} = -2.5 \log \frac{I_{\parallel}}{I_{\perp}} = 1.086(\tau_{\parallel} - \tau_{\perp}) \quad . \tag{5.159}$$

The total extinction of the radiation in all vibration directions, in the case of weak polarisation ($|\tau_{\parallel} - \tau_{\perp}| \ll 1$), is given by

$$A = -2.5 \log \frac{I}{I_0} = 1.086 \left(\frac{\tau_{\parallel} + \tau_{\perp}}{2} \right) \quad , \tag{5.160}$$

since we now have

$$\frac{1}{2}(e^{-\tau_\parallel} + e^{-\tau_\perp}) = \exp\left(-\frac{\tau_\parallel + \tau_\perp}{2}\right) \cosh\left(\frac{\tau_\parallel - \tau_\perp}{2}\right)$$

$$\approx \exp\left(-\frac{\tau_\parallel + \tau_\perp}{2}\right) \quad .$$

5.3.2 Nature of the Interstellar Dust Grains

Albedo and Phase Function: Evidence on these can be obtained from the amount and relative distribution of the Diffuse Galactic Light and also of the light from the reflection nebulae. For this the following approximate formula for the phase function is often useful:

$$p(\Theta) = \frac{1 - g^2}{4\pi}(1 + g^2 - 2g \cos \Theta)^{-3/2} \quad . \tag{5.161}$$

This equation reproduces the essential trends of the phase function calculated from Mie's theory and it depends only on the asymmetry parameter $g = \overline{(\cos \Theta)}$ (5.154 f.).

For a layer of interstellar dust with the geometric thickness Δd and the very small optical thickness $\Delta\tau = \pi a^2 Q_{ext} N_d \Delta d \ll 1$, the intensity of the light scattered at an angle Θ from the incident direction is given by $I' = \Delta\tau I_0 \gamma p(\Theta)$ with $\gamma = $ albedo. This is the case of single scattering: the extinction is so slight, the number of grains so few, that on its passage through the layer the light, once scattered, practically never undergoes a second scatter.

If an interstellar cloud can no longer be considered optically thin, then multiple scattering plays a role and one encounters a complicated radiation transport problem. A successful method of solution is the numerical computation of the intensity of the scattered light I' by the well known Monte Carlo method. In this one simulates the individual track through the particle cloud, with one or more scatter processes, for each individual one of a very large number of photons incident from the same direction. For the scattering angle Θ arising at each individual encounter one selects a random number with the probability distribution (5.161). For each given triad of parameters (γ, g, τ), where τ is the optical thickness of the cloud, for example assumed spherical, one obtains an intensity distribution of the scattered light $I' = I_0 S(\Theta)$, which for $\tau \ll 1$ becomes $I_0 \gamma p(\Theta)$.

After this one can now perhaps calculate the scattered light $I_s(l_c, b_c)$, falling on the observer from a certain cloud in a definite direction (given by the galactic coordinates of the centre of the cloud l_c, b_c), the source of which is the brightness distribution $I_0(l, b)$ of the light of the Milky Way, as seen from the position of the cloud (Fig. 5.21):

$$I_s(l_c, b_c) = \int\int I_0(l, b) S(\Theta) \cos b \, dl \, db \quad . \tag{5.162}$$

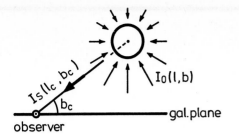

Fig. 5.21. Calculation of the diffuse galactic light

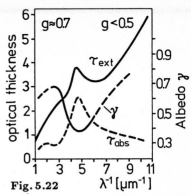

Fig. 5.22

Fig. 5.22. Dependence of the albedo γ of the interstellar dust on the reciprocal wavelength. In the region $\lambda^{-1} > 6\,\mu\mathrm{m}^{-1}$, corresponding to $\lambda < 1700\,\text{Å}$ approximately, the available results are still uncertain. Also shown for comparison are the profiles of the optical extinction depth ($\tau_{\mathrm{ext}} = A_\lambda/1.086$) and the portion of pure absorption (τ_{abs}) for a column density $\mathcal{N}(\mathrm{H\,I}) = 3 \times 10^{21}$ H atoms cm^{-2}. In addition characteristic values of the asymmetry parameter g are given. [After Drapatz (1979)]

Here Θ is the angle between the directions (l_c, b_c) and (l, b). If the optical thickness τ of the cloud or the corresponding extinction A_V is known (for example from star counts), then $S(\Theta)$, and hence the result for $I_s(l_c, b_c)$, depends only on the two parameters γ and g. Observational results for at least two clouds with galactic latitudes as different as possible then allow one to determine both parameters. These considerations can be extended to the Diffuse Galactic Light of the whole cloud layer.

The available results are presented in Fig. 5.22. The visible and UV radiation are evidently scattered forwards relatively strongly: $g \approx 0.6 \ldots 0.8$. From this very small particles with radii $a < 0.01\,\mu\mathrm{m}$, such as rather large complex molecules, are excluded as the cause of the scattered light. It is moreover interesting that the albedo of the dust, relatively high in the visible region ($\gamma \approx 0.6 \ldots 0.7$), sinks to a minimum at the position of the "hump" on the extinction curve at $\lambda \approx 2200\,\text{Å}$, corresponding to $\lambda^{-1} = 4.6\,\mu\mathrm{m}^{-1}$: the increase in the extinction here must therefore be due to genuine absorption. Below 2000 Å the albedo increases again.

Evidence on the scattering characteristics of the interstellar dust grains can be obtained in principle from measurements of typical reflection nebulae. Inadequate knowledge of the geometry (shape of the nebula, position of the illuminating star), however, here often leads to considerable uncertainties.

Interpretation of the Extinction Curve: We have already noticed in Sect. 5.3.1 that the observed wavelength dependence of the interstellar extinction in the visible region of the spectrum can be explained approximately by the assumption of homogeneous dielectric or metallic spheres with similar diameters of the order of the wavelength. Because of the earlier and steeper climb of the Q-curve for metallic spheres, the grain radii assigned to these must simply be correspondingly smaller. The agreement with the observations can be improved almost to perfection if we assume an appropriate frequency distribution for the grain radii $f(a)$ and calculate the normalised reddening law from [Sect. 4.1.1, (4.10)]

$$F(\lambda) = \frac{C(\lambda) - C(\lambda_V)}{C(\lambda_B) - C(\lambda_V)} \qquad (5.163)$$

with the extinction cross-section

$$C(\lambda) = \int\limits_0^\infty Q_{\text{ext}}(a, \lambda)\pi a^2 f(a) da \quad . \qquad (5.164)$$

A good representation of the average observed profile of $F(\lambda)$ between $\lambda \approx 4000\,\text{Å}$ and $\lambda \approx 1\,\mu\text{m}$ is provided *for example* by the assumption of pure water ice grains with a radius distribution $f(a) = \exp(-a/a_0)$ for $a_0 = 0.075\,\mu\text{m}$ (Fig. 5.23) or perhaps the choice of silicate and graphite grains with $f(a){\sim}a^{-3.5}$. The fit also succeeds, however, with iron spheres of $0.02\,\mu\text{m}$ diameter. The nature of the grains is therefore not yet decided in this way.

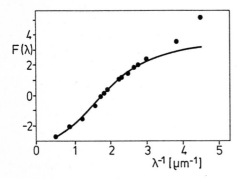

Fig. 5.23. Representation of the observed normalised reddening law (points) in the region $\lambda^{-1} < 3\,\mu\text{m}^{-1}$ with the assumption of spherical water ice grains (real refractive index $n = 1.33$). [After Wickramasinghe, Kahn and Mezger (1972)]

Certain limitations are placed on the acceptable particle matter by the observational results for the behaviour of $F(\lambda)$ in the ultraviolet. In accordance with the results for the albedo in the region of the "hump" of the extinction curve at $\lambda \approx 2200\,\text{Å}$ one needs (also) a material that absorbs strongly at this point but below $2000\,\text{Å}$ becomes a good reflector again. These characteristics can in no way be reproduced by ice grains. On the other hand the absorption capability of graphite has a pronounced maximum just there. However, very small graphite grains are needed, with $a \approx 0.02\,\mu\text{m}$. The behaviour of the profile in the visible, on the other hand, would require considerably larger graphite grains ($\alpha \approx 0.05\,\mu\text{m}$), whose presence must, however, lead to considerable deviations from the observed curve in the UV. Since these small graphite particles can satisfactorily explain neither the strongly forward scattering nor the relatively high albedo in the region $\lambda \gtrsim 3000\,\text{Å}$, the assumption of graphite as the only material is untenable. Even an ice coating round the graphite grains appears to afford no essential change, since it must be very thin if the 2200 Å absorption is to be explained. Accordingly if one attributes this feature to small graphite grains, then one has also to assume another component of the interstellar dust. For example, relatively large ice grains − perhaps with small mixtures of other substances ("impurities") − would give not only the extinction curve, but also the albedo and the asymmetry factor in the region $\lambda \gtrsim 3000$ Å.

A notable indication thereto is given by the interpretation mentioned in Sect. 4.1.2 of the reversal of sign of the observed circular polarisation at a wavelength $\lambda_0 \approx \lambda_{max}$. We restrict ourselves to a statement of the result: $\lambda_0 \approx \lambda_{max}$ is to be expected only for a very small imaginary part of the refractive index $k \approx 0$: for increasing finite values of k the sign reversal would occur at ever increasing wavelengths $\lambda_0 > \lambda_{max}$. For the particles which cause the interstellar extinction and polarisation of the visible stellar radiation, the observations therefore require a small value of k, as possessed by impure ice grains or even silicate grains, whilst graphite for example must be largely excluded.

The identification of the hump in the extinction curve in the UV as the absorption structure of graphite is in no way conclusive. It can also be satisfactorily explained by the assumption of silicate grains, as is shown for example by enstatite $(Fe,Mg)SiO_3$. The position of the absorption, however, depends sensitively on the size and shape of the grains. The marked absorption in the infrared at $\lambda \approx 10\,\mu m$, observed in a few strongly reddened stars (cf. Sect. 4.1.1, Fig. 4.4) can likewise be explained by a mixture of various silicates. When one includes also the visible region, then one has the same situation as for graphite: additional particles of different size and composition are needed. Calculations were therefore carried out of the theoretical extinction curves for numerous mixtures of different sorts of grain: ice with impurities, graphite, silicon carbide, meteoritic silicates, metal oxides, etc. − some also with ice coatings. The interpretation of the observations is possible with various grain combinations, and therefore provides no unambiguous conclusion on the true chemical composition. It is however fairly certain that at least two or three different components are needed: relatively large dielectric grains for the visible and infrared regions, and smaller, in part strongly absorbing grains for the ultraviolet. For the growth of the extinction below $\lambda \approx 2000$ Å very small grains with radii $\lesssim 0.03\,\mu m$ are necessary, which remain practically imperceptible in the visual region.

Interpretation of the Interstellar Linear Polarisation: Let us suppose that all the solid grains have the same shape with only one pronounced axis of symmetry. For ice grains with impurities or perhaps for silicate grains one assumes the greatest extent along the axis of symmetry, and regards them for example as long cylinders or as elongated or needle-shaped ellipsoids of revolution. Graphite crystallises at low pressures into thin flakes, which consist of plane layers of atoms and have highly anisotropic electrical characteristics: they can be regarded approximately as strongly flattened ellipsoids of revolution. Perpendicular to the axis of symmetry, and hence parallel to the flat plane, the electrical conductivity is greater by a factor of about 100 than across the plane. Correspondingly for the extinction cross-sections, which in this case are dominated by the absorption fraction, we have $C_\parallel/C_\perp \gg 1$. Here $C_\parallel$ applies to light with the electric vector parallel to the flat plane. For the calculation of the extinction A_λ through arbitrarily oriented graphite platelets, one can therefore approximate by spherical particles whose diameter is equal to the major diameter of the graphite platelets.

If all the major grain axes are parallel to one another and also oriented perpendicularly to the incident starlight (line of sight), then the difference between the optical thicknesses for light with the electric vector vibrating parallel and perpendicular to the major particle axes is given by $\tau_\parallel - \tau_\perp = (C_\parallel - C_\perp)\overline{N}_d r$. Complete alignment of the grains in the interstellar space is of course very improbable. The amount of resulting polarisation is therefore given in magnitudes by

$$\Delta m_p = 1.086 \times f(C_\parallel - C_\perp)\overline{N}_d r \quad , \tag{5.165}$$

where the factor f is determined by the *degree of alignment*: $f = 1$ corresponds to complete alignment, $f = 0$, and hence $\Delta m_p = 0$, to random orientation of the grains. Using (5.160) it now follows that

$$\frac{\Delta m_p}{A} = 2f \frac{C_\parallel - C_\perp}{C_\parallel + C_\perp} \quad . \tag{5.166}$$

This relation links the degree of alignment and the optical anisotropy of the grains directly with an observed quantity. The determination of the factor f by a theory of the alignment of the grains in the interstellar magnetic field is discussed in Sect. 5.3.3.

The observed dependence of the linear polarisation $P(\lambda) = \Delta m_p(\lambda)/2.172$ on the reciprocal wavelength (Sect. 4.1.2, Fig. 4.7) has a shape similar to the profile of $Q_\parallel - Q_\perp$ with $\alpha = 2\pi a/\lambda$ for long cylinders (Fig. 5.20a) or other non-spherical particles. With appropriate choice of grain size, or better − for the smoothness of the curve − a grain spectrum of finite width, the observed profile $P(\lambda)/P_{max}$ against λ_{max}/λ can be well represented. The value of $\alpha_{max} = 2\pi a/\lambda_{max}$ associated with the maximum of $Q_\parallel - Q_\perp$ then gives evidence on the mean particle radius. For cylinders, for example, it follows from Fig. 5.20a that $\alpha_{max} \approx 2.5$, so that with $\lambda_{max} \approx 5500$ Å we get a mean value of $a \approx 2 \times 10^{-5}$ cm. The conclusion however is not unequivocal, since a satisfactory representation is also possible with other choices of the refractive index and correspondingly altered grain radius. This interpretation assumes that the refractive index does not vary with λ, which is an argument for a dielectric and against, for example, graphite.

For an estimate of the degree of alignment f we read from Fig. 5.20a at $\alpha = \alpha_{max} \approx 2.5$, corresponding to $\lambda \approx \lambda_V$, the ratio $2(Q_\parallel - Q_\perp)/(Q_\parallel + Q_\perp) = 2(C_\parallel - C_\perp)/(C_\parallel + C_\perp) \approx 0.3$. Similar results are obtained for elongated ellipsoids. With the average observed value $\Delta m_p(V)/A_V \approx 0.03$ (see Sect. 4.1.2) we find from (5.166) that $f \approx 0.1$.

The variation mentioned in Sect. 4.1.2 of $\lambda_{\max}$ between 0.45 and 0.8 μm from region to region on the sphere suggests that the characteristics of the grains contributing to the polarisation in the wavelength region under consideration vary somewhat from cloud to cloud. In particular, a larger value of $\lambda_{\max}$ indicates that a is larger, so that the increase of $Q_{\parallel}$ and $Q_{\perp}$ with $1/\lambda$ is steeper: the difference between the extinction in the infrared and in the visual becomes greater, which results in a lower limiting value of $E_{\lambda-V}/E_{B-V}$ for $\lambda \to \infty$ (see Sect. 4.1.1, Fig. 4.3) and with that a larger value for $A_V/E_{B-V} = R$ ($\approx$ 4 to 5 instead of 3). A study by Whittet and Van Breda (1978) led to the relation

$$R \approx 5.6 \lambda_{\max} \, [\mu\text{m}] \quad . \tag{5.167}$$

The Diffuse Interstellar Bands: Two possible interpretations were discussed: (a) production by free molecules and (b) production by solid particles. In case (a) the absorption should not be associated with a polarisation, since there is no mechanism known which could cause a uniform alignment of free molecules in the interstellar magnetic field. Hypothesis (a) could therefore be excluded if the polarisation in the bands could be distinguished from that of the neighbouring continuum. After recent studies of high accuracy, however, this is not the case. Broad absorption structures of free molecules in diffuse interstellar clouds could originate from electronic transitions in polyatomic molecules. For sufficiently large molecular masses the individual rotation lines of a band move so close together that they partially overlap and produce a practically smooth broad band. It is questionable, however, whether these large molecules could withstand the interstellar UV radiation field and not quickly become dissociated − so far only simple molecules have been detected in diffuse clouds.

For hypothesis (b) several arguments apply. We restrict ourselves to recent investigations, according to which the broad absorption bands can be produced by electronic transitions of metal ions such as Fe^{2+}, Ni^{2+}, Cr^{3+} etc., which occur in the crystal lattice of solid interstellar particles. Moreover, broadening by interaction with the lattice can occur. In particular it was shown that the oscillator strengths of the transitions for ions in damaged lattice positions were greater by a factor of 10^2 to 10^3 than in perfect crystals. This made it for the first time seem possible that the relatively sparse metal ions of this sort could afford a quantitative explanation for the diffuse bands − so far as the interstellar grains revealed more lattice defects than the solid bodies studied in the laboratory.

The discovery of the same polarisation in the diffuse bands and in the neighbouring continuum, however, produces a limitation: these bands must be produced by a particle component which does not contribute to the interstellar polarisation, so they cannot consist of the same grains that cause the extinction of the visible starlight. It was therefore proposed to consider very small grains ("clusters" of molecules) with dimensions <0.03 μm = 300 Å as carriers of the "impurities" to which the bands were attributed: only for very small grains does the contour of the band have the same shape as the absorption profile of the embedded material itself. Since, for the explanation of the increase of the extinction in the UV ($\lambda<2000$ Å), one is in any case led to assume a component of very small grains, this suggests that these may be supposed to be the cause of the interstellar bands also.

Origin of the Dust: Ever since the first determinations of the law of interstellar reddening for the visible region, one has sought to obtain information on the nature of the interstellar dust from plausible assumptions as to its origin. In this way grain models under discussion even today play an essential role in attempts at a quantitative interpretation of the entire observed extinction curve. We accordingly here briefly summarise the most important of the proposed hypotheses and a few consequences:

1) *Condensation from the General Interstellar Gas*[4]: Since C, O and to a lesser extent N have the greatest abundances of all the higher elements, the formation first of biatomic molecules such as CH, CO and CN was assumed – consistently with the observations of the diffuse H I clouds. In a second phase large molecules would grow from them, which would consist of about 10 to 20 atoms and are seen as condensation nuclei. From the observations this is of course possible only in the dense cold clouds. The last phase is a slow growth of these condensation nuclei by accretion of interstellar gas atoms through collisions with them. Quantitative estimates allow one to expect, in the low densities of the diffuse clouds, diameters of the order of a light wavelength only after a few 10^8 years. For the grains so created H.C. van der Hulst proposed the following composition: H_2O, H_2, CH_4, NH_3, MgH, ... in the ratios (numbers of molecules) 100:30:20:10:5, ... One therefore expects ice grains with a small amount of "impurities" ("dirty" ice), in which the metallic contribution leads to a small finite imaginary part of the refractive index; one can perhaps set $\tilde{n} = 1.33 - 0.05\,i$.

2) *Condensation in the Atmospheres of Cool Giant Stars*[5]: The difficulty, of actually creating enough condensation nuclei in the extremely low densities of the general interstellar gas, is avoided if one transfers the formation of the solid particles to particular cool and relatively dense stellar atmospheres. From the supersaturated vapour, under certain conditions, graphite, silicon carbide and complex silicates can there condense out and be blown out into space by the radiation pressure.

---

[4] Proposed by B. Lindblad (1935) because of the observed strong correlation of gas and dust densities; quantitatively investigted by J.H. Oort and H.C. van de Hulst (1946).

[5] Proposed and quantitatively discussed by F. Hoyle and N.C. Wickramasinghe (1962). The graphite grains discussed, however, were already used in 1954 by E.L. Schatzman and R. Cayrel for the interpretation of the interstellar polarisation because of their strong anisotropy. A detailed presentation of the theory, here only outlined, is given by Wickramasinghe (1967) in his book "Interstellar Grains".

If, at gas pressures of $10^2 \ldots 10^3\,\mathrm{dyn\,cm^{-2}}$, such as are typical for the atmospheres of late giants, and temperatures a little below 2000 K, the carbon is not all bound in CH, CO, CN etc., but also exists free as C, C_2, C_3 etc., then a certain portion of this free carbon could condense out as small anisotropic graphite particles (C_n with $n \gg 1$) and grow. These conditions are to be expected in the photospheres of the carbon stars of type C5 ... C9 – called N-stars in the Harvard classification – with temperatures of 3000 K to 1500 K, since here the abundance ratio C:O, unlike that in normal stars, is considerably greater than unity. On the other hand, the densities of the photospheres of these stars are still sufficiently low to allow solid particles to be thrown off by the radiation pressure.

Infrared observations have indeed confirmed the presence of warm solid particles in the neighbourhood of late giant stars, but the estimated flow of such grains into the interstellar space is so far very uncertain. The question whether a substantial fraction of the interstellar dust is created in this way still remains open.

3) *Condensation in the Envelopes of Protostars*: Relatively high densities and temperatures below 2000 K can occur in the outer regions of a gravitationally unstable interstellar cloud if the end of the dynamic contraction phase of star formation has been reached. Thus it is assumed today in connection with model computations that, after the formation of the sun, at first a still contracting cloud ("solar nebula") remained, which as a result of the angular momentum conservation became a flat disc, and in which further cooling later created the solid bodies of the planetary system. In such a circumstellar disc conditions exist under which molecules can form in chemical equilibrium. According to thorough investigations, at normal elemental abundances refractory silicates, e.g. Ca_2SiO_4, Al_2SiO_4, Mg_2SiO_4 etc., condense first below 1600 K; iron and nickel are deposited as free metals at about 1300 K; sulphides, potassium and sodium silicates condense at lower temperatures, and finally hydrous silicates below 400 K. The gas outflows observed from Pre-Main Sequence stars of the type T Tauri and other young stars lead one to assume that a stellar wind gradually blows away all the gaseous material as well as small solid particles – up to diameters of perhaps a tenth of a millimetre – from the disc into the interstellar space. Interstellar dust would then be a by-product of star formation and would consist to a considerable extent of silicates and similar compounds.

For the explanation of the observed deficits of several elements in the interstellar gas (Sect. 5.2.5) two different possibilities have been discussed: (1) condensation in a gas with low and falling temperature in chemical equilibrium – in cool stellar atmospheres or in dense interstellar clouds – where elements with high condensation temperatures (Ca, Al, Ti, etc.) pass into the solid phase first, and subsequently to a greater extent than the elements following later with lower condensation temperatures (C, N, O etc.); (2) growth of already existing "condensation nuclei" in the gas through accretion in diffuse clouds – hence no chemical equilibrium! – a scenario which we have already explained in Sect. 5.2.5. Probably both mechanisms play a role.

329

5.3.3 Interactions with the Radiation Field, the Gas and the Magnetic Field

Temperature of the Dust Grains: Processes, such as the accumulation and the evaporation of atoms and molecules, or the formation of molecules on the surfaces of solid particles, depend essentially on the temperature of the particle matter. In particular the temperature determines the spectral profile of the thermal emission from solid particles, which can be directly observed in the infrared.

The temperature of an interstellar grain T_g depends in practice only on the energy imparted to it by radiation (absorption) and on the loss by radiation (emission). The exchange of kinetic energy with the gas particles turns out in the normal case to be negligible; also the energy which can be gained by chemical reactions with adsorbed atoms or ions must be of a lower order of significance.

Let us calculate T_g from the equilibrium of absorbed and emitted radiation. For an individual grain the rates of gain and loss of energy per unit of time in the solid angle element $d\omega$ in a certain direction are given by

$$\text{absorbed energy } = \pi a^2 Q_{abs} I_\lambda d\omega$$
$$\text{emitted energy } = \pi a^2 E_\lambda d\omega \quad .$$

Here I_λ denotes the intensity of the interstellar radiation field and E_λ the emissivity per unit area and unit solid angle (λ wavelength). In thermodynamic equilibrium E_λ is by Kirchhoff's Law equal to $Q_{abs} B_\lambda(T_g)$. We shall require only, however, that the integrals over all wavelengths and solid angle elements of E_λ and of $Q_{abs} B_\lambda(T_g)$ shall be equal to one another. If I_λ is isotropic, then equating the total energies absorbed and emitted by the grain in unit time produces the condition:

$$\int_0^\infty Q_{abs}(\lambda, a) I_\lambda d\lambda = \int_0^\infty Q_{abs}(\lambda, a) B_\lambda(T_g) d\lambda \quad . \tag{5.168}$$

If the irradiation of the grain is caused by only a single star (radius R, distance r), whose disc produces the mean intensity $\overline{I}_\lambda^*$, then radiation is absorbed only from a small solid angle $\Omega_* = \pi R^2/r^2$ and the factor 4π appears only in the emission. The energy balance can then be written as

$$W \int_0^\infty Q_{abs}(\lambda, a) \overline{I}_\lambda^* d\lambda = \int_0^\infty Q_{abs}(\lambda, a) B_\lambda(T_g) d\lambda \quad , \tag{5.169}$$

where W denotes the dilution factor defined by (4.40).

For larger perfectly black bodies, so that $Q_{abs}(\lambda, a) \equiv 1$, it follows from the Stefan-Boltzmann Law together with (5.168) or (5.169), introducing the total radiation fluxes at the position of the grain $\Phi = \pi I$ or $\Phi^*(r) = \overline{I}^* \Omega_*$ (cf. Sect. 4.1.9), that

$$\Phi = \sigma T_{\mathrm{g}}^4 \quad \text{or} \quad \tfrac{1}{4}\Phi^*(r) = \sigma T_{\mathrm{g}}^4 \quad . \tag{5.170}$$

For the value of the interstellar radiation flux given in Sect. 4.1.9, viz. $\Phi \simeq 4 \times 10^{-3}\,\mathrm{erg\,cm^{-2}\,s^{-1}}$ the left-hand equation (5.170) gives $T_{\mathrm{g}} \simeq 3\,\mathrm{K}$. The same value for the temperature is also given, for example, with irradiation of the grain by a single A0V star with $T_{\mathrm{eff}} \simeq 10\,000\,\mathrm{K}$ and $R \simeq 2.6 R_{\odot}$, at a distance $r = 1\,\mathrm{pc}$: one then obtains the dilution factor $W = \Omega_*/4\pi = R^2/4r^2 \simeq 10^{-14}$ and the right-hand equation (5.170) becomes

$$\frac{1}{4}\Phi^*(r) = \frac{1}{4}\overline{I}^*\Omega_* = \pi\overline{I}^*\frac{\Omega_*}{4\pi} = \sigma T_{\mathrm{eff}}^4 W = \sigma T_{\mathrm{g}}^4 \quad .$$

The radiation energy received by the grain comes from the visible and the UV regions, whereas its emission — because of the low temperature — lies in the infrared. The right-hand and the left-hand integrals in (5.168) and (5.169) thus extend over quite different wavelength regions. Now, for real interstellar dust grains the emissivity and the absorptivity or Q_{abs} generally decrease in going from the visible to the infrared region, because the wavelength always becomes large compared with the dimensions of the grain. Formally it is a matter of transition into the region $\alpha = 2\pi a/\lambda \ll 1$. Now (see Sect. 5.3.1) $Q_{\mathrm{abs}} = Q_{\mathrm{ext}} - Q_{\mathrm{sca}} \simeq Q_{\mathrm{ext}} \sim \lambda^{-1}$. Accordingly the value of Q_{abs} appearing in the right-hand integral of (5.168) or (5.169) is considerably smaller than that in the left-hand one, and this must result in a higher value of $B_\lambda(T_{\mathrm{g}})$, and hence a higher grain temperature T_{g}, compared with perfectly black grains. Taking account of the detailed behaviour of Q_{abs} with wavelength leads, for example, for cylindrical and also for spherical ice grains with "impurities" and a radius of order $0.1\,\mu\mathrm{m}$, to temperatures $T_{\mathrm{g}} \simeq 10\ldots 20\,\mathrm{K}$. For spherical grains consisting of a graphite core with $a = 0.05\,\mu\mathrm{m}$ and a mantle of impure ice with $a = 0.15\,\mu\mathrm{m}$ $T_{\mathrm{g}} \simeq 20\,\mathrm{K}$, whilst for pure graphite spheres with $a = 0.05\,\mu\mathrm{m}$ grain temperatures are calculated to be $T_{\mathrm{g}} \simeq 30\ldots 40\,\mathrm{K}$.

In H II regions the solid particles present in general assume higher temperatures, because the energy supply there is greater. "Colour temperatures" of 100 to 200 K, such as are found, for example, in the Kleinmann-Low Nebula (see Sect. 4.2.4), are easily explained. Inside dense dark clouds, on the other hand, one expects considerably lower temperatures than in the normal interstellar medium. Detailed calculations, however, show that despite the considerable screening of the interstellar radiation field in dark clouds of high optical thickness, there is only a relatively small reduction of grain temperature. The reason for this is that the dust grains on the outer edge with temperatures of $10\ldots 20\,\mathrm{K}$ emit infrared radiation which can penetrate deep into the cloud, so that for example with an optical thickness in the visual of $\tau_{\mathrm{v}} = 10$ a grain temperature $T_{\mathrm{g}} \simeq 8\,\mathrm{K}$ can still be maintained. Investigations of the highest possible emissivity of the grains in the far infrared lead one to expect that grain temperatures cannot fall short of 5 to 7 K.

Electric Charge of Solid Particles: Free electrons and positive ions in collisions with dust grains can remain stuck to them and so give them their electric charges. Since the velocities of the electrons are much higher than those of the ions — in the ratio $\sqrt{m_i/m_e}$ for equilibrium distribution of the kinetic energies — the result is a negative charge, which increases until the capture rates for electrons and positive ions are equal. This equilibrium state will only be achieved, however, if the UV radiation field is thoroughly screened. Otherwise the photo-effect must be taken into account: through absorption of shortwave radiation quanta electrons will be lost from the grain surface and it may now even acquire a net positive charge. (An effect of the cosmic radiation acting in the same direction turns out to be negligible in comparison.)

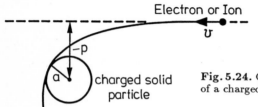

Fig. 5.24. Calculation of the effective cross-section of a charged spherical grain with radius a

Let us calculate first of all the effective cross-section of spherical dust grains with radius a and charge $\pm Z_g e$, whose mass is large compared with the mass of the ions m_i of the surrounding gas. At the grain surface the electric potential amounts to $U(a) = \pm Z_g e/a$ (in electrostatic units). An electron or singly ionized atom that has the velocity v at a great distance from the grain considered at rest can just be deflected to the surface of the positively or negatively charged grain (see Fig. 5.24). Then the effective cross-section of the grain is given by $\sigma = \pi p^2$, where p would be the smallest distance of the charged gas particle from the centre of the solid grain if it were uncharged. The conservation of angular momentum implies that

$$vp = v_a a \tag{5.171}$$

where v_a is the velocity of the gas particle on arrival at the grain surface. In addition, we have the energy law for a gas particle with charge $\pm e$ and mass m, together with a negative or positive grain potential ($U \lessgtr 0$):

$$\tfrac{1}{2}mv^2 = \tfrac{1}{2}mv_a^2 \pm eU(a) \quad . \tag{5.172}$$

Elimination of v_a by means of (5.171) now gives

$$\sigma = \pi p^2 = \pi a^2 \left[1 \mp \frac{2eU(a)}{mv^2} \right] \quad . \tag{5.173}$$

Let us now assume that every gas particle arriving at the surface of the dust

332

grain remains attached, then the total number of negative or positive electron charges acquired by the grain per second is given by $\Gamma = N v \sigma(v)$, where the averaging is carried out with the Maxwell distribution for the kinetic temperature of the gas T [(5.21) in Sect. 5.1.2] as weighting function (integration over v from 0 to ∞, or from the lowest velocity to overcome the grain potential up to ∞). For electrons and single ions (index e and i, respectively) the result is

$$\Gamma_i = \Gamma_i^{(0)} \cdot \begin{cases} (1+y) \\ e^{-y} \end{cases} \quad \text{and} \quad \Gamma_e = \Gamma_e^{(0)} \cdot \begin{cases} e^{-y} \\ (1+y) \end{cases} \tag{5.174}$$

with

$$y = \mp \frac{eU(a)}{kT} \quad , \tag{5.175}$$

where the upper (lower) factor is to be taken in (5.174) for negative (positive) charge of the dust grain. $\Gamma_i^{(0)}$ and $\Gamma_e^{(0)}$ correspond to the case of vanishing grain charge. One finds that

$$\Gamma_i^{(0)} = N_i 2\sqrt{2/\pi}\pi a^2 \sqrt{kT/m_i} \quad \text{and} \quad \Gamma_e^{(0)} = N_e 2\sqrt{2/\pi}\pi a^2 \sqrt{kT/m_e} \quad .$$

The loss Λ_e from the grain of photo-electrons per second is obtained by multiplication of its surface area $4\pi a^2$ by the number of electrons Φ_e lost per second per unit area. For the interstellar radiation field in the neighbourhood of the sun, attenuated by the factor $e^{-\tau_\nu}$, this quantity is estimated [cf. the derivation of (5.112) in Sect. 5.2.3] to be

$$\Phi_e \simeq 2 \times 10^7 y_e Q_{abs} e^{-\tau_\nu} \text{ cm}^2 \text{ s}^{-1} \quad . \tag{5.176}$$

In the equilibrium state we must have

$$\Gamma_i + \Lambda_e = \Gamma_e \quad . \tag{5.177}$$

If, because of a sufficiently strong photoelectric emission, the dust grains have a positive charge, then the number of acquired electrons increases and the number of ion acquisitions Γ_i can be neglected. The condition (5.177) then becomes $\Lambda_e = \Gamma_e = \Gamma_e^{(0)}(1+y)$ or

$$\Phi_e = N_e \sqrt{\frac{8kT}{\pi m_e}} \left(1 + \frac{eU}{kT}\right) \quad . \tag{5.178}$$

Let us take as an example a diffuse cloud with $T = 10^2$ K, $N_e = 10^{-2}$ cm^{-3} and set $Q_{abs} = 1$, $y_e = 0.1$, $\tau_\nu = 1$, then it follows from (5.178) with (5.176) that $U = 3 \times 10^{-4}$ electrostatic units $\simeq 0.1$ Volts, and moreover with $a = 3 \times 10^{-5}$ cm the charge of the dust grain $Z_g e = aU \simeq 20$ positive elementary charges. For the warm H I component one obtains the considerably higher positive charge $Z_g \simeq 10^3 \ldots 10^4$. In the H II regions vigorous emission of photo-electrons can be caused by the strong Lyα-radiation. Because of the relatively high electron density, on the other hand, there may be an over compensation for this loss. Near the exciting star large positive charges are of course always to be expected; in the other region, however, negative potentials to about -1 Volt can occur. More detailed investigation generally produces a frequency distribution of the particle charges $f(Z_g)$ of considerable breadth.

Dynamic Interaction of Dust and Gas: The electric charge of the dust grain increases its effective cross-section for collisions with electrons and ions very strongly. The slowing-down time t_g for the motion of a charged dust grain (mass m_g, charge $Z_g e$, velocity v_g) through gas particles (mass m, charge $Z e$, number density N, kinetic temperature T), defined by $|dv_g/dt| t_g = v_g$, is given in general [after Spitzer (1968; p. 94)] by

$$t_g = \frac{3}{4\sqrt{2\pi}} \frac{m_g(kT)^{3/2}}{\sqrt{m}\, N Z^2 Z_g^2 e^4 q} \quad \text{with}$$

$$q = \ln\left\{ \frac{3}{2 Z Z_g e^3} \left(\frac{k^3 T^3}{\pi N} \right)^{1/2} \right\} . \tag{5.179}$$

Because $t_g \sim 1/\sqrt{m}$, ions as "field particles" produce a considerably shorter slowing-down time than electrons, and only they need therefore be taken into account. For dust particles of radius $a = 3 \times 10^{-5}$ cm and mass density $1\,\mathrm{g\,cm^{-3}}$ we get $m_g \simeq 10^{-13}$ g.

In the cold H I region with $T \simeq 10^2$ K the mass of the ion can be set at about that of C^+ or Si^+. With $N \simeq N_e = 10^{-2}\,\mathrm{cm^{-3}}$, $Z = 1$ and $Z_g = 10$ one then obtains for example $t_g \simeq 10^5$ years. The dust grains here are therefore only weakly linked to the gas; in the neighbourhood of a late supergiant they can be accelerated to considerable velocities relative to the gas by radiation pressure. For normal H II regions ($N \gtrsim 10^2\,\mathrm{cm^{-3}}$, $T \simeq 10^4$ K, $m \simeq m_H$, $Z_g \simeq 10^2 \ldots 10^4$) the relation (5.179) leads us to expect considerably smaller values of t_g, and hence a relatively close coupling between dust and gas.

The electric charge on the solid particles can also lead to a coupling with the interstellar magnetic field (see literature).

Alignment of Solid Particles in the Interstellar Magnetic Field: The interpretation of the interstellar polarisation of the starlight as a consequence of an anisotropy of the interstellar extinction requires nonspherical dust grains whose shape shows a marked preferential direction, and a large scale uniformity of orientation of a significant portion of these grains (Sect. 5.3.2). In what follows we shall base our discussion on rotationally symmetric grains with an elongated shape, such as prolate ellipsoids of rotation.

Under the effect of elastic collisions with gas particles a small solid grain will constantly be given rotational motions. In kinetic equilibrium the energies of the *grain rotation* about a short and the long major axes (indices $i = 1, 2$) are equal: $(1/2)I_i\omega_i^2 = E_{rot}^{(i)} = (1/2)kT$, where I_i is the moment of inertia and ω_i the angular velocity about the axis i; T is the temperature of the gas. Thus $\omega_i = \sqrt{kT}/\sqrt{I_i}$ and the angular momentum is $J_i = I_i\omega_i = \sqrt{kT} \cdot \sqrt{I_i}$. Since the moment of inertia about the short axis is considerably greater than that about the axis of symmetry, the total angular momentum vector $\boldsymbol{J} = \boldsymbol{J}_1 + \boldsymbol{J}_2$ lies close to the direction of the short axis: the rotation of an elongated grain must accordingly be predominantly about an axis perpendicular to the axis of symmetry.

For a rough estimate of the order of magnitude of ω_i let us consider a spherical grain with radius a, mass density ϱ_g and mass $m_g = (4\pi/3)a^3\varrho_g$. In this case $I = (2/5)m_g a^2 = (2/5)(4\pi/3)a^5\varrho_g$ and, from the condition $(1/2)I\omega^2 = (1/2)kT$ with, for example, $a = 5 \times 10^{-5}$ cm, $\varrho_g = 1\,\mathrm{g\,cm^{-3}}$ and $T = 100\,\mathrm{K}$, one finds the considerable angular velocity $\omega = 2 \times 10^4\,\mathrm{rad\,s^{-1}}$.

Colliding gas particles are predominantly H atoms, which probably frequently stick to the surface of the grain and these later depart as components of H_2 molecules. The H_2 molecule retains a portion of its binding energy as kinetic energy of translation; the associated recoil can considerably change the angular momentum of the grain. Since the surface of the grain is presumably not perfectly uniform the result on average is an increase in the angular momentum. Estimates lead one to expect an increase of ω compared with the "thermal" equilibrium value calculated above by a factor $\sim 10^4$ to $\omega \approx 10^8 \ldots 10^9\,\mathrm{s^{-1}}$!

If the elongated dust grains are aligned by the interstellar magnetic field like compass needles in the earth's magnetic field, then the angular momentum produced in the grains must be able to suppress the rotation described. For this the ratio of the magnetic energy of the particle MVB to its rotational energy $(>kT)$ must be greater than unity. Here M denotes the intensity of magnetisation, V the grain volume and B the strength of the interstellar magnetic field [Gauss]. For pure iron the saturation magnetisation $M = 2000$ Gauss and with $V = 10^{-14}\,\mathrm{cm^3}$ and $T = 100\,\mathrm{K}$ one finds $MVB/kT > 1$ only if $B > 10^{-3}$ Gauss. Since the directly observed field is in general at least two orders of magnitude lower (Sects. 4.1.8; 4.1.5), this mechanism can play no part in the alignment of ferromagnetic particles in the direction of the magnetic field. It would moreover require, for that region where the interstellar polarisation shows a predominantly uniform direction, a magnetic field perpendicular to the galactic plane − in contradiction to the evidence of radioastronomical observations (Sect. 4.1.7).

Much lower magnetic field strengths would be needed merely to bring the axis of rotation of the elongated grains gradually into the direction of the field. The theory developed by Davis and Greenstein (1951), already mentioned in Sect. 4.1.2, appeals to the *mechanism of paramagnetic relaxation*: if the axis of rotation of a grain does not coincide with the direction of the magnetic field, then each of its volume elements is subjected to a periodically varying magnetic field and accordingly undergoes an alternating magnetisation M per unit volume. The direction of M always differs somewhat in this unsteady case from the instantaneous direction of the magnetic field B: it corresponds to the field direction of a somewhat earlier point in time, and so follows the grain rotation with a phase delay. This behaviour is described by a complex magnetic susceptibility $\chi = \chi' + i\chi''$, whose imaginary part determines the rotational energy which is converted into heat by the constant remagnetisation (magnetic absorption). This can also be expressed as follows: the imaginary part of χ is associated with an additional component M'' of the magnetisation which is perpendicular to B and the axis of rotation. This is given by $M'' = (\chi''/\omega)(\omega \times B)$, where ω denotes the angular velocity vector of the grain rotation. It produces an angular momentum $D = (V M'' \times B)$ of magnitude

$$D = V|M''| \cdot |B| \sin \Theta = V\chi'' B^2 \sin \Theta \quad . \tag{5.180}$$

Here V denotes the grain volume and Θ the angle between B and ω. D lies perpendicular to B and is directed opposite to the spin component $\omega_\perp$ (perpendicular to B) (Fig. 5.25). The angular momentum of the grain $I\omega_\perp$ about an axis perpendicular to B is thereby decreased and after a retardation time τ_r, determined by $D\tau_r = I\omega_\perp$, eventually reduced to vanishing point. There remains the residual grain rotation about an axis parallel to the magnetic field, for which the magnetisation is not varying.

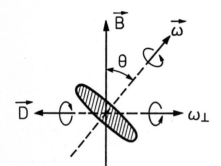

Fig. 5.25. Alignment of elongated grains in the interstellar magnetic field by paramagnetic relaxation. Explanation in text

For many paramagnetic materials we have

$$\chi'' \simeq 2.5 \times 10^{-12} \frac{\omega_\perp}{T_g} \tag{5.181}$$

where T_g denotes the temperature of the solid material. In the expression for the magnetic retardation time $t_r = I\omega_\perp/D$, $\omega_\perp$ cancels out. For the spherical grain considered at the outset with $a = 5 \times 10^{-5}$ cm, $\varrho_g = 1\,\mathrm{g\,cm^{-3}}$, and for example $T_g = 15\,\mathrm{K}$, $B = 3 \times 10^{-6}$ Gauss and $\sin \Theta = 0.5$, one obtains the time $t_r \simeq 4 \times 10^7$ years; choosing $a = 2 \times 10^{-5}$ cm one would get $t_r \simeq 7 \times 10^6$ years. Similar results are also to be expected for elongated particles. On the other hand, if one calculates the length of time taken for the rotation state of a dust grain to be fundamentally altered by collisions with gas particles, i.e. the time

336

t_g needed for the sum of the masses of all the colliding gas particles to equal the grain mass, then we get a period of only a few 10^5 years: the alignment is usually prematurely interrupted and only a few per cent of the dust grains will have their rotation axes approximately aligned with the direction of the magnetic field.

The more precise development of the theory gives in particular the *degree of alignment* f introduced in (5.165) and (5.166). The result is

$$f \simeq 0.3(t_g/t_r) \sim B^2(\chi''/\omega)/N_H a\sqrt{T} \quad , \tag{5.182}$$

where N_H denotes the number density of the neutral H atoms and T the gas temperature. The value $f \simeq 0.1$ estimated in Sect. 5.3.2 in connection with (5.166) would require $t_g/t_r \simeq 0.3$. This will be achieved in a magnetic field of a few 10^{-6} Gauss only by a lengthening of the time t_g given above or by a correspondingly shorter time t_r. Now t_g is known only with relative uncertainty, since the effect of inelastic collisions (with subsequent H_2 formation) depends on unknown properties of the grain surface; t_r could be significantly shortened by a portion of ferromagnetic material in the grain or even by a clumpy distribution of iron atoms. The quantitative interpretation of the observations by the mechanism of magnetic relaxation thus requires, from today's standpoint, assumptions still needing clarification.

5.4 Distribution and Motion of the Interstellar Matter

The main problem in the study of the spatial distribution of the interstellar medium is the determination of distances. This is based in direct or indirect manner, depending upon whether optical or radio phenomena are involved, on distance determinations for stellar objects; in the second case up to now certain assumptions on the large scale velocity field of the gas are necessary. The range achieved, and hence the fraction of the galaxy included, is different according to the nature of the underlying observations.

Evidence on the dust distribution is based essentially on the distance determinations for intrinsically bright stars and is so far limited to about $r < 3\,\text{kpc}$. For the gas of the galactic disc detected by its interstellar absorption lines in the visible stellar spectra this region likewise extends up to about 3 or 4 kpc from the sun, corresponding to the limit up to which the distribution of the OB-stars can be established (Sect. 3.3.1). Detailed information on the cloud structure of the dust can be obtained only for $r \lesssim 1\,\text{kpc}$. This is also true in the main for the gas components. The 21 cm line radiation of neutral hydrogen, detectable at essentially greater distances, makes resolution into individual clouds possible only with observations at middle and high galactic latitudes, so that the relatively thin galactic H I layer is included only in the near neighbourhood.

Observations of radio emission lines provide general information on the interstellar gas up to the most distant regions of the Milky Way system. The derivation of distances is here based, however, essentially on the large scale kinematics of the gas, which is in part still not known with sufficient certainty. A kinematic model of the galaxy, on the other hand, cannot be constructed without the knowledge of certain distance data. The large scale distribution and the large scale motion of the interstellar matter cannot be obtained independently of one another. A satisfactory overall solution is being sought by appropriate combination of optical and radio observations.

5.4.1 Local Distribution and Cloud Structure of the Dust

Mean Distribution: The most important method for deriving the spatial density distribution of the interstellar dust starts from the observed reddening of the stars. We have already mentioned it in Sect. 4.2.5, in connection with the determination of extinction and distance of a dark cloud. The distance r was calculated from (2.43), p. 55, for the largest possible number of stars with well known values of absolute magnitude M_V (from MK spectral type, $H\beta$-index, etc.) and of the reddening E_{B-V} with $A_V = 3 \times E_{B-V}$ and of the apparent magnitudes m_V. For smaller fields on the sphere with centres (l, b) the average profiles of $A_V(r; l, b)$ against r were now obtained (see Sect. 4.2.5, Fig. 4.47). Here of course the selection effect, also already explained in Sect. 4.2.5, must not be forgotten, which causes an apparent decrease in the rate of climb, and eventually a fall, in A_V beyond a certain distance.

Let $\varrho_d(r)$ denote the profile of the mass density of the dust along the line of sight (in $g\,cm^{-3}$) and k_V the *"mass absorption coefficient"* per gramme of dust (dimension $cm^2\,g^{-1}$) at the effective wavelength λ_V of the visual region, then for each fixed direction [Sect. 4.1.1, (4.3) and (4.8)]

$$A_V(r) = 1.086 k_V \int_0^r \varrho_d(r)dr = k_V^* \int_0^r \varrho_d(r)dr \quad . \tag{5.183}$$

Here we have assumed that k_V is independent of position (see Sect. 4.1.1). The profile of the dust density and of the extinction per unit length $a_V(r) = k_V^* \varrho_d(r)$ along the line of sight for the field point (l, b) is given by differentiation: $a_V(r) = dA_V(r)/dr$.

Today the reddening and absolute magnitudes for about 12 000 stars are available for use, with a high proportion of early spectral types. The results obtained up to now for the distribution of the quantity a_V in the galactic plane (for $|z|$ less than about 50 pc) show no clear correlation or anticorrelation between the dust density and the spiral arm filaments found for the young stars. This is true also for the result of a far-reaching new investigation reproduced in Fig. 5.26. Since, on the other hand, photographs of spiral galaxies, such as M 51 (Fig. 1.2), allow one to perceive a clear, albeit often irregular, arrangement of

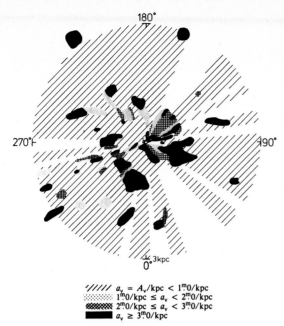

Fig. 5.26. Distribution of the interstellar dust in the galactic plane up to $r = 3$ kpc, derived from the visual extinction values for about 11 000 stars. For the sectors left free no conclusions can be drawn. [After Neckel and Klare (1980)]

the dust on the insides of the spiral arms, one is led to the present conclusion that (1) only the region up to distances of about 1.5 kpc can be covered with any degree of completeness, (2) the uncertainty in M_V and hence in the distances ($\Delta r/r \simeq 0.5 \Delta M_V$!) gives a distortion and smearing of the true distribution and (3) the dense cold clouds of the star formation regions, because of their extremely high visual extinction, are included only in exceptional cases.

The evidence is clearer on the *concentration of the dust near the galactic plane*. With the assumption of a plane parallel layer the distribution perpendicular to the plane $a_V(z)$ can be obtained from the dependence of the mean value over all galactic longitudes $\overline{A}_V(r; b)$ on the galactic latitude b. For $r < 1$ kpc there is a decrease of $a_V(z)$ to a fraction e^{-1} of the maximal value $a_V(0)$ at the surprisingly small "half thickness" $h \simeq 40$ pc.

Mass Density and Extinction Coefficient per Unit of Mass: For the sake of simplicity let us first assume that for each wavelength under consideration only one grain component needs to be taken into account. Then the average extinction up to the distance r can be represented by

$$\overline{A}_\lambda(r) = 1.086 \times C(\lambda)\overline{N}_\mathrm{d}r \tag{5.184}$$

where $C(\lambda)$ is the extinction cross-section to be calculated from Sect. 5.3.2, (5.164) — for nonspherical grains πa^2 is to be replaced there by the corresponding geometric cross-section. With the expression thereby obtained for the mean number density of the dust grains $\overline{N}_\mathrm{d}$ [for estimate of the order of magnitude, see Sect. 4.1.1 (4.14f.)] the average mass density of the dust is given by

$$\overline{\varrho}_d = \frac{4\pi}{3}\overline{a^3}\varrho_g \overline{N}_d = \frac{4\pi}{3.258}\frac{\overline{a^3}}{C(\lambda)}\varrho_g \frac{\overline{A}_\lambda(r)}{r} \quad . \tag{5.185}$$

Here ϱ_g denotes the mass density in the grain itself. The mean value $\overline{a^3}$ is to be formed with the grain radius distribution $f(a)$. For nonspherical particles one has to use the grain volume instead of $(4\pi/3)\overline{a^3}$. If several grain components are essentially effective, then instead of (5.184) and (5.185) one has sums of these expressions.

For the various dust models appropriate for the interpretation of the observed extinction curves one finds throughout mean mass densities which lie about two orders of magnitude below the mean mass density of the gas. From a general relation between extinction and dielectric constant one can deduce that for a spherical grain at a fixed wavelength $\overline{a^3}/C(\lambda)$ is practically independent of the material and is essentially determined by the value of the dielectric constant when $\lambda \to \infty$: for different kinds of spherical dust grain giving the same extinction, the fraction of space occupied by all the grains $(4\pi/3)\overline{a^3}\overline{N}_d = (4\pi/3.258)[\overline{a^3}/C(\lambda_V)](\overline{A}_V/r)$ must be approximately equal (see Aannestad and Purcell, 1973). In this way one obtains the result

$$\overline{\varrho}_d \simeq 3 \times 10^{-27}\varrho_g \frac{\overline{A}_{V(r)}}{r}\,[\text{g cm}^{-3}] \quad , \tag{5.186}$$

where $\overline{A}_V(r)$ is expressed in mag kpc^{-1}. With $\varrho_g = 3\,\text{g cm}^{-3}$ it then follows for the region of the galactic plane that

$$\overline{\varrho}_d \simeq 2 \times 10^{-26}\,\text{g cm}^{-3} \quad . \tag{5.187}$$

For the visual extinction coefficient per unit mass of dust k_V^*, using the expression

$$\overline{A}_V(r) = 1.086 k_V \overline{\varrho}_d r = k_V^* \overline{\varrho}_d r \quad , \tag{5.188}$$

where r will be measured in cm, with $\overline{\varrho}_d$ from (5.186) and $\varrho_g = 3\,\text{g cm}^{-3}$, we get the value

$$k_V^* \simeq \frac{1}{3 \times 10^{-27}\varrho_g 3 \times 10^{21}} \simeq 4 \times 10^4\,\text{mag cm}^2\,\text{g}^{-1} \quad . \tag{5.189}$$

The numerical coefficient is still uncertain: from the various available evidence it should fall between about 2 and 4. For other wavelengths the extinction coefficient can be obtained from this with the help of the reddening curve $F(\lambda)$ from Sect. 4.1.1 (4.10), since $k_B^* - k_V^* = k_V^*/3$, as

$$k_\lambda^* = k_V^*[1 + \tfrac{1}{3}F(\lambda)] \tag{5.190}$$

or from Table 4.1.

The *mass ratio of dust:gas* can be derived from comparison of the extinction $A_V(r) \doteq 3 \times E_{B-V}(r)$ with the column densities of the neutral hydrogen

(Sect. 5.4.2) obtained from observations of the interstellar Lyman α absorption line − for the same star − in association with (5.186). Here we anticipate (5.195) in Sect. 5.4.2, whence $\overline{N_H} \cdot r / \overline{A_V}(r) \simeq 2 \times 10^{21}$. For the ratio of $\overline{\varrho_d}$ to the mass density of the gas (with ratio of numbers of atoms H:He = 10 : 1) $\overline{\varrho_{gas}} = 1.4\overline{\varrho_H} = 1.4 m_H \overline{N_H}$ [taking into account the factor 3×10^{21} in the conversion of r in (5.186) from kpc into cm] it then follows that

$$\overline{\varrho}_d / \overline{\varrho}_{gas} \simeq 2 \times 10^{-3} \varrho_g \simeq 0.6 \times 10^{-2} \quad . \tag{5.191}$$

Cloud Structure: The question posed here is the statistical nature of the small scale structure of the dust distribution. In mathematical terms one assumes either discrete clouds or a strongly fluctuating continuum. Since only *one* realisation of the structure is under discussion, the assumption of statistical homogeneity in space at least in partial regions is necessary. From the observational point of view essentially two statistical functions are derived: (1) frequency distributions $H(A; r)$ of the extinction value A for stars at specified distances r or within narrow distance intervals $r - \frac{1}{2}\Delta r \ldots r + \frac{1}{2}\Delta r$, which still lie within the galactic layer (Fig. 5.27, left-hand side) and (2) an appropriate function for the statistical description of the distribution $A(l, b; r)$ on the sphere in the region of the Milky Way at a fixed distance r.

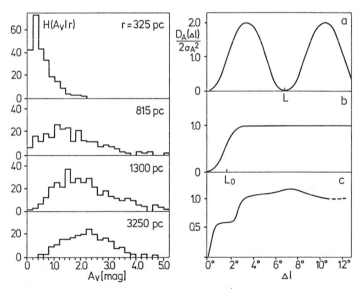

Fig. 5.27. *Left-hand side*: Frequencies $H(A_V | r)$ of observed extinction values A_V in intervals of width $0^{\text{m}} 2$ for stars in narrow distance intervals around $r = 325, 815, 1300$ and 3250 pc, which are at distances less than 75 pc from the galactic plane. The abscissa value 1.0, for example, characterises the A_V interval $0^{\text{m}} 9 \ldots 1^{\text{m}} 1$. *Right-hand side*: Normalised structure functions $D_A(\Delta l)/2\sigma_A^2$ (a) for the strictly periodic profile $A(l) = A_0 \sin(2\pi/L)l + A$, (b) for a completely randomly varying profile $A(l)$, having no waves however with lengths $\Delta l < L_0$, and (c) for the observed distribution $A_V(l) = A_V(l, b \simeq 0° | r \simeq 1$ kpc)

341

For (2) one can perhaps form the *"structure function"*. For the one-dimensional profile of $A(l, b; r)$ at fixed b, denoted in the following by $A(l)$, this is defined by

$$D_A(\Delta l) = \overline{[A(l + \Delta l) - A(l)]^2}$$

$$= \frac{1}{360°} \int_0^{360°} [A(l + \Delta l) - A(l)]^2 dl \quad . \tag{5.192}$$

The averaging in the right-hand side over the single realisation under consideration, viz. the observed profile $A(l)$, has been used to represent the averaging over very many realisations $A(l)$ with the same statistical properties for fixed Δl and l, as is required by the mathematical concept of the structure function. $D_A(\Delta l)$ vanishes at $\Delta l = 0$ and increases for large Δl, since then $A(l + \Delta l)$ and $A(l)$ become statistically independent, to the saturation value $2\sigma_A^2 = 2\overline{(A - \overline{A})^2}$, for we then have

$$\overline{[A(l + \Delta l) - A(l)]^2} = \overline{A^2(l + \Delta l)} + \overline{A^2(l)} - \overline{2A(l + \Delta l) \cdot A(l)}$$
$$= 2\overline{A^2} - 2(\overline{A})^2 = \overline{2(A - \overline{A})^2} \quad .$$

$D_A(\Delta l)$ is linked with the so-called correlation function

$$K_{\Delta A}(\Delta l) = \overline{\Delta A(l + \Delta l) \cdot \Delta A(l)} / \overline{(\Delta A)^2} \quad ,$$

where $\Delta A = A - \overline{A}$, by

$$D_A(\Delta l) = 2\sigma_A^2 [1 - K_{\Delta A}(\Delta l)] \quad .$$

A rise of the curve $D_A(\Delta l)$ with increasing Δl occurs whenever Δl reaches the extent of characteristic structural elements of $A(l)$. Extreme cases are, for one, the strictly periodic profile $A(l) = A_0 \sin(2\pi/L)l + \overline{A}$ with $D_A(\Delta l)/2\sigma_A^2 = 1 - \cos(2\pi \Delta l/L)$ and, for another, a completely random structure of $A(l)$, for which, however, a very small element size, the so-called "microscale" L_0, will occur. In this case one often sets

$$D_A(\Delta l)/2\sigma_A^2 = 1 - \exp[-(\Delta l/L_0)^2] \quad .$$

Both cases are illustrated in the right-hand side of Fig. 5.27, together with a result obtained with the extinction values A_V actually observed at $r \simeq 1$ kpc. This empirical structure function $A_V(l, b \simeq 0°; r \simeq 1$ kpc$)$ shows first of all that very small irregular structures, namely small clouds with apparent extents down to $\Delta l \simeq 0°.25$ are present, which produce about half of the total variance σ_A^2. The further climb occurring beyond $\Delta l = 2°$ must meanwhile be caused by essentially larger clouds or cloud complexes with apparent extents of more than $2°$.

342

Let us turn next to the interpretation of the frequency distributions $H(A; r)$ (Fig. 5.27, left-hand side). They show the typical shapes of probability distributions of random numbers n with mean value $\bar{n}$, viz. the Poisson Distribution:

$$P(n) = \frac{\bar{n}^n}{n!} e^{-\bar{n}} \quad , \tag{5.193}$$

for various values of $\bar{n}$, if one associates larger values of $\bar{n}$ with greater distances: for small r and $\bar{n}$ the distributions are strongly asymmetric; for larger r and $\bar{n}$ they approximate more and more to the symmetrical shape of a Gauss distribution. This suggests assuming a random distribution of clouds in space, in which the line of sight encounters n clouds up to a distance r. If for the sake of simplicity we take the same extinction a_1 for all clouds, so that $A(r) = n(r)a_1$ then the probability distribution of A at fixed r is

$$W(A|r) = P\left(\frac{A}{a_1}\right)$$

with $\bar{n} = \bar{n}(r) = \overline{A}(r)/a_1 = \nu_1 r$. Here ν_1 is the average number of these clouds per unit length [kpc^{-1}]. With this simple model the normalised distribution $H(A; r)/\Sigma H(A; r)$ can be approximately represented if we choose $\nu_1 = 5\,\mathrm{kpc}^{-1}$ and $a_1 = 0\overset{m}{.}25$. Of course, a spectrum of cloud absorptions $f(a)$ with a certain width is to be expected. More detailed discussion shows, in particular, a portion of the clouds with significantly higher extinction values which can be represented by a mean value $a_2 \simeq 1\overset{m}{.}5$ and which lie on average at $\nu_2 \simeq 0.5\,\mathrm{kpc}^{-1}$ along the line of sight. The frequency of the cloud extinction values a falls off roughly as $f(a) \sim a^{-3}$.

The structure functions once again provide information on the fraction of large clouds, since these show up particularly prominently in the scatter σ_A : we have

$$\sigma_n^2 = \overline{(n - \bar{n})^2} = \Sigma (n - \bar{n})^2 P(n) = \bar{n} = \nu r \quad ,$$

and hence

$$\sigma_A^2 = \nu_1 r a_1^2 + \nu_2 r a_2^2 \quad .$$

Chiefly, however, from the profile of $D_A(\Delta l; r)/2\sigma_A^2(r)$ one can obtain statistical conclusions on the *linear dimensions of the clouds*. If one interprets the positions of the first and second increases of the result for $r = 1\,\mathrm{kpc}$ in Fig. 5.27 (lower right) at $\Delta l \simeq 0\overset{\circ}{.}25$ and $\simeq 3°$ as approximate angular dimensions of the two types of cloud at average distances $r/2 = 500\,\mathrm{pc}$, then it follows that the linear diameters $D \simeq 2\,\mathrm{pc}$ and $\simeq 25\,\mathrm{pc}$, respectively. The representation of the profile $D_A(\Delta l; r)/2\sigma_A^2(r)$ under the assumption of a continuous stochastic model of the cloud structure, which we cannot go into more closely here, produces mean values $D_1 \simeq 2\ldots 3\,\mathrm{pc}$ and $D_2 \simeq 70\,\mathrm{pc}$. These numbers correspond roughly to the connection between cloud extinctions and cloud diameters $a \sim D^{0.6}$. Since

$a \sim \varrho_d D$ with ϱ_d = mass density of the dust in the cloud, it follows that the mass of dust in the cloud is $\mathcal{M}_d \sim \varrho_d D^3 \sim (a/D)D^3 - aD^2 \sim a^{4.3}$. The *"mass spectrum"* is given by $F(\mathcal{M}_d) \sim f(a)(da/d\mathcal{M}_d)$ which, with $f(a) \sim a^{-3}$, finally becomes

$$F(\mathcal{M}_d) \sim \mathcal{M}_d^{-1.5} \ .$$

(5.194)

If dust and gas stand in a fixed ratio, this relationship — which is of course only an approximation — is valid also for the total mass of the clouds. This is qualitatively confirmed by direct studies of the gas components. After calculation of the mass density ϱ_d from $\bar{a}_i = k_V^* \varrho_d D_i$ with $k_V^* = (2...4) \times 10^4$ mag cm^2 g^{-1} (5.189) and an assumed gas:dust ratio $\varrho_{gas}/\varrho_d = 100$ for both frequently occurring cloud types, we obtain the mass values $\mathcal{M}_1 = 10...20\mathcal{M}_\odot$ and $\mathcal{M}_2 = (4...8) \times 10^4 \mathcal{M}_\odot$. For the ratio of the dust densities in the two cloud types it follows that $\varrho_{d1}/\varrho_{d2} = (\bar{a}_1/\bar{a}_2)(\overline{D}_2/\overline{D}_1) \simeq 5$. The reason that just the "large clouds" appear to have a low density, is probably that these consist of a loose association of smaller clouds with D about $3...10$ pc, whose densities must correspond at least to those of the "small clouds". For the exponent in the mass spectrum of individual clouds one should expect a value even somewhat greater than 1.5; $\mathcal{M}_1$ perhaps characterises the centre of the region of the "diffuse clouds" detected by the 21 cm data (see Sect. 5.4.2).

5.4.2 Density and Cloud Structure of the Neutral Hydrogen in the Solar Neighbourhood

Analysis of Optical Observations: Evidence on the local density of the neutral hydrogen has been obtained from the extra-terrestrial observations of the interstellar Lyman α absorption lines (Sect. 4.1.4). The curve of growth analysis (Sect. 5.2.5) leads to column densities of the neutral hydrogen atoms $\mathcal{N}(\mathrm{H\,I})$. Because of the cloud structure quite different values are obtained even for stars at almost equal distances; there is however a striking general deficit in the region of the galactic longitudes $l \simeq 180°$ to $l \simeq 270°$. Here one finds up to $r \simeq 1$ kpc a general density of only 0.13 H atoms cm^{-3} compared with the average value $\overline{N}(\mathrm{H\,I}) \simeq 0.7$ atoms cm^{-3} for the other regions included. In the region Scorpius-Ophiuchus there is a rather large mass of gas at distances $r < 200$ pc. For the immediate neighbourhood of the sun ($r < 20$ pc) a particularly low density, between 0.02 and 0.1 atoms cm^{-3}, has been found from Ly α absorption structures which appear inside emission lines from nearby stars of late spectral type, e.g. α Bootis.

From the observed diffuse soft x-ray background with photon energies $h\nu \lesssim 0.3$ keV (see Sect. 4.1.9) it has been concluded that a large fraction of the local interstellar volume must be filled with a thin plasma at about 10^6 K. These x-rays cannot penetrate column densities greater than $\mathcal{N}(\mathrm{H\,I}) \approx 5 \times 10^{19}$ cm^{-2}. Using, for example, $\overline{N}(\mathrm{H\,I}) \approx 0.15$ cm^{-3} the x-rays must come from distances less than about 100 pc.

A comparison of the column densities $\mathcal{N}(\mathrm{H\,I})$ obtained for $r \simeq 100 \ldots 1000\,\mathrm{pc}$ with the individual values of interstellar extinction measured by the reddening $E_{\mathrm{B-V}}$ shows good correlation of the two results. The observational material obtained with OAO-2 and the "Copernicus" satellite — taking into account the molecular hydrogen — leads to the relation

$$\frac{\overline{\mathcal{N}_{\mathrm{H}}}}{E_{\mathrm{B-V}}} = 5.8 \times 10^{21}\,\mathrm{H\ atoms\,cm^{-2}\,mag^{-1}} \quad . \tag{5.195}$$

With $\overline{E}_{\mathrm{B-V}} = 0.6\,\mathrm{mag\,kpc^{-1}}$ (see Sect. 4.1.1) the mean density obtained thence inside $r \simeq 1\,\mathrm{kpc}$ is $N_{\mathrm{H}} = 1.1\,\mathrm{H\ atoms\,cm^{-3}}$. For an atom number ratio H:He=10:1 one therefore obtains for the *mean local mass density of the gas*

$$\varrho_0(\mathrm{gas}) \simeq 3 \times 10^{-24}\,\mathrm{g\,cm^{-3}} \quad .$$

Practically the same value was obtained for the mean *stellar* mass density inside $r = 20\,\mathrm{pc}$ (Sect. 3.2.2)!

With regard to the *cloud structure* it has already been possible to conclude directly from the splitting of the interstellar absorption lines that a line of sight lying close to the galactic plane will encounter on average about 5 to 10 clouds per kpc, and these appear to be overwhelmingly concentrated on the spiral arms (Sect. 4.1.4). With $E_{\mathrm{B-V}} = 0.6\,\mathrm{mag\,kpc^{-1}}$ we find from (5.195), assuming 7 clouds per kpc, an average value for the column density per cloud of $\mathcal{N}_{\mathrm{H}} \simeq 5 \times 10^{20}\,\mathrm{cm^{-2}}$.

Evidence on the densities in these clouds can be obtained in some cases from the equivalent widths of the lines of two or more ionisation stages of the same element, after conversion to the corresponding column densities. For ionisation equilibrium we have in each case a statistical equation of the sort (5.73) in Sect. 5.2.1, and if we select lines whose starting levels (fine structure in the ground term) are excited essentially by collisions with neutral H atoms, then we get, besides the electron density N_{e}, also the density $N(\mathrm{H\,I})$. In this way, from observations in the ultraviolet, values $N(\mathrm{H\,I}) \simeq N_{\mathrm{H}} \simeq 10 \ldots 10^2\,\mathrm{cm^{-3}}$ and $N_{\mathrm{e}} \simeq 10^{-2}\,\mathrm{cm^{-3}}$ were estimated for typical "small" clouds.

In a similar way one can apply the equivalent widths of the various rotational lines of the $\mathrm{H_2}$ molecule observed in the ultraviolet to the determination of the density and the temperature. From the column densities $\mathcal{N}(J)$ for the individual rotation levels $J = 0, 1, 2, \ldots$ excited by photons and by collisions there follow the population ratios $N(J)/N(0)$, which depend on the radiation field and on the density $N(\mathrm{H\,I})$ of the neutral H atoms. Quantitative discussion likewise gives average total densities for normal diffuse clouds (including the hydrogen in molecular form) of about $N_{\mathrm{H}} = 30\,\mathrm{cm^{-3}}$ and a mean temperature $T = 80\,\mathrm{K}$. With the mean column density of a cloud $\mathcal{N}_{\mathrm{H}} \simeq 5 \times 10^{20}\,\mathrm{cm^{-2}}$ found above we deduce an average cloud diameter of around $2\,\mathrm{pc}$.

For the "high velocity components" mentioned at the end of Sect. 4.1.4 and in Sect. 5.2.5 — presumably gas accelerated by supernovae or strong stellar winds — somewhat higher densities and temperatures but lower column densities were obtained. These regions must be thin sheets of gas probably produced by moving shock waves.

Analysis of the 21 cm Line: For an optically thin, homogeneous layer (5.52) can be applied for the representation of the line intensity, where $I_{\nu,0} = 0$, $h\nu/kT \ll 1$, so that $\beta = (N'Q_{21}/A_{21})(h\nu/kT)$, and $N'Q_{21}/A_{21} \gg 1$ can be assumed. One then easily obtains $I_{21}^L = \tau_\nu^L B_\nu(T)$. We shall first of all reckon the expression τ_ν^L appearing for the optical thickness to be free of the assumption of homogeneity.

From (5.15a) in Sect. 5.1.2 it follows, since $h\nu_{nm}/kT \ll 1$ with $m = 0$ and $n = 1$ for the lower and upper states, respectively, that

$$\kappa_\nu^L = \frac{h\nu_{10}}{c}\psi(\nu)B_{01}N_0[1 - \exp(-h\nu_{10}/kT)] \approx \frac{h\nu_{10}}{c}\psi(\nu)B_{01}N_0\frac{h\nu_{10}}{kT} \quad .$$

$$(5.196)$$

The stimulated emission is essential and enters the expression as a strong reduction of the absorption coefficient.

In Sect. 5.1.2 it has already been shown that the excitation temperature occurring here for the 21 cm line is to a satisfactory approximation equal to the kinetic temperature, as in the case of LTE. For the population numbers one therefore has:

$$\frac{N_1}{N_0} = \frac{g_1}{g_0}\exp\left[-\frac{h\nu_{10}}{kT}\right] \simeq \frac{g_1}{g_0} = 3 \tag{5.197}$$

so that for the density of all neutral H atoms

$$N = N_0 + N_1 = 4N_0 \quad . \tag{5.198}$$

If one expresses B_{01} in terms of A_{10} using (5.14), then with $\lambda_0 = c/\nu_{10}$ the optical depth up to distance r is given by

$$\tau_\nu^L(r) = \frac{3}{8\pi}A_{10}\lambda_0\frac{hc}{k}\frac{1}{4}\int_0^r \frac{N(r)\psi(\nu|r)}{T}dr \quad . \tag{5.199}$$

Here we have expressed the fact that the profile of the line absorption coefficient $\psi(\nu)$ will depend on the position of the line of sight as a result of the systematic Doppler displacement by the differential galactic rotation.

It is usual, instead of the frequency ν, to introduce the radial velocity v (we omit the index r for the sake of brevity). We have $v/c = (\nu_{10} - \nu)/\nu_{10}$, and hence $|dv/d\nu| = c/\nu_{10} = \lambda_0$, so that in particular

$$\psi(\nu|r) = \varphi(v|r)\left|\frac{dv}{d\nu}\right| = \varphi(v|r)\lambda_0 \quad .$$

Then (5.199) becomes

$$\tau_v^{\mathrm{L}}(r) = \frac{3}{32\pi} A_{10} \lambda_0^2 \frac{hc}{k} \int\limits_0^r \frac{N(r,v)}{T} dr \quad , \tag{5.200}$$

where $N(r,v)dr = N(r)dr\,\varphi(v|r)$ denotes the number of neutral H atoms in the distance interval $r \ldots r + dr$ with radial velocities in the interval $v - \frac{1}{2} \ldots v + \frac{1}{2}$.

For the relation between the observed line intensity I_v^{L} or I_v^{L} and the density the *case of an optically thin layer*, $\tau_v^{\mathrm{L}} \ll 1$, is of primary significance, since there is then a direct linear connection if we assume in addition constant temperature. Then it follows from (5.9) that the brightness temperature at the velocity v in the line profile is

$$T_{\mathrm{b}}(v) = \tau_v^{\mathrm{L}}(r^*)T \quad . \tag{5.201}$$

Since $\tau_v^{\mathrm{L}}(r^*) \sim T^{-1}$, the brightness temperature T_{b} – and also the intensity itself – is independent of the gas temperature. After substitution of (5.200), (5.201) with $r^* = \infty$ gives

$$\mathcal{N}'(v) = \int\limits_0^\infty N(r,v)dr = 1.835 \times 10^{18} T_{\mathrm{b}}(v) \quad . \tag{5.202}$$

The column density $\mathcal{N}'(v)$ here indicates the number of neutral H atoms per cm^2 whose radial velocities lie in the interval of width $\Delta v = 1\,\mathrm{km\,s^{-1}}$ about the value v [km s^{-1}]. Integration over the line produces $\mathcal{N}(\mathrm{H\,I})$, the column density of all neutral H atoms.

An initial empirical statement on the *temperature* of the H I gas is on the other hand possible in the case of large optical thickness. For constant gas temperature it then follows from (5.9) that $T_{\mathrm{b}}(v) = T$, i.e. the observed brightness temperature assumes a constant highest value over a certain frequency region or velocity region inside the line. This is observed in the direction of the galactic central region. Here the geometric path length attains the greatest value, and at the same time the broadening effect of the differential galactic rotation disappears, so that the profile of the optical thickness τ_v^{L} against v becomes narrower and correspondingly higher. The observations give $T \simeq 130\,\mathrm{K}$.

The case of an optically thin layer occurs as a rule at galactic latitudes $|b| > 20°$. Here the path length inside the gas disc is relatively short, so that there are always only a few clouds on the line of sight. Accordingly the line profiles in emission and also in absorption frequently show a distinct splitting into several components. For purely thermal Doppler broadening the profile of the individual component can be taken as a Gauss distribution with the variance

$$\sigma_v^2 = \overline{(v - \overline{v})^2} = \frac{\mathcal{R}T}{\mu} = \frac{kT}{m} \tag{5.203}$$

($\mathcal{R}$ =gas constant, μ = molecular weight, m = mass of a particle). If one mea-

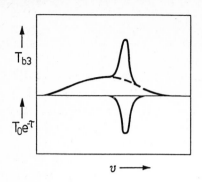

Fig. 5.28. Schematic presentation of observed emission and absorption profiles of the 21 cm line. Explanation in text

sures the radial velocity v or the dispersion σ_v in km s^{-1}, then for hydrogen it follows that the kinetic temperature is given by

$$T = 121\sigma_v^2 \quad . \tag{5.204}$$

The individual components of the emission profile are often superimposed on a broad flat component (Fig. 5.28). The representation of a large number of such profiles from observations at galactic latitudes around $|b|\simeq30°$ and galactic longitudes from 0° to 360° by formal superpositions, each of several Gauss profiles ("Gauss analysis"), produces for the narrow components on average about $\sigma_v^{(1)} = 2\ldots3$ km s^{-1} and for the broad flat components about $\sigma_v^{(2)}\simeq10$ km s^{-1}. For the narrow components (5.204) implies − necessarily allowing for a proportion of internal turbulent motion − a temperature of order 10^2 K, whilst the underlying broad components give a few hundred K up to about 8000 K. This emission was therefore at first ascribed to a "warm" and thin "intercloud gas" that could exist in pressure equilibrium with the cold cloud. After the discovery of a widespread hot component (Sect. 5.2.3), however, it seemed more plausible to move the "warm" gas to the transition regions between the cold clouds and the hot gas, thus primarily as an envelope for the cold clouds.

Absorption observations furnish further information. If one selects for the continuum source an extra-galactic object ($|b|\gtrsim20°$), and observing with a sufficiently narrow directional lobe (interferometer), practically the same gas must be effective in the absorption profile as one can observe directly alongside the source in emission alone (cf. Sect. 4.1.5, Fig. 4.19). Because $\tau_v^{L}\sim T^{-1}$ from (5.200), however, the "warm" gas will appear only very weakly in absorption. The comparison of emission and absorption observations can therefore in principle provide the temperatures of both components. If we denote the brightness temperature of the continuum source by T_0 and there is only one homogeneous cloud of optical thickness τ and of kinetic temperature T on the line of sight, then for the observed brightness temperatures at a certain frequency or velocity we have:

at the source in the line : $\quad T_{b1} = T(1 - e^{-\tau}) + T_0\,e^{-\tau}$

at the source beside the line : $T_{b2} = T_0$

beside the source in the line : $T_{b3} = T(1 - e^{-\tau})$

(interpolation at the position of the source!).

Hence it follows that

$$T_{b1} - T_{b2} - T_{b3} = -T_0(1 - e^{-\tau}) = -T_{b2}(1 - e^{-\tau}) \quad .$$

This equation first of all produces τ for each point of the line profile, whereby from the third of the above equations the temperature T can be calculated. If the warm gas component beside a cold cloud is represented by a broad flat emission, then in the simplest case an additional term $T'\tau'$ appears in the expression for T_{b1}, where T' denotes the temperature and $\tau'(\ll 1)$ the optical thickness of the warm region. Observations beside the narrow line, but still in the broad component (Fig. 5.28), with $T_{b3} = T'\tau'$ produce directly the column density of the warm gas.

Let us summarise important *results* obtained in the manner described on the distribution of the neutral hydrogen gas in the solar neighbourhood: since clouds with reddening $E_{B-V} \simeq 0.3$ already show a significant fraction of molecular hydrogen (Sect. 5.2.5), the evidence of the 21 cm line observations apply primarily to the generally distributed cloudy neutral gas of low density. Observations of small regions of the sphere with high angular resolution ($<10'$) show as the dominant structure cold "diffuse clouds" with distinct, partially filamentary and flat shapes, whose column densities range from $\mathcal{N}(H I) \simeq 1 \times 10^{21}$ cm^{-2} down to about 2×10^{19} cm^{-2}; the smallest of these condensations are often called "cloudlets". The cloud diameters are known only roughly and lie between about 1 pc and 10 pc. The temperatures of the clouds also observed in absorption with optical thicknesses $\tau_0 > 0.2$ in the centre of the line turn out to be 20 K to 120 K. Mean values of the cloud parameters are presented in Table 5.14. From the observed dependence of the total column density – summed over all clouds on the line of sight – on the galactic latitude b, one has derived the effective half-thickness H^* of the galactic H I cloud layer, defined by

$$\int\limits_{-\infty}^{+\infty} \overline{N}(z)dz = 2H^*\overline{N}(0) \tag{5.205}$$

(z = perpendicular distance from the galactic plane, $\overline{N}$ = density averaged over the clouds and over the intervening space). The distances needed for this were obtained from the radial velocities of the line components with the aid of a model of the galactic rotation (see Sect. 5.4.3). The result for H^* is contained in Table 5.14. Assuming on the average four standard clouds per 1 kpc of path

Table 5.14. Characteristic data of the cold diffuse H I clouds in the galactic disc

Mean colum density $\mathcal{N}(H I)$	3×10^{20} cm^{-2}
Mean cloud diameter $\overline{D}$	5 pc
Mean H I density of the cloud $\overline{N}(H I)$	20 cm^{-3}
Mean cloud mass $\mathcal{M}(H I)$	30 $\mathcal{M}_\odot$
Mean temperature of the cloud $\overline{T}$	80 K
Effective half-thickness of the cloud layer H^*	130 pc
Mean magnitude of the cloud velocity	10 km s^{-1}

length in the galactic plane (Sect. 4.1.5), the mean H I density of the included cloud components is found to be $\overline{N}(0) \simeq 4 \times 3 \times 10^{20}\,\mathrm{cm}^{-2}/3 \times 10^{21}\,\mathrm{cm} \simeq 0.4\,\mathrm{cm}^{-3}$.

For the warm gas ($\overline{T} \approx 6000\,\mathrm{K}$) one finds $\mathcal{N}(b) \simeq 1.4 \times 10^{20}\,\mathrm{cosec}\,b$ and an effective half-thickness $H^* \simeq 200\,\mathrm{pc}$, from which the maximum mean density is $\overline{N}(0) \simeq 1.4 \times 10^{20}/H^* \simeq 0.2\,\mathrm{cm}^{-3}$. Together with the corresponding value for the cold clouds this gives a *mean local* H I *density* ($r < 1\,\mathrm{kpc}$) in the galactic plane of $\overline{N}(\mathrm{H\,I}) \simeq 0.6$ atoms cm^{-3}, in satisfactory agreement with the result from Ly α observations given at the beginning of this section. In the immediate solar neighbourhood, say $r < 20\,\mathrm{pc}$, there appears to be only the thin "warm" H I-component present with $N(\mathrm{H\,I}) \simeq 0.1\,\mathrm{cm}^{-3}$ (see above: Ly α absorption) – besides the hot gas. This is confirmed by observations of diffuse Ly α emission, which arises from resonance scattering of the solar Ly α radiation by nearby interstellar H atoms ($r \simeq 100\,\mathrm{AU}$) and whose line widths lead to temperatures of this H I gas around $T \simeq 10\,000\,\mathrm{K}$.

For the individual clouds the density $N(\mathrm{H\,I})$ can be found only for known cloud diameter D, for the determination of which rough distance estimates must also be available. If one associates the individual column densities $\mathcal{N}(\mathrm{H\,I})$ given above with the values of $D \simeq 3\ldots10\,\mathrm{pc}$ suggested by observations of the interstellar extinction (see Sect. 5.4.1), then one obtains the density values $N \simeq 2 \ldots 30$ atoms cm^{-3}. With quite a rough approximation to the cloud shape by a sphere, these numbers give *cloud masses* $\mathcal{M} = (\pi/6)D^3 N m_{\mathrm{H}}$ in the region $\mathcal{M}(\mathrm{H\,I}) \simeq 1\mathcal{M}_{\odot} \ldots 3 \times 10^2\mathcal{M}_{\odot}$. For the *mass spectrum* of the H I clouds in the form $F(\mathcal{M}) \sim \mathcal{M}^{-\beta}$, values of the exponent β between 1.3 and 1.7 were found (Braunsfurth, 1975).

5.4.3 The Neutral Hydrogen on the Large Scale

Observations of radio line emission can give the column densities $\mathcal{N}'(v)$ for a fixed direction (l, b), and if each radial velocity v can be associated unambiguously with a distance r, then the density profile $N(r)$ along the line of sight can be derived from $\mathcal{N}'(v)$. We have $\mathcal{N}'(v)dv = N(r)dr$, so the number density at r is given by

$$N(r) = \mathcal{N}'(v)\frac{dv}{dr} \quad . \tag{5.206}$$

Accordingly, for each direction (l, b) under consideration one needs the derivative of the radial velocity $dv/dr = f(r; l, b)$ as a function of r, and one must therefore know the velocity field $v(r; l, b)$ itself. We first of all apply ourselves to this theme.

Large Scale Kinematics: Taking as the first approximation, that the gas moves round the centre of the system in *circular orbits* in the galactic plane, it is possible in the following way to derive the "rotation curve" $\Theta = \Theta(R)$ for the "inner region" $R < R_0$ from the 21 cm profiles, if the orbital velocity at the

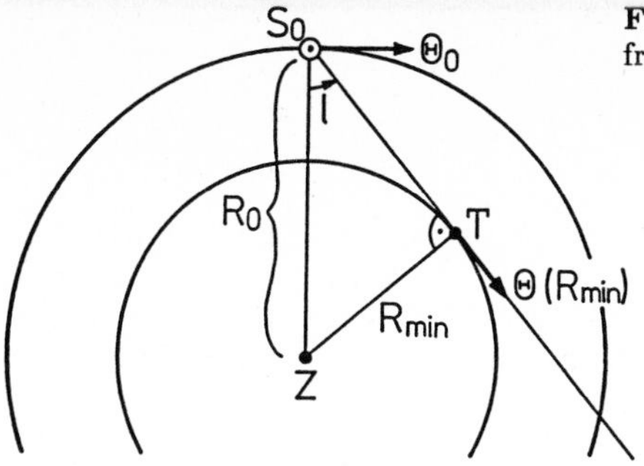

Fig. 5.29. Derivation of the rotation curve $\Theta(R)$ from 21 cm line observations

position of the sun Θ_0 and the distance of the sun from the centre R_0 are given (Fig. 5.29).

On each line of sight which runs in the galactic plane and lies in a longitude l between $0°$ and $90°$ or between $270°$ and $360°$, there is a point labelled T where the smallest distance from the centre

$$R_{min} = R_0 \sin l \tag{5.207}$$

is reached. As we already know from optical observations of the differential galactic rotation of the stars and of the interstellar gas (Sects. 3.4.1f., 4.1.5), the radial velocity relative to the sun along the line of sight increases generally at first with increasing distance r. At the tangent point T of the galactic circular orbit with radius R_{min}, however, the velocity must reach a maximal value v_{max} and thereafter decrease again. From Fig. 5.29 one gets:

$$v_{max} = \Theta(R_{min}) - \Theta_0 \sin l \quad . \tag{5.208}$$

$\Theta(R)$ is often expressed by the angular velocity $\omega(R) = \Theta(R)/R$, and one then obtains, using (5.207),

$$v_{max} = R_0[\omega(R_{min}) - \omega_0] \sin l \quad , \tag{5.209}$$

a relation which also follows immediately from (3.59), Sect. 3.4.1. For $0° < l < 90°$ v_{max} is positive and corresponds to the longwave "edge" of the line profile, so that its magnitude can be taken directly from the observations; for $270° < l < 360°$ the sign of v_{max} is reversed, and one has to take the shortwave boundary of the profile. With the values of R_0 and Θ_0 or ω_0 obtained from optical observations, the quantities $\Theta(R_{min})$ or $\omega(R_{min})$ are obtained from (5.208) or (5.209) for the central distance $R_{min} < R_0$ calculated from (5.207).

Strictly speaking, one obviously obtains only the orbital velocities on the two semicircles with the line segment $S_0 Z$ as diameter, and for this it is necessary that there be an approximately homogeneous large scale distribution of the gas. For example, if there were no or only a little gas in a large region around the point T, one would get a false value of v_{max}. Moreover it is of course impossible in general to obtain information for the longitudes around $0°$, $90°$,

351

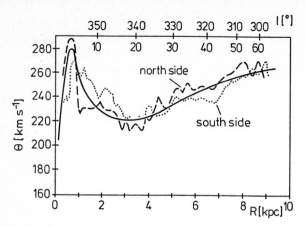

Fig. 5.30. Galactic rotation curves $\Theta(R)$ derived from observations of the 21 cm line emission for the northern and for the southern sides of the galactic centre [after Sinha (1978)]. The *full smooth curve* represents a mean profile

and $270°$, at which $v_{max} \simeq 0$. The rotation curve $\Theta(R)$ obtained extends only over the region $0 < R < 9$ kpc. Results for both quadrants $- l = 0° \ldots 90°$ lying to the north and $l = 270° \ldots 360°$ to the south of the Milky Way $-$ are shown in Fig. 5.30 and approximated by a smooth mean profile.

Obtaining $\Theta(R)$ for $R > R_0$: For a first approximation we set $\Theta(R) = \sqrt{G\mathcal{M}/R}$, corresponding to equality of the centripetal and gravitational accelerations at a distance R from a point mass $\mathcal{M}$ located at the centre, which follows from Θ_0 and R_0 (discussed further in Sect. 6.1.5).

For $|R_{min} - R_0| \ll R_0$ a noteworthy relationship between R_0 and Oort's Constant A can be derived from (5.209). In this case it follows with the first of equations (3.66) that

$$R_0[\omega(R_{min}) - \omega_0] = R_0 \left(\frac{d\omega}{dR}\right)_{R_0} (R_{min} - R_0) = -2A(R_{min} - R_0) \quad ,$$

so that, since $R_{min} = R_0 \sin l$, (5.209) becomes

$$v_{max} = 2AR_0(1 - \sin l) \sin l \quad . \tag{5.210}$$

Observations in directions $l \simeq 90°$, corresponding to $R_{min} \simeq R_0$ then enable one to determine the product AR_0. The values obtained lie between 135 and 150 km s^{-1}. The rounded off value of $A = 15$ km s^{-1} kpc^{-1} given in Sect. 3.4.2 is therefore compatible with a distance to the galactic centre $R_0 \simeq 9 \ldots 10$ kpc.

If the assumption of pure circular orbits of the gas were strictly fulfilled and the local standard of rest were itself also moving on a circular orbit round the galactic centre, then the two observed curves shown in Fig. 5.30 should coincide. The average difference occurring in the region $R \simeq 5 \ldots 9$ kpc could be attributed to an outwardly directed motion of about 5 km s^{-1} of the local standard of rest used in the derivation (see also Sect. 6.2.2, last sentence). The wavelike deviations from the smooth mean profile, with amplitudes of about $\pm 10\,\mathrm{km\,s^{-1}}$, were first of all explained by the assumption that at these galactic longitudes there is almost no gas at the tangent points, and one accordingly drew smooth rotation curves as upper envelopes through the maximal values of the observed profiles. Recent discussion of the line profiles has shown that this interpretation is not tenable. The interpretation of the observed waves — the derived $\Theta(R)$ apply to points each of which has a different galactocentric longitude ϑ! — require the assumption of deviations of the velocity field from pure circular symmetry, if one is to take as basis a not wholly unacceptable density model.

For a purely empirical derivation of the actual, noncircularly symmetric, velocity field, which now depends on R and ϑ, and also shows components directed radially and perpendicularly to the galactic plane, the observed line profiles by themselves are not enough. A tentative way out is offered by the theory of the spiral structure of galaxies (Sect. 6.2.2), giving an improved model structure in comparison with circular symmetry (Fig. 5.31).

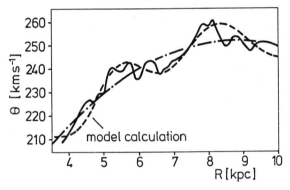

Fig. 5.31. Approximation of the rotation curve derived from observations in the region $l = 22°\ldots70°$ (*continuous*) by a profile (*dashed*) which has deviations from the mean, smooth curve (*chain-dotted*), such as are to be expected from the density-wave theory of the spiral structure. [After Burton (1974)]

Large Scale Density Distribution in the Galactic Plane: $N(r; l, b{\simeq}0°)$ has first been derived from (5.206) on the basis of a mean smoothed profile $\Theta = \Theta(R)$ under the assumption of purely circular orbital motion. The two-valuedness of the relation between v and r appearing in quadrants 1 and 4 — there are always two r-values to a given v-value symmetrically about the tangent point T — can be sorted out by using the emission distribution over galactic latitude b: for constant galactic gas layer thickness the emission from a nearer region on a given longitude extends to a significantly higher latitude b than the emission from a more distant region. We have already presented the result in Fig. 1.12. For the solar neighbourhood $r<4\,\mathrm{kpc}$, primarily in its right-hand (northern) half, it shows similarity with the available results of optical observations (cf. Sect. 3.3.4, Fig. 3.39), namely a local, as well as an inner and an outer arm. Many of the structures here appearing to be oriented radially to the sun, for example at $l \simeq 160°$, must owe their shape to uncertainties of the distance determinations.

The remarkable advance of this chart published in 1958 lay in that for the first time conclusions on the distribution of cosmic matter had been drawn for almost the whole region of the galactic plane outside the centre. The somewhat more recent results mentioned above on the kinematics of the gas show, however, that this picture now needs revision. Model calculations carried out to this end, in which for the first time the influence of deviations from purely circular orbital motion on the line profiles was to be quantitatively investigated, have nevertheless failed to fulfil the hopes of obtaining better results. It appears that the main features of the observed line profiles can already be explained with the assumption of constant density, if one takes as basis a velocity field from the density wave theory of spiral structure. An example is given in Fig. 5.32. The main maximum of this profile at $+60\,\mathrm{km\,s^{-1}}$ arises simply because at this velocity the reversal point of the $r - v$ relation is reached, at which $dv/dr \simeq 0$ and accordingly the intensity $I_v \sim \mathcal{N}'(v) \sim (dv/dr)^{-1}$ must become very large – suitably known as "velocity crowding"! The velocity field is of such considerable influence that conclusions on the large scale structure of the density distribution can be based only on the small differences, partly arising from the cloud

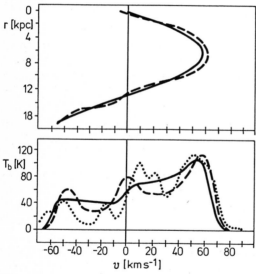

Fig. 5.32. Example of a result of model calculations on the influence of the velocity field on the profile of the 21 cm line. *Above*: two triflingly different profiles of the radial velocity of the gas v (abscissa) with the distance from the observer r (ordinate) for the direction $l = 50°$, $b = 0°$. *Below*: the two corresponding line profiles, calculated under the assumption of *spatially uniform density* $N(\mathrm{H\,I})$, and for comparison the observed line profile (*dotted*). The *full curves* correspond to a model with pure circular orbital motion, the dashed curves make allowance for deviations from this, such as are to be expected from the density-wave theory of the spiral structure. [After Burton (1974)]

structure of the gas, between the observations and the theory for $N = $ constant. Although, for example, the contour chart resulting from Fig. 4.17 by smoothing the small scale irregularities can be reproduced in its main features by the spiral arm model of the density wave theory, it nevertheless does not appear to be possible to deduce reliably the true large scale structure, the so-called "Grand Design", *solely* from the 21 cm observations. For the inner region ($R<R_0$), of course, the significance of the maxima of the radiation flux, integrated over the whole 21 cm line, as tangential directions to the spiral arms provides a sure guideline. These directions are $l = 33°$ (Scutum), $l = 50°$ (Aquila), $l = 283°$ (Carina), $l = 305°$ (Centaurus, Crux) and $l = 327°$ (Norma). They lie on the whole relatively near the maxima of the optical surface brightness of the Milky Way (cf. Sect. 3.3.4, Table 3.13). We return to this in Sect. 5.4.4.

Relatively reliable conclusions can be obtained on the radial distribution of the H I density, referred to the galactic centre, averaged over all galactic longitudes. A recent result is presented in Sect. 5.4.4, Fig. 5.37a. The mean H I density lies at about $0.35\,\mathrm{cm}^{-3}$, with only relatively small variations, between about $R = 4\,\mathrm{kpc}$ and $R = 14\,\mathrm{kpc}$, and falls off rapidly on both sides – in contrast to the number density of the stars, which on the whole increases right up to the centre. For $R>12\,\mathrm{kpc}$ the surface density decreases about exponentially, the H I density appears to show no sharp boundary on the outside.

The total mass of the H I gas in the galactic disc has been estimated at $2.5 \times 10^9\,\mathcal{M}_\odot$.

Thickness and Warping of the Galactic H I Layer: For $R<R_0$ clear conclusions can be reached on the layer thickness in the region of the tangent points T (see Fig. 5.29), using the dependence of the emission at the velocity $v_{\max}(l)$ on the galactic latitude. Outside $R = R_0$ the relation between v and r is single-valued in its main features so that here at fixed l information can be derived on the mean thickness and position of the layer for various distances r. Investigations of this sort have given a systematic increase of the layer thickness for increasing distances form the galactic centre $R>R_0$ and a distortion of the surface of maximal density (Fig. 5.33). One can describe the density distribution perpendicular to the galactic plane approximately by the formula

$$N(z) = N(z_0)\exp\left[-\frac{1}{2}\left(\frac{z - z_0}{h}\right)^2\right] \quad, \tag{5.211}$$

where z_0 denotes the distance of the surface of maximal density from the galactic plane. The effective half-thickness defined by (5.205) is linked with the "scale height" h introduced here by $H^* = \sqrt{\pi/2}\,h = 1.253\,h$. Whereas for $R\leq R_0$ we can in the main take $h\simeq120\,\mathrm{pc}$ and $z_0\simeq0$, for $R>R_0$ we have [after Baker and Burton (1975)] approximately (for $R_0 = 9.5\,\mathrm{kpc}$):

$$h = 0.12 + 0.023(R - R_0)\mathrm{kpc}$$
$$z_0 = 0.12(R - R_0)\cos(\vartheta - 85°)\mathrm{kpc} \quad . \tag{5.212}$$

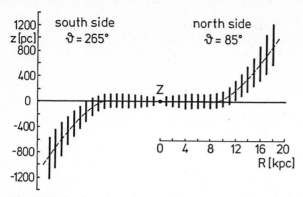

Fig. 5.33. Section through the Milky Way system perpendicular to the galactic plane in the galactocentric longitudes of maximum distortion of the H I layer $\vartheta = 85°$ and $265°$. The ordinate scale (z) is exaggerated by a factor 10 compared with the abscissa scale (R); the lengths of the vertical lines correspond to the thickness $2h$ of the H I layer. The presentation idealises the true proportions: actually the distortions and also the layer thicknesses are somewhat different on the northern and southern sides

Here ϑ denotes the galactocentric longitude, counted in the direction of the galactic rotation, beginning with $\vartheta = 0$ for the direction centre-sun. The layer thickness is already twice as great at $R \simeq 15$ kpc as it is at $R \sim R_0$, and z_0 reaches roughly the values $+600$ pc and -600 pc at $\vartheta \simeq 90°$ and $\vartheta \simeq 270°$, respectively.

The simplest explanation for the bending of the galactic gas layer is based on the assumption of a tidal effect by our neighbour galaxies, the Magellanic Clouds, since the galactocentric longitude of the Large Magellanic Cloud coincides roughly with the direction of maximum distortion on the negative (south) side. Detailed calculations for this are available, which we cannot go into here. In connection with this, some High Velocity Clouds in the line of sight to the Magellanic Clouds have been interpreted as masses of gas which were "torn away" from the galaxy at the closest approach of the Magellanic Clouds, and are now falling back again into the galactic disc ("Magellanic Stream").

H I Distribution in the Galactic Central Region: Direct conclusions from the observations have already been discussed in Sect. 4.1.5 in connection with the schematic Fig. 4.18. The contour charts of $T_b(v, l)$ show in the detail for the central region at $b = 0°$ numerous structural elements, which have been interpreted, for example, as "bars", "spiral arms" or "isolated mass ejections". The picture is considerably simplified, however, if a tilt of the central disc is also taken into consideration. The contour chart is then above all seen to achieve symmetry, if one plots T_b at each longitude l, not for $b = 0°$, but for $b = -l \tan 22°$ (Fig. 5.34). This corresponds to the assumption that we are looking at a disc inclined to the galactic plane at an angle – seen from the sun – of about $22°$ (Fig. 5.35).

The interpretation of this contour chart is possible with two different models (Burton and Liszt, 1978; Liszt and Burton, 1980): (1) with a rotating and

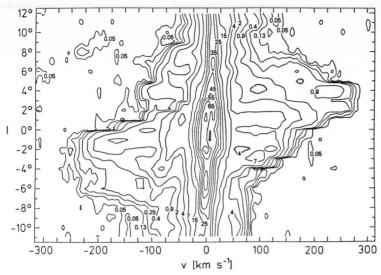

Fig. 5.34. Contour chart of the brightness temperature $T_b(v, l)$ for the 21 cm line emission of the central region from the directions with the galactic latitudes $b = -l \tan 22°$. [By permission of Burton and Liszt (1978)]

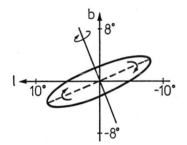

Fig. 5.35. Position of the rotating and expanding disc of neutral hydrogen in the galactic central region, as proposed by Burton and Liszt (1978)

expanding circular disc (radius $\simeq 1.5\,\mathrm{kpc}$, effective half-thickness $H^* \simeq 125\,\mathrm{pc}$ or scale height $h \simeq 100\,\mathrm{pc}$, mean density $\overline{N}(\mathrm{H\,I}) \simeq 0.3\,\mathrm{cm}^{-3}$; rotational velocity $\simeq 175\,\mathrm{km\,s}^{-1}$ and expansion velocity $\simeq 170\,\mathrm{km\,s}^{-1}$, each at a distance from the centre of $1\,\mathrm{kpc}$) or (2) with a nonexpanding, strongly elliptical disc of the same thickness, i.e. a flat cylinder with elliptic base, where the motion follows elliptical streamlines. In both cases the inclination of the disc to the galactic plane must be assumed to be about 25°: the line of intersection of the disc plane and the galactic plane differs somewhat from the line joining the sun to the galactic centre. The assumption of an appropriate bar-like structure of the gas in the galactic central region with elliptical streamlines avoids an improbably high mass flow outwards and has the advantage that it admits a dynamic basis (see Sect. 6.1.4, esp. Fig. 6.5). The total mass of the neutral hydrogen in the region $R < 1.5\,\mathrm{kpc}$ has been estimated at $1 \times 10^7\,\mathcal{M}_\odot$.

5.4.4 Large Scale Distribution of H II Regions and Molecular Clouds

The Region $R > 3$ kpc: From extragalactic stellar systems with spiral structure it is known that the *H II regions* detected optically by their $H\alpha$ emission are often strung like a pearl necklace on rather large arm segments. For the nearest large system, the Andromeda Galaxy M 31, this can be demonstrated by means of nearly 700 emission nebulae. It is therefore to be expected that the spiral structure of our system also is shown up particularly clearly in the distribution of the H II regions. In order to show this one must locate relatively accurately as many as possible of these objects in the whole galaxy.

For the distance determination of a galactic H II region there are essentially two methods. (1) If the exciting star is known, then usually its distance can be derived spectrophotometrically, which in practice gives one the distance of the whole object. (2) If the radial velocity of the H II region can be obtained from measurements of the Doppler displacement of the emission lines (Sects. 4.2.1, 3), then the "kinematic distance" can be derived from foreknowledge of the large scale velocity field. The independence of knowledge of the velocity field gives the first method a fundamental importance. On the other hand only the second method — by observations of radio lines — offers the possibility of advancing significantly beyond distances of about 4 kpc.

Results of the first method produce a similar distribution to that shown in Fig. 3.26 in Sect. 3.3.1 for the Associations and the young open star clusters. For these H II regions with known distances, and a comparable number more, lying outside of $r \simeq 4$ kpc, whose distances were still unknown, the radial velocities have been measured. With a rotation curve — which is not independent of 21 cm

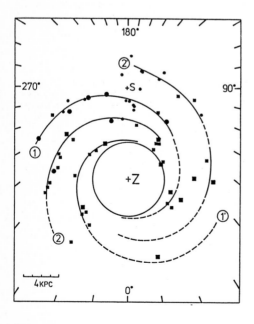

Fig. 5.36. Distribution of 60 "giant H II regions" in the galactic plane. Circular symbols: optically detected H II regions; squares: H II regions with distances from radio observations. Large symbols: excitation parameter $u > 200$ pc cm^{-2}; small symbols: $u = 100\ldots200$ pc cm^{-2}. The four tentatively fitted spiral arms are logarithmic spirals in galactocentric polar coordinates: $R = R^* \exp(a\vartheta)$ with $\vartheta = $ galactocentric longitude. The value chosen for a corresponds to an inclination angle of the spirals of 13.5° from concentric circles (angle ψ in Fig. 6.5). The conventional labels for the spiral arms are: *1* Sagittarius-Carina arm, *2* Scutum-Crux arm, *1'* Norma arm, *2'* Perseus arm. The position of the sun is indicated by S. [After Georgelin, Georgelin and Sivan (1979); by courtesy of the International Astronomical Union]

observations − the distances r of the latter objects can be thereby derived. A result obtained in this way is shown in Fig. 5.36. Since in extragalactic systems the spiral structure is shown up best by the brightest and most extensive H II regions, the objects with excitation parameters $u > 200\,\text{pc cm}^{-2}$ [see (5.95 ff.)] are here given the greatest weight − already with $u > 100\,\text{pc cm}^{-2}$ one speaks of "giant H II regions". The five tangential directions for the inner spiral arms indicated correspond tolerably accurately to the maxima of the total 21 cm emission (see Sect. 5.4.3) and − as is to be expected − also to the maxima of the thermal radio continuum radiation in the decimetre region. Close to these directions, however, also lie the maxima of the surface brightness profile of the optical Milky Way in the constellations Scutum, Aquila, Carina, Centaurus and Norma (cf. Sect. 3.3.4, Table 3.13). The inclination angles of the arm directions to the circular orbit tangents at the same positions amount only to 10° to 12°!

The "local arm" is represented almost only by rather weak H II regions and is therefore not seen here as an independent spiral arm. According to this it could be only a "bridge" between two spiral arms, or a "spur", which runs for a while nearly parallel to the spiral arms and thereby produces the maxima of the optical Milky Way in Cygnus ($l \simeq 73°$) and Vela ($l \simeq 266°$). Such "irregularities" of the large scale structure are well known in many extragalactic spiral systems. We refer to the photograph of the spiral system M 101 = NGC 5457 shown in Fig. 1.3.

Results for the distribution of the H II regions averaged over all galactocentric longitudes, as also for the extended regions of low density ionised hydrogen existing between them − from observations of the "diffuse" or "weak" recombination line emission (Sect. 4.2.3) − are shown in Fig. 5.37b. Average values of the electron density $N_e \simeq N(\text{H II})$ and mass values of the ionised gas can be taken from Table 5.15. The effective half-thickness H^* turns out (at $R \simeq R_0$) to be about 120 pc.

Table 5.15. Radial distribution of the interstellar gas in the galactic disc between $R = 4$ and 12 kpc (mean values for annuli with $\Delta R = 1\,\text{kpc}$). $\mathcal{M}_{\text{neutral}}$ includes the mass of the neutral atomic and molecular hydrogen as well as mass fractions $Y = 0.28$, $Z = 0.02$ for helium and the heavier elements. $\mathcal{M}_{\text{ion}}$ refers to the ionised gas. [After Mezger (1978)]

R-interval [kpc]	$N(\text{H I})$ [cm^{-3}]	$2N(\text{H}_2)$ [cm^{-3}]	N_e [cm^{-3}]	$\mathcal{M}_{\text{neutral}}$ [$10^6\,\mathcal{M}_\odot$]	$\mathcal{M}_{\text{ion}}$ [$10^6\,\mathcal{M}_\odot$]
4–5	0.32	1.56	0.72	475	41
5–6	0.38	1.82	0.62	681	37
6–7	0.35	1.21	0.44	572	22
7–8	0.43	1.13	0.34	664	15
8–9	0.39	0.51	0.28	417	11
9–10	0.40	0.39	0.26	418	11
10–11	0.44	0.24	0.17	396	5
11–12	0.44	0.11	0.14	354	4
4–12				$4.0 \times 10^9\,\mathcal{M}_\odot$	$1.5 \times 10^8\,\mathcal{M}_\odot$

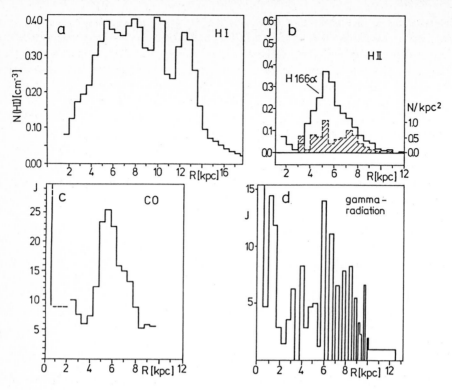

Fig. 5.37a–d. Radial distribution in the galactic plane for various components of the interstellar gas and the diffuse galactic gamma radiation: **(a)** neutral hydrogen [after Burton and Gordon (1978)]; **(b)** emission of the H 166 α line of the diffusely distributed ionised hydrogen (*left scale*: arbitrary units) and number density of "giant H II regions" (*hatched*) per unit surface area in the galactic plane [after Lockman (1975)]; **(c)** CO emission per unit volume (*left scale*: arbitrary units [after Solomon, Sanders and Scoville (1979)]; **(d)** emission per unit volume of photons with energies >100 MeV in arbitrary units [after Caraveo and Paul (1979)]

The large scale distribution of the dense *molecular clouds* was revealed by systematic observations of the 2.6 mm line emission of the CO molecule (Sect. 4.1.6) in an analogous way to the procedure for neutral hydrogen. The $C^{13}O^{16}$ line was used and the optically thin case, $\tau_v \ll 1$, assumed (see Sect. 4.2.7). The further assumption was made of sufficiently high density of the molecular hydrogen, so that from the discussion in Sect. 5.1.2 in connection with (5.32) $T_{ex} = T$, hence LTE may be assumed. Then $\mathcal{N}'_{13}(v)$, the column density of the $C^{13}O^{16}$ molecules with velocity v, in complete analogy with (5.202), is proportional to the brightness temperature $T_b(v)$. If one takes into account the factor of stimulated emission for $T \simeq 10$ K, then the connection reads

$$\mathcal{N}'_{13}(v) = 1.8 \times 10^{15} T_b(v) \quad .$$

The total column density $\mathcal{N}(C^{13}O^{16})$ is given from this by integration over the whole line contour (v in km s^{-1}).

As already mentioned in Sect. 4.1.6, the CO emission is primarily limited to the longitude region $l \simeq 300° \ldots 0° \ldots 60°$, corresponding to galactocentric distances $R \leq R_0$. This also shows up in the fact that the CO-contour chart displayed in Sect. 4.1.6, Fig. 4.20 shows practically no emission with negative velocities between about $l \simeq 10°$ and $40°$, which has to be expected for gas beyond $R = R_0$ (cf. Fig. 5.29) and on the other hand is observed for neutral hydrogen (Sect. 4.1.5, Fig. 4.17). The result shown in Fig. 5.37c for the radial density distribution of the CO confirms these expectations.

The transition to the density of the molecular hydrogen, which is dominant in these clouds, requires knowledge of the ratio $\mathcal{N}(CO)/\mathcal{N}(H_2) \simeq N(CO)/N(H_2)$. For dark clouds with visual extinction values $A_V \simeq 1.5 \ldots 10$ mag one knows that

$$\mathcal{N}(C^{13}O^{16}) \simeq 2.5 \times 10^{15} A_V \, [\text{cm}^{-2}]$$
$$\mathcal{N}(H_2) \quad\quad \simeq 1.25 \times 10^{21} A_V \, [\text{cm}^{-2}] \quad . \tag{5.213}$$

If one selects the isotope ratio $C^{12}/C^{13} = 40$ — derived from observations of the interstellar H_2CO — instead of the terrestrial value ($\simeq 89$), then we get

$$\frac{\mathcal{N}(CO)}{\mathcal{N}(H_2)} \simeq \frac{N(CO)}{N(H_2)} \simeq 8 \times 10^{-5} \quad . \tag{5.214}$$

Mean values of the number density of the hydrogen in molecular form are given in Table 5.15. Investigations of the molecular line emission of typical dense individual clouds lead to a ratio $\mathcal{N}(CO)/\mathcal{N}(H_2)$ which is smaller by a factor of about 4 (Sect. 5.2.4), with a correspondingly higher H_2 density as a result.

The molecular gas is almost completely concentrated in the clouds, which in the region of the galactic plane occupy only about one thousandth of space — for the H I gas, including the warm component, one must allow about 10%. For the *effective half-thickness of the molecular cloud layer* one finds the surprisingly low value $H^* \simeq 65$ pc! The *total mass of the molecular hydrogen* in the galactic disc, about $2 \times 10^9 \mathcal{M}_\odot$, is comparable with the mass of neutral hydrogen atoms given above. The greatest part of this amount is contributed by about 4000 giant molecular clouds with individual masses of about $3 \times 10^5 \mathcal{M}_\odot$.

A prominent characteristic of the radial distribution of the molecular, as of the ionised, gas is the strong *concentration on the annular region* $R \simeq 4 \ldots 8$ kpc, with a pronounced maximum at around $R \simeq 5.5$ kpc, which does not occur in the neutral hydrogen. It is noteworthy moreover that the observations of interstellar gamma radiation described in Sect. 4.1.9 also indicate a radial emission distribution with a maximum at $R \simeq 5$ kpc (Fig. 5.37d). In regions with a high concentration of dense cold clouds the rate of star formation should be increased. This is confirmed, in that the radial distribution of supernova remnants shows a maximum at $R \simeq 5$ kpc.

Molecular Clouds and Ionised Gas in the Galactic Central Region:
The distribution of dense clouds, derived from observations of the molecular

lines, primarily of CO, OH and H_2CO, shows a deep minimum between $R\simeq1$ and 3 kpc. For $R<1$ kpc the average density then increases right up to the centre. Here the ratio $N(CO)/N(H\,I)$ and with it also $N(H_2)/N(H\,I)$ are considerably higher than at $R>3$ kpc: for the mass ratio we have $\mathcal{M}(H_2)/\mathcal{M}(H\,I)>10$.

The contour charts for the brightness temperature $T_b(v,l)$ in the 2.6 mm CO line can be interpreted, like the 21 cm data, with the model of a slightly inclined disc or bar-like structure ($R\lesssim1.5$ kpc). The molecular clouds must therefore follow the same kinematics as the H I gas. The effective half-thickness of the molecular cloud disc must amount at most to about 100 pc; the mean density was estimated at $N(H_2)\simeq10\,\mathrm{cm}^{-3}$. The total mass of the molecular hydrogen in the central region has been estimated at about $10^9\mathcal{M}_\odot$.

Inside $R\simeq150$ pc one has accordingly to reckon with a gas mass (H_2) of about $10^7\mathcal{M}_\odot$, whilst the stars in this region contain around $10^9\mathcal{M}_\odot$ (Sect. 3.3.4, p. 131). The *ratio of gas to star mass* is therefore of the order of 1%, and considerably lower than in the solar region [between about 20 and 30%, see Sect. 6.1.5, (6.77 ff.)].

From the details of recent observations of OH and H_2CO in absorption the following conclusions, amongst others, have been drawn for the nucleus region: (1) the innermost portion of the central rotating gas disc, R less than about 300 pc, seems to be scarcely inclined to the galactic plane and (2) in this region there is a rotating and *expanding* "molecular cloud ring" with the approximate characteristics: radius $\simeq150\ldots200$ pc (probably oval), rotation velocity $\simeq50\ldots60\,\mathrm{km\,s}^{-1}$, expansion velocity $\simeq130\ldots150\,\mathrm{km\,s}^{-1}$. The giant H II region Sgr B2 lies on this ring. The reason for its expansion is still unclear.

Also in the central region the dense molecular clouds – and the "giant H II regions" in part associated with them – occupy only a small fraction of the space near the plane of symmetry. The spaces in between contain ionised gas at relatively low density, observations on which were reported in Sect. 4.2.2 (see especially Fig. 4.40). The model there mentioned of a distribution in the form of a flattened ellipsoid of rotation with semi-axes of 150 and 60 pc implies a total mass $\mathcal{M}_{ion}\simeq2\times10^6\mathcal{M}_\odot$.

5.4.5 Gaseous Galactic Halo

The main mass of the interstellar gas is indeed concentrated in the galactic disc, but there are also considerable quantities of gas between the stars of the galactic halo, with temperatures and densities varying greatly from place to place. The significance of this "gaseous halo" for the dynamics and the energy balance of the gas disc has only recently been recognised. The diffuse clouds in the halo usually appear as high velocity clouds with radial velocities relative to the local standard of rest from $20\,\mathrm{km\,s}^{-1}$ up to about $300\,\mathrm{km\,s}^{-1}$.

The existence of clouds at great distances from the galactic plane was first detected from the appearance of corresponding components of the interstellar Ca II K-lines in the spectra of distant stars at considerable distances from the galactic plane (G. Münch, 1956). Since clouds in empty space would rapidly disperse, or "evaporate", L. Spitzer proposed the assumption of an extremely tenuous, ionised inter-cloud gas at very high temperature ($\simeq 10^6$ K), which produced no interstellar absorption lines in the visible region − at that time the only accessible region − but whose pressure would balance the gas pressure of the cloud. Because of its high temperature such a gas in hydrostatic equilibrium must have a pressure scale height of a few kpc and so form a "galactic corona". The interstellar O VI-lines subsequently discovered in the ultraviolet have confirmed the existence of such a hot gas component, at least in the neighbourhood of the galactic disc. The static picture, however, must be replaced by a dynamic one.

On a larger scale, high velocity clouds at higher galactic latitudes were discovered by the 21 cm emission line components which they produced. An examination of the northern sky revealed nearly 800 clouds. Accordingly the line of sight would encounter H I high velocity clouds over about 10% of the sphere. Since the line of sight passes outside the galactic disc for $r > 1$ kpc if $|b| > 15°$, it must be predominantly halo objects that are involved. A conclusion on the actual distance of a high velocity cloud, however, is not possible from the 21 cm data, for the kinematic method is not applicable here, since the cloud by definition deviates from the rotation law. If one corrects the observed radial velocities v_r for the solar motion around the galactic centre, thus referring them to our local standard of rest, then it appears that both on the northern and the southern sides of the galactic disc the residual velocities are predominantly negative. This result shows that most of the observed cool high velocity clouds are falling in on the galactic disc. The observed distribution of the radial velocities of these clouds on the sphere appears to be readily explained by the following assumptions: participation in the galactic rotation of only 100 km s^{-1}, distances from the galactic plane mainly between 2 and 5 kpc, components of the velocity perpendicularly towards the galactic plane, and also towards the galactic centre, of 100 km s^{-1} (Kaelble, de Boer, and Grewing, 1985).

With high velocity clouds in higher latitudes, which have been detected by their absorption lines in the visible and ultraviolet spectra of halo stars, one has the advantage that upper limits to the distances can be given. Since intrinsically bright stars, such as one needs for obtaining highly resolved spectra, are concentrated on the galactic disc, the number of cases so far investigated is unfortunately still small. Examples are the clouds with velocities of -80 km s^{-1} and -100 km s^{-1}, respectively, found in the UV spectra of bright stars of the globular clusters M 3 ($b = +73°$, $r = 10$ kpc) and M 13 ($b = +41°$, $r = 6$ kpc). Typical is the occurrence of the absorption lines not only of C II, Si II, Mg II, Fe II, which imply low ionisation and predominantly neutral hydrogen, but also of C IV and Si IV, whose occurrence requires temperatures around 10^5 K. The effective height of the clouds above the galactic plane observed by these C IV-lines has been estimated at 3 to 4 kpc. In a few cases high velocity clouds have been detected by 21 cm line emission and also by absorption lines in stellar spectra, as in the directions of the two Magellanic Clouds. Statements on the mean density in the gas halo are uncertain. Values between 10^{-4} and 10^{-2} H atoms cm^{-3} have been estimated.

A simple model representation of the gaseous galactic halo, which can make plausible the co-existence of gas of very different temperatures and the falling of cool clouds towards the galactic plane, takes the formation of the hot corona gas as the starting point. This gas is probably a result of the shock fronts emanating from supernovae, behind which relatively large "bubbles" must form of highly ionised thin gas with temperatures of the order of 10^7 K (see Sect. 6.3.3). In the galactic disc there are primarily this corona gas and the dense cold clouds. It is now to be expected that the thin hot gas is buoyant relative to the gas of the cold clouds and expands. In this way it rises into the halo (Shapiro and Field, 1976). Slight local compressions can here lead to increased cooling and hence through thermal instability to cloud formation in similar manner to that described in Sect. 5.2.3, Fig. 5.10 for the warm H I gas. The further cooled clouds themselves then fall back into the disc. This rising and falling process has been dubbed the "galactic fountain".

6. Dynamics of the Galaxy

6.1 Stellar Dynamics

6.1.1 Posing of the Problem, Fundamental Equations

Stellar dynamics is concerned with the theoretical study of idealised stellar systems: assemblages of point masses which interact with one another according to the laws of classical mechanics and of Newtonian gravitation. The influence of the gravitational field of the interstellar matter or an external field is also taken into account when this appears necessary.

An outstanding problem of stellar dynamics is the construction of a dynamic model of the galaxy: the connection mentioned in Chap. 3 between the spatial distribution and the motion relationships of the stars demands a dynamical interpretation. Whilst the galactic stars themselves in their totality cause the gravitational field, this for its part influences the stellar motions – the interstellar matter is involved in this to an essentially minor extent. If a steady state of the system, unchanging with time in a statistical sense, has already been established, then the motions of the stars are completely governed by the distribution of the gravitational field and certain relationships should exist between the distributions of matter, of the gravitational field and of the stellar velocities.

Further questions of stellar dynamics concern the structure and the cohesion of the open star clusters and the globular clusters. An interesting ancillary problem of galactic dynamics is the explanation of the spiral structure: the spiral arms can be interpreted as density waves in the galactic disc of stars, caused by gravitational interactions.

In the star clusters the assumption of a dynamically steady state turns out to be untenable. Even for the galaxy as a whole, however, this assumption can be fulfilled only approximately, since in the present structure of the system clear indications still survive of development from a protogalactic gas cloud (Sect. 3.5.2). Full understanding of the complicated structure of the galaxy can only be obtained from consideration of the general, time dependent dynamical problem, in which the gasdynamic development of the protogalactic cloud and the subsequent star forming processes are also included.

The basis of stellar dynamics is formed by the mechanics of a system of many point masses. At a certain time t each star of any stellar system under consideration is characterised by its position vector $r(x, y, z)$, its velocity vector $v(u, v, w)$ and its mass $\mathcal{M}$ (first of all considered constant). A system of N stars is represented by N points in seven-dimensional phase space $(x, y, z, u, v, w, \mathcal{M})$. The number density of these points, referred to a quite definite size of volume element in the phase space, is called the *distribution function* $f(r, v, \mathcal{M}, t)$.

The corresponding stellar mass density in three dimensional space is obtained from f by summation or integration over the velocity components and over the stellar masses after multiplication by $\mathcal{M}$:

$$\varrho_s(r, t) = \int \int \int \int f(r, v, \mathcal{M}, t) \mathcal{M} \, du \, dv \, dw \, d\mathcal{M} \quad . \tag{6.1}$$

The corresponding stellar gravitational potential is given by

$$V_s(r, t) = -G \int \int \int \frac{\varrho_s(r', t)}{|r - r'|} dx' dy' dz' \quad . \tag{6.2}$$

This relation is equivalent to Poisson's Equation

$$\Delta V_s = -4\pi G \varrho_s \tag{6.3}$$

with the boundary condition $V \to 0$ for $|r| \to \infty$.

If one chooses the volume element entering the basis of the definition of f sufficiently large compared with the average stellar separation distance, then ϱ_s and V_s have "smooth" mean profiles, similar to the macroscopic quantities of a gas. However, because of the composition of the system from a finite number of discrete individual stars, the actual profiles of f, ϱ_s and V_s show "microscopic" fluctuations in detail from place to place (Fig. 6.1). Correspondingly, by a close encounter with another star, the actual path of any star can be altered strongly in comparison with its behaviour in the case of a smooth potential distribution. Taking into account these encounters complicates the problem considerably.[1]

In the collisionless case[2] the distribution function f satisfies an important partial differential equation. We assume that all stars have the same mass $\mathcal{M}$.

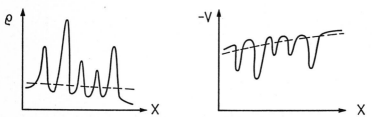

Fig. 6.1. Actual and smoothed profiles (*continuous* and *dashed curves*) of the mass density ϱ and the potential V in a stellar system

---

[1] Genuine collisions between stars, involving actual contact, are in most stellar systems extremely rare, since the radii of stars are always very small in comparison with their mutual separation distances.

[2] One speaks indeed of encounters rather than collisions, but employs the not quite correct terms "collisionless stellar dynamics" and "collisional stellar dynamics".

Furthermore the stars shall undergo only conservative forces, i.e. forces which can be represented by the gradient of a potential function V. If, for the sake of brevity, the components of the position vector $\boldsymbol{r}$ and of the velocity vector $\boldsymbol{v}$ are denoted by x_i and u_i $(i = 1, 2, 3)$, respectively, then the canonical equations of motion for an arbitrarily selected star can be written in the form

$$\dot{x}_i = \frac{\partial H}{\partial u_i} \;, \qquad \dot{u}_i = -\frac{\partial H}{\partial x_i} \tag{6.4}$$

with the Hamiltonian Function

$$H = \frac{1}{2} \sum_{i=1}^{3} u_i^2 + V(x_i, t) \;. \tag{6.5}$$

Here a dot over the symbol indicates differentiation with respect to time. The gravitational potential $V = V_s + V_i$ produced by all the masses of the system, both of stellar and interstellar matter, is linked with the total mass density $\varrho = \varrho_s + \varrho_i$ by Poisson's Equation (6.3).

In the six-dimensional phase space each star, with the starting conditions $x_i = x_i^{(0)}$, $u_i = u_i^{(0)}$ for $t = 0$, describes a "trajectory", corresponding to the solution of the system (6.4).

$$x_i = x_i(t; x_i^{(0)}, u_i^{(0)})$$
$$u_i = u_i(t; x_i^{(0)}, u_i^{(0)}) \;. \tag{6.6}$$

Let us *follow the motion* up to the time $t = t_1$ of the stars contained in a certain volume element of the phase space. With a "smooth" potential function $V(x_i, t)$ the volume of phase space under consideration will indeed be displaced and deformed, but the number of stars contained therein must remain constant, if no encounters take place, by which individual stars are "ejected"!

With the generalised "velocity vector" $\boldsymbol{v} = (x_i, u_i)$ we than have a continuity equation for the density in the phase space, and hence for the distribution function $f(x_i, u_i, t)$:

$$\operatorname{div}(f \times \boldsymbol{v}) = -\frac{\partial f}{\partial t} \tag{6.7}$$

(Liouville's Law). Written out more fully, after differentiation of the products $(f x_i)$ and $(f u_i)$,

$$\sum_{i=1}^{3} \left[\frac{\partial f}{\partial x_i} \dot{x}_i + f \frac{\partial \dot{x}_i}{\partial x_i} \right] + \sum_{i=1}^{3} \left[\frac{\partial f}{\partial u_i} \dot{u}_i + f \frac{\partial \dot{u}_i}{\partial u_i} \right] + \frac{\partial f}{\partial t} = 0 \;. \tag{6.8}$$

From the canonical equations of motion (6.4) we have

$$\frac{\partial \dot{x}_i}{\partial x_i} = \frac{\partial^2 H}{\partial u_i \partial x_i} = -\frac{\partial \dot{u}_i}{\partial u_i} \quad ,$$

so that (6.8) becomes

$$\sum_{i=1}^{3}\left[\frac{\partial f}{\partial x_i}\dot{x}_i + \frac{\partial f}{\partial u_i}\dot{u}_i\right] + \frac{\partial f}{\partial t} = 0 \quad . \tag{6.9}$$

The left-hand side represents the change in f if one follows the motion of the star in the phase space, the so-called total derivative Df/Dt. The assumption of conservative forces means that

$$\dot{u}_i = -\frac{\partial V}{\partial x_i} \quad . \tag{6.10}$$

According (6.9) may be rewritten as

$$\frac{\partial f}{\partial t} + \sum_{i=1}^{3} u_i \frac{\partial f}{\partial x_i} - \sum_{i=1}^{3} \frac{\partial V}{\partial x_i}\frac{\partial f}{\partial u_i} = 0 \quad . \tag{6.11}$$

This relation is often called the *fundamental equation of stellar dynamics*; it corresponds to the collison-free Boltzmann Equation of the kinetic theory of gas or the Vlasov Equation of plasma physics.

The three equations (6.1), (6.3) and (6.11) allow in principle the calculation of the distribution function $f(x_i, u_i, t)$ of the collision-free system, in particular the development from a given starting state $f(x_i, u_i, 0)$.

If a *steady state* has been established, then the derivative of f with respect to time vanishes:

$$\frac{\partial f}{\partial t} = 0 \quad . \tag{6.12}$$

General considerations and numerical calculations lead one to expect that this is already approximately true in the collision-free case if the individual stars have traversed the system a few times or made a few circuits of their orbits around the centre of total mass. If L denotes the linear extent of the system and v an appropriate mean value of stellar velocity, then the *crossing time* is approximately given by

$$t_{\rm cr} = \frac{L}{v} \quad . \tag{6.13}$$

In what follows $t_{\rm cr}$ will be used as the characteristic time constant for the attainment of a steady state. With the help of the virial theorem of classical mechanics $E_{\rm pot} + 2E_{\rm kin} = 0$ we can express v in terms of the total mass of the system $\mathcal{M}_{\rm tot} = N\mathcal{M}$ and L : with the potential energy $E_{\rm pot} \simeq -G\mathcal{M}_{\rm tot}/L$

and the kinetic energy $E_{kin} = \frac{1}{2}\mathcal{M}_{tot}v^2$ it follows that

$$v^2 \approx G\frac{\mathcal{M}_{tot}}{L} = G\frac{N\mathcal{M}}{L} \quad . \tag{6.14}$$

Hence (6.13) becomes

$$t_{cr} \approx \left(\frac{L^3}{GN\mathcal{M}}\right)^{1/2} \quad . \tag{6.15}$$

Values of t_{cr} for typical open clusters and globular clusters, as well as for a large galaxy, are given in Sect. 6.1.2, Table 6.1. In all three cases t_{cr} is small compared with the ages of these systems, if one ignores the youngest open clusters. This means that a steady state, if it is at all possible, is reached relatively rapidly.

If encounters in the sense explained above, play a role, then the stars affected by them do not follow the stream in the phase space, but are "scattered" from it. In the right-hand side of (6.7), and hence also of (6.11), there then appears an additional term $(\partial f/\partial t)_{enc}$, which takes account of the stars "ejected" per unit time, and now in general $(\partial f/\partial t)\neq0$. This can be expressed as follows: the strongly fluctuating portion of the gravitational field has the effect that an arbitrarily selected star describes an irregular zig-zag motion in velocity space. The quantitative theory of this stochastic process leads to an expression for $(\partial f/\partial t)_{enc}$, containing the mean values of the timewise changes of velocity due to close encounters:

$$a_i = \lim_{\Delta t\to 0} \frac{\overline{[u_i(t) - u_i(t + \Delta t)]}}{\Delta t} \quad ,$$

$$b_i = \lim_{\Delta t\to 0} \frac{\overline{[u_i(t) - u_i(t + \Delta t)]^2}}{\Delta t} \quad . \tag{6.16}$$

The a_i are usually known as *coefficients of dynamical friction* and b_i as *diffusion coefficients*. The result is here given without proof

$$\left(\frac{\partial f}{\partial t}\right)_{enc} = -\sum_{i=1}^{3} \frac{\partial(a_i f)}{\partial u_i} + \frac{1}{2}\sum_{i=1}^{3} \frac{\partial^2(b_i f)}{\partial u_i^2} \quad . \tag{6.17}$$

This relation is identical with the Fokker-Planck Equation of classical diffusion theory. It was first applied to treatment of the stellardynamic problem by S. Chandrasekhar (1942).

In the next section we study first of all the question of the conditions under which encounters between stars can have important effects on the distribution function.

6.1.2 Close Stellar Encounters, Relaxation

The distance r_e from a star of mass $\mathcal{M}$, at which this makes a contribution to the gravitational field equal in magnitude to the smoothly varying field produced there by all the other stars of the system, can be estimated roughly from the equation $\mathcal{M}/r_e^2 = \mathcal{M}_G/R_0^2$. Here $\mathcal{M}_G$ denotes the total mass of the galaxy and R_0 the distance of the sun from the galactic centre, here taken as characteristic radius of the system. With $\mathcal{M} = 1\mathcal{M}_\odot$, $\mathcal{M}_G = 10^{11}\mathcal{M}_\odot$ and $R_0 = 10\,\mathrm{kpc}$ it follows that $r_e \simeq 0.03\,\mathrm{pc}$. In the solar neighbourhood the average separation distance of two stars $(\simeq D_0^{-1/3})$ is about 50 times as great. Approaches to distances smaller than or equal to r_e must therefore be relatively rare occurrences. If we consider $r_e = 0.03\,\mathrm{pc}$ as the "encounter radius" and $\sigma_e = \pi r_e^2$ as the corresponding "encounter cross-section", then, e.g. for a relative velocity $v = 20\,\mathrm{km\,s^{-1}}$ and the number density of stars in the solar neighbourhood $D_0 = 0.1\,\mathrm{pc^{-3}}$ (cf. Fig. 5.2), the *"mean encounter time"* τ_e is given by:

$$\tau_e = \frac{1}{\sigma_e v D_0} \approx 2 \times 10^8 \text{ years} \quad , \tag{6.18}$$

and hence a galactic solar orbital period.

Change of the Direction of Motion: The effect of the encounter with a star of mass $\mathcal{M}'$ on the star under consideration of mass $\mathcal{M}$, measured by the deflection produced (or the change of kinetic energy) increases with the duration of the interaction, and so with decreasing relative velocity v. We therefore calculate first of all the deflection angle ψ, through which the direction of motion of the star $\mathcal{M}'$ has changed after the encounter. We assume that $\mathcal{M}'$ previously had velocity v_1 at a great distance $(r = \infty)$, and would, in the absence of all gravitational force, have passed at a minimum distance d ("impact parameter") from the position of $\mathcal{M}$ (cf. Fig. 6.2). We make use of the following relations occurring in the treatment of the two-body problem: The orbit equation of $\mathcal{M}'$ relative to $\mathcal{M}$ (centre of mass located at the origin = focal point F) is generally given in polar coordinates (r, ϕ) by

$$r = \frac{a(1 - e^2)}{1 + e \cos \phi} \tag{6.19}$$

(e = eccentricity, a = major semi-axis of the orbit in the case of an ellipse, $|a| = \overline{\Pi M}$ in Fig. 6.2 in the case of a hyperbola). The conservation of angular momentum per unit mass $\boldsymbol{h}$ can be stated thus:

$$r^2 \dot{\phi} = |\boldsymbol{h}| = \sqrt{\mu a(1 - e^2)} \quad \text{with} \quad \mu = G(\mathcal{M} + \mathcal{M}') \tag{6.20}$$

(G = gravitational constant). For very large r the (constant) angular momentum $|\boldsymbol{h}|$ is given by $v_1 d$ so, from (6.20),

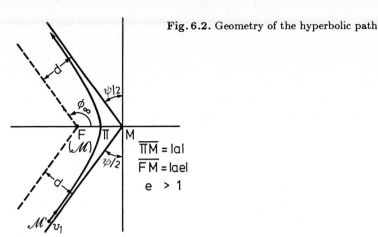

Fig. 6.2. Geometry of the hyperbolic path

$\overline{\Pi M} = |a|$
$\overline{FM} = |ae|$
$e > 1$

$$v_1 d = \sqrt{\mu a (1 - e^2)} \quad . \tag{6.21}$$

The energy law can be written in the form

$$v^2 = \mu \left(\frac{2}{r} - \frac{1}{a} \right) \quad , \tag{6.22}$$

where $v = v(r)$ denotes the magnitude of the orbital velocity ("vis viva integral"). In the case here under consideration with a finite velocity at infinity $v(\infty) = v_1$ the orbit is a hyperbola and one has $e > 1$, so that a must be negative [see (6.20)].

When $r = \infty$, (6.22) gives

$$-\mu/a = v_1^2 \quad .$$

Hence, substituting for a in (6.21), we get

$$\sqrt{e^2 - 1} = (v_1^2 d)/\mu \quad . \tag{6.23}$$

By means of (6.19) the half angle of deflection $\psi/2$ can be expressed in terms of e : when $r = \infty$, $1 + e \cos \phi_\infty = 0$ or $\cos \phi_\infty = -1/e$, whilst from Fig. 6.2 evidently $\psi/2 = \phi_\infty - 90°$. One accordingly obtains

$$\tan \frac{\psi}{2} = -\cot \phi_\infty = -\frac{\cos \phi_\infty}{\sqrt{1 - \cos^2 \phi_\infty}} = \frac{1}{\sqrt{e^2 - 1}} \quad , \tag{6.24}$$

which, using (6.23) and (6.20), gives

$$\tan \frac{\psi}{2} = \frac{\mu}{v_1^2 d} = \frac{G(\mathcal{M} + \mathcal{M}')}{v_1^2 d} \quad . \tag{6.25}$$

371

For the case $M = M' = 1\mathcal{M}_\odot = 2 \times 10^{33}$ g, $d = r_e = 0.03\,\mathrm{pc} = 10^{17}$ cm, and $v_1 = 20\,\mathrm{km\,s^{-1}} = 2 \times 10^6\,\mathrm{cm\,s^{-1}}$ this relation gives $\tan \psi/2 = 6.7 \times 10^{-4}$, corresponding to $\psi \simeq 5$ arc minutes.

In conjunction with the result for the mean "encounter time" τ_e (6.18), this shows that in the solar neighbourhood significant deflections are to be expected only after repeated encounters and many revolutions of the galaxy. The deflection $\psi = 90°$ in a single encounter would under the same assumptions require $d = 4.5\,\mathrm{AU}$. In the case of an approach to the sun, the star would accordingly need to penetrate as close as the orbit of Jupiter. For this close approach (6.18) gives $\tau_e \simeq 3 \times 10^{14}$ years!

Relaxation: On its path through the field of the other stars, a given star after n encounters experiences a total deflection, compared with its original direction of motion, of

$$\psi = \sum_{i=1}^n \psi_i \; . \tag{6.26}$$

Even if the individual deflections ψ_i are mainly small, after a sufficiently long period of observation of a "test star" ψ will no longer be negligible. The accumulated effect of the "encounters" between individual stars upon their orbits and hence upon the state of the whole system is known as relaxation. We shall consider the ψ_i as random variables with magnitudes $\ll 1$ and a mean value $\overline{\psi_i} = 0$. Then also $\overline{\psi} = 0$, but the mean squared deviation $\overline{(\psi - \overline{\psi})^2} = \overline{\psi^2} = \sum \overline{\psi_i^2}$ has a finite value. The time t_r, after which $(\sum \overline{\psi_i^2})^{1/2} = 1$ radian is attained, is called the *relaxation time*.[3]

If all the stars of the system have the same mass $M' = M$, then (6.25) gives $\psi_i = \psi(d) \simeq 4GM/v^2 d$. The number of encounters in the time t with an encounter cross-section πd^2 in a number density of encounter partners D (cf. Fig. 5.2) is given by $n(d) = \pi d^2 Dvt$. The relaxation time $t = t_r$ can then be estimated from the condition

$$\sum \overline{\psi_i^2} = \psi^2(d) = 1$$

with the result

$$t_r \approx \frac{v^3}{16\pi G^2 \mathcal{M}^2 D} \; . \tag{6.27}$$

The impact parameter d cancels out in this approximation. The more exact treatment introduces into the denominator of this expression the additional factor $\simeq \frac{1}{4} \ln N \simeq 0.6 \log N$, which however does not reach the value 3 before $N =$ total number of stars in the system $\simeq 10^5$ (Henon, 1973).

---

[3] A similar definition starts from the variations in the kinetic energy ΔE_i, and requires that $[\sum (\Delta E_i)^2]^{1/2} = E_0 =$ the original kinetic energy.

By elimination of v with the help of (6.14) in Sect. 6.1.1 and with $D \simeq N/ (\pi/6)L^3$, where L = linear extent (diameter) of the system, it follows from (6.27) that

$$t_r \approx \frac{1}{24} \left(\frac{NL^3}{GM} \right)^{1/2} \frac{1}{\ln N} \quad . \tag{6.28}$$

Here we have taken into account the correction factor $\frac{1}{4} \ln N$ in the denominator mentioned above. A numerical example for t_r is contained in Table 6.1. The relaxation times for open clusters, as also for globular clusters, are quite short compared with the stellar evolution times. Each cluster member therefore experiences numerous encounters during its evolution. The orbits of the "field stars" in the galaxy, on the other hand, should have been scarcely affected since the formation of the system.

Table 6.1. Crossing time t_{cr} and relaxation time t_r for three different stellar systems consisting of stars of equal mass $\mathcal{M} = 1\mathcal{M}_\odot$. N = total number of stars, $\mathcal{M}_{tot} = N \cdot \mathcal{M}$ total mass, L = linear dimension (diameter)

Quantity	Open cluster	Globular cluster	Galaxy
N	10^3	2×10^5	10^{11}
$\mathcal{M}_{tot} [\mathcal{M}_\odot]$	10^3	2×10^5	10^{11}
L [pc]	2	10	10^4
t_{cr} [years]	2×10^6	1.5×10^6	7×10^7
t_r [years]	8×10^6	7×10^8	7×10^{15}

An important quantity is the ratio formed from (6.28) and (6.15)

$$\frac{t_r}{t_{cr}} \approx \frac{N}{24 \ln N} \quad . \tag{6.29}$$

For large N the relaxation time t_r is thus much greater than the crossing time t_{cr}.

One should not simply conclude from this and from the numbers for t_r itself that a steady state with a Maxwellian velocity distribution must have been reached in the star clusters. Especially in the open clusters with low member populations and correspondingly low total mass, close encounters can lead relatively frequently to the exceeding of the escape velocity and thereby to significant deviations from the behaviour of an enclosed gas. In the smaller systems the encounters cause a slow variation with time of the distribution function, which thus passes through a sequence of quasi-steady states. In large stellar systems, on the other hand, the time scale of the effect of encounters is so great that its influence can be neglected.

From the standpoint of stellar dynamics one can therefore divide stellar systems into two classes: "collisionless" systems and systems in which encounters play an important role. The first class comprises chiefly the galaxies, in which collective effects dominate. To the second class belong the stellar associations, the open clusters and the globular clusters.

6.1.3 Stellar Dynamics Taking Account of Encounters

The determination of the distribution function $f(\boldsymbol{r}, \boldsymbol{v}, \mathcal{M}, t)$ for a self-gravitating system of many stars using analytical methods comes up against formidable, partially still unclarified, mathematical difficulties. This is also true to a certain extent for the fundamentally simpler collisionless case. In order to obtain general theoretical statements for comparison with the observational results, various simplifications are necessary, whereby the validity of the result may be considerably curtailed. Since the availability of fast computers with large storage capacity, however, it has become possible to integrate numerically the equations of motion for the stars of a whole system, and thus to simulate the system. One speaks of *numerical experiments* in contrast to analytical theory. Since this method is simpler in principle than the analytical treatment and is today coming increasingly to the fore, we give a discussion of it at the outset. We restrict ourselves therein to the open star clusters.

Numerical Simulation of Star Clusters: The treatment of the dynamics of star clusters by direct numerical integration of the equations of motion of the N-body problem:

$$\ddot{\boldsymbol{r}}_k = -\frac{\partial V}{\partial \boldsymbol{r}_k} = -\sum_{\substack{j=1 \\ j \neq k}}^{N} \frac{\mathcal{M}_j(\boldsymbol{r}_k - \boldsymbol{r}_j)}{|\boldsymbol{r}_k - \boldsymbol{r}_j|^3} \qquad k = 1, \ldots, N \tag{6.30}$$

without smoothing of the potential profile was carried out for the first time in 1960 by von Hoerner and in recent years has become possible for numbers of cluster members up to $N \simeq 10^3$, and thus for typical open clusters. For this purpose one usually writes these second order differential equations in the equivalent form

$$\dot{\boldsymbol{r}}_k = \boldsymbol{v}_k, \quad \dot{\boldsymbol{v}}_k = -\frac{\partial V}{\partial \boldsymbol{r}_k} \quad , \qquad k = 1, \ldots, N \quad . \tag{6.31}$$

These $6N$ equations of the first order can be solved numerically by integration in sufficiently short steps. For the starting values one chooses the most realistic configuration possible.

For large N the calculation primarily of the expression for the acceleration in the right-hand side of (6.30) requires much computation time. The same is true for the calculation of the outcome of close encounters, which become particularly frequent if a relatively dense nucleus of the system develops. To overcome these numerical problems many methods have been developed, which we cannot go into here. We refer to Wielen (1974) who also investigated the problem of the "microscopic" instability of individual numerical solutions, which we ignore in our presentation. Realistic model calculations take into account also the influence of an external gravitational field (tidal influence of the galactic field; passage of interstellar clouds) and the effects of stellar evolution (mass loss of evolved stars!).

The available *results* of numerical integrations for systems with $N \leq 10^3$ can be qualitatively summarised as follows:

1) The development of the simulated open star clusters is dominated by encounters. Stars are continually exceeding the escape velocity and being lost from the system: the number of these "escapes" per crossing time, $\dot{N}t_{cr}$, is from 1 to 10, where the larger values occur when the external galactic field is taken into account (lowering of the escape velocity!). The establishment of a long-term steady state is thereby prevented.

2) A cluster develops a nucleus-halo structure. The approximately radial motions of the "halo stars" (in strongly eccentric orbits) are nearly "collision-less", whilst encounters between the stars of the nucleus region are relatively frequent.

3) If instead of equal masses for all stars one chooses a mass spectrum of the form $F(\mathcal{M}) \sim \mathcal{M}^{-\beta}$, with, say, $\beta = 2.35$ [Sect. 3.2.2, (3.38 f.)], then the dynamic development of the cluster is accelerated and its "evaporation time" is shortened. In the encounters between unequal masses the kinetic energies are not equalised; the larger masses preferentially stay in the nucleus, whilst more small masses are found outside. The mean velocity of the stars decreases from the inside to the outside, i.e. the star clusters are not "isothermal".

4) The escapers take kinetic energy with them, so that the remainder of the system is more strongly bound. In the nucleus a few binary stars develop, which take up the greater part of the binding energy of the system. Since no long-term stable system with more than two components of comparable mass is known, the end state of the evolution appears to be a single close binary system. Several binary stars will not persist in the nucleus region for long, since triple encounters are relatively probable there. Massive stars with abnormally high spatial velocities, so-called runaway stars, can also be formed in this way.

5) A slow rotation of the whole system does not change these results appreciably.

The calculated "evaporation" time t_{ev} for given values of the total mass $\mathcal{M}_{tot}$ and fixed stellar masses $\mathcal{M} = 0.5\mathcal{M}_\odot$ (corresponding to the mean mass of the stars of the Hyades) are displayed graphically in Fig. 6.3 as a function of the cluster radius R. The smaller R is, so much greater is the density in the cluster for a given $\mathcal{M}_{tot}$, and so much more quickly the encounters between the stars effect its evaporation. Above a certain radius (dependent on $\mathcal{M}_{tot}$), which for $\mathcal{M}_{tot} = 500\mathcal{M}_\odot$ is about $R = 0.3\,\mathrm{pc}$, the external gravitational field (for the region of the solar neighbourhood) causes a marked shortening of t_{ev} compared with the behaviour (shown by a dashed curve) in the absence of the external field. Finally, for systems with $R > 1\,\mathrm{pc}$, a reduction of the lifetime is

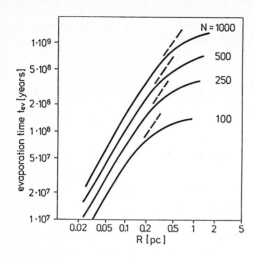

Fig. 6.3. Evaporation time t_{ev} of open star clusters as function of the cluster radius R for various member numbers N, when the mass of the individual star is taken as $\mathcal{M} = 0.5\mathcal{M}_{\odot}$ (schematic, after Wielen, 1974)

also to be expected from encounters with interstellar clouds as well as from the differential rotation.

We restrict the *comparison with observations* of open star clusters to a discussion of the lifetimes. From the frequency distribution of the ages of galactic open star clusters (as derived from their CM-diagrams), in conjunction with the assumption of a constant formation rate of clusters with time, the distribution of actual lifetimes can be inferred. In this way one finds that generally about 50 % of all open clusters must dissolve within 2×10^8 years; only 2% survive longer than 1×10^9 years (Wielen, 1971).

This indirect observational result can be explained at least qualitatively by the results of the numerical calculations displayed in Fig. 6.3: if one takes for the average open cluster the plausible values $L = 2R \simeq 0.6\,\mathrm{pc}$, $N = 250$, $\mathcal{M} = 0.5\mathcal{M}_{\odot}$, then it follows that $t_{ev} \simeq 2 \times 10^8$ years.

Analytical Theory: The fundamental equation (6.11) formulated in Sect. 6.1.1, supplemented on the right-hand side by an encounter term of the form (6.17), proves so complicated that a solution for the time-dependent distribution function $f(x_i, u_i, \mathcal{M}, t)$ even in the case of spherical symmetry and using modern numerical methods, encounters great difficulties (computation time!). It is therefore customary to impose various simplifications. Appropriate assumptions are, for example: (1) all star masses are equal, (2) the density in the cluster is homogeneous, (3) the velocity distribution is isotropic. In view of the results of the numerical simulations described above, all three assumptions are extremely unrealistic. If one ignores special cases, results obtained in this way can therefore scarcely be suitable for detailed comparison with observations.

We can, however, draw some conclusions on the general outcome of the dynamic development of a star cluster, such as we have already sketched in Sect. 6.1.2 in connection with (6.29). The statements agree with the results of the numerical experiments and can be summarised as follows: every gravitationally bound star cluster, which is not originally in a state of dynamical equilibrium, passes into a quasi-steady state after a crossing time t_{cr}. An equilibrium statistically invariant with time cannot be established by the relaxation effect of encounters with a time scale $t_r > t_{cr}$. The cluster develops towards an end state in which nearly all the stars have escaped, down to at least one binary system.

6.1.4 Collisionless Dynamics of the Galactic Stellar System

Numerical Simulation: The starting point is again provided by the system of $6N$ ordinary differential equations of the first order (6.31), where now, however, a smoothed profile is used for the potential function $V(x_i, t)$ to be calculated from Sect. 6.1.1, (6.2). For the gravitational field, therefore, a relatively coarse-meshed representation can be chosen, with fewer lattice points than stars present. The computational labour is so much reduced thereby, that systems with total star numbers up to a few times 10^5 can nowadays be numerically simulated.

The problem had first been treated for a two-dimensional disc ($\sim$ 1970: R.H. Miller, K.H. Prendergast, W.J. Quirk, F. Hohl and others). Meanwhile results were also obtained for the three-dimensional case and even a spherical or ellipsoidal component was taken into account. We mention in the following a few important results without, however, going into the numerical methods used in obtaining them:

1) If one chooses as the starting state of the system a rotating disc uniformly filled with stars, all following circular orbits in equilibrium with the gravitational force, then the system very quickly becomes unstable. First an irregular small-scale structure develops with strong density variations. The system almost disintegrates into individual elements. Within about one rotation thereafter there forms an eddy-like overall structure with a longlived large central "bar" (Fig. 6.4).

2) If one assumes at the start a sufficiently large (radial) velocity dispersion of the stars compared with the circular orbital motion, then the small-scale strong density fluctuations do not appear, and a certain stabilisation of the system occurs. Later, however, the bar structure appears. After several rotations this takes the form of an oval or bar-shaped central concentration, which is superimposed on a stationary disc with approximately exponential density fall off. In the galactic disc simulated in these calculations there often resulted a velocity dispersion greater by at least a factor of two than the corresponding observed values summarised in Sect. 3.5.1, Table 3.14 − the system was too "hot".

3) The formation of a large central bar in the disc can be prevented if one assumes in addition a spherical or ellipsoidal stellar mass distribution with a density increasing towards the centre (central bulge or halo). The stabilising effect can be clearly traced to the fact that each star is now linked by a

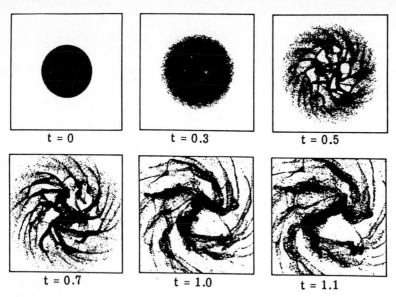

| t = 0 | t = 0.3 | t = 0.5 |

| t = 0.7 | t = 1.0 | t = 1.1 |

Fig. 6.4. Numerical simulation of the development of a disc-shaped rotating stellar system. Time t in units of rotation period. (After Hohl, 1975; by courtesy of the International Astronomical Union)

stronger force to its equilibrium orbital radius − a step in the direction of the case of planetary motion under the dominating influence of the solar mass, as opposed to the situation in a pure disc-galaxy, whose gravitational field is not dominated by a single central superstar! The least mass required for the whole approximately spherical component depends on the mass distribution in this, as also in the disc, and is found to be between 60 % and 70 % of the total mass.

Examples of results of this are shown in Fig. 6.5. The system includes 25 000 stars in the disc and 25 000 stars in the halo. For the calculation of the gravitational field $30 \times 30 \times 14$ lattice points were chosen, whose mutual distances apart of 2 kpc correspond to a galactic total mass of $1.84 \times 10^{11} \mathcal{M}_\odot$. The shape of the halo corresponded at the start of the calculation to a spheroid with flattening ratio of 1:1.4, whilst the disc was flattened in the ratio 1:30. The halo had a diameter of 30 kpc in the galactic plane; the diameter of the disc was of similar size (see Fig. 6.5). For the initial rotation curve a similar profile to that shown in Fig. 5.30 was chosen, without, however, taking account of the peak at $R \simeq 0.7$ kpc.

Since our Milky Way system appears to show no barred spiral, the mass of the bulge-halo component should make up more than 60 % of the total mass. In the direct determinations up to now, however, the halo stars contribute only about 20 % (Sect. 3.3.3). Possibly, however, the central (bulge-)mass is so high that the result altogether is over 60 % (cf. Sect. 6.1.5, Table 6.4). This question is still today unresolved.

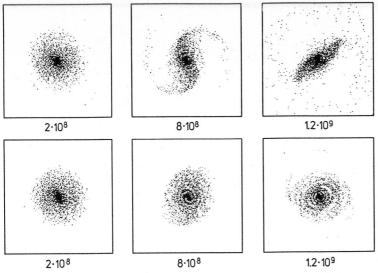

Fig. 6.5. Numerical simulation of a stellar system that consists of a rotating disc and a stellar halo strongly concentrated at the centre. Mass ratio of disc: halo is $0.5:0.5$ (*top*); $0.3:0.7$ (*bottom*). Only the disc stars are shown. Frame side length 60 kpc. Rotation counter-clockwise. (After Sellwood, 1980)

Approach of the Analytical Theory: If one treats the potential function $V(x_i, t)$ as given, then the fundamental equation (6.11) is a linear homogeneous partial differential equation of the first order for the distribution function $f(x_i, u_i, t)$. The characteristics are just the six equations of motion (6.31) for a selected star:

$$\dot{x}_i = u_i , \qquad \dot{u}_i = -\frac{\partial V}{\partial x_i} . \tag{6.32}$$

This system of ordinary differential equations has in general six independent first integrals, which are constant along an individual stellar trajectory in phase space: $\phi_1(x_i, u_i, t) = c_1, \ldots, \phi_6(x_i, u_i, t) = c_6$. The functions with these properties are called integrals of the motion. The six constants ("orbital elements") can each be obtained by substituting the initial values of x_i and u_i in these functions; the x_i and u_i of the star under consideration are thereby determined as functions of t and the constants $c_1, \ldots, c_6$. From the theory of first order partial differential equations the general solution of (6.11) is given by

$$f = F(\phi_1, \ldots, \phi_6) , \tag{6.33}$$

where F is an arbitrary positive function of the six arguments. Also f is accordingly an integral of the motion (Jeans's Theorem).

A limitation on the included integrals is the requirement that for a physically meaningful solution arbitrarily close trajectories in the phase space must

379

not be associated with arbitrarily different constants or values of f (Kuzmin, 1953; Lynden-Bell, 1962).

For a dynamically *steady state*: $\partial f/\partial t = 0$, $V = V(x_i)$ the variable t disappears, and there are only five first integrals. One integral is the energy law

$$\phi_1(x_i, u_i) = \frac{1}{2} \sum_{i=1}^{3} u_i^2 + V(x_i) = E = c_1. \tag{6.34}$$

In *the case of an axially symmetric mass distribution* three further integrals ϕ_2, ϕ_3, ϕ_4 correspond to the conservation law for angular momentum about the axis of symmetry. Transformation into cylindrical coordinates reduces them to *one* function I_2 with *one* parameter C_2. The remaining integral is therefore known as the *Third Integral I_3*. It was first shown by numerical calculations of three-dimensional stellar trajectories that such a Third Integral exists, and so the distribution function does not in general have the form $f = F(I_1, I_2)$ with $I_1 \equiv \phi_1$.[4] According to the numerical studies, I_3 primarily imposes limitations on the velocity components perpendicular to the plane of symmetry. For the characterisation of the galactic orbit of a star one therefore needs, as well as the total energy $E = c_1$ and the angular momentum C_2, a third independent parameter C_3; the complete solution of the axially symmetric problem must have the form $f = F(I_1, I_2, I_3)$. The force field determining the motions is given by the potential $V(x_i)$ with (6.1) and (6.2) or (6.3) given in Sect. 6.1.1.

In this general form, however, the problem cannot be solved. Solutions are obtained only for special cases. For extremely flattened axially symmetric systems, for example, we may set $f = F(I_1, I_2)$. Conversely, therefore, one puts plausible assumptions for $f(x_i, u_i)$ into the fundamental equation (6.11) with $\partial f/\partial t = 0$, in order to discover which special forms are dynamically permissible. From the empirical results displayed in Sect. 3.3.4, the assumption suggested itself of a general Gaussian distribution of the peculiar velocities $u_i - \overline{u_i}$:

$$f = D(x_i)\exp\left\{-\frac{1}{2} \sum_{i,k=1}^{3} a_{ik}(u_i - \overline{u}_i)(u_k - \overline{u}_k)\right\} \quad . \tag{6.35}$$

For steady axially symmetric systems, in which the potential $V(x_i)$ is derived solely from the stars (self consistency), G.L. Camm (1941) and more thoroughly W. Fricke (1951) showed that the requirement $V \sim 1/R$ for arbitrarily increasing central distance $R \to \infty$ cannot be fulfilled with ellipsoidal distribution functions of the type (6.35). Such an assumption is a possible solution only for infinite systems with infinite mass. This is understandable, since a Gaussian distribution allows arbitrarily high velocities.

An ellipsoidal distribution function of the type (6.35) is however appropriate as a local approximation to the solution of the fundamental equation. We

---

[4] This assumption was made by Jeans in 1916, after no further integrals had been found by analytical methods.

380

discuss this as an example of the *interpretation of the observed velocity distribution in the solar neighbourhood*. Here it is convenient to transform at once to cylindrical coordinates (R, ϑ, z) by the transformation[5] $x_1 = x = R \cos \vartheta$, $x_2 = y = R \sin \vartheta$, $x_3 = z$ with the introduction of the velocity components in the radial, tangential and z-directions (unit vectors e_R, e_ϑ, e_z):

$$\Pi = \dot{R} \; , \qquad \Theta = R\dot{\vartheta} \; , \qquad Z = \dot{z} \; . \tag{6.36}$$

Then the fundamental equation (6.11), under the steady state assumption, assumes the form:

$$\Pi \frac{\partial f}{\partial R} + \frac{\Theta}{R} \frac{\partial f}{\partial \vartheta} + Z \frac{\partial f}{\partial z} + \dot{\Pi} \frac{\partial f}{\partial \Pi} + \dot{\Theta} \frac{\partial f}{\partial \Theta} + \dot{Z} \frac{\partial f}{\partial Z} = 0 \; . \tag{6.37}$$

Here $\dot{\Pi}, \dot{\Theta}$ and $\dot{Z}$ are given by the equations of motion

$$\dot{\Pi} = \frac{\Theta^2}{R} - \frac{\partial V}{\partial R} \; , \qquad \dot{\Theta} = -\frac{\Pi \Theta}{R} - \frac{1}{R}\frac{\partial V}{\partial \vartheta} \; , \qquad \dot{Z} = -\frac{\partial V}{\partial z} \; . \tag{6.38}$$

Then the acceleration components b_R and b_ϑ in the directions of the unit vectors e_R and e_ϑ are given in general by (see textbooks on theoretical physics)

$$b_R = \ddot{R} - R\dot{\vartheta}^2 \quad = -\frac{\partial V}{\partial R}$$

$$b_\vartheta = R\ddot{\vartheta} + 2\dot{R}\dot{\vartheta} \quad = -\frac{1}{R}\frac{\partial V}{\partial \vartheta}, \tag{6.39}$$

where we have introduced $\dot{\Pi} = \ddot{R}$ and $\dot{\Theta} = d(R\dot{\vartheta})/dt = \dot{R}\dot{\vartheta} + R\ddot{\vartheta} = b_\vartheta - \dot{R}\dot{\vartheta} = b_\vartheta - \Pi\Theta/R$.

We now use the expression (6.35) in the form of an ellipsoidal distribution with arbitrarily oriented principal axes in cylindrical coordinates:

$$f(R, z, \Pi, \Theta, Z) = D_s(R, z)\frac{hkl}{\pi^{3/2}}\exp[-h^2\Pi^2 - k^2(\Theta - \overline{\Theta})^2$$

$$- l^2 Z^2 - m\Pi(\Theta - \overline{\Theta}) - n\Pi Z - p(\Theta - \overline{\Theta})Z] \; . \tag{6.40}$$

Here $\overline{\Theta}$ denotes the local mean value of the tangential velocity component ($\overline{\Pi} = 0$, $\overline{Z} = 0$). The coefficients $h^2 = [2\overline{\Pi^2}]^{-1}$, $k^2 = [2\overline{(\Theta - \overline{\Theta})^2}]^{-1}$, $l^2 = [2\overline{Z^2}]^{-1}$ and m, n, p may be functions of R and z. If one substitutes (6.40) in the fundamental equation (6.37) with (6.38) and assumes *axial symmetry*, thus setting all derivatives with respect to Θ equal to zero, then after some rather lengthy elementary manipulation and division by the common factor f one obtains an expression which can be brought into the form

---

[5] Instead of R, the symbol ϖ (a "curly" π), originating with Jeans, is often used.

$$A + B\Pi + C\Theta + DZ + E\Pi^2 + F\Pi\Theta + \ldots + T\Pi\Theta Z = 0 \quad .$$

Since this equation must be satisfied for arbitrary values of Π, Θ and Z, it follows that $A = 0$, $B = 0$, $C = 0$ etc. From this the condition equations for the coefficients h, k, l, m, n and p can be obtained, of which we shall here need only the following:

$$m = 0 \ , \qquad p = 0 \ , \qquad \frac{\partial k^2}{\partial z} + \frac{n}{R} = 0 \tag{6.41}$$

$$\frac{\partial h^2}{\partial R} = \frac{\partial l^2}{\partial z} = 0 \ , \qquad \frac{\partial h^2}{\partial z} + \frac{\partial n}{\partial R} = 0 \ , \qquad \frac{\partial l^2}{\partial R} + \frac{\partial n}{\partial z} = 0 \tag{6.42}$$

$$\frac{\partial k^2}{\partial R} - \frac{2(k^2 - l^2)}{R} = 0 \ , \qquad \frac{\partial(k^2\overline{\Theta})}{\partial R} - \frac{k^2\overline{\Theta}}{R} = 0 \quad . \tag{6.43}$$

Since $z = 0$ will be the plane of symmetry, it must be generally true of $k(R, z)$ that $\partial k^2/\partial z = 0$ when $z = 0$. We thus show that in the galactic plane $m = n = p = 0$, i.e. the principal axes of the velocity ellipsoid must lie in the directions of the unit vectors e_R, e_ϑ, e_z – in qualitative agreement with observation.

The first equation (6.42) states that $\overline{\Pi^2} = [2h^2]^{-1}$ must be independent of R and $\overline{Z^2} = [2l^2]^{-1}$ independent of z.

Equations (6.43) permit an important statement on the ratio of the velocity dispersions in the radial and tangential directions Σ_Π and Σ_Θ : differentiation of the product $k^2\overline{\Theta}$ in the right-hand equation and substitution of the expression obtained above for $\partial k^2/\partial R$ in the left-hand equation gives

$$\frac{h^2}{k^2} = \frac{1}{2}\left[1 + \frac{R}{\overline{\Theta}}\frac{\partial\overline{\Theta}}{\partial R}\right] \quad . \tag{6.44}$$

For the position of the sun $R = R_0$, $z = 0$ we can express the right-hand side in terms of Oort's Constants A and B by means of (3.66). Since $\omega(R) = \overline{\Theta}(R)/R$ we have

$$\omega_0 = \frac{\overline{\Theta}_0}{R_0} \ , \qquad \omega_0' = \left(\frac{\partial\omega}{\partial R}\right)_{R_0} = \frac{1}{R_0}\left(\frac{\partial\overline{\Theta}}{\partial R}\right)_{R_0} - \frac{\overline{\Theta}_0}{R_0^2} \tag{6.45}$$

and it is easily verified that the right-hand side of (6.44) is equal to $-B/(A-B)$. We write the result as

$$\frac{\Sigma_\Theta^2}{\Sigma_\Pi^2} = \frac{\overline{(\Theta - \overline{\Theta})^2}}{\overline{\Pi^2}} = \frac{-B}{A-B} \quad . \tag{6.46}$$

For rigid rotation $\overline{\Theta} \sim R$ it follows from (6.44) that $\Sigma_\Theta^2/\Sigma_\Pi^2 = 1$, whence $A = 0$, whilst if $\overline{\Theta} = $ constant then $\Sigma_\Theta^2/\Sigma_\Pi^2 = \frac{1}{2}$. The observations give values which lie between these two extremes: with the numerical values (3.75) one obtains $\Sigma_\Theta^2/\Sigma_\Pi^2 = 0.40$, essentially in accordance with the mean of the empirically determined values Σ_V^2/Σ_U^2 of 0.42 ± 0.06[6] for Main Sequence stars taken from Table 3.9.

Epicyclic Stellar Orbits and Velocity Ellipsoid: The fact that the major axis of the velocity ellipsoid of the stars with peculiar velocities $< 60\,\mathrm{km\,s^{-1}}$ lies approximately in the radial direction, i.e. $\Sigma_\Theta/\Sigma_\Pi < 1$, was explained by B. Lindblad (1927) in a less formal way, namely as a direct consequence of small deviations from purely circular orbital motion.

Let us therefore study the motion of a star in the galactic plane in relation to a rectangular coordinate system (ξ, η) with origin at $R = R_0$, which partakes in the circular orbital motion with the angular velocity ω_0, where the ξ-axis points continuously towards the anticentre, and the η-axis in the direction of motion (Fig. 6.6a). Let $\Theta_K(R)$ denote the velocity of the circular orbital motion as a function of the distance R from the centre, so that $\Theta_K^2(R)/R = -K_R$. In particular, let us write $\Theta_K(R_0) = \Theta_{K0}$.

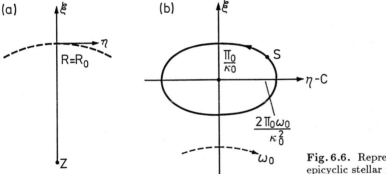

Fig. 6.6. Representation of epicyclic stellar motion

At time $t = 0$ the star is at $\xi = 0$, $\eta = 0$, but it has a finite *radial velocity component* $\dot\xi = \Pi_0 \ll \Theta_{K0}$, whilst initially $\dot\eta = \Theta_0 - \Theta_{K0} = 0$. Then after a finite time we have $\xi = R - R_0 \neq 0$ and also $\dot\xi = \Pi \neq \Pi_0$ and $\dot\eta = \Theta - \Theta_{K0} \neq 0$. The radial acceleration at time t is given by

$$\ddot\xi = \ddot R = \frac{\Theta^2}{R} - K_R = \frac{\Theta^2}{R} - \frac{\Theta_K^2(R)}{R} \quad . \tag{6.47}$$

Using the conservation of angular momentum $R\Theta = R_0\Theta_0$ with $R^{-3} = (R_0 + \xi)^{-3} = R_0^{-3}[1 + (\xi/R_0)]^{-3} \approx R_0^{-3}[1 - (3\xi/R_0)]$ to the first order in ξ/R_0, it follows for the first term that:

---

[6] Equation (6.46) should not be applied to the interstellar gas without further consideration. Accordingly the result found for the OB-stars, viz. $\Sigma_V^2/\Sigma_U^2 \simeq 1$ (Sect. 3.2.3) does not immediately mean that $A \simeq 0$, since these young objects still reflect primarily the velocity ratios of the interstellar gas, for which itself nevertheless $A \neq 0$.

$$\frac{\Theta^2}{R} = \frac{\Theta_0^2 R_0^2}{R^3} \approx \frac{\Theta^2}{R_0}\left(1 - \frac{3\xi}{R_0}\right) \quad.$$

Also since $\Theta_K(R) \simeq \Theta_K(R_0) + \xi\Theta_K'(R_0) = \Theta_0 - \xi(A+B)$ [see (3.66)] and $R^{-1} \simeq R_0^{-1}[1 - (\xi/R_0)]$ the second term in (6.47), again to the first order in ξ/R_0, becomes

$$\frac{\Theta_K^2(R)}{R} \approx \frac{\Theta_0^2}{R_0} - \frac{\Theta_0^2}{R_0^2}\xi - \frac{2\Theta_0}{R_0}(A+B)\xi \quad.$$

Substituting this in (6.47) we obtain

$$\ddot{\xi} = \frac{2\Theta_0}{R_0}\left[(A+B) - \frac{\Theta_0}{R_0}\right] \quad.$$

Since $\Theta_0/R_0 = \omega_0 = A - B$ we can write

$$\ddot{\xi} = -\kappa_0^2 \xi \quad \text{with} \quad \kappa_0^2 = -4B(A-B) \quad. \tag{6.48}$$

The solution of this equation, with the initial values $\xi(0) = 0$, $\dot{\xi}(0) = \Pi_0$, is:

$$\xi(t) = \frac{\Pi_0}{\kappa_0}\sin\kappa_0 t \quad. \tag{6.49}$$

In particular, it follows that

$$\dot{\xi}(t) = \Pi(t) = \Pi_0 \cos\kappa_0 t \quad. \tag{6.50}$$

This result means that the star describes a harmonic oscillation radially about $R = R_0$. With the numerical values of (3.75) − after conversion into units of $\mathrm{km\,s^{-1}/km}$ − the period is found to be

$$P_0 = \frac{2\pi}{\kappa_0} = \frac{\pi}{\sqrt{-B(A-B)}} \approx 2 \times 10^8 \,\text{years} \quad. \tag{6.51}$$

Accordingly P_0 is of the same order of magnitude as the rotation period T_0 [Sect. 3.4.2, (3.77)]. The amplitude of the oscillation amounts, for example if $\Pi_0 = 30\,\mathrm{km\,s^{-1}}$, simply to $\Pi_0/\kappa_0 = 1\,\mathrm{kpc}$.

Let us consider now the *tangential component* of the motion. From the conservation of angular momentum $R^2\dot{\vartheta} = R\Theta = R_0\Theta_0$ it follows first of all, again to the first order in ξ/R_0, that

$$\dot{\vartheta} = \frac{R_0}{R^2}\Theta_0 \approx \frac{\Theta_0}{R_0}\left(1 - \frac{2\xi}{R_0}\right) = \omega_0 - \frac{2\omega_0}{R_0}\xi \quad.$$

For $t > 0$ we have a finite value

$$\dot{\eta} = \Theta - \Theta_{K0} = R\dot{\vartheta} - \omega_0 R_0 \approx R_0(\dot{\vartheta} - \omega_0) = -2\omega_0\xi \tag{6.52}$$

384

or

$$\dot{\eta} = -\frac{2\Pi_0\omega_0}{\kappa_0} \sin \kappa_0 t \quad . \tag{6.53}$$

Integration gives, taking into account (3.67) and (6.48) and with $\eta(0) = 0$,

$$\eta(t) = \frac{2\Pi_0\omega_0}{\kappa_0^2} \cos \kappa_0 t + C = \frac{\Pi_0}{2(-B)} \cos \kappa_0 t - \frac{\Pi_0}{2(-B)} \quad . \tag{6.54}$$

Equations (6.49) and (6.54) state that the star describes an ellipse with semi-axes Π_0/κ_0 and $2\Pi_0\omega_0/\kappa_0^2$ in the radial and tangential directions, respectively, in the rotating ξ, η-system, and therefore in practice relative to the local standard of rest (Fig. 6.6b). This ellipse – by analogy with the circle introduced by Ptolemy for the description of planetary motion – is called the "epicycle" and κ_0 the *epicyclic frequency*. Using the definitions of A and B one generally writes (arbitrary choice of R_0)

$$\kappa^2 = 4\omega^2 \left(1 + \frac{1}{2} \frac{R}{\omega} \frac{d\omega}{dR} \right) \quad , \tag{6.55}$$

where $\omega = \omega(R)$ is the angular velocity of rotation ("rotation frequency").

The *axis ratio of the epicycle* is given by the ratio $\kappa/2\omega$. In the case of the solar neighbourhood under consideration it follows that

$$\kappa_0/2\omega_0 = \sqrt{(A - B)/(-B)} \simeq 1.6 \quad ,$$

i.e. the tangential component of the motion is greater than the radial. The sense of motion in the epicycle is counterclockwise (retrograde). In relation to a coordinate system fixed in space with the galactic centre as origin it gives a rosette-shaped orbit.

In order to formulate a statement on the *velocity ellipsoid* let us find appropriate mean values of Π and $\Delta\Theta = \Theta - \Theta_K$ for many stars located in all possible phases of the epicyclic motion, considered equally probable. We take this mean over a volume around $R = R_0$, which is small compared with the semi-axis of the epicycle ($\simeq 1\,\mathrm{kpc}!$). That is to say, we integrate over a complete epicyclic period $P_0 = 2\pi/\kappa_0$ and take Θ_K for the same position as the peculiar velocity (correction for differential rotation, see Sect. 3.2.3). For the tangential component we must therefore not apply the definition (6.52), but must set $\Delta\Theta = \Theta(R) - \Theta_K(R)$. Again using the conservation of angular momentum $R\Theta = R_0\Theta_0$ as well as $\Theta_K(R) \simeq \Theta_0 + \xi\Theta_0' = \Theta_0 - \xi(A + B)$ and $R = R_0[1 + (\xi/R_0)]$, it then follows from (6.49) that, to the first order in ξ/R_0,

$$\Delta\Theta(t) = \frac{2B\Pi_0}{\kappa_0} \sin \kappa_0 t \quad . \tag{6.56}$$

Since $\overline{\cos^2 \kappa_0 t} = \overline{\sin^2 \kappa_0 t}$ over a complete period, (6.50) and (6.56) give

385

$$\frac{\Sigma_\Theta^2}{\Sigma_\Pi^2} = \frac{\overline{(\Delta\Theta)^2}}{\overline{\Pi^2}} = \frac{4B^2}{\kappa_0^2} = -\frac{B}{A-B} \quad , \tag{6.57}$$

in agreement with the result (6.46), which was obtained from the fundamental equation of stellar dynamics. The statistical correction of the epicyclic motion on the effect of the differential rotation diminishes the tangential component of the peculiar velocity, and indeed so strongly that Σ_Θ becomes smaller than Σ_Π (cf. Sect. 3.4.1, Fig. 3.41 top: flow field in local standard of rest). That the major axis of the observed velocity ellipsoid for stars with small peculiar velocities points approximately towards the centre of the galaxy, is according to this derivation evidently a result of the (differential) galactic rotation.

Fluctuations of the Gravitational Field and Age Dependence of the Velocity Dispersion: In Sect. 3.5.2 the different kinematic behaviour of objects of different ages (and different chemical compositions) was examined in terms of the scheme of stellar populations. The high velocity dispersion of the halo objects and the strong inclinations of their galactic orbits prompted the statement that they scarcely took any part in the galactic rotation. According to the qualitative interpretation given there these properties are attributable to the conditions during the collapse of the protogalaxy. The increase with age, shown in Table 3.10, of the velocity scatter of stars whose orbits run inside the galactic disc, however, may have another cause.

The effect of close stellar encounters is excluded as a possible explanation because of the long relaxation time for the galactic disc found in Sect. 6.1.2. It can be shown, however, that extensive local fluctuations of the gravitational field, such as are produced by massive interstellar clouds – also possibly through rather large accumulations of stars – lead to significant variations of the kinetic energies of the stars, because of the strength and the long duration of the disturbance (L. Spitzer and M. Schwarzschild, 1951). If the field perturbations have a statistical nature, then the effect on a selected group of stars in the phase space is a "diffusion of the stellar orbits" compared with the undisturbed mean orbits. This effect can be described by the second member of the encounter term in Sect. 6.1.1 (6.17). If one writes the distribution function as $f = f_0 + f_1$, where f_1 describes *only* the diffusion in velocity space, then, since f_0 satisfies (6.11), we have approximately

$$\frac{\partial f_1}{\partial t} = \frac{D}{2} \sum_{i=1}^{3} \frac{\partial^2 f_1}{\partial u_i^2} \quad . \tag{6.58}$$

Here $D = b_1 + b_2 + b_3$ denotes the total diffusion coefficient, and we have assumed for the sake of simplicity that the b_i are equal to one another and independent of the u_i. Equation (6.58) is of the heat diffusion type. For the initial condition $u_i = 0$ for $t = 0$, or more exactly $f_1(u_i, 0) = \delta(u_1)\delta(u_2)\delta(u_3)$, the solution is a Gauss distribution:

$$f_1(u_i, t) = \frac{1}{(2\pi Dt)^{3/2}} \exp\left\{ -\frac{u_1^2 + u_2^2 + u_3^2}{2Dt} \right\} \quad . \tag{6.59}$$

The square of the total velocity dispersion Σ increases in proportion to the time: $\Sigma^2 = Dt$. With a finite initial dispersion Σ_0 we get

$$\Sigma^2 = \Sigma_0^2 + Dt \quad . \tag{6.60}$$

The observed age dependence of the velocity dispersion (Sect. 3.2.3, Table 3.10) can in fact be represented by a relation of this form. The interpretation of the observed values requires that $\Sigma_0 \simeq 10\,\mathrm{km\,s^{-1}}$ and $D \simeq 6 \times 10^{-7}(\mathrm{km\,s^{-1}})^2$ year^{-1} (Wielen, 1977). The diffusion coefficient in this case clearly indicates an increase of kinetic energy which is transferred to the stars in encounters with massive condensations. There is a change in Σ of about $10\,\mathrm{km\,s^{-1}}$ per galactic revolution. The distance of the position of the star from the undisturbed orbit then amounts on average to about 1.5 kpc! The dissolution of star groups is accordingly accelerated. Another important consequence is that the mathematical tracing back of individual stars over rather long periods of time to the place of their formation, perhaps for evidence of origin in spiral arms, may not be possible in practice.

Vertex Deviation: In contrast to what was found for axially symmetric systems, the principal axis of the velocity ellipsoid according to the observations is not exactly in the direction of the centre, but deviates in galactic longitude by an angular amount up to about 30° (Sect. 3.2.3, esp. Table 3.9). Since this deviation becomes smaller from early to late spectral types, and hence with the average age of the star, the following possible interpretation was first mooted. If the stars are created in the spiral arms and these move relative to the stars and the interstellar matter (density waves, see Sect. 6.2.1), then newly formed stars must gradually move further away from the spiral arm. The kinematic adjustment of these stars now follows only slowly, so the young objects will still reflect the characteristics of their birth region. For the spiral arm model mentioned in Sect. 6.2.2 the vertex deviations of the stars of the solar neighbourhood with spectral types earlier than A0 could be explained in this way (Yuan, 1971).

The effect explained above of fluctuations of the gravitational field on the kinematic characteristics of the stars, however, leads one to expect a much stronger decrease of the effect with mean age (or spectral type) than is observed. The principal cause of the vertex deviation is therefore sought today in perturbations of the axisymmetric gravitational field, such as would appear in the presence of a spiral structure, but also with local density fluctuations (Mayor, 1972).

Hydrodynamical Approach: Important simplifications occur, if instead of the distribution function $f(R, z, \Pi, \Theta, Z)$ one studies only the first and second moments $\overline{\Pi}, \overline{\Theta}, \overline{Z}, \overline{(\Pi - \overline{\Pi})^2}, \overline{(\Pi - \overline{\Pi})(\Theta - \overline{\Theta})}, \ldots$. The mean values $\overline{\Pi}, \overline{\Theta}, \overline{Z}$

correspond to the components of the stream velocity of the "stellar gas"; the matrix of the second order moments characterises the shape of the velocity distribution and reduces in the isotropic case to the scalar pressure.

From the fundamental equation (6.37), by appropriate averaging (integration over the velocity components), relations of the form of Euler's Equations of hydrodynamics (in cylindrical coordinates) can be derived. Multiplying (6.37) by $\Pi \, d\Pi \, d\Theta \, dZ$ and integrating over the whole velocity space, the first term, for example, applying (6.1) (normalisation!), becomes

$$\int\limits_{-\infty}^{+\infty}\!\!\int\!\!\int \Pi^2 \frac{\partial f}{\partial R} d\Pi \, d\Theta \, dZ = \frac{\partial}{\partial R} \int\limits_{-\infty}^{+\infty}\!\!\int\!\!\int \Pi^2 f \, d\Pi \, d\Theta \, dZ$$

$$= \frac{\partial}{\partial R}(\varrho_s \overline{\Pi^2}) \quad . \tag{6.61}$$

For the terms containing $\dot{\Pi}$, $\dot{\Theta}$, $\dot{Z}$, appropriate partial integrations can be carried out, where these quantities can be expressed by (6.38). One then obtains an equation in which, instead of f, the various mean values $\overline{\Pi^2}$, $\overline{\Pi\Theta}$ etc. occur. Two further equations are obtained by multiplication of (6.37) by $\Theta \, d\Pi \, d\Theta \, dZ$ and $Z \, d\Pi \, d\Theta \, dZ$, respectively, and subsequent integration. We give the results directly:

$$\frac{\partial(\varrho_s\overline{\Pi^2})}{\partial R} + \frac{1}{R}\frac{\partial(\varrho_s\overline{\Pi\Theta})}{\partial\vartheta} + \frac{\partial(\varrho_s\overline{\Pi Z})}{\partial z} - \frac{\varrho_s\overline{\Theta^2}}{R} + \frac{\varrho_s\overline{\Pi^2}}{R} + \varrho_s\frac{\partial V}{\partial R} = 0$$

$$\frac{\partial(\varrho_s\overline{\Pi\Theta})}{\partial R} + \frac{1}{R}\frac{\partial(\varrho_s\overline{\Theta^2})}{\partial\vartheta} + \frac{\partial(\varrho_s\overline{\Theta Z})}{\partial z} + \frac{2\varrho_s\overline{\Pi\Theta}}{R} + \varrho_s\frac{1}{R}\frac{\partial V}{\partial\vartheta} = 0 \tag{6.62}$$

$$\frac{\partial(\varrho_s\overline{\Pi Z})}{\partial R} + \frac{1}{R}\frac{\partial(\varrho_s\overline{\Theta Z})}{\partial\vartheta} + \frac{\partial(\varrho_s\overline{Z^2})}{\partial z} + \frac{\varrho_s\overline{\Pi Z}}{R} + \varrho_s\frac{\partial V}{\partial z} = 0 \quad .$$

These equations are of fundamental importance for stellar dynamics. We also add the corresponding continuity equation, which is given by integration of (6.37) over the velocity space:

$$\frac{\partial(\varrho_s\overline{\Pi})}{\partial R} + \frac{1}{R}\frac{\partial(\varrho_s\overline{\Theta})}{\partial\vartheta} + \frac{\partial(\varrho_s\overline{Z})}{\partial z} + \frac{\varrho_s\overline{\Pi}}{R} = 0 \quad . \tag{6.63}$$

The Poisson Equation (6.3) assumes in cylindrical coordinates the form:

$$\frac{1}{R}\frac{\partial}{\partial R}\left(R\frac{\partial V}{\partial R}\right) + \frac{1}{R^2}\frac{\partial^2 V}{\partial\vartheta^2} + \frac{\partial^2 V}{\partial z^2} = -4\pi G\varrho \quad . \tag{6.64}$$

If in the derivation of (6.62) and (6.63) the basic distribution function is interpreted as the number density of a certain star type S, then these four equations are obviously also valid if the mass density ϱ_s is replaced by the number density of this star type $D_s = D_s(R, \vartheta, z)$. We shall consider under this assumption the case of axial symmetry: $V = V(R, z)$, $f = f(R, z, \Pi, \Theta, Z)$, in which all derivatives with respect to ϑ vanish. Then the system (6.62) becomes:

$$\frac{1}{D_s}\frac{\partial(D_s\overline{\Pi^2})}{\partial R} + \frac{1}{D_s}\frac{\partial(D_s\overline{\Pi Z})}{\partial z} + \frac{\overline{\Pi^2} - \overline{\Theta^2}}{R} = K_R$$

$$\frac{1}{D_s}\frac{\partial(D_s\overline{\Pi\Theta})}{\partial R} + \frac{1}{D_s}\frac{\partial(D_s\overline{\Theta Z})}{\partial z} + \frac{2\overline{\Pi\Theta}}{R} = 0 \tag{6.65}$$

$$\frac{1}{D_s}\frac{\partial(D_s\overline{\Pi Z})}{\partial R} + \frac{1}{D_s}\frac{(D_s\overline{Z^2})}{\partial z} + \frac{\overline{\Pi Z}}{R} = K_z \quad.$$

Here we have introduced the "force components" K_R and K_z:

$$K_R = -\frac{\partial V}{\partial R}\ , \quad K_z = -\frac{\partial V}{\partial z}\ . \tag{6.66}$$

Asymmetry of the Stellar Motions: As an example of an application of the "moment equations" (6.65), let us discuss the velocity component Θ in the direction of the galactic rotation. According to the observational results (Sects. 3.3.2 and 4) there are systematic differences between the local mean values $\overline{\Theta}$ for different sub-systems of similar objects, in which an increasing asymmetry of the distribution of Θ appears with increasing velocity scatter. In particular, from case to case there are varying differences between $\overline{\Theta}$ and the circular orbit velocity Θ_K. The theoretical connection between $\overline{\Theta}$ and Θ_K can be obtained by means of the first equation (6.65).

For each star of a certain sub-system we set $\Theta = \overline{\Theta} + \Theta_1$, where Θ_1 denotes its velocity component relative to the centroid of the objects under consideration. From the property $\overline{\Theta_1} = 0$ it follows that $\overline{\Theta^2} = \overline{\Theta}^2 + \overline{\Theta_1^2}$. Substituting this expression in the first equation (6.65), and applying the formula

$$\frac{1}{D_s}\frac{\partial(D_s\overline{\Pi^2})}{\partial R} = \frac{1}{D_s}\left(\frac{\partial D_s}{\partial R}\overline{\Pi^2} + D_s\frac{\partial\overline{\Pi^2}}{\partial R}\right)$$

$$= \overline{\Pi^2}\left(\frac{\partial\ln D_s}{\partial R} + \frac{\partial\ln\overline{\Pi^2}}{\partial R}\right)$$

we obtain the relation

$$\overline{\Pi^2}\left[\frac{\partial\ln D_s}{\partial R} + \frac{\partial\ln\overline{\Pi^2}}{\partial R} + \frac{1}{R}\left(1 - \frac{\overline{\Theta_1^2}}{\overline{\Pi^2}}\right) + \frac{1}{D_s\overline{\Pi^2}}\frac{\partial(D_s\overline{\Pi Z})}{\partial z}\right]$$

$$= \frac{\overline{\Theta}^2}{R} + K_R = \frac{\overline{\Theta}^2 - \Theta_K^2}{R}\quad. \tag{6.67}$$

For the object groups with small asymmetry of the velocity distribution we can set

$$\overline{\Theta}^2 - \Theta_K^2 = (\overline{\Theta} + \Theta_K)(\overline{\Theta} - \Theta_K) \approx 2\Theta_K(\overline{\Theta} - \Theta_K)$$

and for the solar neighbourhood $R \simeq R_0$ with $\Theta_{K0}/R_0 = \omega_0 = A - B$ (3.67) we obtain directly the *asymmetric drift*

$$(\overline{\Theta} - \Theta_K)_{R_0} = \frac{\overline{\Pi^2}}{2(A - B)} \left[\frac{\partial \ln D_s}{\partial R} + \frac{1}{R} \left(1 - \frac{\overline{\Theta_1^2}}{\overline{\Pi^2}} \right) \right]_{R_0} . \tag{6.68}$$

In the large brackets we have neglected the radial gradient of $\overline{\Pi^2}$ and the term with the derivative of $D_s \cdot \overline{\Pi Z}$ with respect to z, corresponding to the assumption of a very flat system. If it is assumed that the velocity distribution is ellipsoidal and that the major axis, even when $|z| > 0$, always points towards the centre — hence in this case inclined to the galactic plane — then the neglected last term in the large brackets in (6.67) can be expressed as follows:

$$\frac{1}{D_s \overline{\Pi^2}} \frac{\partial (D_s \overline{\Pi Z})}{\partial z} = \frac{1}{R} \left(1 - \frac{\overline{Z^2}}{\overline{\Pi^2}} \right) . \tag{6.69}$$

The second term in the brackets in (6.68) can be directly expressed in terms of the rotation constants A and B by means of the result (6.46).

Equation (6.68) provides the dynamical basis for the procedure (explained in Sect. 3.2.3) for *establishing a definite local standard of rest*. It shows that the asymmetric drift $\overline{\Theta} - \Theta_K$ increases, as observed, with increasing velocity dispersion $(\overline{\Pi^2})^{1/2}$ (denoted in Sect. 3.2.3 by Σ_U) and as $\overline{\Pi^2} \to 0$, $\overline{\Theta} \to \Theta_K$. By plotting the observed solar motion $V_\odot = \Theta_\odot - \overline{\Theta}$, relative to various object groups, against $\overline{\Pi^2} \equiv \Sigma_U$, and then extrapolating to $\overline{\Pi^2} = 0$, we obtain $V_\odot = \Theta_\odot - \Theta_K$. In other words, we obtain the solar velocity relative to a local standard of rest moving with the circular orbital velocity around the galactic centre. This dynamically defined solar motion has been derived by Jahreiss (1974) using recent observational material with the result:

$$U_\odot = +9 \, \text{km s}^{-1} , \quad V_\odot = +10 \, \text{km s}^{-1} , \quad W_\odot = +6 \, \text{km s}^{-1} .$$

The deviation from the "Basic solar motion" (3.40) falls within the limits of error ($\simeq \pm 1 \, \text{km s}^{-1}$).

Equation (6.68) shows that the asymmetry of the stellar motions is closely connected with the radial density gradient $\partial \ln D_s / \partial R$ of the objects under consideration: *the stronger the density gradient towards the centre, the greater the asymmetric drift*. The quantitative relationship is displayed in Table 3.14 in Sect. 3.5.1. The density gradients given there are calculated in part — where observational values are not yet available — using the relations (6.67) or (6.68).

6.1.5 The Mass Distribution in the Galaxy

If one knows the spatial distribution of the potential field $V(R, z)$ for the system, regarded up to now as steady and axially symmetric, then the mass distribution $\varrho(R, z)$ can be obtained from the Poisson Equation (6.64). If the stellar system is self-consistent, i.e. if the gravitational field is produced by the stars alone — ignoring the interstellar matter! — then $V(R, z)$ can be derived from the best possible fit of an appropriate solution $f(R, z, \Pi, \Theta, Z)$ of the stellar dynamic fundamental equation (6.37) with (6.38) to the available observational results for the number density D_s and the velocity field. The difficulties in the solution of the fundamental equation have so far permitted the execution of this concept only in outline.

The evidence available today on the mass distribution in the galaxy is based primarily on empirical determinations of the force components $K_z(R_0, z)$ and $K_R(R, 0)$ from the motions of the stars perpendicular to the galactic plane and the rotation curve $\Theta = \Theta(R, 0)$, respectively. Let us discuss first the analysis of the force law perpendicular to the galactic plane in the solar neighbourhood, which makes possible a determination of the total mass density at $z = 0$. Afterwards we shall turn to the derivation of a model for the mass distribution in the entire galaxy.

The Local Mass Density in the Galactic Plane: Because of the assumption of axial symmetry the term with $\partial^2 V / \partial \vartheta^2$ drops out of the Poisson Equation (6.64):

$$\frac{1}{R}\frac{\partial}{\partial R}(R K_R) + \frac{\partial K_z}{\partial z} = -4\pi G \varrho \quad , \tag{6.70}$$

where $\varrho = \varrho_s + \varrho_i$ denotes the total mass density of the stellar and interstellar matter. For the solar neighbourhood $R \simeq R_0$ and $|z| < 1\,\text{kpc}$ we can express the first term by the constants of the galactic rotation. With $K_R = -\Theta^2/R$ it follows [Sect. 3.4.1, (3.66) and (3.67) as well as Sect. 3.4.2, (3.75)] that:

$$\frac{1}{R}\frac{\partial}{\partial R}(R K_R) = -\frac{2\Theta_0}{R_0}\left(\frac{d\Theta}{dR}\right)_{R_0} = 2(A - B)(A + B)$$

$$\approx 250\,\text{km}^2\,\text{s}^{-2}\,\text{kpc}^{-2} = 2.6 \times 10^{-31}\,\text{s}^{-2} \quad . \tag{6.71}$$

For comparison with this value, evaluating the right-hand side of (6.70) using the mass density $\varrho_{s0} = 3 \times 10^{-24}\,\text{g cm}^{-3}$ given in Sect. 3.2.2 for the observed stars in the immediate solar neighbourhood, we obtain $4\pi G \varrho_{s0} \simeq 25 \times 10^{-31}\,\text{s}^{-2}$. Since ϱ is certainly significantly larger than ϱ_{s0}, the gradient $\partial K_z / \partial z$ in the left-hand side of (6.70) must completely dominate the first term, as one would expect for a widely extended plane parallel layer. The term (6.71) thus only plays the role of a small correctional term in (6.70), whose value we know sufficiently accurately. An independent determination of $(\partial K_z / \partial z)_{z=0}$ enables one to derive immediately the total mass density ϱ for $z = 0$.

First of all the following qualitative statements can be made on the behaviour of K_z with increasing z. Since any realistic symmetric density distribution $\varrho = \varrho(z)$ must be practically constant near $z = 0$, it follows from (6.70) – neglecting the first term – that $|K_z(z)|$ at first increases linearly with $|z|$ (see Fig. 6.7). On the other hand for z-values of the order of magnitude of the distance of the sun from the galactic centre there must be a transition to $|K_z| \sim z^{-2}$. In the region in between, therefore, at a distance of a few kpc, there

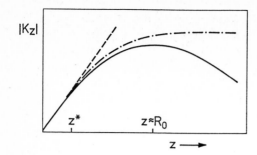

Fig. 6.7. Profile of force component perpendicular to the galactic plane $K_z(z)$ with increasing distance z. For $z<z^*$ the density is practically constant and equal to $\varrho_0 = \varrho(0)$

must be a maximum or an approximately constant profile, which is determined by the mass distribution in the disc *and* in the halo.

A possible way of determining $K_z(R_0, z)$, and hence also $\partial K_z/\partial z$, is provided by the third equation (6.65). In the outer regions of the galactic disc the motions perpendicular to the plane of symmetry and the motions parallel to this plane should not be coupled in dynamical equilibrium, so that $\overline{\Pi Z} = 0$. Then it follows that

$$\frac{1}{D_s}\frac{\partial(D_s\overline{Z^2})}{\partial z} = K_z \quad . \tag{6.72}$$

Whilst $K_z(z)$ represents the total gravitational acceleration perpendicular to the galactic plane, $D_s(z)$ and $\overline{Z^2(z)}$ may here also relate to homogeneous sub-groups of stars. In the Milky Way system the assumption $\overline{\Pi Z} = 0$ may be permissible as an approximation only for distances $|z|<z_1$, where z_1 is of the order of 1 kpc.

If Z and Π are decoupled and as a result the total energy of the stellar motion in the z-direction is constant:

$$\frac{1}{2}Z^2 = \frac{1}{2}Z_0^2 + \int\limits_0^z K_z(z)dz \quad ,$$

then it can be shown that with a Gauss distribution for Z_0 (= velocity at $z = 0$) $Z = Z(z)$ also has a Gauss distribution for $|z|>0$ and moreover $\overline{Z^2}$ is independent of z, being therefore equal to $\overline{Z_0^2}$. In this case (6.72) becomes

$$\overline{Z_0^2}\frac{\partial \ln D_s}{\partial z} = K_z \quad . \tag{6.73}$$

Since $K_z(z)$ is always the same profile for different sub-groups of stars, a larger velocity dispersion $\overline{Z_0^2}$ must be associated with a correspondingly smaller density gradient: larger values of Z_0 lead on average to longer excursions of the stars from the galactic plane.

The determination of a reliable value for $(\partial K_z/\partial z)_{z=0}$ by differentiating the profile of $K_z(z)$ produced by (6.73) requires a relatively exact knowledge

392

of the density profile $D_s(z)$ for small $|z|$ and a representative value of $\overline{Z_0^2}$. The available data (see Tables 3.7 and 3.9, where $\overline{\Sigma_W^2} \equiv \overline{Z_0^2}$) lead to

$$\left(\frac{\partial K_z}{\partial z}\right)_{z=0} \approx 10^4 \, \mathrm{km\,s^{-2}\,kpc^{-2}} \approx 10^{-29}\,\mathrm{s^{-2}} \quad . \tag{6.74}$$

The maximal distance from the galactic plane $|z|_{\max}$, which can be reached by stars with the velocity component $Z(0) = Z_0$, and their period of oscillation about the plane $|z| = 0$, can now be estimated by assuming a linear "force law". The equation of motion

$$\ddot{z} = K_z = -\alpha^2 z \quad \text{with} \quad \alpha^2 = \left(\frac{\partial K_z}{\partial z}\right)_{z=0} \tag{6.75}$$

produces $z = a \sin \alpha t$ and $\dot{z} = Z = a\alpha \cos \alpha t$, so that

$$|z|_{\max} = a = Z_0/\alpha \quad . \tag{6.76}$$

With the approximate value (6.74) it follows that $\alpha \simeq 3 \times 10^{-15}\,\mathrm{s^{-1}}$ giving the period $P = 2\pi/\alpha \simeq 6 \times 10^{14}\,\mathrm{s} \simeq 2 \times 10^7$ years and the values given in Table 6.2 for $|z|_{\max}$.

Table 6.2. Maximal distances from the galactic plane $|z|_{\max}$ for stars with velocity components Z_0 in the z-direction at $z = 0$, calculated from (6.76) with $\alpha = 0.1\,\mathrm{km\,s^{-1}\,pc^{-1}}$

Z_0 [km s^{-1}]	5	10	20	40
$|z|_{\max}$ [pc]	50	100	200	400

The simultaneous satisfaction of (6.70) and (6.72), in conjunction with an appropriate, empirically based assumption on the form of the z-dependence of ϱ, permits the derivation not only of $K_z(z)$ but also of the total mass density ϱ_0 in the galactic plane. The application to $D_s(z)$ and $\overline{Z^2}$ for K-giants by Oort gave a profile for $K_z(z)$ which indeed flattened out above $|z| \simeq 500\,\mathrm{pc}$, but which was still climbing gently at $|z| \simeq 1\,\mathrm{kpc}$. The often quoted result for the *total mass density in the solar neighbourhood* reads

$$\varrho_0 = 10 \times 10^{-24}\,\mathrm{g\,cm^{-3}} = 0.15 \mathcal{M}_\odot \mathrm{pc^{-3}} \quad . \tag{6.77}$$

The gradients of K_z from a few more recent studies yield values between 0.09 and $0.16 \mathcal{M}_\odot \mathrm{pc^{-3}}$.

The mass density of the known stars according to Sect. 3.2.2 contributes less than a third of the value (6.77). For the interstellar matter in atomic form, with 0.5 to 1.0 H atoms cm^{-3} (see Sect. 5.4.2), assuming the mass fraction of helium and the higher elements to be about 40 % of the total atomic gas density, we get a value between 1.4 and 2.8 × 10^{-24} g cm^{-3}. The average contribution of molecular hydrogen (and other molecules) lies about an order of magnitude lower than these numbers in the solar neighbourhood (Sect. 5.2.5); the contribution from interstellar dust is still lower. The mean density of the interstellar matter will scarcely be overestimated at $\varrho_0 \simeq 3 \times 10^{-24}$ g cm^{-3}, so that the total stellar and interstellar matter known up to now accounts for

$$\varrho_0 \approx 6 \times 10^{-24} \, \text{g cm}^{-3} = 0.09 \mathcal{M}_\odot \, \text{pc}^{-3} \quad .$$

We thus reach only the lowest of the various dynamically determined values for ϱ_0. According to more recent studies, however, it cannot be ruled out that the stellar contribution on the part of white dwarfs and of faint M-dwarfs has been underestimated. Since a dynamic value for ϱ_0 a little over $0.10 \mathcal{M}_\odot \, \text{pc}^{-3}$ is today entirely within the realm of possibility, the "missing mass problem" existing up to now seems to be resolvable without dramatic assumptions concerning the missing mass. The far extended "corona" of the galaxy, discussed in a later part of this section, would cause an increase of ϱ_0 in the solar neighbourhood by only the order of 1 %.

Derivation of a Mass Model from the Rotation Curve: Because of the dominating significance of the radial force equilibrium for the dynamics of the galactic disc, there is a direct connection between the radial potential profile and the profile of the mean value $\overline{\Theta} = \Theta_K$ [for vanishing velocity dispersion in the first equation (6.65) and in particular $\overline{\Theta^2} = \overline{\Theta}^2$]:

$$-\left(\frac{\partial V}{\partial R}\right)_{z=0} = K_R(R, 0) = -\frac{\Theta_K^2}{R} \quad . \tag{6.78}$$

In conjunction with the results mentioned above for $K_z = -\partial V/\partial z$, conclusions can be drawn, with the help of the Poisson Equation, on the *spatial distribution* of the total mass density $\varrho = \varrho_s + \varrho_i$.

In practice one starts − neglecting the spiral structure − from a convenient mathematical model for $\varrho(R, z)$, calculates the pertinent potential function $V(R, 0)$ or the "force law" $K_R(R, 0)$, and then the resulting profile of the circular orbital velocity $\Theta_K(R)$. One selects the parameters of the model so that, not only the observed values $\Theta_K(R)$, but also as many as possible of the other known facts (mass density $\varrho_0 = \varrho(R_0, 0)$, z-dependence of ϱ etc.) are approximated optimally. The numerical result for $\varrho(R, z)$ is known as "mass model" of the galaxy.

The representation which suggests itself is a galactic disc made up of "similarly oriented" oblate spheroidal shells of uniform density. Because of the possibility of arbitrary choice of oblateness and perhaps of superimposing several inhomogeneous spheroidal bodies of different oblateness, this concept is very flexible. Thereby one can take advantage of the property already known to

Newton, that inside a spheroidal shell the gravitational force is zero and the potential is constant.

The derivation of the potential and of the *force components for a spheroidal mass distribution* is a classic problem of potential theory, which we shall not here go into in detail. We start from the results for a spheroid uniformly filled with mass (Schmidt, 1956). Let a denote the major (equatorial) semi-axis, c the minor (polar) semi-axis, $e = \sqrt{1-(c/a)^2}$ the numerical eccentricity and ϱ the constant mass density, then the force component in the equatorial plane $(z = 0)$ is given by

$$K_R(R,0;a,e) = -\frac{2\pi G}{e^3}\sqrt{1-e^2}\,\varrho R\left[\arcsin\left(\frac{ae}{R}\right)\right.$$
$$\left. -\frac{ae}{R}\sqrt{1-\left(\frac{ae}{R}\right)^2}\right]\;. \tag{6.79}$$

A thin shell, which is bounded by two spheroids of equal eccentricity e with major semi-axes a and $a+da$ (Fig. 6.8) and in which the mass density is $\varrho = \varrho(a)$, produces at an external point $(R>a, z = 0)$ the radial acceleration

$$K_R(R,0;a+da,e) - K_R(R,0;a,e) = (\partial K_R/\partial a)da \;.$$

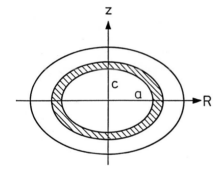

Fig. 6.8. Derivation of (6.80)

The total effect $K_R(R,0)$ of a continuous series of nested shells, each with constant density $\varrho(a)$, is given by integration of $(\partial K_R/\partial a)$, where this derivative for fixed ϱ is to be formed from (6.79). We obtain

$$K_R(R,0) = -4\pi G\sqrt{1-e^2}\,\frac{1}{R}\int\limits_0^R \frac{\varrho(a)a^2 da}{\sqrt{R^2-a^2e^2}} \;. \tag{6.80}$$

The integral only needs extend to $a = R$, since spheroidal shells with $a>R$ make no contribution.

Because of the extreme oblateness of the galactic disc the limiting case $e\to 1$ is of interest as an approximation. One then conveniently introduces the circularly symmetric "surface mass density" in the plane $z = 0$ by

$$\lim_{e \to 1} \sqrt{1 - e^2}\, a\varrho(a) = \lim_{e \to 1} c\varrho(a) = \sigma(a) \, [\mathcal{M}_\odot \, \text{pc}^{-2}] \quad . \tag{6.81}$$

Equation (6.80) then becomes

$$R K_R(R,0) = -4\pi G \int\limits_0^R \frac{\sigma(a) a \, da}{\sqrt{R^2 - a^2}} \quad . \tag{6.82}$$

With the given profile for $\Theta_K^2(R)/R = K_R(R,0)$ obtained from observations, (6.80) represents an integral equation for $\varrho(a)$. Several different methods have been developed for the inversion of this equation: the simplest is to approximate $\varrho(a)$ by a power series. Integration term by term then gives a power series for $K_R(R,0)$ also. Thus the series

$$\varrho(a) = \frac{p_{-2}}{a^2} + \frac{p_{-1}}{a} + p_0 + p_1 a + \ldots \tag{6.83}$$

leads to the result

$$\frac{\Theta_K^2(R)}{R} = -K_R(R,0) = R\left\{ \frac{c_{-2}}{R^2} + \frac{c_{-1}}{R} + c_0 + c_1 R + \ldots \right\} \tag{6.84}$$

where the c_n are related in elementary fashion to the p_n and e. From the c_n the p_n and hence $\varrho(a)$ can be calculated without difficulty.

In the limiting case $e \to 1$ the inversion of the resulting integral equation (6.82) for the surface density $\sigma(a)$ is possible with complete generality (Abel's Integral Equation, transformation $x = R^2$, $y = a^2$). The result reads [see, e.g. Trumpler and Weaver (1953; Chap. 1.4)]

$$\sigma(a) = \frac{1}{2\pi^2 G} \frac{1}{a} \frac{d}{da} \int\limits_0^a \frac{K_R(R) R^2 \, dR}{\sqrt{a^2 - R^2}} \quad . \tag{6.85}$$

This method has frequently been applied to the derivation of the mass distribution in other spiral galaxies.

To obtain a mass model of our galaxy let us return to the case of arbitrary eccentricity $e \leq 1$. After various mass models had been developed, composed of many homogeneous or inhomogeneous ellipsoidal bodies, the Dutch astronomer M. Schmidt in 1965 proposed a simple approximation to the accuracy of the measured observational data. He found that the mean profile of the rotation curve derived from the observations then available in the northern and southern parts of the Milky Way could be approximated by just two terms of the series (6.84) and a point mass $\mathcal{M}_0$ at the centre:

$$\frac{\Theta_K^2}{R^2} = G\frac{\mathcal{M}_0}{R^3} + \frac{c_{-1}}{R} + c_1 R \quad , \tag{6.86}$$

where for $R = R_0 = 10\,\text{kpc}$ in particular the values of the constants of the galactic rotation are $A = 15$, $B = -10$ $[\text{km}\,\text{s}^{-1}\,\text{kpc}^{-1}]$. If one reckons Θ_K in $\text{km}\,\text{s}^{-1}$ and R in kpc, then the coefficients $G\mathcal{M}_0 = 30\,000$, $c_{-1} = 10\,120$, $c_1 = -41.722$. The eccentricity e was now fixed so that for $a = R_0 = 10\,\text{kpc}$ (6.83) gave the local density value as $\varrho_0 = 0.145\mathcal{M}_\odot\,\text{pc}^{-3}$, with the result that $c/a = \sqrt{1 - e^2} = 0.05$ or $e = 0.99875$. For $a < 10\,\text{kpc}$ the density law then becomes

$$\varrho(a) = 3.930a^{-1} - 0.02489a \; [\mathcal{M}_\odot\,\text{pc}^{-3}] \quad,$$

where a is expressed in kpc. For the central point mass the value of the first coefficient in (6.86) gives $\mathcal{M}_0 = 7 \times 10^9\mathcal{M}_\odot$. The mass model obtained is described numerically in Table 6.3. For illustration Fig. 6.9 shows the density profile for a similar model consisting of four inhomogeneous spheroidal bodies.

The mass of the inhomogeneous spheroid is found by the following method. From the formula for the volume of a spheroid $V = (4\pi/3)a^3\sqrt{1 - e^2}$ one obtains by differentiation the volume of a shell corresponding to an increment da and then the following integral for the total mass of an inhomogeneous spheroid:

$$\mathcal{M}_E = 4\pi\sqrt{1 - e^2}\int_0^{R_b} \varrho(a)a^2\,da \quad.$$

Table 6.3. Dynamical mass model of the galaxy by M. Schmidt (1965). For the derived density profile, including a central point mass, we give: the calculated values of the circular orbital velocity $\Theta_K(R)$, the angular velocity $\omega(R)$, the epicyclic frequency κ according to (6.55), the equivalent surface density $\sigma(R)$ obtained by integrating the density $\varrho(R, z)$ over z from $-\infty$ to $+\infty$, and the mass $\mathcal{M}(R)$ lying inside a cylinder of radius R. [From Lin (1970)]

R [kpc]	Θ_K [km s^{-1}]	ω [km s^{-1} kpc^{-1}]	κ [km s^{-1} kpc^{-1}]	$\sigma(R)$ [$\mathcal{M}_\odot$ pc^{-2}]	$\mathcal{M}(R)$ [$10^9\,\mathcal{M}_\odot$]
1	200	200	245	1097	11
2	187	93	136	817	20
3	198	66.0	103	646	31
4	213	53.2	85.0	521	44
5	227	45.3	72.6	421	57
6	238	39.7	62.8	338	70
7	247	35.2	54.4	267	82
8	252	31.4	46.7	206	93
9	253	28.1	39.2	155	103
10	250	25.0	31.6	114	111
11	244	22.2	26.5	86	117
12	238	19.8	22.8	65.9	123
13	231	17.8	20.0	51.8	127
14	224	16.0	17.7	41.5	131
15	218	14.6	15.9	33.7	135
16	213	13.3	14.3	27.8	138
17	207	12.2	13.0	23.2	140
18	202	11.2	11.9	19.5	143
19	197	10.4	10.9	16.6	145
20	193	9.6	10.1	14.2	147

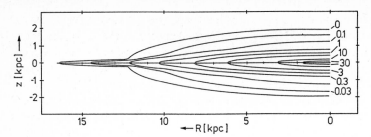

Fig. 6.9. Distribution of total mass density ϱ in the R, z plane for a model derived by Schmidt (1956), in which four inhomogeneous spheroidal components are superposed. Unit of ϱ is the value of ϱ_0 at the location of the sun

In the application of the series (6.83) for $\varrho(a)$ it is necessary to make the choice of an upper boundary R_{b}, since the mass would in general otherwise be infinite. One can, of course, also use (6.83) only up to $a = R_{\mathrm{b}}$, and for $a > R_{\mathrm{b}}$ postulate a steep fall in density $\sim a^{-n}$ with $n > 3$. M. Schmidt chose for $a > R_{\mathrm{b}} \simeq 10\,\mathrm{kpc}$ the expression $\varrho(a) = 1450 a^{-4}$, which with $a = R_0 = 10\,\mathrm{kpc}$ just gave $\varrho = \varrho_0 = 0.145 \mathcal{M}_{\odot}\,\mathrm{pc}^{-3}$ and hence joined up with the profile given above for smaller values of a. The contributions of the regions $a < R_{\mathrm{b}}$ and $a > R_{\mathrm{b}}$ to $\mathcal{M}_E$ are then $0.82 \times 10^{11} \mathcal{M}_{\odot}$ and $0.92 \times 10^{11} \mathcal{M}_{\odot}$, respectively, so that the total mass is

$$\mathcal{M} = \mathcal{M}_0 + \mathcal{M}_E = 1.8 \times 10^{11}\,\mathcal{M}_{\odot} \quad .$$

Since half of this mass comes from the only roughly estimated profile of $\varrho(a)$ for $a > R_{\mathrm{b}}$, this result is correspondingly uncertain. An approximately spherical "corona" of the galaxy, consisting of hitherto unobserved faint objects, with a relatively slow decrease in density for $R > 10\,\mathrm{kpc}$, would have only a weak influence on the rotation curve in the region $R < 10\,\mathrm{kpc}$ included in observations, even if its total mass amounted to many times the above result.

Recent Approaches to Mass Models: The knowledge available today on the structure of other galaxies similar to the Milky Way provides additional evidence, which had not previously been available when viewing our own galaxy from the position of the sun; for example, on the relationship between the disc and halo components in the inner region, or on the rotation curve for $R > 10\,\mathrm{kpc}$.

The brightness distribution in spiral galaxies usually makes it possible to distinguish two stellar components: a flat *disc* with rapid radial decrease of the surface brightness, which can be represented by $I_{\mathrm{d}}(R) \sim \exp(-\alpha_{\mathrm{d}} R)$, and a nearly spherical *halo* with slower density decrease, which was represented, for example, by the formula $I_{\mathrm{h}}(R) \sim \exp[-(\alpha_{\mathrm{h}} R)^{1/4}]$ or $I_{\mathrm{h}}(R) \sim (1 + \alpha_{\mathrm{h}} R)^{-2}$. The inner part of this halo can be perceived as a central "bulge" of the stellar distribution, it is therefore also called the "bulge component" and sometimes considered separately. The disc component appears to fall significantly below the exponential expression for $I_{\mathrm{d}}(R)$ in this inner region ("central hole").

398

Under the assumption of a constant mass-luminosity relationship these expressions are valid also for the projected surface density $\sigma_d(R)$ and $\sigma_h(R)$. If one represents the mass distribution of each component by appropriate spheroids with constant eccentricity, then the mass density profiles $\varrho_d(R,z)$, $\varrho_h(R,z)$ are also thereby fixed. The parameters appearing are the oblatenesses (or eccentricities), the scale factors α^{-1} and the total masses of the components. By repeated calculations of the "observable" quantities, such as the rotation curve for $R<10\,\mathrm{kpc}$, local mass densities ϱ_0 for each component, etc., under different assumptions concerning these parameters, the most appropriate values can be decided. The reproduction of a minimum in the observed rotation curve between $R = 1$ and $4\,\mathrm{kpc}$ (see Sect. 5.4.3, Fig. 5.30) can be achieved by the assumption of a "central hole" in the disc, where even the sum $\varrho_d(R,0)+\varrho_h(R,0)$ passes through a minimum value. Results from studies of the sort described for the masses and effective radii of the components are contained in Table 6.4.

The observed rotation curves typical of large spiral galaxies usually show in the outer region, corresponding to, say, $R>10\,\mathrm{kpc}$, a *flat* profile, falling considerably more slowly than $R^{-1/2}$. Such a result requires the assumption of an additional, very extended, dark "corona" with rather a large total mass. It is at least probable that this is true also for our galaxy. Ostriker and Caldwell (1979) have put forward a further argument, that available data on high velocity stars lead one to infer an escape velocity from the whole galactic system of around $500\,\mathrm{km\,s^{-1}}$, which requires a total mass of about $10^{12}\mathcal{M}_\odot$ (see Table 6.4).[7]

Table 6.4. Mass fractions and effective radii of the various components of the galaxy according to recent studies (Ostriker and Caldwell, 1979; Einasto, 1979). The effective radius used here is approximately equal to 3.5 times the "scale factor" α^{-1} (see text)

Component	Mass [$\mathcal{M}_\odot$]	Effective radius [kpc]
"Nucleus"	$1\ldots2 \times 10^8$	$\lesssim 0.005$
Disc	8×10^{10}	$5\ldots8$
Bulge	$\Big\}\;2\ldots8 \times 10^{10}$	0.2
Halo		$1\ldots2$
"Corona"	$1\ldots2 \times 10^{12}$	$50\ldots75$

Stability of the Gravitative Equilibrium of the Stellar System: If the equilibrium between the radial gravitational force and the centrifugal force is disturbed, then the system can either return of its own accord to the original state of equilibrium (stability) or move towards a different state, where the case of continuing contraction is of primary concern. In order to derive a simple criterion of the conditions under which the one or the other possibility will occur, let us consider a thin disc with constant surface mass density σ, which has radius R and rotates with constant angular velocity ω. A unit mass *on the*

---

[7] Such a very extended "corona" would not contribute to the stabilising of the central region of the system discussed in Sect. 6.1.4.

edge of the disc experiences there the gravitational force $K_R = G(\pi R^2 \sigma)/R^2$ and the centrifugal force is $Z_R = \omega^2 R$. Starting from the equilibrium state $K_R = Z_R$, we now enquire into the changes of K_R and Z_R with R at *fixed total mass* $\pi R^2 \sigma$ whilst conserving angular momentum: ωR^2 = constant, so that $d\omega/dR = -2\omega/R$. One then has

$$\frac{dK_R}{dR} = -\frac{2\pi G\sigma}{R} \quad \text{and} \quad \frac{dZ_R}{dR} = 2\omega\frac{d\omega}{dR}R + \omega^2 = -3\omega^2 \quad .$$

For a decrease in $R(dR<0)$, if the gravitational force increases more strongly than is needed to maintain the new circular orbit, i.e. if $2\pi G\sigma/R > 3\omega^2$ or

$$R < \frac{2\pi G\sigma}{3\omega^2} \quad , \tag{6.87}$$

then the contraction will continue and one speaks of *gravitational instability*. Otherwise the system will revert to its initial state. Stabilisation through rotation thus occurs only if the radius of the disc reaches or exceeds a certain limiting value. For the Milky Way system, for example, with $\sigma = 3 \times 10^2 M_\odot \, \mathrm{pc}^{-2}$ $\simeq 2 \times 10^{-2} \, \mathrm{g\,cm}^{-2}$ for the region $R<10$ kpc (Table 6.3) and $\omega = \omega_0 = 25 \, \mathrm{km\,s}^{-1}$ $\mathrm{kpc}^{-1} \simeq 8 \times 10^{-16} \, \mathrm{cm\,s}^{-1}\,\mathrm{cm}^{-1}$ one obtains as the minimal radius for stability about 5 kpc. In comparison with this, the effective radius of the galactic disc (Table 6.4) is not very much higher.

According to the results of the numerical simulations of stellar systems (Sect. 6.1.4), however, the stability is even further increased by deviations from circular orbital motion, which appear as statistically distributed radial motions of the individual stars. According to studies in the framework of analytical theory (Toomre, 1964), radial perturbations of extent L, which we shall characterise by the radial wave number $k = 2\pi/L$, lead to a *local* gravitational collapse, if

$$\frac{k^2 \Sigma_{II}^2}{\kappa^2} < 0.2857 \quad . \tag{6.88}$$

Here Σ_{II} denotes the radial velocity scatter and κ the epicyclic frequency, which can be expressed in terms of ω by (6.55). The stability limit following from (6.87) for the whole disc, $R = 2\pi G\sigma/3\omega^2$, corresponds to a critical value of the wave number $k = \kappa^2/2\pi G\sigma$, from which it follows that the condition for stability of the system for disturbances of all wavelengths $L<2R$ is given by

$$\Sigma_{II} \geq \sqrt{0.2857}\frac{2\pi G\sigma}{\kappa} = 3.36\frac{G\sigma}{\kappa} \quad . \tag{6.89}$$

With the numbers for σ and κ at $R = 10$ kpc from Table 6.3 one obtains for the critical value on the right-hand side about $50 \, \mathrm{km\,s}^{-1}$. This number is reduced to about $35 \, \mathrm{km\,s}^{-1}$ by taking into account the finite thickness of the disc, and then agrees approximately with the value (Σ_U in Table 3.9) observed for the main mass of the disc stars in the solar neighbourhood (G- to M-dwarfs). The galactic disc therefore still just fulfils the condition for gravitational instability.

6.2 Gravitational Theory of the Spiral Structure

6.2.1 Statement of the Problem and General Review

Observations of Extragalactic Systems: The question of the origin of the spiral structure of the Milky Way system is suggested by observations of other galaxies. About 60 % of the extragalactic stellar systems with total apparent magnitudes $m_{pg} < 13^m$ show spiral arms. These are generally galaxies with a flat rotating disc of stars and interstellar matter, such as we find also for the Milky Way system; the spiral arms are marked primarily by intrinsically bright young objects (OB-stars, supergiants, H II regions). In detail the observed spiral structure is often irregular or only fragmentary in form, the arms are partially deformed, and between them there quite often exist "bridges" (Figs. 1.2, 3). One can therefore generally speak only of a "basic spiral structure", a so-called *Grand Design*, and it is this interpretation with which theory must be concerned.

For extragalactic systems these inferences on the large scale spiral structure follow from a mere inspection of photographic exposures of the sky, but reliable evidence of spiral arms in our own galaxy can be obtained in general only from an accurate location of suitably selected objects (Sects. 3.3, 5.4.3, 5.4.4). It now appears that parts of spiral arms can be detected with the help of typical "spiral arm indicators", but one cannot yet claim an undisputed empirical comprehension of the Grand Design of the spiral arm structure of our system. Theoretical pronouncements on the spiral basic structure of flat, rotating galaxies are therefore regarded with particular interest.

The Winding Up Dilemma: Like our Milky Way system, the other flat galaxies as a rule rotate differentially, with an angular velocity $\omega = \omega(R)$ which decreases outwards. Noncircular structures, therefore, such as a straight band lying in the main plane through the centre of rotation, will be drawn out with time into a spiral configuration. The development of spiral structure accordingly appears at first easy to explain. However a spiral arm which always consisted of the same stars and interstellar clouds, thus representing a structure fixed in the material, would be already fully "wound up" (like a clock spring) after a few rotation periods and thereafter disappear. For a galaxy of the size of the Milky Way system this would require only a few 10^8 years. To explain the high observed frequency of galaxies with spiral arms, therefore requires either the assumption that a new spiral structure be created after only about 10^8 years, or that the observed spiral arms are not structures fixed in the material, but represent a relatively longlived wave-like perturbation of the original rotationally symmetric galaxy.

Density Wave Concept: The second assumption had already been discussed in the twenties by B. Lindblad in connection with calculations of the orbits of individual test stars. A successful quantitative treatment of the collective effect, however, was first achieved by the work of C.C. Lin and fellow workers since 1963 and it has now become the most fully developed theory of the spiral structure of galaxies, which we shall accordingly examine more closely in the next section. Here the spiral arms are regarded as spatial maxima of a density wave moving relative to the material in the disc of a galaxy: not only the star field but also the interstellar medium are compressed in a wave crest curved into a spiral form, and correspondingly rarefied in the neighbouring wave trough. Since a density wave can move relative to the material, it does not need to take part in the differential rotation and so is not wound up by it. Here we must of course discuss the possibility of the density wave being damped out, which could place the advantage of this concept in question.

The propagation of a density wave in the collision-free star field is based simply on the fact that the raising of the mass density in the wave causes a corresponding wave-like perturbation in the profile of the gravitational potential, which then reflects back on the motions of the stars and hence the stellar distribution. It is a purely gravitational process, in which the pressure effect – as in acoustic waves – plays no part. This is at first assumed also true of the interstellar gas (linear theory).

Stars, like interstellar matter, must spend longer in the density maxima, and hence in the spiral arms, than in the minima. There are systematic deviations from the average flow (circular orbital motion!), consisting of periodic convergences on the places of maximum density and divergences from the regions between the arms.

The observational fact, that young stellar objects and interstellar matter are particularly concentrated on the spiral arms, can be explained on the basis of the density wave concept in broad outline as follows. If an extensive rigid spiral-form potential wave moves relative to the interstellar gas with supersonic velocity, then this has the result of developing a shock front (see Sect. 6.3.1). The quantitative treatment of this nonlinear process shows, for example, that for a potential wave amplitude of only 5 % the gas in the shock front is concentrated to five to eight times the mean density. Sufficiently dense interstellar clouds could thereby be compressed up to the limit of gravitational instability with consequent star formation.

Following the discovery of interstellar magnetic fields there appeared also to be a possible magnetohydrodynamic interpretation of spiral structure. The observational results so far available, however, give only a few microgauss for the strength of the mean magnetic field (Sects. 4.1.5, 4.1.8). The magnetic energy density $B^2/8\pi$ is therefore somewhat smaller than the energy density $\frac{1}{2}\varrho v^2$ of the irregular motions of the interstellar clouds (ϱ = mass density in the cloud, v = peculiar velocity of the cloud). In particular, however, $B^2/8\pi$ is negligible compared with the energy density of the galactic rotational motion. One must therefore assume that the magnetic field can neither prevent the winding up of material spiral arms, nor cause appreciable deviations from the circular orbital motion of the gas.

Generation of Spiral Structure: A possible cause is an encounter with another galaxy. Model calculations have been carried out on this question. In the case of the Milky Way system the production of spiral arms by "tidal interactions" with the two Magellanic Clouds could play a role. In the neighbourhood of galaxies with well formed spiral structures, however, one often finds no other system which might be a collision partner. Nevertheless, the high frequency of spiral systems permits this explanation if necessary for certain peculiar galaxies, from which even material bridges to neighbouring systems are often detectable.

Various suggestions have been made of internal stimulation or generation mechanisms, which might lead directly to the formation of spiral arms: for example, explosive gas ejection from the rotating galactic kernel region. However, one can manage without dramatic assumptions of this sort: the numerical simulation of the development of disc-like, differentially rotating systems composed of many gravitating point masses (Sect. 6.1.4) shows that, assuming the dimensions of the Milky Way system, a spiral structure arises after only a few 10^8 years, which, for example, disappears again after half a rotation and later gets superseded by a new one. The deeper reason for this lies in that the rotationally symmetric stellar disc selected as starting point finds itself in marginally stable equilibrium, which can be perturbed by trivial random accumulations of stars. The internal *stimulation* of spiral-form density waves could therefore be a completely natural process. A quite different possibility for the creation of spiral arms is described in Sect. 6.3.4.

6.2.2 Spiral-Form Density Waves

"Hydrodynamic Equations" for a Disc of Vanishing Thickness: Let us restrict ourselves to consideration of the hydrodynamic approximation for a plane rotating disc, since for this case relatively simple relations are obtained, yet all the important conclusions of the general theory can be obtained. We start from (6.62) and (6.63), but must now consider the unsteady case. In the left-hand side of (6.62) there now appear the additional terms $\partial \overline{\Pi}/\partial t$, $\partial \overline{\Theta}/\partial t$ and $\partial \overline{Z}/\partial t$; in (6.63) $\partial \varrho_s/\partial t$ appears. Integrating these equations with respect to z from $-\infty$ to $+\infty$, and letting the thickness of the disc go to zero, then $\overline{\Pi}$, $\overline{\Theta}$, $\overline{Z}$, $\overline{\Pi\Theta}$ etc. between $z = 0 - \varepsilon$ and $0 + \varepsilon$ must be seen to be independent of z and we obtain the same expressions as in (6.62) and (6.63), without, however, the (vanishing) derivatives with respect to z and with

$$\sigma_s = \sigma_s(R, \vartheta; t) = \int_{-\infty}^{+\infty} \varrho_s(R, \vartheta, z; t)dz \qquad (6.90)$$

instead of ϱ_s. For example, the continuity equation reads

$$\frac{\partial \sigma_s}{\partial t} + \frac{\partial (\sigma_s \overline{\Pi})}{\partial R} + \frac{\sigma_s \overline{\Pi}}{R} + \frac{1}{R}\frac{\partial (\sigma_s \overline{\Theta})}{\partial \vartheta} = 0 \quad . \qquad (6.91)$$

For the sake of simplicity we shall further assume an isotropic velocity dispersion, so that in the plane $z = 0$ a scalar "surface pressure" can be introduced by

$$p_s = \sigma_s \overline{(\varPi - \overline{\varPi})^2} = \sigma_s \overline{(\varTheta - \overline{\varTheta})^2} \quad \text{or}$$

$$\sigma_s \overline{\varPi^2} = \sigma_s (\overline{\varPi})^2 + p_s \quad \text{and} \quad \sigma_s \overline{\varTheta^2} = \sigma_s (\overline{\varTheta})^2 + p_s \quad ;$$

further let $\varPi$, $\varTheta$ and Z be uncorrelated with one another:

$$\overline{(\varPi - \overline{\varPi})(\varTheta - \overline{\varTheta})} = \overline{\varPi \varTheta} - \overline{\varPi} \, \overline{\varTheta} = 0$$

and hence $\overline{\varPi \varTheta} = \overline{\varPi} \, \overline{\varTheta}$ etc. Accordingly we obtain for the first two terms in the first equation (6.62)

$$\frac{\partial (\sigma_s \overline{\varPi^2})}{\partial R} = \frac{\partial p_s}{\partial R} + 2\overline{\varPi} \frac{\partial \overline{\varPi}}{\partial R} \sigma_s + \frac{\partial \sigma_s}{\partial R} (\overline{\varPi})^2 \quad \text{and}$$

$$\frac{1}{R} \frac{\partial (\sigma_s \overline{\varPi \varTheta})}{\partial \vartheta} = \frac{1}{R} \frac{\partial \overline{\varPi}}{\partial \vartheta} \sigma_s \overline{\varTheta} + \frac{\overline{\varPi}}{R} \frac{\partial (\sigma_s \overline{\varTheta})}{\partial \vartheta} \quad .$$

By means of (6.91) the last term can be expressed in terms of derivatives with respect to R and t, and eventually – including the temporal derivative $\partial (\sigma_s \overline{\varPi})/\partial t$ – one gets

$$\frac{\partial \overline{\varPi}}{\partial t} + \overline{\varPi} \frac{\partial \overline{\varPi}}{\partial R} + \frac{\overline{\varTheta}}{R} \frac{\partial \overline{\varPi}}{\partial \vartheta} - \frac{(\overline{\varTheta})^2}{R} + \frac{1}{\sigma_s} \frac{\partial p_s}{\partial R} + \frac{\partial V}{\partial R} = 0 \quad . \tag{6.92}$$

In similar manner the second equation (6.62) follows:

$$\frac{\partial \overline{\varTheta}}{\partial t} + \overline{\varPi} \frac{\partial \overline{\varTheta}}{\partial R} + \frac{\overline{\varTheta}}{R} \frac{\partial \overline{\varTheta}}{\partial \vartheta} + \frac{\overline{\varPi} \, \overline{\varTheta}}{R} + \frac{1}{\sigma_s} \frac{1}{R} \frac{\partial p_s}{\partial \vartheta} + \frac{1}{R} \frac{\partial V}{\partial \vartheta} = 0 \quad . \tag{6.93}$$

A third equation no longer appears. Equations (6.92) and (6.93) are identical with Euler's Equations of Hydrodynamics for a thin rotating disc and can therefore be applied also to the interstellar gas. In what follows we write in general σ and p instead of σ_s and p_s.

Equations for Small Perturbations: Let us consider the spiral structure as a small perturbation of the rotationally symmetric solution (index 0) with pure circular orbital motion $[\overline{\varPi} = 0, \overline{\varTheta} = \varTheta_K = R\omega(R)]$ and accordingly set

$$
\begin{aligned}
V(R, \vartheta, z, t) &= V_0(R, z) + V_1(R, \vartheta, z, t) \text{ with } |V_1| \ll |V_0| \\
\sigma(R, \vartheta, t) &= \sigma_0(R) + \sigma_1(R, \vartheta, t) \text{ with } |\sigma_1| \ll |\sigma_0| \\
p(R, \vartheta, t) &= p_0(R) + p_1(R, \vartheta, t) \text{ with } |p_1| \ll |p_0| \\
\overline{\varPi}(R, \vartheta, t) &= u(R, \vartheta, t) \\
\overline{\varTheta}(R, \vartheta, t) &= R\omega(R) + v(R, \vartheta, t) \quad .
\end{aligned}
\tag{6.94}
$$

Let us substitute these in (6.91) and also in (6.92) and (6.93), both multiplied

404

throughout by σ_s, taking account of the fact that v_0, σ_0, p_0, $\overline{\varPi} = 0$ and $\Theta = Rw(R)$ satisfy the corresponding equations for the steady case.

Neglecting all terms of higher than the first order in V_1, σ_1, p_1, u, v, and therefore also all terms with products of these quantities, such as uv or even $(\partial V_1/\partial R)\sigma_1$ etc., we are left with the linearised *perturbation equations*

$$\frac{\partial \sigma_1}{\partial t} + w\frac{\partial \sigma_1}{\partial \vartheta} + \frac{\partial(\sigma_0 u)}{\partial R} + \frac{\sigma_0 u}{R} + \frac{1}{R}\frac{\partial(\sigma_0 v)}{\partial \vartheta} = 0 \tag{6.95}$$

$$\frac{\partial u}{\partial t} + w\frac{\partial u}{\partial \vartheta} - 2wv + \frac{1}{\sigma_0}\frac{\partial p_1}{\partial R} + \frac{\partial V_1}{\partial R} = 0 \tag{6.96}$$

$$\frac{\partial v}{\partial t} + w\frac{\partial v}{\partial \vartheta} + \frac{\kappa^2}{2w}u + \frac{1}{\sigma_0}\frac{1}{R}\frac{\partial p_1}{\partial \vartheta} + \frac{1}{R}\frac{\partial V_1}{\partial \vartheta} = 0 \quad . \tag{6.97}$$

Here κ is the epicyclic frequency from (6.55). V_1 also satisfies the Poisson Equation

$$\frac{1}{R}\frac{\partial}{\partial R}\left(R\frac{\partial V_1}{\partial R}\right) + \frac{1}{R^2}\frac{\partial^2 V_1}{\partial \vartheta^2} + \frac{\partial^2 V_1}{\partial z^2} = -4\pi G\sigma_1\delta(z) \quad , \tag{6.98}$$

where $\delta(z)$ denotes the delta function.

Statement and Procedure of Lin's Theory: Let us consider a spiral-form wave rotating as a rigid entity with the angular velocity w_p (p for *pattern*) with m arms, which can be expressed, e.g. for density, in the form

$$\sigma_1(R,\vartheta,t) = \mathrm{Re}\{\sigma_1^* e^{i\chi}\} \quad \text{with} \tag{6.99}$$

$$\chi = mw_p t - m\vartheta + \varPhi(R) \quad . \tag{6.100}$$

In order to explain this statement let us first of all take a fixed time t, hence a spiral model at rest, and set $m = 2$. The profile of the maximal density σ_1 with R and ϑ is obtained by setting $\chi = 0$, i.e. $-2\vartheta + \varPhi(R) = $ constant. This relation describes a spiral $R = f(\vartheta)$ (Fig. 6.10). The *pitch angle* ψ of the *spiral* compared with a circle $R = $ constant through the point (R,ϑ) is given by

$$\tan\psi = \frac{df(\vartheta)}{R\,d\vartheta} = \frac{1}{R}\frac{1}{(d\vartheta/dR)} = \frac{2}{R\varPhi'(R)} \quad . \tag{6.101}$$

In the case of a logarithmic spiral $R = R^* e^{a\vartheta}$, for example, one has $\varPhi(R) = (2/a)\ln(R/R^*)$ and hence $\tan\psi = a$. If, at fixed time, we move along a circle of radius R with increasing ϑ, we obtain a double wave $\sigma_1 \sim \exp(-2i\vartheta)$, since we have postulated two arms.

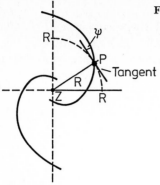

Fig. 6.10. Geometry of the spiral arms

Let us now consider the time dependence of σ_1. At a fixed point (R, ϑ) (6.99) represents an oscillation with the frequency $2\omega_{\mathrm{p}}$: with each circuit two arms pass the point. At fixed ϑ a perturbation with amplitude σ_1^* runs radially outwards: $2\omega_{\mathrm{p}}t + \Phi(R) = \mathrm{constant}$, defining the functional relationship $R = R(t)$. The magnitude of the phase velocity of this radial wave motion is therefore given by

$$v = \left| \frac{dR}{dt} \right| = \left| \frac{dt}{dR} \right|^{-1} = \frac{2\omega_{\mathrm{p}}}{\Phi'(R)} \quad . \tag{6.102}$$

With the notation

$$\Phi'(R) = \frac{d\Phi(R)}{dR} = k \tag{6.103}$$

the *radial wavelength* is given by

$$\lambda(R) = \frac{v}{(\omega_{\mathrm{p}}/2\pi)} = \frac{2\pi}{k} \quad . \tag{6.104}$$

$\lambda(R)$ is approximately equal to the radial distance between neighbouring arms ΔR. Since for consecutive arms in a two-armed spiral structure at fixed ϑ we must have $\chi(R + \Delta R, \vartheta + \pi) = \chi(R, \vartheta)$, it follows from (6.100) that $\Phi(R + \Delta R) - \Phi(R) = 2\pi$ or approximately $\Delta R \, \Phi'(R) = \Delta R \, k = 2\pi$. Here k denotes the radial wave number of the perturbation.

Let us now ask whether and under what conditions a steady spiral structure in the form of a rigidly rotating wave of the type (6.99) for V_1, σ_1, u, v satisfies the equations (6.95) to (6.97) and Poisson's Equation (6.98). The solution shall thus be self-consistent: the density distribution σ_1 produces the potential V_1 and this causes the waves in u, v and σ_1. The simultaneous fulfillment of the "hydrodynamic equations" *and* Poisson's Equation is achieved by proceeding in steps. This is explained in the following scheme by Lin.

First of all one postulates

$$V_1 = V_1(R, \vartheta, t) = V_1^*(R) \exp[i\chi(R, \vartheta, t)] \tag{6.105}$$

406

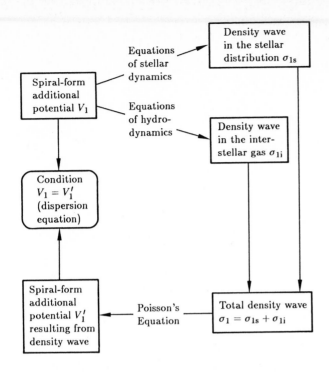

and finds the resulting density wave $\sigma_1(R, \vartheta, t)$ in the star field and in the interstellar gas. Now one puts the solution into Poisson's Equation and recovers a potential perturbation V_1', which can produce this wave in the mass density $\sigma_1(R, \vartheta, t)$. Self-consistency means that the potential perturbation V_1' must be equal to the originally postulated expression (6.105). This identity represents a condition equation on the parameters appearing: radial wavelength $\lambda = 2\pi/k$ and the frequency of the orbital motion ω or ω_p ("dispersion equation" – see below).

Let us now follow this procedure in detail. Because of the formal analogy of the "hydrodynamic equations" of stellar dynamics with Euler's Equations of hydrodynamics we can simply treat the stars and the gas as a single package and then indicate any differences only where necessary. Attention is first of all directed to the solutions σ_1, u, v of (6.95) to (6.97), which are set in the form

$$
\begin{aligned}
\sigma_1 &= \sigma_1^*(R)e^{i\chi}\\
u &= u^*(R)e^{i\chi}\\
v &= v^*(R)e^{i\chi} \ .
\end{aligned}
\tag{6.106}
$$

Here and in the *postulated* potential perturbation (6.105) we shall now assume a general spiral wave with m arms. Finally for simplicity we shall assume *tightly wound* spirals, showing a small angle ψ. This means that $\tan \psi \ll 1$ or according to (6.101) and (6.103) (in practice $m \leq 4$)[8]

$$k = \Phi'(R) \gg 1/R \quad . \tag{6.107}$$

Each of equations (6.95) to (6.97) then gives a simple relation between two of the quantities σ_1^*, u^* and v^*. We enlarge on this for (6.95): one first of all obtains

$$\sigma_1^* e^{i\chi} i m \omega_p + \omega \sigma_1^* e^{i\chi} i(-m) + \frac{\partial \sigma_0}{\partial R} u^* e^{i\chi} + \sigma_0 u^* e^{i\chi} i k$$

$$+ \frac{\sigma_0 u^*}{R} e^{i\chi} + \frac{\sigma_0 v^*}{R} e^{i\chi} i(-m) = 0 \quad .$$

Because $\partial \sigma_0 / \partial R \simeq 0$ and $1/R \ll |ik|$ according to (6.107) it follows that

$$m(\omega_p - \omega) \sigma_1^* + k \sigma_0 u^* = 0 \quad . \tag{6.108}$$

Here we introduce the *local relative frequency* of the wave:

$$\nu = \frac{m(\omega_p - \omega)}{\kappa} \quad , \tag{6.109}$$

where κ denotes the epicyclic frequency according to (6.55). $|\nu|$ is the frequency of passage of the spiral arms as seen by an observer rotating with the material, measured in units of the epicyclic frequency κ. (The significance of κ for the present problem is explained below.) Now (6.108) takes the form

$$\nu \sigma_1 + \frac{k \sigma_0}{\kappa} u^* = 0 \quad . \tag{6.110}$$

In similar fashion (6.96) and (6.97) with (6.105) and (6.101) lead to the relations

$$i\nu u^* - \frac{2\omega}{\kappa} v^* = \frac{ik}{\kappa} V_1^* \tag{6.111}$$

$$\frac{\kappa}{2\omega} u^* + i\nu v^* = 0 \quad . \tag{6.112}$$

Here for the sake of simplicity we have first of all set $p_1 = 0$ and then $\partial p_1 / \partial R = 0$ and $\partial p_1 / \partial \vartheta = 0$. This assumption means that the velocity scatter is neglected.

Although the observations show that the flattest sub-systems show the smallest velocity dispersions, this approximation goes too far: according to the condition (6.88), at vanishing velocity scatter the gravitational effect between the stars of the disc would cause a local collapse. The agreement between the

---

[8] This approximation means the application of the so-called WKB method (after Wenzel, Kramers and Brillouin): one assumes that the amplitudes σ_1^*, u^* and v^* vary only slowly with R : the first term in $\partial \sigma_1 / \partial R = (\partial \sigma_1^* / \partial R) e^{i\chi} + \sigma_1^* e^{i\chi} i \Phi'$ is small compared with the second term, and with $\partial \sigma_1 / \partial R \simeq \sigma_1 / R$ it follows that $1/R \ll |\Phi'| = |k|$.

observed and the critical velocity dispersion established in connection with (6.89) means on the contrary that the galaxy is stable against gravitational collapse, but leaves open the possibility of instability in the nature of periodic oscillations. A significantly higher radial velocity dispersion was however associated with a greater stability of the system against periodic spiral-form perturbations. Only in flat systems with relatively small radial velocity dispersion can a spiral structure form. This can also be otherwise expressed thus: if the stars pass with high velocity through a potential perturbation, they have too little time to be able to undergo a marked change in their kinetic energy. Too low a velocity dispersion, on the other hand, would lead to contraction and an accompanying increase of $\overline{(\Pi - \overline{\Pi})^2}$ up to a critical value,[9] corresponding to the optimal response to wave-form periodic perturbations. Lin and his co-workers therefore assume that $\overline{(\Pi - \overline{\Pi})^2}$ in the spiral galaxies has just reached its critical value, and set

$$p_1 = \sigma_1 \overline{(\Pi - \overline{\Pi})^2_{\mathrm{crit}}} \quad \text{with}$$

$$\overline{(\Pi - \overline{\Pi})^2_{\mathrm{crit}}} = 0.2857 \frac{\kappa^2}{k^2_{\mathrm{crit}}} \quad .$$

Compared with the case $p_1 = 0$ then, the left-hand side of (6.111) is supplemented by the term $-ik^2\overline{(\Pi - \overline{\Pi})^2}u^*/\kappa^2\nu$. The solution of the equations (6.110) to (6.112) for the amplitudes of the waves in the density and in the velocity components then gives[10]

$$\sigma_1^* = -\frac{k^2 V_1^* \sigma_0}{\kappa^2} \frac{1}{1 - \nu^2 + x}$$

$$u^* = \frac{k V_1^*}{\kappa} \frac{\nu}{1 - \nu^2 + x}$$

$$v^* = \frac{ik V_1^*}{2\omega} \frac{1}{1 - \nu^2 + x} \qquad \text{with} \qquad (6.113)$$

$$x = \frac{k^2 \overline{(\Pi - \overline{\Pi})^2}}{\kappa^2} \quad . \qquad (6.114)$$

This quantity characterises the susceptibility of the system to density waves and is called the *stability number*. According to the above statements x is less than about $1/4$ for gravitationally unstable systems. For very stable systems, in which no density waves can form, $x \gg 1/4$, and in spiral systems we probably have $x \simeq 1/4$.

---

[9] The gaining of kinetic energy by the stars from gravitational energy by means of encounters, as these become more frequent, is analogous to the heating of a gravitationally contracting sphere of gas.

[10] The hydrodynamic approximation for the collision-free "star gas", according to studies by means of the fundamental equation of stellar dynamics, also requires certain "reduction factors" in the expressions (6.113), which we cannot go into here.

We have so far shown that a spiral shaped perturbation of the potential profile of the form (6.105) produces a density wave which is in phase with the potential perturbation. Now, conversely, we ask whether this density wave has the potential wave (6.105) as direct consequence, and thus seek V_1 as the solution of Poisson's Equation (6.98). One must here be satisfied with an asymptotic solution for tightly wound arms found by Lin and Shu − a general solution appears to be so far unknown. Since a readable derivation would claim a disproportionate amount of space here, we refer to the literature (e.g. Rohlfs, 1977) and give the result directly: one finds that (6.105) is a solution with

$$V_1^* = -2\pi G \frac{\sigma_1^*(R)}{|\Phi'(R)|} = -2\pi G \frac{\sigma_1^*}{|k|} \quad . \tag{6.115}$$

Self-consistency means that this expression is to be inserted in (6.113). The first equation (6.113) then gives with (6.115) the required condition equation for the existence of a steady rigidly rotating spiral-form density wave, the so-called *dispersion equation*

$$1 - \nu^2 + x = \frac{2\pi G \sigma_0 |k|}{\kappa^2} \tag{6.116}$$

or, with (6.109) and (6.114),

$$m^2 (\omega_p - \omega)^2 = \kappa^2 + k^2 \overline{(\Pi - \overline{\Pi})^2} - 2\pi G \sigma_0 |k| \quad . \tag{6.117}$$

This equation connects the local wavelength $\lambda = 2\pi/k$ with the local frequency of the density wave for an observer rotating with the material $\omega_p - \omega$ and thereby produces the form of the spiral structure. This is a quadratic equation for λ, and its solution for each frequency $\omega_p - \omega$ or ν gives in general two different values for the wavelength: the dispersion relation curve consists of two branches, one for short and one for long waves.

Models of the Spiral Structure: Whereas $\omega = \omega(R)$ and hence $\kappa = \kappa(R)$ are given, ω_p is a free parameter of the theory, which however is subject to important restrictions. Equations (6.113) show that the amplitude of the density wave increases without limit when $\nu^2 = 1 + x$, i.e., using (6.109), when

$$m(\omega_p - \omega) = \pm \kappa \sqrt{1 + x} \approx \pm \kappa \quad . \tag{6.118}$$

Here the linear theory breaks down. These are the so-called *Lindblad Resonances*, which appear at that central distance R, at which an observer rotating with $\omega(R)$ sees the individual spiral arms pass with a frequency equal to the local epicyclic frequency κ.

For each chosen value of ω_p there is an *inner* and an *outer* Lindblad Resonance, which are determined by $\omega_p = \omega - (\kappa/m)$ or $\omega_p = \omega + (\kappa/m)$.

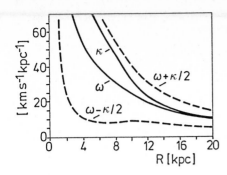

Fig. 6.11. Variations of angular velocity of galactic rotation and the epicyclic frequency with distance R from the galactic centre for Schmidt's mass model. Two-armed spiral structure is to be expected only between $\omega - (\kappa/2)$ and $\omega + (\kappa/2)$. (After Lin, Yuan and Shu, 1969; by permission)

For the rotation model of Table 6.3 and $m = 2$ the corresponding R-values can be taken from Fig. 6.11, being the abscissae of the intersection points of the two curves for $\omega - (\kappa/2)$ and $\omega + (\kappa/2)$ with the horizontal straight line $\omega_p =$ constant. Only in the region in between, which also contains the distance of the "co-rotation" with $\omega = \omega_p$, is spiral structure to be expected. For $m = 3$ or 4 this annulus is essentially narrower than for $m = 2$. Accordingly two-armed spirals are the most common.

In the framework of "epicycle theory" treated in Sect. 6.1.4 the Lindblad resonances can be illustrated in the following way. We restrict ourselves to the case $m = 2$. A star with motion deviating slightly from the circular orbit traces out, relative to a coordinate system rotating with the angular velocity of the circular orbital motion $\omega(R)$, a small ellipse (epicycle) with the epicycle frequency $\kappa(R)$ (cf. Fig. 6.6). If $\kappa/2 = \omega$, then after half a galactic rotation the star will have returned exactly to its starting point on the epicycle. The same is true for $\kappa/2 = \omega - \omega_p$, if we refer to a coordinate system that rotates with the angular velocity ω_p, in which the spiral structure is accordingly at rest. The perturbation of the motion of the star by the density wave at each point of the epicycle is in this case exactly the same after each rotation. The perturbations must therefore continually accumulate ("resonance"). With regard to the significance of the resonances in the works of B. Lindblad (from 1925) on the theory of the spiral structure, we refer to the article by Wielen (1974) mentioned in the literature survey.

Lin, Yuan and Shu (1969) selected $\omega_p = 11\,\mathrm{km\,s^{-1}\,kpc^{-1}}$ in conjunction with $m = 2$ for a first attempt at finding a density wave model of the galactic spiral structure. Then the inner Lindblad resonance fell roughly on the "3 kpc arm", inside which there appeared to be no spiral arms. The outer Lindblad resonance in this case lay at $R \simeq 25\,\mathrm{kpc}$. With $\omega(R), \kappa(R)$ as well as σ_0 (= mean surface density of the galactic disc) taken from the Schmidt mass model (Table 6.3), and $x = 1/4$ or an appropriate value for $\overline{(\Pi - \overline{\Pi})^2}$ the dispersion equation produces $k(R)$ or $\lambda(R)$ ($\simeq 4\,\mathrm{kpc}$). Then $\Phi(R)$ can be obtained by numerical integration of (6.103) and hence the profile of the spirals from (6.100). Lin, Yuan and Shu used the "short wave mode" of the dispersion equation and obtained the result displayed in Fig. 6.12. The amplitudes of the waves in the gravitational field and in the surface density of the material (stars + gas)

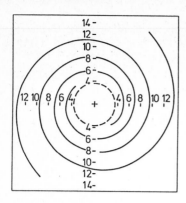

Fig. 6.12. Example of a model for the spiral structure of the Milky Way system with $\omega_p = 11\,\mathrm{km\,s^{-1}\,kpc^{-1}}$. Distances from centre in kpc. (After Lin, Yuan and Shu, 1969; by permission)

amounted in this model to 5 % and 10 % of the mean values. The radial velocity dispersion of the stars was set at $\pm 40\,\mathrm{km\,s^{-1}}$, the scatter of the velocities of the gas was taken as $10\,\mathrm{km\,s^{-1}}$. The deviations of the velocity field from the pure circular orbital motion are $< 15\,\mathrm{km\,s^{-1}}$.

Other authors found a similar spiral structure with the solution leading to "long waves", with considerably larger values for ω_p. Since there are several free parameters, the adaptability of the density wave theory is really considerable.

Lifetime of Density Waves: Lin's waves have a finite radial group velocity of about $8\,\mathrm{km\,s^{-1}}$, with which they run out of the system (Toomre, 1969). This involves a loss of energy, and hence a damping, so that the lifetime of the whole phenomenon is limited. There is a similar effect from the formation, established in Sect. 6.3.3, of spiral shaped shock fronts in the interstellar gas with the passage of the potential waves, since this causes a dissipation of mechanical energy which is not negligible. The density waves used as a way of escape from the winding-up dilemma (Sect. 6.2.1) survive for only about 10^9 years and are therefore only about five times as long lived as material spiral arms. One therefore needs a mechanism which produces new density waves after only about 10^9 years. This problem cannot yet be considered completely cleared up: however, the internal stimulation of spiral-form density waves, mentioned in Sect. 6.2.1, by deviations from rotational symmetry − especially in the central region − seems to promise a quite plausible solution.

Comparison Between Theory and Observation: The formulae of the linear theory of density waves assume a steady dynamic equilibrium which can only become established after a few rotations. The density wave described up to now is therefore to be expected in the field of those stars which are older than a few 10^8 years, and hence in the older disc population. The *amplitude* of this wave probably amounts to *about 5 to 10 % of the mean value of the star density* and is therefore scarcely detectable with certainty in our galaxy. In a few extragalactic systems such a relatively uniform spiral structure, as opposed to arms formed of young objects, appears to be indicated by surface photometry in the visible and in the near infrared. A considerable fraction of younger stars could, of course, still be contained therein.

The formulae obtained are not directly applicable to *young objects*. The distribution and motion of young stars is determined essentially by the non-linear effect of the potential wave on the interstellar medium from which these stars have been created, and they accordingly start out with the current velocity of the gas and may leave the original spiral wave. The theoretical representation of the spiral phenomenon best documented by observation (in extragalactic systems), namely, the arrangement (often necklace-like) of the young stars and the H II regions in spirals, thus also requires assumptions on the process of star formation and turns out to be a very involved exercise. Since knowledge of a few results of gasdynamics is required, we go into this first of all in the following section. Let it be remarked now, however, that in the case of our own galaxy, here of primary interest, we unfortunately still do not have reliable knowledge of the large scale configuration of the spiral arms represented by the young objects.

The best evidence of the existence of a density wave in our galaxy is provided by the wave-form deviations from smooth profiles of the *rotation curves* derived for the H I gas (Sect. 5.4.3). By appropriate choice of the parameters of the density wave model the main features of the curves can be satisfactorily approximated. An example of this is given in Sect. 5.4.3, Fig. 5.31, but this, of course, still does not provide a compelling corroboration of the density wave theory. This is true particularly in view of the fact that here only the linear theory has been applied, in which the effect on the velocity field of the shock waves in the gas has still not been taken into account.

As has already been explained, the *kinematics* of the linear density wave should be accurately reflected only in the motions of the older stars, whereas for the young stars and for the interstellar matter a velocity field deviating from this is to be expected. For the solar neighbourhood in particular the mean velocities of these various object groups can be predicted relative to the circular orbit velocity. For the old disc population, for example, with the model of Lin et al. explained above, one obtains a radial component $u \simeq +1 \, \mathrm{km \, s^{-1}}$. Typical objects of this group are the dwarf stars near the sun, relative to which the radial component of the solar motion turns out to be $U_\odot \simeq +5 \, \mathrm{km \, s^{-1}}$ (see Sect. 3.2.3). Relative to the local circular orbital velocity, therefore, the sun would move radially with the velocity $u_\odot \simeq +4 \, \mathrm{km \, s^{-1}}$, in contradiction to the result (3.40) according to which this is $+9 \, \mathrm{km \, s^{-1}}$. A way out is to assume a radially outwards motion of the local standard of rest of about $5 \, \mathrm{km \, s^{-1}}$. In Sect. 5.4.3 it has already been mentioned that this could be an explanation for the average difference between the two rotation curves derived from observations of the northern and of the southern parts of the Milky Way (Wielen, 1979).

6.3 Dynamics of the Interstellar Gas

6.3.1 Posing of the Problem and Fundamentals

In the two previous sections of this chapter we have been concerned in essence with the dynamical interpretation of the motions and the distribution of the stars; the interstellar gas has indeed been included, formally, in the linear theory of density waves, but the critical differences compared with the "collision-free" star field have not been taken into account. We turn now to the dynamics of the gas, with the aim of gaining a thorough physical understanding of the motion and the structure in the interstellar gas. After setting out the fundamentals we shall discuss a selection of individual problems recognised as important with the aid of relatively simple theoretical models. Where the space needed would be excessive within the present framework we shall limit ourselves to qualitative considerations.

Fundamental Equations: Since even in the H I region there is a weak ionisation and a correspondingly high conductivity, we have in general to deal with problems of plasma physics. In the present case the fundamental equations can usually be put in the approximation of macroscopic *magnetohydrodynamics*: one combines the equations of hydrodynamics, including allowance for the mutual interactions of motion and magnetic field, with the equations of electrodynamics for slow processes (velocity of the material and phase velocity of change of field quantities $\ll$ velocity of light). The modified Euler's equation for the macroscopic velocity of the gas v, neglecting the viscosity term, then becomes

$$\varrho \frac{Dv}{Dt} \equiv \varrho \frac{\partial v}{\partial t} + \varrho(v \cdot \nabla)v = -\nabla P - \varrho \nabla V - \frac{1}{8\pi}\nabla B^2$$
$$+ \frac{1}{4\pi}(B \cdot \nabla)B \quad . \tag{6.119}$$

Here D/Dt denotes the Lagrangean time derivative, i.e. the rate of change with time for a point moving with the material; ϱ and P denote density and pressure, V is the potential of the total gravitational force. The two last terms on the right-hand side represent the force of the magnetic field, where we use these quantities in vacuo $B = H =$ magnetic field (magnetic flux density) with the unit 1 Gauss $= 1\,\mathrm{g}^{1/2}\,\mathrm{cm}^{-1/2}\,\mathrm{s}^{-1}$: for the force field in vacuo $k = (1/c)(j \times B)$, where the electric current density j can be expressed by Maxwell's first equation as $j = (c/4\pi)(\nabla \times B)$ for $\dot{E} = 0$. From the identity $(\nabla \times B) \times B = -\frac{1}{2}\nabla B^2 + (B \cdot \nabla)B$ we get the terms indicated. For a homogeneous magnetic field [e.g., $B_x = 0$, $B_y = 0$, $B_z = B_z(x,y)$] it follows that $(B \cdot \nabla)B = 0$ and there remains only the gradient of the magnetic pressure $(1/8\pi)B^2$ acting perpendicular to the field direction.

Between ϱ and v there exists also the continuity equation

$$\frac{\partial \varrho}{\partial t} + \nabla \cdot (\varrho v) = 0 \tag{6.120}$$

and the gravitational potential V is connected with the total mass density (stars and interstellar matter) by Poisson's Equation (6.3) formulated in Sect. 6.1.1 for the stars only.

Taking the electric field E from Ohm's Law for material at rest $j = \sigma E$ (σ = conductivity) and j from Maxwell's first equation already used above, we can transform Maxwell's second equation (the induction law) $\nabla \times E = -(1/c)\dot{B}$ into

$$\dot{B} = -\frac{c^2}{4\pi\sigma}(\nabla \times (\nabla \times B)) = \frac{c^2}{4\pi\sigma}\Delta B \quad . \tag{6.121}$$

The right-hand side follows from $(\nabla \times (\nabla \times B)) = \nabla(\nabla \cdot B) - \Delta B$ and the absence of sources in the magnetic field $\nabla \cdot B = 0$. Equation (6.121) is of the heat conduction type and describes the diffusion of the magnetic field, as the magnetic energy is gradually transformed into heat by the ohmic dissipation in the electric currents produced. If for a rough estimate we put $\dot{B} \simeq B/\tau$, $\Delta B \simeq B/L^2$, where L is a measure of the original extent of the field, then it follows that the "decay time"

$$\tau \approx \frac{4\pi\sigma}{c^2}L^2 \quad .$$

For the interstellar gas in the H I region σ is of the order of magnitude of $10^{10}\,\mathrm{s}^{-1}$. With $L > 1\,\mathrm{AU}$ it then follows that $\tau > 3 \times 10^{16}\,\mathrm{s} \simeq 10^9$ years, and for $L = 1\,\mathrm{pc}$ we already have $\tau \simeq 2 \times 10^{14}$ years!

If $v \neq 0$, then in the above derivation we have $j = \sigma[E + (1/c)(v \times B)]$ and on the right-hand side of (6.121) the term $\nabla \times (v \times B)$ appears. Because of the smallness of the diffusion coefficient $c^2/4\pi\sigma \simeq 0$ we can then write

$$\dot{B} - \nabla \times (v \times B) = 0 \quad . \tag{6.122}$$

If one follows a finite surface F (element df) fixed in the material and moving with it, with a boundary curve C (line element dl), on a small part of the path of the flow $ds = v\,dt$ (Fig. 6.13), then the change of magnetic flux Φ through this surface is given by

$$d\Phi = \iint_F (\dot{B}\,dt) \cdot df + \oint_C B \cdot (ds \times dl) \quad .$$

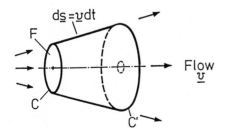

Fig. 6.13. Derivation of (6.123)

415

The first integral is the change in Φ as a result of the purely timewise variation of B at a fixed point, the second integral results from the motion of the boundary curve from C to C' with the flow and represents the additional flux through the strip between the boundary curves C and C'. Remembering that $B \cdot (ds \times dl) = -(ds \times B) \cdot dl$ and $ds = v\, dt$, and applying Stokes's Theorem to the line integral we find, using (6.122), than

$$\frac{d\Phi}{dt} = \iint_F [\dot{B} - \nabla \times (v \times B)] \cdot df = 0 \quad . \tag{6.123}$$

This means that, under interstellar conditions ($\sigma \to \infty$), the field lines will move with the material: they are "frozen in".

Shock Fronts: Pressure perturbations with small relative amplitude $|\Delta P|/P \ll 1$ propagate in a homogeneous gas (density ϱ) with the sonic speed c_s, which is given by

$$c_s^2 = \frac{dP}{d\varrho} \quad . \tag{6.124}$$

If the duration of the perturbation is short compared with the cooling time of the compressed volume, then changes may be assumed to be adiabatic: $P = K\varrho^\gamma$ with $\gamma = c_p/c_v =$ ratio of the specific heats at constant pressure and constant volume, respectively, K is a constant, and one obtains

$$c_s^2 = \gamma K \varrho^{\gamma-1} = \gamma \frac{P}{\varrho} = \gamma \frac{\mathcal{R}T}{\overline{\mu}} \quad . \tag{6.125}$$

The right-hand expression follows for the ideal gas with the equation of state $P = \varrho \mathcal{R}T/\overline{\mu}$, where $\mathcal{R} =$ gas constant per mole, $T =$ temperature, $\overline{\mu} =$ mean molecular weight = mean mass per particle in units of the mass of the H atom. Outside the dense molecular clouds we can in the interstellar gas set $\gamma = 5/3$ (monatomic gas).

If it can be assumed that the energy gained in the compression is immediately radiated away, then the temperature remains constant and the right-hand side of (6.124) is formed for isothermal changes of state: instead of (6.125) we obtain, with $dT/d\varrho = 0$,

$$c_s^2 = \frac{\partial P}{\partial \varrho} + \frac{\partial P}{\partial T} \frac{dT}{d\varrho} = \frac{\mathcal{R}T}{\overline{\mu}} \quad . \tag{6.126}$$

The "isothermal sonic velocity" is therefore obtained from the "adiabatic sonic velocity" by setting $\gamma = 1$. Assuming one He atom to 10 H atoms, the values obtained for the isothermal sonic velocity in the various regions of the interstellar medium are displayed in Table 6.5 (p. 420).

For pressure perturbations in the interstellar gas, for example, such as those connected with the propagation of the ejected shell of a supernova or the

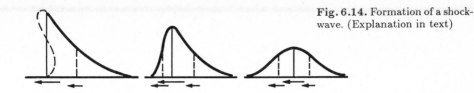

Fig. 6.14. Formation of a shock-wave. (Explanation in text)

expansion of an H II region into the surrounding medium, velocities often occur which far exceed the sonic velocity of this medium, and it can no longer be assumed that $|\Delta P|/P \ll 1$. The treatment of this general case of strong pressure perturbations on the basis of the hydrodynamic equations shows that one can no longer speak of a propagation velocity for the whole perturbation: c_s increases with increasing density, as is already clear from (6.125) for $|\Delta P|/P \ll 1$. Accordingly points of the perturbation profile at which ϱ is higher move more quickly (Fig. 6.14) and the part of the profile signalling a rise in pressure therefore becomes gradually steeper. In the mathematical formulation a physical impossibility is eventually reached − multi-valuedness of the density function − whereas what actually occurs is a jump, known as a shockwave or *shock front*.

The greater the initial supply of energy (e.g. a strong explosion), the higher the resulting jump in the pressure and density profiles, and the more the velocity v of the front exceeds the sonic velocity in the undisturbed medium: the Mach number, defined as $m \equiv v/c_s$, is greater than unity. For a simple treatment of the development and propagation of shockwaves, such as occur in the above well known astrophysical phenomena, one often starts from the following scenario: at a certain time a plane piston moves with constant velocity into homogeneous gas at rest. A shock front develops, which in general moves with a velocity different from that of the piston.

Quantitative relations between P, ϱ and T in front of and behind the shock front can be obtained, without going into the complicated processes in the immediate frontal region itself, by means of the conservation laws for mass, momentum and energy. One chooses a coordinate system in which the shock front is at rest (Fig. 6.15). In this system we have in the undisturbed medium in front of the shock a one-dimensional flow with the constant velocity v_1, whose magnitude is equal to the velocity of the shockwave relative to the undisturbed gas in front of it; the values there of pressure P_1, density ϱ_1 and temperature T_1 are also constant. Behind the front, after a "transition region", these parameters assume different values v_2, P_2, ϱ_2, T_2.

In the transition to the region behind the shock front the following must stay constant: the mass flow ϱv, the momentum flow $P + \varrho v^2$ and, if there are no radiation losses, the energy flow[2] $\varrho v[\frac{1}{2}v^2 + U] + Pv$, the kinetic energy $\frac{1}{2}v^2$

Fig. 6.15. Derivation of "shock relations" for the physical parameters at a shock front

417

plus the internal energy U transported by the mass, plus the work done by the pressure, per unit area per unit time.

The *"shock relations" for adiabatic shockwaves* therefore read:

$$\varrho_1 v_1 = \varrho_2 v_2 = j \tag{6.127}$$

$$P_1 + \varrho_1 v_1^2 = P_2 + \varrho_2 v_2^2 \tag{6.128}$$

$$\varrho_1 v_1 \left[\frac{v_1^2}{2} + U_1 \right] + P_1 v_1 = \varrho_2 v_2 \left[\frac{v_2^2}{2} + U_2 \right] + P_2 v_2 \quad . \tag{6.129}$$

Using (6.127) one can rewrite (6.129) as

$$\frac{v_1^2}{2} + h_1 = \frac{v_2^2}{2} + h_2 \tag{6.130}$$

introducing the enthalpy $h = U + P/\varrho$. We assume an ideal gas, so that

$$U = \frac{1}{\gamma - 1} \frac{P}{\varrho} \quad , \qquad h = \frac{\gamma}{\gamma - 1} \frac{P}{\varrho} \tag{6.131}$$

(from $U = c_v T$ with $T = P\overline{\mu}/\varrho R$ and $R/\overline{\mu} = c_p - c_v$). To derive the ratio ϱ_2/ϱ_1 one substitutes the mass flow j from (6.127) into (6.128), obtaining

$$j^2 = \frac{P_2 - P_1}{V_1 - V_2} \quad . \tag{6.132}$$

Here the specific volume $V = 1/\varrho$. Substituting (6.127) in (6.130) and using (6.132) then gives

$$h_2 - h_1 = \tfrac{1}{2}(P_2 - P_1)(V_1 + V_2) \quad . \tag{6.133}$$

Eliminating h from (6.131) and (6.133) and solving the resulting equation for V_1/V_2 then gives:

$$\frac{\varrho_2}{\varrho_1} = \frac{V_1}{V_2} = \frac{(\gamma - 1)P_1 + (\gamma + 1)P_2}{(\gamma + 1)P_1 + (\gamma - 1)P_2} \quad . \tag{6.134}$$

This relation between P and V or ϱ is known as the shock adiabat or *Hugoniot's Adiabat*. For the *density jump at a strong shock front* ($P_2 \gg P_1$) it follows from (6.134), with $\gamma = 5/3$ for a monatomic gas, that

$$\frac{\varrho_2}{\varrho_1} \approx \frac{\gamma + 1}{\gamma - 1} \approx 4 \quad . \tag{6.135}$$

For the velocities v_1 and v_2, one can obtain from (6.127), using (6.132) and (6.134), expressions which, after division by the square of the sonic velocity

$c^2 = \gamma P/\varrho = \gamma PV$, become

$$\left(\frac{v_1}{c_{s1}}\right)^2 = \frac{(\gamma - 1) + (\gamma + 1)(P_2/P_1)}{2\gamma} \quad ,$$

$$\left(\frac{v_2}{c_{s2}}\right)^2 = \frac{(\gamma - 1) + (\gamma + 1)(P_1/P_2)}{2\gamma} \quad . \tag{6.136}$$

Since $P_2 > P_1$ the front moves with supersonic speed $v_1 > c_{s1}$ relative to the undisturbed gas, whereas behind the discontinuity the shock speed is subsonic $v_2 < c_{s2}$ relative to the compressed gas: in the limiting case of the "strong shock" $v_2 = c_{s2}\sqrt{(\gamma - 1)/2\gamma}$. For a very weak shockwave with $P_2/P_1 \simeq 1$, $v_1 \simeq c_{s1} \simeq c_{s2} \simeq v_2$, corresponding to the case of normal sound waves.

We shall consider now the case when the compressed gas can radiate away the thermalised kinetic energy. Instead of the condition of constancy of energy flow (6.129), we then have the conditions for thermal equilibrium between energy supply and radiation on the two sides of the front, by which the temperatures T_1 and T_2 are determined. If the cooling time of the gas is shorter than the duration of the transition of the shock front and the ionisation state of the gas does not change, then we can assume that behind the transition region $T_2 = T_1$. One speaks of an isothermal shockwave. The ratios ϱ_2/ϱ_1, v_1/c_{s1} and v_2/c_{s2} can be determined approximately from (6.134) and (6.136) with $\gamma = 1$. One finds that $\varrho_2/\varrho_1 = P_2/P_1$ and $(v_1/c_{s1})^2 = P_2/P_1$, so it follows that

$$\frac{\varrho_2}{\varrho_1} = \frac{v_1^2}{c_{s1}^2} = m^2 \quad . \tag{6.137}$$

Physically, this means that the gas behind the shock front can be compressed by an arbitrary amount, as a result of the rapid energy loss by radiation, if only the velocity of the front v_1 is sufficiently high compared with the sonic velocity c_{s1}. For example, in the isothermal collision of two normal H I clouds ($c_s \simeq 1\,\mathrm{km\,s^{-1}}$) with relative velocity $v_1 = 10\,\mathrm{km\,s^{-1}}$ we find that already $\varrho_2 = 10^2 \varrho_1$. This capability for producing high compressions suggests that the origin of diffuse interstellar clouds is to be sought in the formation of shock fronts in the interstellar gas.

Whereas the *thickness* of adiabatic shock fronts in neutral gas is only of the order of the mean free path length of the atoms ($\simeq 10^{-5}\,\mathrm{pc}$ for a density of $10\,\mathrm{cm}^{-3}$), one finds in the isothermal case a very much greater extent for the additional emitting transition region, corresponding roughly to the free pathlength of the photons. For the more detailed structure of shockfronts, the reader must consult the literature.

A *magnetic field* running everywhere parallel to the shock front introduces into (6.128) the additional magnetic pressures $B_1^2/8\pi$ and $B_2^2/8\pi$, respectively. Since the field may be considered "frozen in" the material, we also have $B_2/B_1 = \varrho_2/\varrho_1$. In the adiabatic case B will thereby be strengthened by at most a factor of 4, so that a magnetic field of a few microgauss is negligible in the first approximation. For an isothermal shock wave, with $P_1 = \varrho_1 c_{s1}^2$, $P_2 = \varrho_2 c_{s2}^2$, (6.128) becomes

$$\varrho_1 c_{s1}^2 + \varrho_1 v_1^2 + B_1^2/8\pi = \varrho_2 c_{s2}^2 + \varrho_2 v_2^2 + (\varrho_2/\varrho_1)^2 B_1^2/8\pi \quad .$$

We now introduce the *Alfven Velocity* V_{A1} ahead of the shock defined by $V_{A1}^2 = B_1^2/4\pi\varrho_1$. This is the velocity of a transverse wave along the magnetic field lines (analogous to waves on a stretched string). Substituting V_{A1} into the last equation we find for the limiting case where $V_{A1} \gg c_{s1}, c_{s2}$ and $\varrho_2 \gg \varrho_1$:

$$\frac{\varrho_2}{\varrho_1} = \sqrt{2}\frac{v_1}{V_{A1}} \quad . \tag{6.138}$$

For an interstellar cloud with $B = 10^{-5}$ Gauss and 30 H atoms cm^{-3} we find, for example, $V_A \simeq 4\,\text{km}\,\text{s}^{-1}$. Then, with $v_1 = 10\,\text{km}\,\text{s}^{-1}$, we find from (6.138) that $\varrho_2 \simeq 3.5\varrho_1$.

Initial Data for the Theory: Table 6.5 gives representative values for important physical parameters of the individual components of the interstellar gas, which are usually taken as a basis in the following sections.

Table 6.5. Mean physical parameters of the components of the interstellar gas. T temperature, N particle density, P/k static pressure in units $[\text{K}\,\text{cm}^{-3}]$, f fraction of space occupied by the component in the region of the spiral arms, $\bar{\mu}$ mean molecular weight for $N(\text{He})/N(\text{H}) = 0.1$ (higher elements not taken into account, He in the normal H II regions singly ionised), c_s isothermal sonic velocity

Parameter	Hot component	Diffuse H II low density gas	Normal H II regions	Warm H I gas	Diffuse H I clouds	Molecular clouds
T [K]	6×10^5	8×10^3	8×10^3	6×10^3	80	20
N [cm^{-3}]	3×10^{-3}	3×10^{-2}	5×10^2	0.3	20	10^4
$P/k = NT$ [$\text{K}\,\text{cm}^{-3}$]	2×10^3	2.4×10^2	4×10^6	2×10^3	2×10^3	2×10^5
f	0.7	0.1	10^{-4}	0.2	0.02	5×10^{-4}
$\bar{\mu}$	0.6	0.6	0.64	1.3	1.3	2.3
c_s [$\text{km}\,\text{s}^{-1}$]	100	10	10	6	0.7	0.3

6.3.2 Dynamical Equilibrium of the Gas

Magneto-Hydrostatic Model: In the relatively thin rotating galactic gas disc we can assume that the radial component of the gravitational force is practically equal to the centrifugal force, whilst the radial velocity dispersion from thermal and macroscopic motions are of subordinate importance. Perpendicular to the galactic plane, however, in the z-direction, the existing steady state with approximately plane parallel stratification is essentially maintained by pressure forces and macroscopic motions. The component $K_z(z)$ of the gravitational acceleration dominated by the stars is here opposed by the following individual partial pressures:

1. gas pressure of the thermal motion $P_{th} = \frac{1}{3}\overline{\varrho}\,\overline{v_{th}^2} = \overline{N}kT$,
2. "turbulent" pressure associated with the motions of the clouds $P_c = \frac{1}{3}\overline{\varrho}\,\overline{v_c^2}$,
3. magnetic pressure $P_m = B^2/8\pi$,
4. pressure produced by the cosmic ray particles $P_{cr} = U_{cr}/3$.

Here v_{th} denotes the magnitude of the thermal velocity, $\overline{N}$ the mean number density of the gas particles, $\overline{\varrho}$ the mean gas density, v_c the magnitude of the cloud velocity, whose direction is assumed distributed isotropically, and U_{cr} the energy density of the cosmic rays. We assume a homogeneous magnetic field, whose field lines run parallel to the galactic plane so that the term $(\boldsymbol{B}\cdot\nabla)\boldsymbol{B}$ in (6.119) vanishes.

We shall describe the influence of the macroscopic motions only by the partial pressure P_c. Then no finite quantity appears on the left-hand side of (6.119) and we have the fundamental hydrostatic equation

$$\frac{d}{dz}(P_{th} + P_c + P_m + P_{cr}) = \overline{\varrho}(z)K_z(z) \quad . \tag{6.139}$$

For a simple model treatment we select the linear law $K_z(z) = -az$, assume that $\overline{v_{th}^2}$ and $\overline{v_c^2}$ are independent of z, and set $P_m = \alpha(P_{th}+P_c)$, $P_{cr} = \beta(P_{th}+P_c)$ with the factors α and β also independent of z (Parker, 1966). Then, using the abbreviation

$$\frac{(1+\alpha+\beta)(\overline{v_{th}^2} + \overline{v_c^2})}{3a} = \frac{1}{2}H^2 \tag{6.140}$$

we obtain from (6.139) a differential equation for the density

$$\frac{1}{\varrho}\frac{d\overline{\varrho}}{dz} = -\frac{2}{H^2}z \quad , \tag{6.141}$$

whose solution is

$$\overline{\varrho}(z) = \overline{\varrho}(0)\exp\left[-\left(\frac{z}{H}\right)^2\right] \quad . \tag{6.142}$$

The quantity H has the interpretation of a layer half-thickness: for the effective layer thickness $2H^*$ defined in Sect. 5.4.2 by (5.205) it follows that

$$2H^* = \sqrt{\pi}H \quad . \tag{6.143}$$

We apply (6.140) first of all to the *galactic HI cloud layer* in the solar neighbourhood. In this case the thermal velocities of the gas particles (in the clouds) are essentially smaller than the velocities of the clouds themselves and we may neglect $\overline{v_{\text{th}}^2}$ in comparison with $\overline{v_{\text{c}}^2}$. The observations of the synchrotron radiation lead one to expect that the gradient of the pressure of the cosmic rays dP_{cr}/dz in the thin H I cloud layer is very small, so we set $\beta = 0$. With mean gas density $\overline{\varrho} = 3 \times 10^{-24}\,\text{g cm}^{-3}$ (Sect. 6.4.2) and also $\sqrt{\overline{v_{\text{c}}^2}} = 10\,\text{km s}^{-1}$ (Sect. 5.4.2, Table 5.14 or Sect. 4.1.5) and $B = 3 \times 10^{-6}$ Gauss (Sect. 4.1.8), one obtains $\alpha = P_{\text{m}}/P_{\text{c}} = 0.36$. With the approximation $a = |\partial K_z/\partial z|_{z=0} \simeq 10^4$ $\text{km}^2\,\text{s}^{-2}\,\text{kpc}^{-2} \simeq 10^{-29}\,\text{s}^{-2}$ (see Sect. 6.1.5), usable for small distances z, we obtain $H \simeq 100\,\text{pc}$ and $H^* \simeq 90\,\text{pc}$. The deviation from the empirical value for H^* given in Table 5.14 can be removed by assuming higher temperatures for some of the clouds ($\overline{v_{\text{th}}^2} \neq 0$).

For homogeneous stratified *warm neutral gas in between the clouds* let us set $\overline{v_{\text{c}}^2} = 0$, whereas with the values of N and T from Table 6.5 the mean thermal velocity is $\sqrt{\overline{v_{\text{th}}^2}} \simeq 11\,\text{km s}^{-1}$. Retaining the assumptions that $B = 3 \times 10^{-6}$ Gauss and $dP_{\text{cr}}/dz = 0$ one gets for H^* practically the same value as for the cold H I clouds. If, however, one takes as basis the one-dimensional velocity dispersion $\sqrt{\overline{v_{\text{r}}^2}} \simeq 10\,\text{km s}^{-1}$ derived from the line widths for the warm gas component [Sect. 5.4.2, (5.204 f.)], and accordingly replaces $\sqrt{\overline{v_{\text{th}}^2}}$ by $\sqrt{3\overline{v_{\text{r}}^2}} \simeq 17\,\text{km s}^{-1}$, then one gets (with $\alpha = 0.36$, $\beta = 0$) $H^* \simeq 150\,\text{pc}$, a value which agrees better with the result of about $200\,\text{pc}$ (Sect. 5.4.2), derived from observations of the warm gas. The higher velocity scatter can result from a cloudy structure of the warm component with corresponding macroscopic motions, but can also be explained qualitatively by the assumption that the warm gas is located predominantly in the neighbourhood of the cold clouds and to a certain degree partakes in their motions (see below: Global Model).

In the *hot gas component* the electrons contribute about as much to P_{th} as do the protons. We take account of this by putting $\overline{v_{\text{th}}^2} = 3\overline{N}kT/\overline{\varrho} = 3\mathcal{R}T/\mu$ with $\mu = 0.5$. For $T = 10^6\,\text{K}$ it then follows that $\sqrt{\overline{v_{\text{th}}^2}} = 224\,\text{km s}^{-1}$. The magnetic field here can at the most be $B \simeq 1 \times 10^{-6}$ Gauss, so we assume that $\alpha = 0$. Since according to the observations the hot gas extends to great distances, one finds, if this is also true for the cosmic rays, that with $P_{\text{th}} \simeq P_{\text{cr}} \simeq 4 \times 10^{-13}$ dyn cm^{-2} and $P_{\text{c}} \simeq 0$, then approximately $\beta = 1$. Using here tentatively also the value of the gradient $\partial K_z/\partial z$ given above, then it follows that $H^* \simeq 2.4\,\text{kpc}$. Since the gradient of $|K_z|$ flattens out at $|z| > 500\,\text{pc}$ (Sect. 5.1.4, Fig. 6.7), a model with $K_z = \text{constant}$ would give a better approximation to the truth. As the nominal value for the layer half-thickness of the hot component, in this case one obtains – neglecting the pressure of the cosmic rays – the equivalent height

$$H^* = \frac{\mathcal{R}T}{\mu g_z}\,. \tag{6.144}$$

If one uses the value for the gravitational acceleration $g_z = -K_z = az_0 \simeq 10^{-8}$

cm s^{-2}, valid at $z = z_0 = 300$ pc, which corresponds to a mass loading of the galactic plane of $\varrho(0)2z_0 \simeq 100 \mathcal{M}_\odot$ pc^{-2} (integral over z), then it follows that $H^* \simeq 5 \times 10^{-3} T$ [pc], and so $H^* \simeq 5$ kpc for $T = 10^6$ K. As already shown from the empirical point of view in Sect. 5.4.5, the hot gas component must therefore form an extensive galactic halo.

Parker Instability: The foregoing estimates show that the interstellar magnetic field in the galactic disc makes a considerable contribution to the pressure of the gas. If the field lines run parallel to the galactic plane, and so perpendicular to the gravitational force, there appears in the horizontal homogeneous gas layer an instability first recognised by Parker (1966), which is analogous to the Rayleigh-Taylor instability of a heavy fluid lying on top of a light one.

The magnetic field is in practice fixed to the gas; conversely, its strength is sufficient to hinder considerably any motion of the gas perpendicular to the field lines. Any selected small volume of gas is therefore limited, as a bead on a necklace, to motions along the field lines. A perturbation of the plane parallel stratification which consists of small local displacements of plasma and magnetic field upwards or downwards (Fig. 6.16), is first of all opposed to the tendency of the field lines to contract. Nevertheless, wherever the fieldlines are now no longer horizontal, the gas will follow them "down". The perturbation thereby becomes more marked: in a gas volume pushed upwards the density becomes lower, the field has less to carry and expands upwards; in a gas volume pushed downwards the field is loaded more heavily and so it sinks still further. The pressure contribution of the cosmic rays reinforces this process: since P_{cr} remains constant along the perturbed field lines, this quantity is above the equilibrium value in rising gas volumes, and below it in sinking gas volumes, causing even more rising or sinking, respectively.

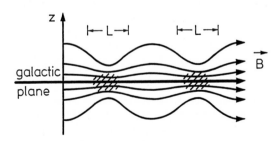

Fig. 6.16. Instability of a plane parallel layer of interstellar gas with homogeneous magnetic field $\boldsymbol{B}$ in the homogeneous gravity field K_z

The instability described can only arise if the variation of the potential energy of the vertically rising parcels of gas is greater than the variation of the energy of the magnetic field, since otherwise the tension in the field lines would re-establish the plane parallel stratification. Since the mass of the displaced gas for a fixed amplitude increases with the spatial extent (half wavelength) of the perturbation L, one must expect that L must exceed a certain critical value L^*. The quantitative theory due to Parker (neglecting P_{cr}) leads to

$$L^* \approx \frac{4\sqrt{\pi}H^*}{\sqrt{1+2\alpha}} \quad . \tag{6.145}$$

With $H^* = 130\,\mathrm{pc}$ and $\alpha \simeq 1$ it follows that $L^* \simeq 500\,\mathrm{pc}$. The timescale of the development of the instability is of the order of 10^7 years. The formation of material accumulations in this way, perhaps induced by the galactic density waves, could be of significance for the structure of the spiral arms and the formation of interstellar clouds. By detailed calculations it has been shown that a new stable state with curved field lines over a length $L \simeq L^*$ is possible, in which the tension in the field lines and the outwards directed magnetic pressure are in equilibrium. Perhaps in this way the curved filaments observed in the distribution of dust and gas can be explained.

Global Dynamical Model of the Interstellar Medium: In the galactic disc by far the largest part of the mass of the interstellar gas is concentrated in clouds; at least half of the total cloud mass is contained in the diffuse H I clouds. The origin and development of the clouds have still not been satisfactorily explained. Grossly simplified model representations have so far been used for the description of the global dynamics of the cloud system.

Let us sketch first of all an *extremely idealised model*, treating the diffuse clouds as molecules of a gas, between which inelastic collisions occur. In these, the clouds will, at least partially, merge with one another. Gradually more and more massive clouds will be formed, in which eventually stars will be formed. Then the remaining gas will be swept up by the shockwaves emitted by the most massive of these stars (supernovae, H II regions, stellar winds) and new small diffuse clouds will be formed: the cycle can begin anew.

To assess the lifetime of this process let us first of all calculate the mean collision time τ of the diffuse clouds. Suppose that these are spherical, with radius R [pc] and their number per unit volume is N [pc^{-3}]. Then the mean cloud number on the line of sight $\nu\,[\mathrm{kpc}^{-1}] = \pi R^2 \times 10^3 \times N$ (= number of clouds in a cylinder of cross-section πR^2 and length 10^3 pc). With $\nu = 5\,\mathrm{kpc}^{-1}$ and $2R = 3\,\mathrm{pc}$ it follows that $N = 7 \times 10^{-4}\,\mathrm{pc}^{-3}$. A cloud undergoes on average one collision with another cloud in the time $\tau = (\pi R^2 N v_\mathrm{c}')^{-1}$. Here v_c' denotes the relative velocity of the two clouds. For the three-dimensional dispersion $\sqrt{\overline{v_\mathrm{c}^2}} \simeq 10\,\mathrm{km\,s}^{-1}$ it follows that $\overline{v_\mathrm{c}'} \simeq 14\,\mathrm{km\,s}^{-1}$ and one obtains, with $\pi R^2 N = \nu \times 10^{-3} = 5 \times 10^{-3}\,\mathrm{pc}^{-1}$ the mean collision time $\tau \simeq 10^7$ years. The merging into massive clouds, in which stars can be created, therefore requires about 10^8 years. As the formation of massive stars requires at most the same time, the whole cycle lasts only a few 10^8 years, and since the formation of the galactic disc, about 10^{10} years ago, the cycle must have taken place many times. A certain "equilibrium" mass spectrum of the clouds must therefore have been established. Model calculations lead one to expect $F(\mathcal{M}) \sim \mathcal{M}^{-\gamma}$ with $\gamma = 1.5 \ldots 2$, and are thereby in satisfactory agreement with available empirical evidence (Sects. 5.4.1, 5.4.2).

The loss of kinetic energy in inelastic collisions between the diffuse clouds appears to be about equal to the energy gain from the explosively expanding H II regions and supernova shells:

Assuming tentatively that within a cloud collision time $\tau \simeq 10^7$ years a quarter of the kinetic energy of all the diffuse clouds in the volume unit $\varepsilon = N \cdot \frac{1}{2} M v_c^2$ is lost, then it follows with $M = 30 M_\odot$ that the rate of energy loss is $\varepsilon/4\tau \simeq 2 \times 10^{-27} \, \mathrm{erg\,cm^{-3}\,s^{-1}}$.

The energy gain can be estimated in the following manner:

1) In an H II region around an O-star the energy gain by photo-ionisation per electron and per proton is of the order of $\Gamma_{\mathrm{ph}}/N_e N_p \simeq 2 \times 10^{-24}$ $\mathrm{erg\,cm^{-3}\,s^{-1}}$ (Sect. 5.2.3, Fig. 5.8). According to radio continuum observations, for the total interstellar medium near the galactic plane, the average value of $N_e N_p \simeq N_e^2 \simeq 0.1 \, \mathrm{cm^{-6}}$, and we find that $\Gamma_{\mathrm{ph}} \simeq 2 \times 10^{-25} \, \mathrm{erg\,cm^{-3}\,s^{-1}}$. According to more detailed studies, as a result of high radiation losses the kinetic energy gained by the gas will probably be only a few per cent of this, but should reach at least $2 \times 10^{-27} \, \mathrm{erg\,cm^{-3}\,s^{-1}}$.

2) The expelled shell of a supernova has a kinetic energy of around 10^{51} erg. With a frequency of appearance of one supernova every 30 years in the whole galactic gas disc (radius $\simeq 10 \, \mathrm{kpc}$, thickness $\simeq 250 \, \mathrm{pc}$) one obtains about $10^{-24} \, \mathrm{erg\,cm^{-3}\,s^{-1}}$. More exact treatment shows here also that the efficiency amounts to about one per cent (Sect. 6.3.3), resulting in an order of magnitude of $10^{-26} \, \mathrm{erg\,cm^{-3}\,s^{-1}}$.

A satisfactory theory of the interstellar medium must take account of the fact that there are not only clouds, but actually at least three components of the gas with temperatures in the ranges $T \leq 100 \, \mathrm{K}$, $T \simeq 10^4 \, \mathrm{K}$ and $T \simeq 10^6 \, \mathrm{K}$, and they must be explained (cf. Sect. 5.2.3). It is made more difficult by the fact that observations still cannot provide a reliable picture of the mutual boundaries between these components. A consistent *three-component model*, which enables one to explain the most important observed features in broad outline, was constructed in 1977 by McKee and Ostriker. Therein the explosion waves from supernovae play a dominant role. We reproduce here important results of this study.

It is supposed that originally the interstellar space was filled only with the warm medium between the clouds. The shockwave of a supernova attains pressure equilibrium with the surrounding gas when it reaches a radius of about $r_s \simeq 50 \, \mathrm{pc}$ (see Sect. 6.3.3). The volume behind the shock front, $V_0 = (4\pi/3) r_s^3 \simeq 5 \times 10^5 \, \mathrm{pc^3}$, is then essentially filled with a very thin gas at temperature $T \simeq 10^6 \dots 10^7 \, \mathrm{K}$. The lifetime of this hot "gas bubble" corresponds roughly to the time $t = r_s/c_{s0}$ taken by the surrounding warm intercloud gas with sonic velocity c_{s0} to penetrate to the middle of the spherical volume. With $c_{s0} = 6 \, \mathrm{km\,s^{-1}}$ (Table 6.5, p. 420) it follows that $t \simeq 10^7$ years. Since in the galactic disc a supernova occurs about every 30 years there are constantly about $10^7/30 = 3 \times 10^5$ hot gas bubbles of this type present, with a

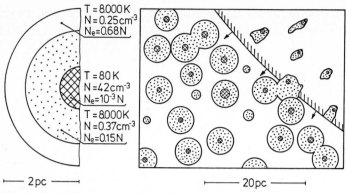

| T = 8000 K |
| N = 0.25 cm^{-3} |
| Ne = 0.68 N |

| T = 80 K |
| N = 42 cm^{-3} |
| Ne = 10^{-3} N |

| T = 8000 K |
| N = 0.37 cm^{-3} |
| Ne = 0.15 N |

├── 2 pc ──┤   ├──── 20 pc ────┤

Fig. 6.17. McKee and Ostriker's (1977; by permission) three-component model of the interstellar medium. *Left*: cross-section of a diffuse cloud. *Right*: effect of the passage of a supernova explosion wave on the diffuse clouds

total volume of $3 \times 10^5 V_0 \simeq 10^{11}$ pc^3. This corresponds to the volume of a disc with radius $R_0 = 10$ kpc and thickness $2H^* = 250$ pc. The space-filling factor f of the hot component is therefore expected to be close to unity.

The cold clouds are embedded in this thin hot "corona gas". Quantitative discussion shows that, at the "cloud surfaces", an extended warm transition region with $T \simeq 10^4$ K must arise (Fig. 6.17): the cooling by radiation causes this in a manner similar to that which we have explained for the two-phase model in Sect. 5.2.3. The outer layer of the warm cloud shell is ionised by soft x-rays from the hot component. For very small cold nuclei the appearance is dominated by the warm clouds. Consistent values for the space-filling factors f of the typical cold cloud nuclei, the warm neutral and the warm ionised gas are about 0.02, 0.1 and 0.2, respectively. The warm shells form the greater part of the cloud volume, whereas the cold nuclei are decisive for the cloud mass. Clouds without cold nuclei are completely "vaporised" by the passage of the shockwave from a supernova.

McKee and Ostriker assume a steady cycle, in which equilibrium exists: (1) in the mass balance between the "vaporisation" of clouds by the passage of supernova explosion waves and the enlargement of the cloud masses by the uptake of the corona gas swept up by the shock fronts; and also (2) in the energy balance, in which the kinetic energy of the supernova shells (see Sect. 6.3.3) is just sufficient for the heating and the replacement of the losses by radiation of the galactic gas corona, the replacement of the energy dissipation in the penetration of the shock fronts into the clouds and the radiation losses in the transition regions between the clouds and the corona gas. The spectrum of the cloud diameters D is put in the form $f(D) \sim D^{-4}$. Clouds with $D < 1$ pc vaporise quickly in the hot corona gas, so the spectrum is curtailed here. The O VI absorption lines (Sect. 4.1.4) probably arise in the outer shells of the clouds. The diffuse soft x-rays (Sect. 4.1.9) come from the thin hot gas inside the supernova shells; the emission amount required for the interpretation of the radiation observed at 0.3 keV, of the order of 10^{-2} cm^{-6} pc at $T = 10^6$ K, can be attained by the assumption of a few such "bubbles" of hot corona gas ($N_e \simeq 3 \times 10^{-3}\,\mathrm{cm}^{-3}$) on the line of sight.

6.3.3 Shock Fronts in the Interstellar Gas

Dynamics of H II Regions: In Sect. 5.2.1 the ionisation of the H II regions was treated statically. In actual fact the heated gas expands explosively with velocities of around $10\,\mathrm{km\,s}^{-1}$, and only when the ionising O-star becomes weaker after at most 10^7 years are the dimensions observed in typical developed H II regions attained. It is thus a matter of a relatively short-lived creation. In the following sections we discuss a theory of the development of these objects with time.

For the initial phase we can assume that the ionising star is suddenly "switched on": the calculation of the Pre-Main Sequence development of O-stars shows that the increase in brightness up to Main Sequence values occurs within only about 10^4 years. If the hydrogen surrounding the star is already ionised, then the short mean free path length of the Lyman continuum photons in neutral hydrogen ($l \simeq 1/\kappa_{\nu 0} \simeq 0.04\,\mathrm{pc}/N(\mathrm{H\,I})$, see Sect. 5.2.1), has the effect that the inside unhindered photons are absorbed in a thin layer of the neutral shell. This causes ionisation there also, and the next photons can penetrate unhindered this further thickness of newly won H II layer, and so on. One therefore speaks of the propagation of an ionisation front.

As in the consideration of shock fronts, one also introduces here a reference system moving with the ionisation front, in which (6.127) must be satisfied. The mass flow j is determined by the flux of ionising photons $L_c(R_*)/4\pi r^2$ [photons $\mathrm{cm}^{-2}\,\mathrm{s}^{-1}$] reaching the ionisation front (at a distance r from the star):

$$\varrho_1 v_1 = \varrho_2 v_2 = j = \mu_i m_{\mathrm{H}} \frac{L_c(R_*)}{4\pi r^2} \quad , \tag{6.146}$$

$\mu_i m_{\mathrm{H}}$ is the mean mass per ion created. If $\mathrm{H : He} = 10 : 1$, then it follows that $\mu_i = (10+4)/10 = 1.4$, if the He is not ionised. With the value of $L_c(R_*)$ given in Table 5.6 for an O7-star, and also $r = 3\,\mathrm{pc}$ and $\varrho_1 = 100 m_{\mathrm{H}}\,\mathrm{g\,cm}^{-3}$, (6.146) gives $v_1 \simeq 1000\,\mathrm{km\,s}^{-1}$ for the velocity of the ionisation front! The Strömgren radius r_s (Sect. 5.2.1) is therefore reached after a time $<10^4$ years. As the beginning of the expansion of the heated gas one takes the point in time when the radius of the ionisation front r_i is equal to the Strömgren radius r_s.

The subsequent expansion phase is considered as a sequence of equilibrium states. Because $r_s \sim N^{-2/3}$ and the density is decreasing, r_s is constantly increasing. The total mass of the ionised gas $\mathcal{M}_{\mathrm{H\,II}} \sim N_{\mathrm{H}} r_m^3 \sim N_{\mathrm{H}}^{-1}$ is therefore also increasing: with the same number of photons and the lower density more H atoms can be held in the ionised state, since the number of recombinations per unit volume $\sim N_{\mathrm{H}}^2$, but the number of ionisations per unit volume is only $\sim N_{\mathrm{H}}$.

427

The ionisation front moves on, but with a lower velocity $v_1 = dr_s/dt$ determined by the expansion process. Whereas in the passage of the ionisation front for $r_i < r_s$ there was practically no material motion, now it is a matter of the propagation of the "warm" H II gas into the previously undisturbed neutral gas of the cloud in which the ionising star was created. A shock front will therefore form, whose position we can take as $r = r_s$.

For an internal point of the H II region at a distance r_2 from the star, lying immediately behind the ionisation front, with spacewise constant density $\varrho = \varrho_2$ in the H II region we have $r_2^3 \varrho_2 = $ constant, whereas for the boundary of the expanding gas r_s according to Sect. 5.2.1 (5.95) we can assume that $r_s^3 \varrho_2^2 = $ constant. If one forms $v_2 = dr_2/dt$ and $v_1 = dr_s/dt$, the comparison of the two results shows that $v_2 = \frac{1}{2}v_1$. Solution of the condition equation (6.128) for the velocity of the shock front in the *isothermal* case with $P_2 = \varrho_2 c_{s2}^2 \gg P_1 \simeq 0$ and $v_2 = \frac{1}{2}v_1$ gives

$$v_1^2 = c_{s2}^2 \frac{(\varrho_2/\varrho_1)}{1 - (\varrho_2/4\varrho_1)} \ . \tag{6.147}$$

If the ratio ϱ_2/ϱ_1 falls as a result of expansion to 0.1, then it follows from this that even now $v_1 \simeq 0.3 c_{s2} \simeq 3 \text{ km s}^{-1}$: the front moves with supersonic velocity into the surrounding denser neutral gas of the cloud; the H II region is in this phase still ionisation bounded and is therefore often detectable only by radio or infrared observations.

The propagation of the thin H II medium into the denser neutral gas is unstable against perturbations in a similar fashion as one finds with a heavy fluid lying on a light fluid in a homogeneous gravity field (Rayleigh-Taylor instability). In the region of the boundary surface trivial inhomogeneities can lead to major deformations. The so-called "Elephant's Trunks" (Sect. 4.2.1, Fig. 4.32), in which the cold gas projects deeply into the H II region, and in certain cases is quite enclosed by it (globules), hereby finds an explanation.

In recent studies the development of H II regions has been followed numerically using the general hydrodynamic equations. Here one starts out from the assumption suggested by the observations, that the exciting O-star is formed in a dense molecular cloud with about 10^3 gas particles per cm^3. Interesting results, exceeding the foregoing considerations, followed especially in the case when the ionisation front of the expanding H II region passed the edge of the cloud and broke through into the inter-cloud medium (Fig. 6.18). This is the first time, as a rule, that the H II region becomes detectable in the visible region.

According to the calculations the velocity of the ionisation front increases strongly at this instant and produces a strong pressure gradient behind it, between the ionised part of the cloud and the contiguous, now ionised inter-cloud medium. This jump in the pressure profile has as a result the so-called "champagne effect": it creates a strong isothermal shock front, which moves with velocities $\geq 30 \text{ km s}^{-1}$ (Mach number ≥ 3) into the ionised inter-cloud

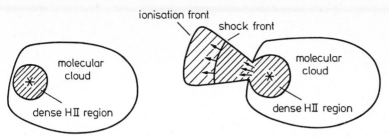

Fig. 6.18. Development of an H II region. *Left*: compact H II region still optically invisible in a molecular cloud. *Right*: after breakthrough of the H II region into the inter-cloud medium: "champagne effect". (After Tenorio-Tagle, 1979)

medium and now becomes the boundary of the nebula. At the same time a "rarefaction wave" moves back in the direction of the star towards the other, still ionisation bounded, edge of the H II region (the bottom of the champagne bottle!). Expressed simply, this means that the gas of the H II region pours out into the inter-cloud medium, with constant thinning in its interior; whilst, however, the H II region is thereby expanding in the direction of the molecular cloud, the latter is providing further material for this outpouring [Tenorio-Tagle (1979) and others].

The significance of these calculations lies primarily in that, in contrast to earlier model studies, they correspond more to reality. The older theory had considered first of all a spherically expanding H II region, which was surrounded on all sides by an optically thin medium. The systematic Doppler displacements and broadenings of the optical emission lines to be expected in this case are quite at variance with the observational results (cf. Sect. 4.2.1). The new model, according to the results available up to now, seems to be able to produce an interpretation of the velocities observed in developing H II regions. The frequently observed bright rims of the nebula must coincide with the shock front itself.

Propagation of Supernova Shells: The mutual interaction here follows a fundamentally different course from that accompanying the advance of an expanding H II region into the surrounding neutral gas. Because of the high initial velocity of the supernova shell, typically about $v = 10\,000\,\mathrm{km\,s^{-1}}$ for Type II supernovae, corresponding to an energy (per H atom) of 1 MeV, the dominant protons have mean free paths in the interstellar gas of the order of $l \simeq 10^3\,\mathrm{pc}$. This follows from taking into account that the proton loses only a fraction of about 10^{-4} of its energy in an ionisation, so that $l \simeq 10^4 l_i$, where l_i is the free path length up to the first ionisation: $l_i = 1/N_\mathrm{H}\sigma_\mathrm{cr}$ with σ_cr from Sect. 5.2.1 (5.77). Shock fronts of the hydrodynamic sort described up to now are therefore not to be expected. The interstellar magnetic field can however establish a momentum transfer in very short distances: with $E = 1\,\mathrm{MeV} = 1.6 \times 10^{-6}\,\mathrm{erg}$ and even a weak magnetic field with $B = 1 \times 10^{-6}$ Gauss, (4.43) in Sect. 4.1.9 produces the small gyration radius $r_B \simeq 1 \times 10^{-9}\,\mathrm{pc}$. The motion of the pro-

429

tons and other charged particles perpendicular to the magnetic field is thereby prevented and, since the field is frozen in, the momentum of the particle stream is transferred to the interstellar gas. The resulting discontinuity in this case is called a *hydromagnetic shock front.*

One distinguishes three phases in the dynamic development of a supernova shell [Spitzer (1978; Chap. 12)].

In the *first phase* the density of the supernova shell is large compared with the density of the interstellar matter and the mass of the gas swept up by the shock front $(4\pi/3)r^3\varrho$ is smaller than the mass ΔM of the supernova shell. The expansion therefore maintains approximately constant velocity up to a radius r_1, which is given approximately by $(4\pi/3)r_1^3\varrho = \Delta M$. For $\Delta M = 0.5\mathcal{M}_\odot$ and $\varrho = 2 \times 10^{-24}\,\mathrm{g\,cm}^{-3}$, for example, we find that $r_1 = 1.6\,\mathrm{pc}$.

During the *second phase* the effect of the swept up interstellar gas is dominant: the shock front slows down. The temperature reaches a few $10^7\,\mathrm{K}$. Cooling then occurs only by thermal Bremsstrahlung and is therefore very slow [see Sect. 5.2.3, (5.124)]. One can therefore assume that changes are approximately adiabatic. To estimate the velocity of the shock front v_1 and the temperature T_2 directly behind the density jump, one needs an approximate value for the pressure P_2. The mean pressure $\overline{P}$ in the region $r<r_1$ (= radius of the shock front) is equal to 2/3 of the mean thermal energy per unit volume $E_{\mathrm{th}}/(4\pi/3)r_1^3$. Since P_2 is considerably larger than $\overline{P}$, but E_{th} will be only a fraction of the total energy E of the ejected supernova shell, let us set simply

$$P_2 \approx \frac{E}{(4\pi/3)r_1^3} \quad , \tag{6.148}$$

where the factor 2/3 has been omitted. If one solves the first equation (6.136) for P_2/P_1 and puts $P_1 = (1/\gamma)\varrho_1 c_{\mathrm{s1}}^2$, it follows, with $v_1^2/c_{\mathrm{s1}}^2 \gg 1$, that

$$P_2 = \frac{2\varrho_1 v_1^2}{\gamma + 1} \quad . \tag{6.149}$$

With $\gamma = 5/3$, it then follows from (6.148) that

$$v_1^2 = \frac{E}{\pi \varrho_1 r_1^3} \quad . \tag{6.150}$$

The temperature behind the shock front is given by the equation of state for an ideal gas, with (6.149) and (6.148), and $\varrho_2/\varrho_1 \simeq 4$ (6.135), as

$$T_2 = \frac{\overline{\mu}P_2}{\mathcal{R}\varrho_2} \approx \frac{3\overline{\mu}}{16\mathcal{R}}v_1^2 \approx \frac{3\overline{\mu}}{16\pi\mathcal{R}}\frac{E}{\varrho_1 r_1^3} \quad . \tag{6.151}$$

For example, with $\varrho_1 = 2 \times 10^{-24}\,\mathrm{g\,cm}^{-3}$ and $E = 10^{51}\,\mathrm{erg}$ these relations give for $r_1 = 10\,\mathrm{pc}$ the numerical value $v_1 \simeq 8 \times 10^2\,\mathrm{km\,s}^{-1}$ and $T_2 \simeq 2 \times 10^7\,\mathrm{K}$.

According to more detailed calculations the density falls off steeply behind the shock front to the supernova. About half of the mass inside r_1 is concentrated in the region between $r = 0.94r_1$ and $r = r_1$. The temperature indeed increases towards the supernova, but not so strongly as the density decreases. The thermal x-rays between 0.2 and 10 keV observed with supernova remnants must come from this relatively thin layer (exception: the Crab Nebula, for which it is also a matter of synchrotron radiation in the x-ray region).

The *third development phase* is characterised by the dominance of the radiation: the temperature behind the shock front has fallen off so far that electrons can be bound to the common sorts of atom and therefore also excited. The cooling by subsequent radiation, predominantly in the ultraviolet, of the excitation energy taken from the thermal energy of the gas leads finally to almost isothermal expansion with a temperature of $10^7 \ldots 10^6$ K in the interior and $T \leq 100$ K in the now very dense layer immediately at the shock front (6.137). The velocity v_i of the mass of interstellar gas $\mathcal{M}_i \simeq (4\pi/3)r_1^3 \bar{\varrho}$ swept outwards by the front is now no longer determined by thermal expansion, but by the outward directed momentum of the gas present inside $r = r_1$ (like a snowplough!). Most of the known supernova remnants are not yet in this stage. When $r_1 \simeq 50$ pc is reached the velocity v_1 has fallen to about 10 km s^{-1}. The order of magnitude of the kinetic energy $\frac{1}{2}\mathcal{M}_i v_i^2$ transferred to the interstellar medium is, with these numbers, around 10^{49} erg, and thus amounts to about one per cent of the original energy $E \simeq 10^{51}$ erg.

The actual development in detail is more complicated than here described. When the cold high density layer has developed at the shock front and is accelerated by the very hot gas of the internal region at lower density, then one must expect, for example, the Rayleigh-Taylor instability already mentioned for the H II regions. Deformations will probably develop, which may assume the form of filaments because of the magnetic field, as is observed.

Theory of the Spiral Structure in the Interstellar Gas: The linear density wave theory of the spiral structure (Sect. 6.2.2) gives similar waves of small amplitude in the density profile for both stars and interstellar gas. It has already been seen, however, that the reaction of the gas to the passage of the wave in the gravitational potential can be stronger than that of the star field: according to (6.113) and (6.114) the density amplitude σ_1^* is the greater, the smaller the velocity dispersion $\sqrt{(\Pi - \overline{\Pi})^2}$. In the homogeneous gas this quantity is represented by the scatter of the one-dimensional thermal velocity $\sqrt{v_x^2} = \sqrt{RT/\bar{\mu}}$, which is equal to the isothermal sonic velocity c_s. In the H I region c_s is (Table 6.5, p. 420), like the mean cloud velocity (Sect. 5.4.2, Table 5.14), considerably smaller than the dispersion of the stellar velocities of 40 to 50 km s^{-1} (in the solar neighbourhood).

For the frequently assumed angular velocity of the spiral model $\omega_p = 13.5$ km s^{-1} kpc^{-1} the interstellar gas in the solar neighbourhood moves with the velocity $R(\omega_0 - \omega_p) \simeq 115$ km s^{-1} relative to the potential wave in the direction of rotation. Perpendicular to the potential wave, with a spiral arm angle of $\psi = 7°$, one has a velocity of around $14\,\mathrm{km\,s^{-1}}$. This value lies considerably above the sonic velocity for the warm gas component. If the space between the clouds were practically filled with this medium, then *spiral-form shock fronts must form* in it; the simple relations (6.135) and (6.137) derived in Sect. 6.3.1 produce increases in the density by factors of 4 to 5. If the inter-cloud gas consists essentially of the hot component (Sect. 6.3.2), however, whose sonic velocity is $100\,\mathrm{km\,s^{-1}}$, then no shock fronts can form in this way. We mention first of all a few important results of the theory for the first case and then discuss a cyclic model, in which the existence of a very extensive hot component is taken into account.

The starting point is formed by the hydrodynamic equations for a thin gas disc (Sect. 6.2.2), whose potential is taken from the first equation (6.94). Here it is appropriate to transform to a coordinate system which rotates with the spiral pattern and whose ξ-axis lies parallel to the spiral arm, whilst the η-axis is oriented perpendicularly thereto. It can be shown directly that, under the prevailing conditions, in addition to the unperturbed velocity of the gas perpendicular to the spiral arm $v_{\eta 0} = R(\omega_0 - \omega_p)\sin\psi$, in the case $v_{\eta 0} > c_s$ there will be a component $v_{\eta 1}$, which is directed towards the potential minimum (and hence towards the maximum of the density wave), and so causes a strengthening of the density wave (Roberts, 1969). The development of the flow perpendicular to the spiral arm has been studied by numerical integration of the time-dependent equations. Typical results for $c_s = 8.6\,\mathrm{km\,s^{-1}}$ are illustrated in Fig. 6.19: the parameter F of the curves for $v_\eta = v_{\eta 0} + v_{\eta 1}$ and for $\varrho/\overline{\varrho}$ is the

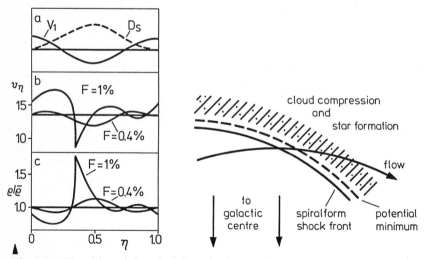

Fig. 6.19. Transition of the spiral-form density wave into a shockwave. Abscissa: coordinate η perpendicular to spiral arm in units of wavelength of the density wave. (**a**) Qualitative profiles of the gravitational potential V_1 and the star density D_s. (**b**) Velocity component perpendicular to spiral arm v_η in $\mathrm{km\,s^{-1}}$. (**c**) Density ϱ relative to the mean density $\overline{\varrho}$. (After calculations by Woodward, 1975; by permission)

Fig. 6.20. Position of the galactic shock front and the newly formed stars in the spiral arm

ratio of the spiral-form portion of the gravitational field to the rotationally symmetrical field of the galactic disc, and hence the amplitude of the gravitational field of the density wave in units of the mean field. Whereas, for example, with $F = 0.4\,\%$ the result is a smooth wave similar to the potential or the stellar density, with $F \geq 1\,\%$ a shock wave appears. For the spiral arm model explained in Sect. 6.2.2 with $F = 5\,\%$ there is a density jump $\varrho_2/\varrho_1 \simeq 6$. The spiral-form shock front always lies on the inner side of the density wave. The compression of the interstellar gas within and behind the spiral-form shock front leads one to expect the following interesting consequences (Fig. 6.20), which are amenable to observation:

1) The compression of the inter-cloud medium perturbs the heat equilibrium (Sect. 5.2.3) and brings this gas partially into a thermally unstable state, from which the transition to a stable cold and dense phase takes place within about 10^6 years. Directly behind the shock front new interstellar clouds form.

2) Sufficiently large and massive clouds already in existence, which are not far removed from the edge of gravitational instability, could be concentrated by the compression to the point of collapse (Sect. 6.3.4). The subsequent star formation would require of the order of 10^7 years. During this time the gas moves relative to the shock front by only about the thickness of the transition layer. The newly born OB-stars and the H II regions associated with them will thus appear in a relatively narrow region (behind the shock front), as one in fact observes in many extragalactic spiral systems and as appears to be the case also in our own galaxy.

3) The strengthening of the magnetic field running roughly parallel to the spiral arms immediately behind the shock front will lead to a local increase there of the synchrotron radiation from electrons of the cosmic rays. Observations of the non-thermal continuous radio emission from the spiral arms of the galaxy M 51 (Fig. 1.2) have confirmed this expectation. The distance found in M 51 between the spiral arms defined "optically" by early star types and H II regions on the one hand, and the lines of maximal non-thermal radio emission identified with the shock front on the other hand, correspond to a time difference of 6×10^6 years, which suffices for the star formation. It seems further that evidently the dust density as well as the gas density is considerably increased in the region of the shock front, as one would expect. Studies of the spatial distribution of the dust can thus provide evidence on the positions of the large scale spiral-form shock fronts.

How can this picture be reconciled with the three-component model of the interstellar gas mentioned in Sect. 6.3.2, in whose hot inter-cloud medium with $c_s \simeq 100\,\mathrm{km\,s^{-1}}$ (Table 6.5) the galactic potential wave would not give rise to a shock front? If one starts from the point that the heating of the hot gas component in the galactic disc is caused by supernovae of Type II, this opens up the following possibility of a model reconciling both concepts (Reinhardt and Schmidt-Kaler, 1979): after the passage of the potential wave the hot gas can cool in between the spiral arms, where there are practically no supernovae. From $T \simeq 10^6$ K down to $T \simeq 10^4$ K the cooling time amounts to about $t_C = (10^5/N_e)$ [years] [cf. Sect. 5.2.3, (5.124)]. After about 10^8 years one therefore has in the inter-arm region predominantly only the still ionised gas at $T \simeq 10^4$ K and the warm phase of the neutral gas with about the same temperature. In both media the sonic velocity is about $c_s \simeq 10\,\mathrm{km\,s^{-1}}$. The next potential wave, with the two-arm model discussed in Sect. 6.2.2, only comes along after half a cycle time $\pi(\omega_0 - \omega_p) \simeq 3 \times 10^8$ years (for $\omega_0 = 25\,\mathrm{km\,s^{-1}\,kpc^{-1}}$, $\omega_p = 13.5\,\mathrm{km\,s^{-1}\,kpc^{-1}}$) and finds there at least approximately the conditions postulated as the basis for the results of the hydrodynamic calculations described above. The picture sketched in this connection can therefore be retained in essence, if one assumes a cycle of continual reiteration of the sequence of processes: compression, star formation, heating and cooling.

6.3.4 Condensation of the Gas and Galactic Evolution

Since a star is held together by gravitation, this must also play a role in the process of the local condensation of interstellar gas to stellar densities. The existence of an approximate dynamic equilibrium in the Milky Way system as a whole, in conjunction with the existence of numberless galaxies in a similar state, suggests in similar fashion the assumption that these great building stones of the cosmos have also been formed by gravitational condensation from a pre-galactic medium. Let us therefore first address the question of the conditions under which gas distributed more or less homogeneously in space begins to contract under its own intrinsic gravitation.

Gravitational Instability: The starting point is provided by the fundamental hydrodynamic equation [(6.119) for $B \equiv 0$], the continuity equation (6.120) and Poisson's Equation [Sect. 6.1.1, (6.3)]. We give here for a homogeneous gas an analysis due to J.H. Jeans (1902, 1928). One starts from an equilibrium state corresponding to the static solution $\varrho = \varrho_0 = $ constant, $P = P_0 = $ constant, $v = v_0 = $ constant, $V = V_0 = $ constant, and enquire into the development with time of a closely neighbouring solution $\varrho = \varrho_0 + \varrho_1$, $P = P_0 + P_1$, $v = v_0 + v_1$, $V = V_0 + V_1$ with $|\varrho_1| \ll |\varrho_0|$, ..., $|V_1| \ll |V_0|$, in complete analogy with the procedure in Sect. 6.2.2 for the derivation of (6.95) to (6.97). Equations (6.119) for $B \equiv 0$, (6.120) and (6.3), with retention only of terms of the first order, and under the additional assumption $v_0 = 0$, lead in this way to the linearised equations

$$\frac{\partial v_1}{\partial t} = -\frac{1}{\varrho_0}\nabla P_1 - \nabla V_1 \quad , \tag{6.152}$$

$$\frac{\partial \varrho_1}{\partial t} = -\varrho_0 \nabla \cdot v_1 \quad , \tag{6.153}$$

$$\Delta V_1 = -4\pi G \varrho_1 \quad . \tag{6.154}$$

We assume that $\varrho_0 \neq 0$: this means that our procedure represents an approximation, insofar as, according to Poisson's Equation (6.3), $V_0 = $ constant strictly implies that $\varrho_0 = 0$, and hence a completely empty space. Assuming that $c_s = $ constant, and hence isothermal conditions [see (6.126)], the pressure perturbation is given, according to (6.124), by

$$P_1 = c_s^2 \varrho_1 \quad . \tag{6.155}$$

If the functions ϱ_1, P_1, $\boldsymbol{v}_1$, V_1 representing the solution of this system of equations increase with time the medium is called *unstable*; otherwise it is called *stable*. To obtain the solution we form the divergence of (6.152), with P_1 from (6.155), and the time derivative of (6.153):

$$\nabla \cdot \frac{\partial \boldsymbol{v}_1}{\partial t} = -\Delta \left(c_s^2 \frac{\varrho_1}{\varrho_0} + V_1 \right) \tag{6.156}$$

$$\frac{\partial^2 \varrho_1}{\partial t^2} = -\varrho_0 \nabla \cdot \frac{\partial \boldsymbol{v}_1}{\partial t} \quad . \tag{6.157}$$

Elimination of $\nabla \cdot (\partial \boldsymbol{v}_1 / \partial t)$ from these equations, and using (6.154), leads to the wave equation

$$\frac{\partial^2 \varrho_1}{\partial t^2} = 4\pi G \varrho_0 \varrho_1 + c_s^2 \Delta \varrho_1 \quad . \tag{6.158}$$

Of the well known solutions we consider the plane wave propagating in the x-direction

$$\varrho_1 = \varrho_1^* \exp\{i[\omega t - kx]\} \quad . \tag{6.159}$$

Substituting this in (6.158), since $\ddot{\varrho}_1 = -\omega^2 \varrho_1$ and $\Delta \varrho_1 = -k^2 \varrho_1$, we obtain

$$\omega^2 = k^2 c_s^2 - 4\pi G \varrho_0 \quad . \tag{6.160}$$

For ϱ_1 to increase (exponentially) with time at fixed x it is necessary to have a real coefficient of t in (6.159), i.e. $\omega^2 < 0$. Accordingly instability can only occur for $k^2 < 4\pi G \varrho_0 / c_s^2 = k_g^2$, i.e. for wavelengths $l = 2\pi/k$ greater than a limiting wavelength $l_g = 2\pi/k_g$. Since the extent of the compressed region L corresponds to about half a wavelength, one sets $L = l/2$ and obtains the Jeans Criterion for instability

$$L > L_J = \sqrt{\frac{\pi c_s^2}{4 G \varrho_0}} = \sqrt{\frac{\pi k T}{4 G \varrho_0 \bar{\mu} m_H}} = \frac{7.822}{\bar{\mu}} \sqrt{\frac{T}{N_H}} [\text{pc}] \quad . \tag{6.161}$$

435

L_J is often called the *Jeans length*. The two right-hand expressions are valid for isothermal behaviour of the perturbation [c_s according to (6.126) with $\mathcal{R} = k/m_H$, where k = the Boltzmann constant, N_H = number density of H atoms = $\varrho_0/\bar{\mu}m_H$].

With the specified density ϱ_0 the length L_J corresponds to a certain *Jeans Mass* $\mathcal{M}_J = L_J^3\varrho_0$ and one can also write the instability condition as

$$\mathcal{M} > \mathcal{M}_J = \left(\frac{\pi kT}{4G\bar{\mu}m_H}\right)^{3/2}\varrho_0^{-1/2} = \frac{11.74}{\bar{\mu}^2}\sqrt{\frac{T^3}{N_H}}[\mathcal{M}_\odot] \quad . \tag{6.162}$$

In this case the thermal pressure of the gas, characterised by T and N, can no longer withstand the gravitational force. These results are confirmed by detailed studies of self-gravitating spheres. In particular, the Virial Theorem for the radius R_J and the mass $\mathcal{M}_J$ of a sphere at the limit of gravitational instability produces expressions which differ from the right-hand sides of the inequalities (6.161) and (6.162) only by a factor lying close to unity. Moreover it is noteworthy that even the feasible exact analysis of gravitational instability of an infinitely extended plane parallel gas layer leads to a limiting value for the layer thickness differing only by a factor $\sqrt{2}$ from the right-hand side of (6.161) [see e.g. Spitzer (1978; pp. 283 ff.)].

Table 6.6. Jeans length L_J in pc and Jeans mass $\mathcal{M}_J$ in solar masses ($\mathcal{M}_\odot$) for a selection of values of number density of H atoms N_H [cm^{-3}] and kinetic temperature T [K], calculated from (6.161) and (6.162) with $\bar{\mu} = 1.3$ (corresponding to one He atom to 10 H atoms)

N_H \ T	L_J [pc]			$\mathcal{M}_J$ [$\mathcal{M}_\odot$]		
	10	100	1000 K	10	100	1000 K
0.1	60	190	602	700	22000	7.0×10^5
1	19	60	190	220	7000	2.2×10^5
10	6.0	19	60	70	2200	7.0×10^4
100	1.9	6.0	19	22	700	2.2×10^4
1000	0.6	1.9	6	7	220	7.0×10^3

Table 6.6 gives a review of the values of L_J and $\mathcal{M}_J$ obtained from (6.161) and (6.162) for the densities and temperatures occurring in the H I region of the interstellar space. Further values for $N_H = 10^4, 10^5, \ldots$ and $T = 10^4, 10^5, \ldots$ are obtained directly from the obvious behavioural regularity of the numbers given. We can now test for which of the known interstellar structures gravitational instability is to be expected. Small diffuse clouds with $T \simeq 100$ K, $N_H \simeq 10\ldots30$ cm^{-3} and dimensions $L \simeq 3$ pc or masses $\simeq 5\ldots20\mathcal{M}_\odot$ are obviously far removed from it. On the other hand, for the typical molecular clouds listed in Sect. 5.2.4, Table 5.9, with $N_H \simeq 2N(H_2) \simeq 10^4\ldots10^5$ cm^{-3} and $T \simeq 10\ldots50$ K the average values of L_J and $\mathcal{M}_J$ lie considerably below the actual dimensions and masses. These clouds should therefore collapse.

If the cloud has a *magnetic field,* this can have a stabilising effect similar to that of the thermal pressure. Numerical calculations have shown that collapse then first occurs with a cloud mass

$$\mathcal{M} > \mathcal{M}_{mag} \approx 2 \times 10^4 B^3 N_{H}^{-2} [\mathcal{M}_\odot] \qquad (6.163)$$

(Mouschovias, 1978), where the magnetic field is expressed in microgauss. $\mathcal{M}_{mag}$ is the analogue of the Jeans Mass for the case of the magnetic counter-force to gravitation.

Here the further question arises, whether the massive dense clouds can arise by gravitative condensation from the generally distributed interstellar medium. In a "warm" inter-cloud gas with $N_H \simeq 0.1\,\mathrm{cm}^{-3}$, $T \simeq 10^3\,\mathrm{K}$, regions will first become unstable when $L \geq 1\,\mathrm{kpc}$ and $\mathcal{M} \geq 10^6 \mathcal{M}_\odot$; for the hot component considerably larger values are obtained. In condensations with such large dimensions the differential galactic rotation will play a significant role: the dynamic time scale of the collapse for $N_H = 0.1\,\mathrm{cm}^{-3}$, corresponding to $\varrho \simeq 2 \times 10^{-25}\,\mathrm{g\,cm}^{-3}$, is $\tau \simeq 1/\sqrt{G\varrho} \simeq 3 \times 10^8$ years. However, when $L = 1\,\mathrm{kpc}$ even a shear of the velocity field of only $5\,\mathrm{km\,s}^{-1}\,\mathrm{kpc}^{-1}$, obtained in Sect. 3.4.2 (3.80) for the solar neighbourhood, will already after 2×10^8 years produce a tangential displacement of $1\,\mathrm{kpc}$ between the edges of the cloud which are nearest to, and furthest from, the galactic centre. In the inner region $R \simeq 3 \ldots 7\,\mathrm{kpc}$ the differential rotation is still more effective (see Sect. 6.1.5, Table 6.3). We arrive at the conclusion, therefore, that the condensation of the interstellar medium into compact clouds cannot be caused by gravitational instability alone. The formation of the dense molecular clouds must be initiated by other processes, which will be discussed below.

Origin of Gravitationally Unstable Clouds and Star Formation: As a possible mechanism the *Parker Instability* mentioned in Sect. 6.3.2 has been repeatedly suggested. Because of the large dimensions, about $L = 0.5 \ldots 1\,\mathrm{kpc}$, of the "magnetic pockets" thereby created, in which interstellar gas accumulates, the objections raised above must be considered first of all. Since the magnetic field lines near the galactic plane run roughly parallel to it, the gas can flow out of the "pockets" at $z \simeq 0$. More detailed studies show that maximal condensations to be expected therefrom are only about a factor of 3 (Mouschovias, 1974).

This is not sufficient for the creation of a gravitationally collapsing cloud. However, diffuse HI clouds already existing (formed as a result of galactic density waves, see below) could be assembled under mild compression in the large "magnetic valleys". Fusion of some of these clouds would possibly lead to densities at which the hydrogen goes over into molecular form. The associated increase in shielding from ultraviolet radiation could lower the temperature, causing the density to climb further under pressure equilibrium. In individual condensations the self-gravitation could produce a further increase in density to a few hundred per cm^3, without immediately leading to collapse. The resulting complex of small molecular clouds would probably be embedded in a complex of H I clouds, as one actually observes in various cases (region of the Orion Nebula, the Taurus cloud, etc.). The observed velocity scatter inside the cloud complex, a few km s^{-1}, would lead one to expect the individual condensations to merge with one another after a few 10^7 years (Blitz and Shu, 1980).

A further mechanism for the creation of condensations is offered by the *thermal instability* of the inter-cloud medium already discussed in Sect. 5.2.3. In Sect. 6.3.3 we have already explained that the compression of gas associated with the spiral-form galactic density waves can lead to the formation of cold clouds. Diffuse H I clouds could be created in this way. A compression of large quantities of thin gas to the state of gravitationally unstable molecular clouds, however, is extremely improbable. To obtain a total mass of about $10^5 \mathcal{M}_\odot$, a very large region would have to contract as a whole in an essentially uniform manner, in spite of the presence of interstellar clouds and shock fronts. Another objection (Kwan, 1979) is that in the time between two passages of the density wave, about 3×10^8 years, even with a relatively high initial density of 0.3 H atoms cm^{-3}, the mass convertible into molecular clouds would amount at most to $0.3 \times m_{\mathrm{H}}$ cm^{-3}. On the other hand between $R = 4$ and 8 kpc the average density produced by the large molecular clouds is about 3 H$_2$ molecules cm^{-3}. The time is therefore not available for the formation of the large molecular clouds, if one does not assume for this the quite improbably long lifetime of at least 6×10^9 years.

The actual *lifetimes of molecular cloud complexes* have repeatedly been estimated. For densities of, for example, 5×10^3 H$_2$ molecules per cm^3, such as occur in particularly compact clouds, one first of all obtains as the purely gravitational collapse time $\tau \simeq 1/\sqrt{G\varrho}$, i.e. only about 10^6 years. For a large complex (typically: total mass $\simeq 2 \times 10^5 \mathcal{M}_\odot$, extent $\simeq 50$ pc, mean density $\simeq 10^2 \times m_{\mathrm{H}}$ cm^{-3}) one finds 10^7 years. The true lifetime, however, must be longer. If the total mass of these great clouds in the galaxy, about $2 \times 10^9 \mathcal{M}_\odot$ (Sect. 5.4.4) were to collapse into stars in 10^7 years, then the star formation rate for the whole galactic disc would be $2 \times 10^2 \mathcal{M}_\odot$ per year. On the other hand, with the value of $5 \times 10^{-12} \mathcal{M}_\odot$ pc^{-3} year^{-1} for the solar neighbourhood derived in Sect. 3.2.2 (3.38), it follows that a disc of radius 10 kpc and effective thickness 600 pc will have a total rate of only $1 \mathcal{M}_\odot$ per year. Other estimates, which relate directly to the region $R \simeq 4\ldots8$ kpc, give values up to about $4 \mathcal{M}_\odot$ per year. It has been concluded from this contradiction that the actual lifetime of a molecular cloud complex is about 30 to 50 times higher than the collapse time of 10^7 years calculated above. To explain a lifetime of about $(3\ldots5) \times 10^8$ years the following grounds are cited: (1) the time required for star formation in the nucleus of a dense condensation is greater, on account of magnetic effects (magnetic pressure), angular momentum (centrifugal force, turbulence, etc.), than the collapse time τ corresponding to free fall, (2) because of the internal motions the whole complex would not in general collapse towards its mass centre, (3) only a relatively small portion of the cloud mass will be transformed into stars.

The formation of large molecular cloud complexes is explained in a few recent papers by the process of *cloud growth by inelastic collisions* already outlined in Sect. 6.3.2 [among others, Kwan (1979), Norman and Silk (1980)]. The key assumption here is that two colliding clouds as a rule merge with one another. Numerical calculations produce quite complicated results in detail. For example, for the head-on collision of two small clouds a strongly compressed thin disc is first of all produced, in whose subsequent expansion to an approximately spherically symmetric form about 25 % of the gas escapes. In summary one can say that collisions between small diffuse clouds lead rather to decay than to growth. The situation is different, however, if one of the two collision partners is larger and more massive. In this case the smaller cloud can be absorbed by the larger. Since the cloud masses extend over a certain range from the start, the more massive clouds will constantly grow at the expense of the less massive. The large molecular cloud complexes which are the eventual result will have an inhomogeneous, clumpy structure. Model treatments indicate that the *formation time* of these is about 10^8 years, if one takes into account the effects of the galactic density waves and the Parker instability. Since this time is roughly comparable with the lifetime of the complexes, an equlibrium between formation and disruption of the clouds (after the formation of massive stars, see Sect. 6.3.3) seems plausible.

Shock fronts, impinging on stable clouds from outside, can accelerate the development of high densities and − for sufficiently high initial density − can also induce cloud collapse with attendant star formation. The development of a cloud with the passage of a spiral arm shock front has been simulated numerically (Woodward, 1976). The starting values were: density $1.5 m_H \, cm^{-3}$, radius 15 pc (spherical), mass $524 \, \mathcal{M}_\odot$, and amongst the results were the following. The shock front (assumed velocity $28 \, km \, s^{-1}$) first of all caused on the near side of the cloud a compression which by cooling and condensation in a relatively thin layer reached a factor of 60. The cloud thus became strongly flattened in the shock direction. A further compression − by about a factor of 3 − was caused by the increase of pressure in the inter-cloud medium behind the front.

In detail hydrodynamic instabilities occur, leading to very irregular deformations (Fig. 6.21). Where the inter-cloud medium flows roughly parallel to the "cloud surface", Helmholtz Instability (analogous to water waves created by the wind) comes into play, and where the denser cloud gas is accelerated by the thin inter-cloud medium Rayleigh-Taylor Instability has its effect (cf. Sect. 6.3.3: dynamics of H II regions). In front of the cloud, in addition to long dense tongues, a compact condensation emerges which can become gravitationally unstable after a time, although its share of the mass is only 5 %. The flattening makes a collapse of the cloud as a whole improbable, but a few of the fragments formed by hydrodynamic instability could collapse and form stars.

The continuation of the computations showed that, after the passage of the front (duration $\simeq 10^7$ years), the flattened cloud began to expand into the region of lower pressure behind the shock front. Now, by contrast a conical form

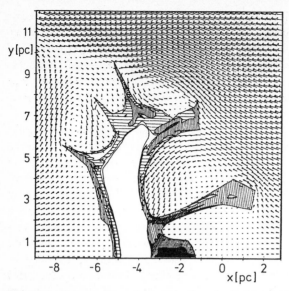

Fig. 6.21. Cross-section of an originally spherical interstellar cloud (midpoint at $x = y = 0$) 6×10^6 years after entry into a spiral arm shock wave (axial symmetry about the x-axis). The various degrees of hatching correspond to the density values $N_H = 20, 40, 100,$ and $200\,H$ atoms cm^{-3}. In the white inner region the gas is still undisturbed. The arrows indicate the instantaneous velocity field. (After Woodward, 1976; by permission)

elongated in the shock direction arose: the cloud was extended in the direction of the gas flow. Perhaps the elongated shapes of a few molecular clouds, such as the CO cloud in the region of the M 17 H II region, can be interpreted in this way.

If massive O-stars are created at the strongly compressed end of a large molecular cloud, then the shock fronts of the expanding H II regions produced by these stars can trigger the formation of a second generation of stars (cf. Sect. 4.2.9). In an elongated cloud complex this process will proceed in the direction of the major axis of the cloud and one can speak of a *chain reaction of star formation*. Model computations show that the accumulation of gas in the narrow layer between the ionisation and shock fronts leads successively to a break-up of the cloud into smaller gravitationally unstable condensations.

As a mechanism for the formation of stars of lower mass, a similar reaction of low mass stars already created on the remaining gas in the cloud has been proposed (Silk and Norman, 1980): numerous T Tauri stars ($\geq 10\,pc^{-3}$) are always embedded in the dense dark clouds. Intense stellar winds stream out from these low-mass Pre-Main-Sequence stars and sweep up the thinner gas present between the clumpy condensations. New zones of compression are thus produced by superposition, which later partially merge with one another and eventually can exceed the Jeans Mass. In this way stars with a certain mass spectrum will be formed. Amongst the low-mass stars there must again be T Tauri stars, which could keep the process going. The cycle ends when O-stars are created, whose H II regions eventually disrupt the clouds.

Very massive stars can also lead to a second generation of stars if they explode as supernovae. The supernova shockwave has a similar effect to that of the spiral arm shockwave. A possible way of deciding empirically in an actual individual case whether the star formation was triggered by a supernova is offered by the non-thermal radio continuum radiation of the expanding shell. If stars have already been formed, this radiation is of course only very weak, so convincing examples are rare.

This picture of the continuing development of new star generations from stable interstellar clouds by the shock fronts emanating from the most massive of the already created stars has been tentatively taken as the basis of a simple stochastic model for the development with time of the star distribution in the whole galactic disc (Gerola and Seiden, 1978). The galaxy — treated as a thin disc — was divided into N rings, which rotate differentially with angular velocities $\omega_i = \omega(R_i)$ (R_i = radius of the centreline of the annulus); each ring was further divided into a sufficient number of equal segments. Starting from an initial population of massive stars in the segments, the probability P_{st} was specified that such a star would stimulate the creation of an additional star in the nearest neighbouring segment in a discrete time step τ. If a new star had already been created in a segment a waiting time τ_w was imposed before new massive stars could be formed. Finally even spontaneous star formation will be possible with a small probability P_{sp}. In an example with $N = 49$, a total of 7350 segments, P_{st} and P_{sp} were set equal to 0.28 and 2×10^{-4}, respectively; a time step was $\tau = 1.5 \times 10^7$ years and the waiting time $\tau_w = 11\tau$. As the starting state 1 % of the segments outside $R = 3$ kpc were occupied with stars. The stepwise numerical computation of the development led to the distributions of young massive objects shown in Fig. 6.22. As one can see, a slightly irregular spiral structure develops very early, which is similar in appearance to many spiral galaxies. Computations with $P_{sp} = 0$ and a starting population restricted to the region $R < 3$ kpc led to qualitatively similar results.

In contrast to the density wave theory, according to which spiral arms are a global characteristic of the system, here a large scale spiral structure is

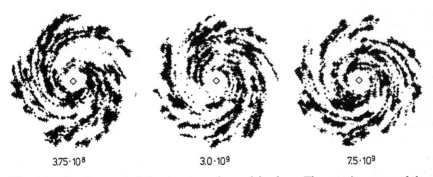

3.75·10⁸   3.0·10⁹   7.5·10⁹

Fig. 6.22. Development of the structure of a model galaxy. The rotation curve of the galaxy M 81 was taken as the basis. The attached numbers relate to the elapsed time in years since the start of the calculation. (After Gerola and Seiden, 1978; by permission)

induced by a purely local process. This concept has the advantage that the mechanism for creating new spiral arms is available from the beginning. The details of the stimulated star formation are of course still largely unclarified and the results obtained concern initially only the spiral structure distribution of young, massive objects.

Observations of the 21 cm emission of the spiral galaxy M 81 with a high resolution radio interferometer have shown, for example, that the neutral hydrogen gas follows the spiral arms, and its motion in broad outline shows the deviations from pure circular orbital motion predicted by the density wave theory. On the other hand galaxies are known with multi-armed and very fragmentary spiral structure, whose interpretation by the density wave theory appears difficult (example: NGC 7793).

Early Development of the Galaxy: The picture outlined in Sect. 3.5.2 of the creation of the present form of the Milky Way system from a very large spherical proto-galaxy can be supported by the following dynamic considerations. Since the stellardynamic relaxation time for the galaxy as a whole is considerably greater than the age of the oldest objects of the system (Sect. 6.1.2, Table 6.1), we can establish first of all that the relationship of the halo (including the bulge) and the disc can scarcely have changed since the formation of the disc. For the state of the system which preceded this process, let us choose an extended spherical gas cloud with radius a and homogeneous mass density ϱ, rotating rigidly with angular velocity ω. The mass inside a concentric sphere with radius r will be denoted by $\mathcal{M}_r$.

At a selected point at a distance r from the centre and at latitude angle β, so that its axial distance $R = r \cos \beta$ (Fig. 6.23), the ratio of the centrifugal and gravitational accelerations perpendicular to the axis of rotation is given by

$$\frac{Z_R}{K_R} = \frac{\omega^2 R}{(G\mathcal{M}_r/r^2) \cos \beta} = \frac{j^2}{G\mathcal{M}_r} \frac{1}{r \cos^4 \beta} \tag{6.164}$$

where $j = \omega R^2$ = angular momentum per unit mass about the axis of rotation. Since $\mathcal{M}_r = (4\pi/3)r^3 \varrho$ one can also write the first expression in the form

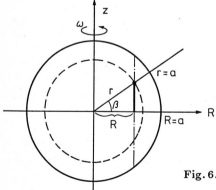

Fig. 6.23. Model of the protogalaxy

$$\frac{Z_R}{K_R} = \frac{3}{4\pi G} \frac{\omega^2}{\varrho} \quad . \tag{6.165}$$

First of all let $Z_R/K_R \ll 1$, so that gravitation is completely dominant. A glance at Table 6.6 shows that the Jeans Mass $\mathcal{M}_J$ is only about $10^6 \mathcal{M}_\odot$, even with a relatively high gas temperature of $1000\,\mathrm{K}$ and, for example, a number density of H atoms $N_H = 0.1\,\mathrm{cm}^{-3}$, corresponding to $\varrho \simeq 2 \times 10^{-25}\,\mathrm{g\,cm}^{-3}$. With a total mass of the proto-galactic cloud of the order of $10^{11} \mathcal{M}_\odot$ one must therefore expect gravitational collapse even for considerably lower densities and for somewhat higher temperatures.

We shall assume that as r grows smaller the angular momentum j and the mass $\mathcal{M}_r$ remain constant. According to (6.164) $Z_R/K_R \sim 1/r$ will then increase. The primeval sphere will contract approximately spherically ($\beta = $ constant) until the increasing centrifugal force halts the collapse perpendicular to the axis of rotation, so that $Z_R/K_R = 1$. The values of density and angular velocity existing at this instant $t = t_1$, regarded as independent of R and z, will be denoted by ϱ_1 and ω_1, respectively, and the radius of the gas sphere by a_1. From (6.165) it follows in particular that

$$\omega_1^2 = \frac{4\pi}{3} G\varrho_1 = G\frac{\mathcal{M}}{a_1^3} \quad . \tag{6.166}$$

The component of the gravitational force parallel to the axis of rotation must now produce a flattening of the sphere into a disc. The *stars of the galactic halo* must be created before the instant t_1. To explain the strongly eccentric orbits of these stars, Eggen, Lynden-Bell and Sandage (1962) assume that the collapse took place very quickly (free fall: duration a few 10^8 years). The "sudden" concentration of mass towards the centre now attracted the existing stars much more strongly or (and perhaps both processes occurred) stars were formed during the free fall phase. The contraction could, however, have taken place considerably more slowly (Yoshii and Saio, 1979).

The finally resultant mass density loading $\sigma_1(R)$ in the disc is given by integration of the density ϱ_1 over z at fixed R in a sphere of radius a_1. Since ϱ_1 is constant it follows that

$$\sigma_1(R) = \varrho_1 \times 2\sqrt{a_1^2 - R^2} \quad . \tag{6.167}$$

In a thin *disc* with this density distribution the component of the gravitational force K_R is greater than that required by the centrifugal acceleration: (6.82) with (6.167) and for example $R = a_1$ easily leads to $K_R(R = a_1) = 4\pi G\varrho_1 a_1$, whereas (6.166) gives $Z_R(R = a_1) = \omega_1^2 a_1 = (4\pi/3)G\varrho_1 a_1$, so that $Z_R(R = a_1) = (1/3)K_R(R = a_1)$. The disc must therefore contract radially. If one assumes from now on that the angular momentum of each element is conserved, then for every specified law of contraction there will be a corresponding law of rotation $\omega(R)$. However, the fraction of the total mass $d\mathcal{M}/\mathcal{M}$ with an angular momentum per unit mass in the interval $j \ldots j + dj$ will remain the same throughout the contraction. This offers the possibility of testing the hypothesis of the development of the galactic disc as a result of a proto-galactic collapse by comparison with the *present functional relationship between mass and angular momentum* (Freeman, 1975).

The fraction of the total mass $\mathcal{M} = (4\pi/3)a_1^3 \varrho_1$, lying between R and $R + dR$ in our disc model, is given by

$$\frac{d\mathcal{M}(R)}{\mathcal{M}} = \frac{2\pi R \, dR \, \sigma_1(R)}{\mathcal{M}} = \frac{3R \, dR}{a_1^2} \sqrt{1 - \frac{R^2}{a_1^2}} \quad .$$

If one expresses R and dR in terms of $j(R) = \omega_1 R^2$ amd $dj = \omega_1 2R \, dR$, one obtains

$$\frac{d\mathcal{M}(j)}{\mathcal{M}} = \frac{3}{2} \sqrt{1 - \frac{j}{\omega_1 a_1^2}} \frac{dj}{\omega_1 a_1^2} \quad . \tag{6.168}$$

The fraction of the mass, for which the angular momentum per unit mass is smaller than j, is given by integration

$$\frac{\mathcal{M}(j)}{\mathcal{M}} = \int_0^j \frac{d\mathcal{M}(j)}{\mathcal{M}} = 1 - \left(1 - \frac{j}{\omega_1 a_1^2}\right)^{3/2} \quad . \tag{6.169}$$

For comparison with the corresponding relationship for the galaxy at present, one can use the exponential law already explained in Sect. 6.1.5

$$\sigma(R) = \sigma_0 \exp(-\alpha R) \tag{6.170}$$

with the "scale factor" $\alpha^{-1} \simeq 2 \, \mathrm{kpc}$. The rotation law, obtained by equating the gravitational force per unit mass and the centrifugal acceleration is according to Freeman

$$\omega^2(R) = \pi G \sigma_0 \alpha [I_0(\tfrac{1}{2}\alpha R) K_0(\tfrac{1}{2}\alpha R) - I_1(\tfrac{1}{2}\alpha R) K_1(\tfrac{1}{2}\alpha R)] \quad , \tag{6.171}$$

where I_0, K_0, I_1, K_1 are the modified Bessel Functions. This agrees approximately with the observed profile. Carrying out the calculation of $\mathcal{M}(j)/\mathcal{M}$ with these profiles of $\sigma(R)$ and $\omega(R)$, one obtains the continuous curve plotted in Fig. 6.24, whereas (6.169) gives the plotted points. The good agreement is a weighty argument for the collapse hypothesis.

The radius a_1, density ϱ_1 and angular velocity ω_1 of the spherical proto-galaxy at the beginning of the disc formation can be estimated in the following way. The total angular momentum of the exponential disc is given by (6.170) and (6.171) according to Freeman as

$$J = 1.109 \sqrt{G \mathcal{M}^3 / \alpha} \quad . \tag{6.172}$$

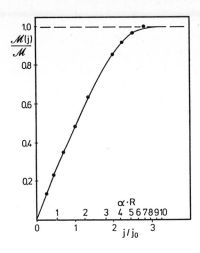

Fig. 6.24. Mass fraction of the galactic disc with angular momentum per unit mass $< j$, where j is expressed in units of an arbitrary magnitude j_0, for an exponential disc with scale parameter α (*continous curve*) and for the disc resulting from the collapse of a rigidly rotating homogeneous sphere (*points*). The unit of angular momentum was taken to be $j_0 = \sqrt{GM/\alpha} \simeq (1/3)\omega_1 a_1^2$. (After Freeman, 1975)

On the other hand, for the homogeneous sphere with radius a_1 and angular velocity ω_1, using (6.166),

$$J = \tfrac{2}{5}Ma_1^2\omega_1 = \tfrac{2}{5}\sqrt{GM^3 a_1} \quad . \tag{6.173}$$

Equating the right-hand sides of (6.172) and (6.173), with $\alpha^{-1} \simeq 2\,\mathrm{kpc}$, gives

$$a_1 = 7.69\alpha^{-1} \approx 15\,\mathrm{kpc} \quad .$$

The disc mass $\mathcal{M} \simeq 8 \times 10^{10}\,\mathcal{M}_\odot$ (Table 6.4) then gives a homogeneous mass density $\varrho_1 \simeq 4 \times 10^{-25}\,\mathrm{g\,cm^{-3}}$. Moreover (6.166) now gives an angular velocity $\omega_1 \simeq 3.4 \times 10^{-16}\,\mathrm{s^{-1}}$, corresponding to a linear velocity $\Theta = 100\,\mathrm{km\,s^{-1}}$ at $R = 10\,\mathrm{kpc}$. For the dynamic timescale of the collapse leading to disc formation (beginning at $t = t_1$) one obtains $\tau \approx 1/\sqrt{G\varrho_1} \simeq 2 \times 10^8$ years, a relatively short time.

As an example of a considerably earlier stage still, let us take the spherical radius $a = 5a_1 \simeq 75\,\mathrm{kpc}$. Conservation of mass and angular momentum then gives, according to (6.173), $a^2\omega = a_1^2\omega_1$, so that $\omega = (a_1/a)^2\omega_1 = 0.04\omega_1$, and also $\varrho = (a_1/a)^3\varrho_1 = 0.008\varrho_1 \simeq 3 \times 10^{-27}\,\mathrm{g\,cm^{-3}}$. Equations (6.165) and (6.166) then give a ratio $Z_R/K_R = a_1/a = 0.2$, so the collapse is still nearly spherical.

Detailed model calculations of the development of a rotating spherical proto-galaxy consisting of many clouds have shown that a thin disc results only if the initial high star formation rate − in the spherical collapse phase − decreases strongly after the formation of the stellar halo, since otherwise no gas is left over for the disc (Larson, 1976). One therefore assumes that between the clouds of the proto-galaxy there has been a thin gas, which remained after the formation of the stars of the halo and the central region from the clouds. This gas finally fell into the principal plane and formed the disc. The star formation here followed more slowly and continues up to the present time.

In the denser *central region* the star formation will have proceeded further than in the outer parts of the disc. In fact near the galactic centre, in contrast to the solar neighbourhood, (1) the gas content is considerably lower (Sect. 5.4.4, p. 362), (2) the frequency of supernovae is lower (fewer massive stars, because less gas), (3) the frequency of planetary nebulae, which arise from low-mass stars, is higher, and (4) the relative frequency of heavier elements is higher (Sect. 5.2.5). We shall go further into the last point below.

Chemical Development of the Galaxy: Essential indices of galactic evolution are contained in the observed spatial variation of the relative abundances of the elements in the galaxy. A part of the "heavy" elements (heavier than hydrogen) synthesised in the stars are given up to the interstellar medium in the course of stellar evolution. In this way there is a continual *enrichment of the interstellar medium and the later generations of stars with heavy elements*. This process continues until star formation ceases.

For the interpretation of the available empirical results on the element abundances in the various sub-systems and populations of galactic objects and the spatial frequency gradients inside the same, numerous model calculations have been carried out. The essential basis for these consists of the knowledge of (a) the formation of the elements in the stars and (b) the processes which lead to the transfer of synthesised matter to the interstellar medium.

For (a): heavy elements are formed during the quiet times of stellar evolution, as well as in the explosive phases (supernovae). In the massive stars element synthesis proceeds essentially more quickly than in low-mass stars with $M \lesssim 1 M_\odot$. Assessments of the production rates for the individual elements in stars of different mass, in conjunction with the initial mass function, lead to statements on the total mass of each of the heavy elements newly formed in each generation of stars. Here one has to distinguish between "primary" elements, which can be formed directly from H and He (O, C, Fe group), and "secondary" elements, whose formation requires the pre-existence of heavy elements (e.g. Ba and N).

For (b): The mass outflow from stars in the various stages of evolution is so far known only relatively inexactly. Quite broadly it can be said that stars with $M \lesssim 1 M_\odot$, even if they belong to the first generation of stars, have still lost practically no material, because they develop too slowly. The mass supply to the interstellar medium is provided predominantly (in order of star mass) by the stellar wind of red giants ($M \simeq 2 \ldots 8 M_\odot$), the planetary nebulae (produced by stars with $M \simeq 2 \ldots 8 M_\odot$), the wind from massive stars ($M > 8 M_\odot$) and the supernovae (especially Type II with initial mass $M \gtrsim 10 M_\odot$).

In the model calculations one further deals with the formation rate $s(M, t)$ of stars of various masses, $s(M, t) dM\, dt$ = number of stars with mass in the interval $M \ldots M + dM$, created in the time interval $t \ldots t + dt$. The time dependence arises primarily from the variation with time of the density of the interstellar medium $\varrho = \varrho(t)$. Often one simply sets

$$s(M, t) \sim F_0(M)[\varrho(t)]^n \quad,$$

where $F_0(\mathcal{M})$ denotes the initial mass function introduced in Sect. 3.2.2 and n lies between 1 and 2 (Schmidt, 1959). The density variation with time $\varrho(t)$ is determined from the dynamic development of the system.

Relatively detailed calculations of the development for a typical spiral galaxy have been carried out by Larson (1976), amongst others. Without going in detail into the calculations, we give here some interesting results of this work for the spatial and temporal variation of the mass fraction Z of the heavy elements − often referred to as "metals". The starting model was a spherical proto-galaxy (clouds + inter-cloud gas) of 10^{11} solar masses with a radius of 100 kpc, whose chemical composition is characterised by $Z = 0$.

It is shown first of all that in the developing dense kernel region with rapid star creation there is a steep increase of the content of heavy elements, which later flattens out. The lower star formation rate in the outer regions leads there to an essentially slower increase of Z. When the disc has formed, Z goes on rising gradually. In this way, the *development of two different stellar populations* can be understood, a metal-poor halo population and a relatively metal-rich disc population (Fig. 6.25). The enrichment of the inflowing proto-galactic gas with heavy elements from the already formed massive stars could also play a role here. These theoretical conclusions agree at least qualitatively with the results of observations (cf. Sect. 3.5.2 and particularly Sects. 3.3.2, 3.3.3).

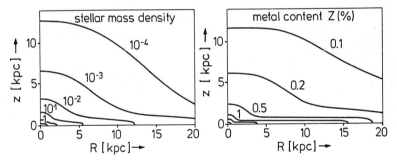

Fig. 6.25. Development model for the present state of the galaxy in radial sections perpendicular to the principal plane. *Left*: lines of equal mass density of the star distribution $[\mathcal{M}_\odot \mathrm{pc}^{-3}]$. *Right*: lines of equal mass fraction Z per cent of the heavy elements ("metals") in the stellar matter. (After Larson, 1976; by permission)

Appendix

A. Fundamental Formulae for the Transformation of Astronomical Coordinates

Formulae of Spherical Trigonometry: The most useful formulae in astronomy, relating the angles α, β, γ and the sides a, b, c of the spherical triangle (depicted in Fig. A.1 and defined by arcs of great circles) are obtained by introducing two rectangular coordinate systems with a common x-axis. The coordinates of the point B in these two rectangular systems are, respectively:

$$
\begin{aligned}
x &= r \sin a \sin \gamma & x' &= r \sin c \sin \alpha \\
y &= -r \sin a \cos \gamma & y' &= r \sin c \cos \alpha \\
z &= r \cos a & z' &= r \cos c
\end{aligned} \qquad \text{(A.1)}
$$

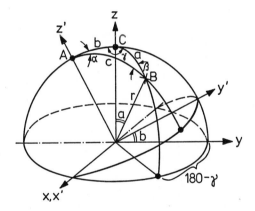

Fig. A.1. Derivation of the basic formulae of spherical trigonometry

The transformation equations between the two rectangular systems are

$$
\begin{aligned}
x &= x' & x' &= x \\
y &= y' \cos b - z' \sin b & y' &= y \cos b + z \sin b \\
z &= y' \sin b + z' \cos b & z' &= -y \sin b + z \cos b
\end{aligned} \qquad \text{(A.2)}
$$

The first equation of system (A.2), using the first equations of (A.1), gives

$$
\sin a \sin \gamma = \sin c \sin \alpha \qquad \text{(A.3)}
$$

449

(the Sine Rule for the spherical triangle), and two more corresponding formulae follow by cyclic permutation. The second equations of system (A.2), using the second and third equations of system (A.1), yield

$$\sin a \cos \gamma = \sin b \cos c - \cos b \sin c \cos \alpha \quad \text{and} \tag{A.4}$$

$$\sin c \cos \alpha = \cos a \sin b - \sin a \cos b \cos \gamma. \tag{A.5}$$

The two last equations of system (A.2) similarly give

$$\cos a = \cos b \cos c + \sin b \sin c \cos \alpha \quad \text{and} \tag{A.6}$$

$$\cos c = \cos a \cos b + \sin a \sin b \cos \gamma \quad . \tag{A.7}$$

Tangential and Radial Components of the Motion of an Object: We again start from the rectangular coordinates of the object. The representation by polar coordinates (r, L, B) with the x-axis as the zero direction for L (Fig. A.2) gives:

$$x = r \cos B \cos L \ , \quad y = r \cos B \sin L \ , \quad z = r \sin B \quad . \tag{A.8}$$

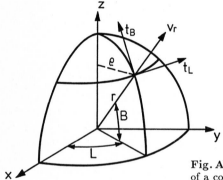

Fig. A.2. Tangential and radial components of motion of a cosmic object

Let us find the components of the velocity vector $\boldsymbol{v}$ of an object in relation to a rectangular coordinate system whose base plane is the plane tangential to the sphere at the point (L, B) and whose axes lie in the direction of the parallel (latitude) circle, the vertical (longitude) circle, and the line of sight (radial direction) (Fig. A.2). The components of $\boldsymbol{v}$ in this system are

$$t_L = r \dot{L} \cos B \ , \quad t_B = r \dot{B} \ , \quad v_r = \dot{r} \quad . \tag{A.9}$$

The dot indicates differentiation with respect to time. The factor $\cos B$ arises from the fact that L is measured on the great circle (the "equator") in the $x - y$ plane, and a change dL there means an angular change about the z-axis. The distance of the object from this axis, however, is $r \cos B$ (not r), so that the corresponding arc through which the object travels is $r \cos B \, dL$ (not $r \, dL$). Put in another way, the line of sight moves through an angle $dL \cos B$ (not dL).

In order to express t_L, t_B, v_r in terms of $\dot{x}$, $\dot{y}$, $\dot{z}$, one can differentiate the coordinate expressions (A.8) and solve the resulting system of equations, together with (A.9), for t_L, t_B, v_r. The result can also be obtained directly and clearly from Fig. A.2, by noticing that t_L, t_B, v_r are the scalar products of the velocity vector $\boldsymbol{v}(\dot{x}, \dot{y}, \dot{z})$ with the unit vectors $\boldsymbol{e}_L$, $\boldsymbol{e}_B$, $\boldsymbol{e}_r$ on the axes of the tangential system. Denoting the unit vectors in the x, y, z directions by $\boldsymbol{e}_x$, $\boldsymbol{e}_y$, $\boldsymbol{e}_z$, we obtain

$$v_r = (\boldsymbol{v} \cdot \boldsymbol{e}_r) = \dot{x} \cos(\boldsymbol{e}_x, \boldsymbol{e}_r) + \dot{y} \cos(\boldsymbol{e}_y, \boldsymbol{e}_r) + \dot{z} \cos(\boldsymbol{e}_z, \boldsymbol{e}_r)$$
$$= \dot{x} \cos L \cos B + \dot{y} \sin L \cos B + \dot{z} \sin B \qquad \text{(A.10)}$$

and correspondingly

$$t_L = (\boldsymbol{v} \cdot \boldsymbol{e}_L) = -\dot{x} \sin L + \dot{y} \cos L \qquad \text{(A.11)}$$

$$t_B = (\boldsymbol{v} \cdot \boldsymbol{e}_B) = -\dot{x} \cos L \sin B - \dot{y} \sin L \sin B + \dot{z} \cos B \quad . \qquad \text{(A.12)}$$

To invert this system of equations we write

$$\dot{x} = \gamma_{11} t_L + \gamma_{12} t_B + \gamma_{13} v_r$$
$$\dot{y} = \gamma_{21} t_L + \gamma_{22} t_B + \gamma_{23} v_r$$
$$\dot{z} = \gamma_{31} t_L + \gamma_{32} t_B + \gamma_{33} v_r \quad . \qquad \text{(A.13)}$$

Since it is only a matter of a rotation of the coordinate system, the matrix (γ_{ik}) is orthogonal. It follows that this inverse matrix is equal to the transpose of the matrix (γ_{ki}), whose elements are given by the coefficients of the equations (A.10) to (A.12). One therefore obtains

$$(\gamma_{ik}) = \begin{pmatrix} -\sin L & -\cos L \sin B & \cos L \cos B \\ \cos L & -\sin L \sin B & \sin L \cos B \\ 0 & \cos B & \sin B \end{pmatrix} \quad . \qquad \text{(A.14)}$$

Changes in Spherical Coordinates for Small Rotations of the System: Let us now consider the case when the change of apparent position of the object considered above arises purely from a rotation of the coordinate system. The changes in the spherical coordinates L and B will then be described by the relations (A.11) and (A.12) – in conjunction with (A.9); for the radial component $v_r \equiv 0$.

We resolve the rotation of the system into rotation components about the x, y and z axes: the angles ω_1, ω_2 and ω_2 (Fig. A.3). One imagines these three

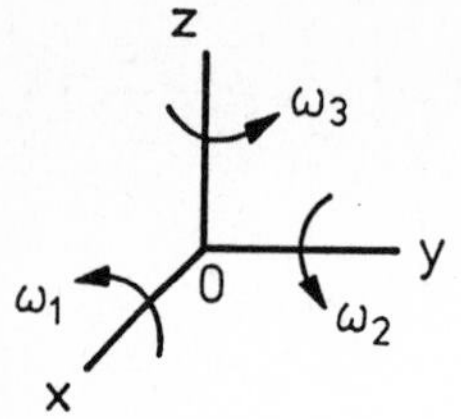

Fig. A.3. Resolution into components of the rotation of the coordinate system

rotations carried out one after the other. Under the assumption that it is only a matter of *small* angles of rotation, one obtains for the coordinate changes of a field point, with the help of Fig. A.3,

$$dx = y\omega_3 - z\omega_2$$
$$dy = z\omega_1 - x\omega_3$$
$$dz = x\omega_2 - y\omega_1 \quad . \tag{A.15}$$

If one considers changes per unit time, and hence the angular velocities ω_i and coordinate changes $\dot{x}$, $\dot{y}$, $\dot{z}$, then by substituting (A.15) in (A.11) and (A.12) and using (A.8) and (A.9) $- r$ cancels out $-$ we obtain the following relations

$$\dot{L} \cos B = \omega_1 \sin B \cos L + \omega_2 \sin B \sin L - \omega_3 \cos B$$
$$\dot{B} = -\omega_1 \sin L + \omega_2 \cos L \quad . \tag{A.16}$$

B. Energy States and Transitions of Molecules

We shall assume that the reader is familiar in broad outline with the notation conventions of the energy states of atoms. We shall consider first of all the diatomic molecule with the aid of a schematic picture similar to the Bohr model of the atom: the valence electrons essential for the chemical bonding move preferentially in closed orbits round the line joining the two atoms as axis, whilst the "inner" electrons form a charged cloud around the respective nuclei (see Fig. B.2a).

Electron Terms: The energy of the system depends critically on the configuration of the outer (valence) electrons. For each electron state there is a certain profile of the potential energy of the diatomic molecule with variable separation distance R of the two atoms: a certain potential curve (Fig. B.1). The orbital angular momenta of the electrons are strongly linked to the field between the two atoms and precess about the molecular axis, whose direction is denoted by the unit vector $\boldsymbol{k}$. Instead of the orbital angular momentum quantum number L of the electrons of an individual atom there now appears a quantum number Λ, which refers to the projection of the total orbital angular momentum vector of all the electrons $\boldsymbol{L}$ on the molecular axis: $\Lambda = |\boldsymbol{k} \cdot \boldsymbol{L}|$.

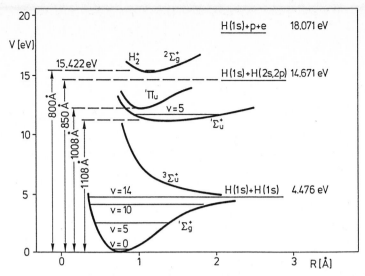

Fig. B.1. Potential energy of the H_2 molecule $V(R)$ as a function of the mutual separation R of the two protons for a few important electronic states. A few vibrational levels (quantum number v) are also indicated by horizontal lines, as well as the dissociation energy and the energy required for direct photodissociation and photo-ionisation. (After Field, Somerville and Dressler. Reproduced, with permission, from the Annual Review of Astronomy and Astrophysics, Vol. 4. Copyright © 1966 by Annual Reviews, Inc.)

The term symbols S, P, D, ... for the internal quantum numbers of the atom $L = 0, 1, 2, \ldots$ have an analogue for the diatomic molecule in the electron term symbols Σ, Π, Δ, ... which are associated with the values $\Lambda = 0, 1, 2, \ldots$. From the sum of the electron spins, which will also be denoted by S for the valence electrons of the molecule, one likewise takes the component in the direction of the line joining the nuclei $\Sigma = |k \cdot S|$. This quantity must not be confused with the electron term notation Σ for $\Lambda = 0$. Λ and Σ are summed algebraically to give $\Omega = \Lambda + \Sigma$. Just as for the individual atoms, one term, characterised by certain values Λ and Σ, is split into $r = 2S + 1$ fine structure levels. Σ-states are simple, so long as the molecule does not rotate. As in the energy terms for the atom, the multiplicity r is placed above and to the left of the term symbol. For the Σ-terms ($\Lambda = 0$) one has to distinguish between Σ^+- and Σ^- states ("positive" and "negative" terms), whose quantum mechanical eigenfunctions either keep or reverse sign with reflection of the electrons in a plane containing the line joining the nuclei.

The energy states of a diatomic molecule can be deduced from the energy terms of the two individual atoms, if these are well separated from one another. Then Λ and Σ are given by simple addition from the corresponding quantum numbers L and S, respectively, for the individual atoms. If one starts, say, from two H atoms in the ground state ^2S, then it follows that $\Lambda = L_1 + L_2 = 0$, hence a Σ-state. The two electrons may have parallel or anti-parallel spins, so that the spin quantum number $S = S_1 + S_2$ is either 1 or 0, respectively, thus giving the two terms $^3\Sigma$ and $^1\Sigma$. With different starting values L_1, L_2, S_1, S_2 one gets further terms. In similar fashion, for example, from a C atom in state ^3P and an H atom in state ^2S one gets a resulting CH molecule with Σ- and Π-states (doublets and quartets, since C has two electrons).

If one forms the molecule from two atoms with equal nuclear charge number Z, then two sorts of term can arise: the quantum mechanical treatment produces two eigenfunctions, one symmetric and one antisymmetric (reflection of the electron coordinates in the centre of symmetry leaves the sign of the eigenfunction unchanged or reversed, respectively). The respective terms are called even or odd, and are distinguished by the letters g and u below and to the right of the term symbol, e.g. $^3\Sigma_\mathrm{g}$ and $^3\Sigma_\mathrm{u}$. The two possibilities were first discovered in the H_2 molecule. Figure B.1 shows the potential curves of the odd and even Σ-states of the H_2 molecule, which correspond respectively to repulsion and attraction with stable molecule formation. With a symmetric eigenfunction the quantum numbers of the two electrons coincide, so that according to the Pauli Principle the two spins must be anti-parallel. Only the approach of two H atoms with anti-parallel spins can therefore lead to the direct formation of an H_2 molecule.

For the transitions between electronic states of a diatomic molecule, which are associated with the emission or absorption of radiation, the following selection rules hold. Transitions with $\Delta\Lambda = 0$, ±1 and $\Delta S = 0$ are allowed. Intercombinations between terms of different multiplicity are therefore forbidden. Accordingly one can only combine even with odd terms, but not positive with negative. The lines arising from transitions between electronic states fall in the visible and the ultraviolet regions of the spectrum. For molecular hydrogen these rules mean, for example, that transitions between the ground state $^1\Sigma_\mathrm{g}^+$ (Fig. B.1) and the 11.2 eV higher state $^1\Sigma_\mathrm{u}^+$ (Lyman bands) or to the $^1\Pi_\mathrm{u}$ state at 12.3 eV (Werner bands) are allowed. On the other hand the intercombining transitions between $^3\Sigma_\mathrm{u}^+$ (repulsion) and $^1\Sigma_\mathrm{g}^+$ (attraction) are forbidden.

Vibrational and Rotational Levels: Deviations from the analogy with the relations of individual atoms arise from the fact that a molecule can rotate and vibrations are possible between the two nuclei.

Correspondingly, each of the electron terms described consists of a number of vibrational energy states, whose separations are essentially smaller than those of the electron states and accordingly produce spectra in the near infrared (up to about $10\,\mu$m); the vibrational states are further split into various rotational levels with still smaller separations, so that transitions between these lead to spectra in the far infrared or in the radio frequency region.

The various discrete vibrational states of a molecule are characterised by the vibrational quantum number v. An electron state of the stable molecule, which corresponds to a certain potential curve, is arranged in a sequence of vibrational levels with the quantum numbers $v = 0$, 1, 2, ... In Fig. B.1 a few of these vibrational levels of the electronic ground term $^1\Sigma^+$ are indicated. In the limiting case $v\to\infty$ both nuclei are completely separated from one another.

The energy difference between $v = 0$ and $v \to \infty$ (practically $v > 14$) is the dissociation energy. In transitions between different electronic terms there are no selection rules for the vibrational quantum numbers v.

The numbers and locations of the rotational states are given by the coupling between $\boldsymbol{S}$, $\boldsymbol{L}$ and the axial field. In the simplest case the total angular momentum $\boldsymbol{J}$ is given by the vectorial addition $\boldsymbol{J} = \boldsymbol{R} + \boldsymbol{k}\Omega$ from the mechanical molecular spin $\boldsymbol{R}$ and the components of the electron orbital and spin momenta in the direction of the molecular axis $\Omega = \Lambda + \Sigma$, multiplied by $\boldsymbol{k}$, the unit vector in the molecular axis (see Fig. B.2b). This is the Hund's coupling case (a): $\boldsymbol{L}$ and $\boldsymbol{S}$ are both strongly coupled to the axial field; there is no interaction between the rotation of the molecule and the electron motion.

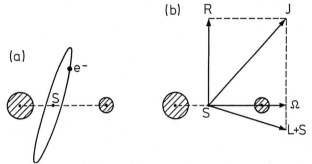

Fig. B.2. (a) Conceptual model of a diatomic molecule with one outer electron. S is the position of the mass centre. (b) Vector model of the diatomic molecule for Hund's case a (explanation in text)

The quantum number J of the different rotational levels can take the values $J = \Omega$, $\Omega + 1$, $\Omega + 2$, ..., where $\Omega = \Lambda + S$, $\Lambda + S - 1$, ..., $\Lambda - S$. For each value of Ω one obtains a sequence of rotational levels with different J-values. The whole term is characterised by the attachment of the value of Ω below and to the right of the term symbol. For an outer electron, and hence $S = 1/2$, for example, with $\Lambda = 1$ we have the values $\Omega = 1/2$ and $\Omega = 3/2$, which lead to the sequences of levels $^2\Pi_{1/2}$ with $J = 1/2, 3/2, 5/2, \ldots$ etc. and $^2\Pi_{3/2}$ with $J = 3/2, 5/2, \ldots$.

There are further coupling cases. With increasing rotation energy the interaction neglected in the previous case between the electron motion, characterised by Λ, and the molecular rotation plays a role; it has the effect that all the energy levels with $\Lambda \neq 0$ split into two components. One speaks of Λ-doubling. The amount of the split is generally very small. Astronomically the Λ-doubling has been primarily of interest because of the observation of the transition between the two individual levels of the rotational state $^2\Pi_{3/2}$, $J = 3/2$ of the interstellar OH molecule with one outer electron (see Fig. 4.48), whose energy difference corresponds to a wavelength $\lambda \simeq 18$ cm.

For the *energy values* of the individual rotational levels the quantum mechanics gives in the simplest case $S = \Lambda = 0$, which occurs with two outer electrons (e.g. CO, CS), with the assumption of a rigid rotator,

$$E_J = hB_0 J(J+1) \quad \text{with} \quad B_0 = \frac{h}{8\pi^2 I_0} \quad (J = 0, 1, 2, \dots) \quad . \tag{B.1}$$

Here I_0 denotes the moment of inertia of the molecule about an axis perpendicular to the line joining the two nuclei through the mass centre; B_0 is called the rotational constant of the molecule. For transitions the selection rule is $\Delta J = \pm 1$, so that the expression for the frequency of the individual lines of the pure rotational spectrum is

$$\nu = \frac{E_{J+1} - E_J}{h} = 2B_0(J+1) \quad . \tag{B.2}$$

For $J = 0, 1, 2, \dots$ one obtains a sequence of equidistant lines with the frequency separation $2B_0$. If $\Lambda \neq 0$, the rigid rotator is transformed into a symmetrical top. In Hund's coupling case (a) one then obtains, for example,

$$E_J = hB_0[J(J+1) - \Omega^2] \quad . \tag{B.3}$$

Polyatomic Molecule: If the atoms are arranged linearly, as for example in HCN (hydrocyanic acid) or HCCCN (cyanacetylene), then the moment of inertia I_0 is formed in similar fashion as for the diatomic molecule and one obtains the one-dimensional sequence of rotation energies from (B.1). In the general case of an arbitrary unsymmetrical molecule, however, there appear in the description of the molecular rotation three different moments of inertia, $I_a < I_b < I_c$, which correspond to the three rotational constants given by the specification in (B.1). For nonlinear symmetric molecules, such as CH_3 or NH_3, two of the three moments of inertia are equal: $I_a = I_b < I_c$. The rotation energy depends on two parameters. Besides J, the quantum number of the total angular momentum, one then introduces another quantum number K for the projection of the total angular momentum on the axis of symmetry of the molecule, and instead of (B.1) we get

$$E_{J,K} = hB_0 J(J+1) + (A_0 - B_0)K^2 \tag{B.4}$$

with $J \geq K$. The selection rule for transitions is $\Delta K = 0$. To a transition $J + 1 \rightarrow J$ there then belongs a whole sequence of lines with $K = 0, 1, 2, \dots, J$. Examples are the lines for $K = 0, 1, \dots, 5$ of the transition $J = 6 \rightarrow 5$ of CH_3CN (methyl cyanide) observed in the direction of the radio source Sgr B2.

An additional possibility of transitions which lead to microwave lines is the inversion doubling of energy levels. We shall explain this case using the example of NH_3. Here the three H atoms form an equilateral triangle and the N atom can be placed either above or below the triangular surface (Fig. B.3), so that it forms the peak of a three-sided pyramid, which in the second case

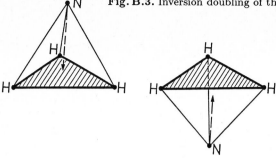

is "standing on its head". The frequency of the line produced in the transition between these two energetically somewhat different states can be explained as the pendulum frequency of the N atom in an oscillation between the two positions − over and under the triangular basis of H atoms. Here, the quantum numbers J and K do not change.

With unsymmetrical molecules (asymmetric tops, e.g. H_2O and H_2CO) a rotational state is characterised by three independent quantum numbers. Besides J one defines quantum numbers K_b, K_c for the projections of the total angular momentum on the two principal axes with the highest moments of inertia. The notation for the individual levels then usually takes the form $J_{K_bK_c}$, hence, for example, 2_{12} for the rotational state determined by $J = 2$, $K_b = 1$, $K_c = 2$.

C. Charts for the Graphical Determination of Approximate Galactic Coordinates

Using the following charts the galactic coordinates l [°] and b [°] can be directly determined from the equatorial coordinates α[h] and δ[°] for the equinox 1950.0, and conversely. [From J.D. Kraus: *Radio Astronomy*, 2nd ed. (1986); by permission].

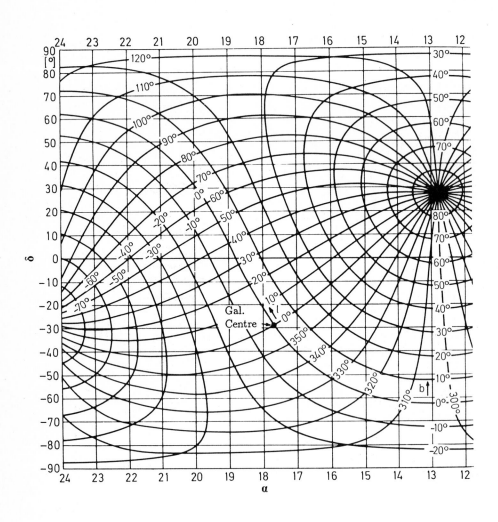

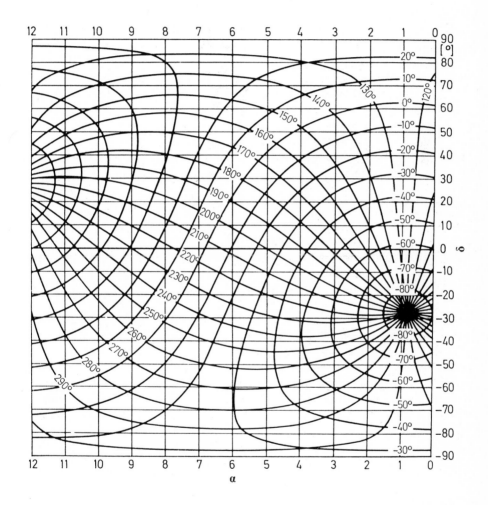

D. Excitation and Ionisation in the Thermodynamical Equilibrium

With sufficiently high densities, for example even in the lower photosphere of the sun, the collisional processes are completely dominant over the radiative processes in the attainment of an equilibrium distribution of the gas particles over their various energy states. The excitation and ionisation ratios are therefore determined solely by the temperature of the gas. Statistical thermodynamics produces a population of the discrete energy states of atoms or ions, characterised by the index n, proportional to $\exp(-\chi_n/kT)$. Here χ_n denotes the excitation energy from the ground state. The relative population accordingly decreases exponentially at higher levels. At low temperatures nearly all the gas particles are in their lowest states.

In an external magnetic field each individual level is split in general into $g = 2J + 1$ levels, where J is the amount of the total electron angular momentum (Fine Structure splitting). In the absence of a magnetic field each level must nevertheless be counted g times ("degeneracy") in the statistics of the population ratios. One calls g the statistical weight of the state. The ratio of the number densities of the atoms (ions) in the energy state n and the ground state $n = 1$ is then given by the Boltzmann formula:

$$\frac{N_n}{N_1} = \frac{g_n}{g_1}\exp(-\chi_n/kT) \quad .$$

For the levels n and m involved in a spectral line, with $\chi_n - \chi_m = h\nu_{nm}$, it follows in particular that

$$\frac{N_n}{N_m} = \frac{g_n}{g_m}\exp(-h\nu_{nm}/kT) \quad .$$

The ratio between the numbers of singly ionised and neutral atoms can be obtained by extension of the Boltzmann formula to the region of continuously distributed energy states above the ionisation energy χ_0. The distance of such a state from the ground state amounts to $\chi_0 + \frac{1}{2}m_e v^2$, where m_e denotes the mass and v the velocity of the liberated electron. Following the quantum theory we "dissect" the continuum of these states in the 6-dimensional phase space, with the position and momentum coordinates x, y, z, $p_x = m_e v_x$, $p_y = m_e v_y$, $p_z = m_e v_z$, into quantum cells with volume h^3, where h is the Planck's constant.

The number of the atoms per cm^3 each with one liberated electron, located in the phase space region $x\ldots x + dx$, $y\ldots y + dy$, $z\ldots z + dz$, $p_x\ldots p_x + dp_x$, $p_y\ldots p_y + dp_y$, $p_z\ldots p_z + dp_z$, calculated per quantum cell (division by h^3!), in proportion to the number of the neutral atoms per cm^3 in the ground state, is given by the Boltzmann formula as:

$$\frac{dN_{1,1}}{N_{0,1}} = 2\frac{g_{1,1}}{g_{0,1}}\exp\left\{-\frac{1}{kT}\left[\chi_0 + \frac{1}{2m_e}(p_x^2 + p_y^2 + p_z^2)\right]\right\}$$
$$\times \frac{dx\,dy\,dz\,dp_x dp_y dp_z}{h^3} \quad .$$

Here $N_{1,1}$ denotes the singly ionised atoms (first index 1) in the ground state (second index 1) and $N_{0,1}$ the neutral atoms (first index 0) in the ground state. The factor 2 (the statistical weight of the free electron) takes account of the fact that the free electron has two possibilities for its spin in an external field, which doubles the number of states.

Integration over the momentum coordinates (from $-\infty$ to $+\infty$ in each case) gives the factor $(2\pi m_e kT)^{3/2}$; integration over the position coordinates is restricted to just the volume in which an electron finds itself (since only one electron will escape from each ion), and accordingly produces the factor $1/N_e$, where N_e is the number of free electrons per cm^3 ("electron density"). One obtains

$$\frac{N_{1,1}}{N_{0,1}} N_e = 2 \frac{g_{1,1}}{g_{0,1}} \frac{(2\pi m_e)^{3/2} (kT)^{3/2}}{h^3} \times \exp\left(-\frac{\chi_0}{kT}\right) \quad ,$$

where the indices of the statistical weights are analogous to those of the densities. By application of the Boltzmann formula one can proceed to the higher energy states and finally sum over all the excited states. The result then becomes in the general case of ionisation equilibrium between the i-fold and the $(i+1)$-fold ions (Saha Equation):

$$\frac{N_{i+1}}{N_i} N_e = 2 \frac{u_{i+1}}{u_i} \frac{(2\pi m_e)^{3/2} (kT)^{3/2}}{h^3} \times \exp\left(-\frac{\chi_i}{kT}\right)$$

with the so-called partition function

$$u_i(T) = \sum_n g_{i,n} \exp[-(\chi_{i,n}/kT)] \quad .$$

Here $\chi_{i,n}$ denotes the excitation energy of the state n of the i-fold ions and $g_{i,n}$ the corresponding statistical weight.

Astronomical and Physical Constants

Primary Astronomical Constants

(System 1976 of the International Astronomical Union, introduced in the astronomical year-books from 1984.)

The astronomical unit of mass is the solar mass ($\mathcal{M}_\odot$). The astronomical unit of time is the time interval of one day of 86 400 seconds; 36 525 days are one Julian century. The astronomical unit of length (AU) is the unit of length in which the major orbital semi-axis a must be expressed in Kepler's Third Law for the two-body problem $a^3/P^2 = (k^2/4\pi^2)(1 + \mathcal{M})$, where the so-called Gaussian gravitational constant takes the value $k = 0.017\,202\,098\,950$.

Velocity of light (in vacuo)	c	$= 2.997\,924\,58 \times 10^8\,\mathrm{m\,s^{-1}}$
Time for light to travel 1 AU	τ_{AU}	$= 499.004\,782\,\mathrm{s}$
Radius of earth (equatorial)	R_δ	$= 6.378\,140 \times 10^6\,\mathrm{m}$
Gravitation constant	G	$= 6.672 \times 10^{-11}\,\mathrm{m^3\,kg^{-1}\,s^{-2}}$
		$= 6.672 \times 10^{-8}\,\mathrm{cm^3\,g^{-1}\,s^{-2}}$
Mass ratio moon/earth	$\mathcal{M}_{\mathbb{C}}/\mathcal{M}_\delta$	$= 0.012\,300\,02$

General precession in longitude per Julian century at epoch 2000	p	$= 5029\rlap{.}''0966$
Obliquity of the ecliptic at epoch 2000	ε	$= 23°26'21\rlap{.}''448$
Constant of nutation at epoch 2000	N	$= 9\rlap{.}''2055$

Derived Astronomical Constants

Astronomical unit of distance ($= c\tau_{AU}$)	AU	$= 1.495\,978\,7 \times 10^{11}\,\mathrm{m}$
Parsec ($=\mathrm{AU}/\sin 1''$)	pc	$= 3.0857 \times 10^{16}\,\mathrm{m}$
Solar mass	$\mathcal{M}_\odot$	$= 1.9891 \times 10^{30}\,\mathrm{kg}$
Solar radius	$R_\odot$	$= 6.960 \times 10^8\,\mathrm{m}$
Solar luminosity	$L_\odot$	$= 3.85 \times 10^{26}\,\mathrm{W}$
Aberration constant at epoch 2000	A	$= 20\rlap{.}''49552$
Flattening factor for the earth	f	$= 1/298.257$
Mass ratio sun/(earth + moon)	$\mathcal{M}_\odot/\mathcal{M}_{\delta+\mathbb{C}}$	$= 328\,900.6$

462

Mass ratio sun/earth $\qquad \mathcal{M}_\odot/\mathcal{M}_\oplus \quad = 332\,946.0$

Mass of earth $\qquad\qquad \mathcal{M}_\oplus \qquad\ = 5.9742 \times 10^{24}\,\text{kg}$

Physical Constants

Planck's constant	h	$= 6.6262 \times 10^{-27}\,\text{erg s}$
		$= 6.6262 \times 10^{-34}\,\text{J s}$
Rest mass of electron	m_e	$= 9.1095 \times 10^{-31}\,\text{kg}$
Rest mass of proton	m_p	$= 1.6726 \times 10^{-27}\,\text{kg}$
Atomic unit of mass $= (1/12)\,\text{C}^{12}$	u	$= 1.6605 \times 10^{-27}\,\text{kg}$
Elementary electric charge	e	$= 4.8032 \times 10^{-10}\,\text{esu}$
		$= 1.6022 \times 10^{-19}\,\text{C}$
Rydberg constant	R_∞	$= 1.09737312 \times 10^{5}\,\text{cm}^{-1}$
Boltzmann constant	k	$= 1.3806 \times 10^{-16}\,\text{erg K}^{-1}$
		$= 1.3806 \times 10^{-23}\,\text{J K}^{-1}$
Gas constant	$\mathcal{R}$	$= 8.3143 \times 10^{7}\,\text{erg K}^{-1}\,\text{mol}^{-1}$
		$= 8.3143 \times 10^{3}\,\text{J K}^{-1}\,\text{kmol}^{-1}$
Avogadro's number	L	$= 6.0221 \times 10^{23}\,\text{mol}^{-1}$
Stefan-Boltzmann radiation law	σ	$= 5.6696 \times 10^{-5}\,\text{erg cm}^{-2}\,\text{s}^{-1}\,\text{K}^{-4}$
constant (radiation flux)		$= 5.6696 \times 10^{-8}\,\text{W m}^{-2}\,\text{K}^{-4}$
Stefan-Boltzmann radiation law	a	$= 4\sigma/c = 7.5647 \times 10^{-15}\,\text{erg cm}^{-3}\,\text{K}^{-4}$
constant (radiation density)		$= 7.5647 \times 10^{-16}\,\text{W m}^{-3}\,\text{K}^{-4}$
Bohr radius	a_0	$= 5.291 \times 10^{-11}\,\text{m}$
Classical electron radius	r_0	$= e^2/m_ec^2 = 2.8179 \times 10^{-15}\,\text{m}$
Compton wavelength	λ_c	$= h/m_ec = 2.4263 \times 10^{-12}\,\text{m}$

Units Often Used in Astrophysical Literature

Wavelength	$1\,\text{Å}$	$= 0.1\,\text{nm} = 10^{-10}\,\text{m}$
Force	$1\,\text{dyn}$	$= 10^{-5}\,\text{N}$
Pressure	$1\,\text{dyn cm}^{-2}$	$= 10^{-6}\,\text{bar} = 10^{-1}\,\text{Pa}$
Energy	$1\,\text{erg}$	$= 10^{-7}\,\text{J} = 6.2414 \times 10^{11}\,\text{eV}$
Magnetic field	$1\,\text{Gauss}$	$= 10^{-4}\,\text{T}$

Latin Names of Constellations
(Nominative and Genitive) and Their Abbreviations
According to the IAU Convention

Andromeda, Andromedae	And	Lacerta, Lacertae	Lac
Antlia, Antliae	Ant	Leo, Leonis	Leo
Apus, Apodis	Aps	Leo Minor, Leonis Minoris	LMi
Aquarius, Aquarii	Aqr	Lepus, Leporis	Lep
Aquila, Aquilae	Aql	Libra, Librae	Lib
Ara, Arae	Ara	Lupus, Lupi	Lup
Aries, Arietis	Ari	Lynx, Lyncis	Lyn
Auriga, Aurigae	Aur	Lyra, Lyrae	Lyr
Bootes, Bootis	Boo	Mensa, Mensae	Men
Caelum, Caeli	Cae	Microscopium, Microscopii	Mic
Camelopardalis, Camelopardalis	Cam	Monoceros, Monocerotis	Mon
Cancer, Cancri	Cnc	Musca, Muscae	Mus
Canes venatici, Canum Venaticorum	CVn	Norma, Normae	Nor
Canis Major, Canis Majoris	CMa	Octans, Octantis	Oct
Canis Minor, Canis Minoris	CMi	Ophiuchus, Ophiuchi	Oph
Capricornus, Capricorni	Cap	Orion, Orionis	Ori
Carina, Carinae	Car	Pavo, Pavonis	Pav
Cassiopeia, Cassiopeiae	Cas	Pegasus, Pegasi	Peg
Centaurus, Centauri	Cen	Perseus, Persei	Per
Cepheus, Cephei	Cep	Phoenix, Phoenicis	Phe
Cetus, Ceti	Cet	Pictor, Pictoris	Pic
Chamaeleon, Chamaeleonis	Cha	Pisces, Piscium	Psc
Circinus, Circini	Cir	Piscis Austrinus, Piscis Austrini	PsA
Columba, Columbae	Col	Puppis, Puppis	Pup
Coma Berenices, Comae Berenices	Com	Pyxis, Pyxidis	Pyx
Corona Austrina, Coronae Austrinae	CrA	Reticulum, Reticuli	Ret
Corona Borealis, Coronae Borealis	CrB	Sagitta, Sagittae	Sge
Corvus, Corvi	Crv	Sagittarius, Sagittarii	Sgr
Crater, Crateris	Crt	Scorpius, Scorpii	Sco
Crux, Crucis	Cru	Sculptor, Sculptoris	Scl
Cygnus, Cygni	Cyg	Scutum, Scuti	Sct
Delphinus, Delphini	Del	Serpens, Serpentis	Ser
Dorado, Doradus	Dor	Sextans, Sextantis	Sex
Draco, Draconis	Dra	Taurus, Tauri	Tau
Equuleus, Equulei	Equ	Telescopium, Telescopii	Tel
Eridanus, Eridani	Eri	Triangulum, Trianguli	Tri
Fornax, Fornacis	For	Triangulum Australe, Trianguli	
Gemini, Geminorum	Gem	Australis	TrA
Grus, Gruis	Gru	Tucana, Tucanae	Tuc
Hercules, Herculis	Her	Ursa Major, Ursae Majoris	UMa
Horologium, Horologii	Hor	Ursa Minor, Ursae Minoris	UMi
Hydra, Hydrae	Hya	Vela, Velorum	Vel
Hydrus, Hydri	Hyi	Virgo, Virginis	Vir
Indus, Indi	Ind	Volans, Volantis	Vol
		Vulpecula, Vulpeculae	Vul

References

Aannestad, P.A., Purcell, E.M. (1973): Annu. Rev. Astron. Astrophys. **11**, 325
Appenzeller, I. (1975): Astron. Astrophys. **38**, 313

Baker, P.L., Burton, W.B. (1975): Astrophys. J. **198**, 281
Barnard, E.E. (1927): *A Photographic Atlas of Selected Regions of the Milky Way*, Part I: Photographs and Descriptions, Part II: Charts and Tables (Carnegie Institution of Washington, Washington, D.C.)
Bethe, H. (1933): *Quantenmechanik der Ein- und Zwei-Elektronen-Probleme*, Handbuch der Physik, Vol. 24/1 (Springer, Berlin) p. 518
Blitz, L., Shu, F.H. (1980): Astrophys. J. **238**, 148
Boss, B. (1936; 1937): *General Catalogue*, Carnegie Inst. Washington Publ., No. 468, Vols. I–V
Braunsfurth, E. (1975): "Zur Wolkenstruktur des neutralen Wasserstoffs"; Dissertation, University of Bonn
Burton, W.B., Liszt, H.S. (1978): Astrophys. J. **225**, 815
Butler, D. (1975): Publ. Astron. Soc. Pac. **87**, 559

Cannon, A.J., Pickering, E.C. (1918–24): *Henry Draper Catalogue of Stellar Spectra*, Ann. Harvard, Vols. 91–99
Cannon, A.J., Pickering, E.C. (1925–49): *Henry Draper Extension*, Ann. Harvard, Vols. 100, 105, 112

Downes, D. et al. (1980): "Observations of Masers in Regions of Star Formation", in *Interstellar Molecules*, ed. by J. Andrew, IAU Symposium, No. 87 (Reidel, Dordrecht) p. 565
Dreyer, J.L.E. (1888): Mem. R. Astron. Soc. **49**
Dreyer, J.L.E. (1895): Mem. R. Astron. Soc. **51**
Dreyer, J.L.E. (1908): Mem. R. Astron. Soc. **59**

Eggen, O.J., Lynden-Bell, D., Sandage, A.R. (1962): Astrophys. J. **136**, 748
Einasto, J. (1979): "Galactic Mass Modelling", in *The Large-Scale Characteristics of the Galaxy*, ed. by W.B. Burton, IAU Symposium, No. 84 (Reidel, Dordrecht) p. 457

Feitzinger, J.V., Stüwe, J.A. (1984): Astron. Astrophys. Suppl. **58**, 365
Freeman, K.C. (1975): "Stellar Dynamics and the Structure of Galaxies", in *Galaxies and the Universe*, ed. by A. Sandage, M. Sandage, J. Kristian, Stars and Stellar Systems, Vol. IX (The University of Chicago Press, Chicago) Chap. 11, p. 460
Fricke, W. et al. (1963): *Fourth Fundamental Catalogue*, Veröff. Astron. Rechen-Inst. Heidelberg, Vol. 10
Fricke, W. (1977): *Basic Material for the Determination of Precession and of Galactic Rotation and a Review of Methods and Results*, Veröff. Astron. Rechen-Inst. Heidelberg, No. 28

Genzel, R. et al. (1981): Astrophys. J. **244**, 884
Gerola, H., Seiden, P.E. (1978): Astrophys. J. **223**, 129
Ginzburg, V.L., Syrovatskii, S.I. (1965): Annu. Rev. Astron. Astrophys. **3**, 312

Gliese, W. (1969): *Catalogue of Nearby Stars*, Veröff. Astron. Rechen-Inst. Heidelberg, No. 22
Gliese, W., Jahreiss, H. (1979): Astron. Astrophys. Suppl. **38**, 423

Henon, M. (1973): "Collisional Dynamics of Spherical Stellar Systems", in *Dynamical Structure and Evolution of Stellar Systems*, ed. by G. Contopoulos, M. Henon, D. Lynden-Bell, Saas-Fee, pp. 182f.

Jahreiss, H. (1974): "Die räumliche Verteilung, Kinematik und Alter der sonnennahen Sterne (The Spatial Distribution, Kinematics and Ages of the Stars near the Sun)"; Dissertation, University of Heidelberg
Jenkins, L.F. (ed.) (1952): *General Catalogue of Trigonometric Stellar Parallaxes* (Yale University Observatory, New Haven, USA)

Kaelble, A., de Boer, K.S., Grewing, M. (1985): Astron. Astrophys. **143**, 408
Kegel, W.H. (1976): Astron. Astrophys. **50**, 293
Khavtassi, J. (1958): Bull. Abastumani Obs. No. 18
Kraus, J.D. (1986): *Radio Astronomy*, 2nd ed. (Cygnus-Quasar Books, Powell, Ohio)
Kukarkin, B.V. (1975): *General Catalogue of Globular Star Clusters*, NASA Technical Translation F-16, 157, Washington D.C.
Kuzmin, G.G. (1953): Publ. Tartu Astron. Obs. **32**, 332
Kwan, J. (1979): Astrophys. J. **229**, 567

Larson, R.B. (1976): Mon. Not. R. Astron. Soc. **176**, 31
Lilley, A.E., Palmer, P. (1968): Astrophys. J. Suppl. **16**, 143
Lin, C.C., Yuan, C., Shu, F.H. (1969): Astrophys. J. **155**, 721
Liszt, H.S., Burton, W.B. (1980): Astrophys. J. **236**, 779
Lynden-Bell, D. (1962): Mon. Not. R. Astron. Soc. **124**, 1
Lynds, B.T. (1962): Astrophys. J. Suppl. **7**, 1
Lynds, B.T. (1965): Astrophys. J. Suppl. **12**, 163

Maihara, T. et al. (1978): Publ. Astron. Soc. Jpn. **30**, 1
Mattila, K. (1970): Astron. Astrophys. **9**, 53
Mattila, K., Scheffler, H. (1978): Astron. Astrophys. **66**, 277
Mayor, M. (1972): Astron. Astrophys. **18**, 97
McCray, R., Snow, T.P. (1979): Annu. Rev. Astron. Astrophys. **17**, 213
McKee, C.F., Ostriker, J.P. (1977): Astrophys. J. **218**, 148
Mezger, P.G. (1972): "Interstellar Matter: An Observer's View", in *Interstellar Matter*, ed. by N.C. Wickramasinghe, F.D. Kahn, P.G. Mezger (Geneva Observatory, Sauverny)
Mouschovias, T. (1974): Astrophys. J. **192**, 37
Mouschovias, T. (1978): "Formation of Stars in Magnetic Clouds", in *Protostars and Planets*, ed. by T. Gehrels (University of Arizona Press, Tucson) p. 209

Neckel, T. (1968): Z. Astrophys. **69**, 112
Neckel, T., Klare, G. (1980): Astron. Astrophys. Suppl. **42**, 251
Norman, C., Silk, J. (1980): Astrophys. J. **238**, 158

Oort, J.H., Plaut, L. (1975): Astron. Astrophys. **41**, 71
Osterbrock, D. (1974): *Astrophysics of Gaseous Nebulae* (Freeman, San Francisco)
Ostriker, J.P., Caldwell, J.A.R. (1979): "The Mass and Light Distribution of the Galaxy: A Three-Component Model", in *The Large-Scale Characteristics of the Galaxy*, ed. by W.B. Burton, IAU Symposium, No. 84 (Reidel, Dordrecht) p. 441

Parker, E.N. (1966): Astrophys. J. **145**, 811
Peters, G. (1970): Astron. Astrophys. **4**, 134

Reinhardt, M. Schmidt-Kaler, Th. (1979): Astrophys. Space Sci. **66**, 121
Roberts, W.W. (1969): Astrophys. J. **158**, 123
Rodgers, A.W., Campbell, C.T., Whiteoak, J.B. (1960): Mon. Not. R. Astron. Soc. **121**, 103
Rohlfs, K. (1977): *Lectures on Density Wave Theory*, Lecture Notes in Physics, Vol. 69 (Springer, Berlin, Heidelberg) pp. 95ff.

Schaifers, K., Voigt, H.H. (eds.) (1982): *Interstellar Matter. Galaxy. Universe*, Landolt-Börnstein, Vol. VI/2c (Springer, Berlin, Heidelberg)

Schmidt, M. (1956): Bull. Astron. Inst. Neth. **13**, 15

Schmidt, M. (1959): Astrophys. J. **129**, 243

Schoenberg, E.E. (1964): Veröff. Sternwarte München **5**, No. 21

Serkowski, K. (1973): "Interstellar Polarization", in *Interstellar Dust and Related Topics*, ed. by J.M. Greenberg, H.C. van de Hulst, IAU Symposium, No. 52 (Reidel, Dordrecht) p. 145

Shapiro, P.R., Field, G.B. (1976): Astrophys. J. **205**, 762

Silk, J., Norman, C. (1980): "The Interaction of T Tauri Stars with Molecular Clouds", In *Interstellar Molecules*, ed. by J. Andrew, IAU Symposium, No. 87 (Reidel, Dordrecht) p. 165

Smithsonian Astrophysical Observatory Star Catalog (SAO Catalog) (1966), Smithsonian Institution, Washington, D.C.

Spitzer, L. (1968): *Diffuse Matter in Space* (Wiley, New York)

Spitzer, L. (1978): *Physical Processes in the Interstellar Medium* (Wiley, New York)

Tammann, G.A. (1970): "The Galactic Distribution of Young Cepheids", in *The Spiral Structure of our Galaxy*, ed. by W. Becker, G. Contopoulos, IAU Symposium, No. 38 (Reidel, Dordrecht) p. 236

Tenorio-Tagle, G. (1979): Astron. Astrophys. **71**, 59

Toomre, A. (1964): Astrophys. J. **139**, 1217

Toomre, A. (1969): Astrophys. J. **158**, 899

Torgård, I. (1961): Ann. of the Observatory of Lund, Nos. 15, 16, 17

Trumpler, R.J., Weaver, H.F. (1953): *Statistical Astronomy* (University of California Press, Berkeley) (reprinted in 1962 by Dover Publications Inc., New York)

Voigt, H.H. (ed.) (1965): *Astronomy and Astrophysics*, Landolt-Börnstein, Vol. VI/1 (Springer, Berlin, Heidelberg)

Von Hoerner, S. (1960): Z. Astrophys. **50**, 184

Whittet, D.C.B., van Breda, I.G. (1978): Astron. Astrophys. **66**, 57

Wickramasinghe, N.C. (1967): *Interstellar Grains* (Chapman and Hall, London)

Wielen, R. (1971): Astron. Astrophys. **13**, 309

Wielen, R. (1974): Astron. Astrophys. **15**, 1

Wielen, R. (1974): Publ. Astron. Soc. Pac. **86**, 341

Wielen, R. (1974): "The Gravitational N-Body Problem for Star Clusters", in *Stars and the Milky Way System*, ed. by L.N. Mavridis, Proceedings of the First European Astronomical Meeting, Vol. 2 (Springer, Berlin, Heidelberg) p. 326

Wielen, R. (1977): Astron. Astrophys. **60**, 263

Wielen, R. (1979): "The Density-Wave Theory Confronted by Observations", in *The Large-Scale Characteristics of the Galaxy*, ed. by W.B. Burton, IAU Symposium, No. 84 (Reidel, Dordrecht) p. 133

Wilking, B.A. et al. (1980): Astrophys. J. **235**, 905

Woltjer, L. (1975): Astron. Astrophys. **42**, 109

Woodward, P. (1976): Astrophys. J. **207**, 484

Yoshii, Y., Saio, H. (1979): Publ. Astron. Soc. Jpn. **31**, 339

Yuan, C. (1971): Astron. J. **76**, 664

Supplementary and Advanced Literature

Reviews, important literature sources and compilations of data are contained in the publication series abbreviated as follows:

Annu. Rev. Astron. Astrophys. = Annual Review of Astronomy and Astrophysics (1963), Vol. I... (Annu. Reviews Inc., Palo Alto, California)

Encycl. Physics = Encyclopedia of Physics, ed. by S. Flügge (1958–1962): *Astrophysics I–V*, Vols. 50–54 (Springer, Berlin, Göttingen, Heidelberg)

IAU Symposia = Symposia of the International Astronomical Union (Reidel, Dordrecht). Each volume contains contributions of numerous authors on a particular topic.
Example: Burton, W.B. (ed.) (1979): *The Large-Scale Characteristics of the Galaxy*, No. 84 (Reidel, Dordrecht)

Landolt-Börnstein = Landolt-Börnstein, Numerical Data and Functional Relationships in Science and Technology, New Series, Group VI, Astronomy, Astrophysics, and Space Research:
Voigt, H.H. (ed.) (1965): *Astronomy and Astrophysics*, Vol. 6/1 (Springer, Berlin, Heidelberg);
Schaifers, K., Voigt, H.H. (eds.) (1982): *Methods, Constants, Solar Systems*, Vol. 6/2a (Springer, Berlin, Heidelberg);
Schaifers, K., Voigt, H.H. (eds.) (1982): *Stars and Stellar Clusters*, Vol. 6/2b (Springer, Berlin, Heidelberg);
Schaifers, K., Voigt, H.H. (eds.) (1982): *Interstellar Matter. Galaxy. Universe*, Vol. 6/2c (Springer, Berlin, Heidelberg)

Stars Stellar Systems = Stars and Stellar Systems, ed. by G.P. Kuiper, B.M. Middlehurst (1960–1975), Vols. I–IX (The University of Chicago Press, Chicago)

An entry to the new literature is afforded by the "Reports on Astronomy" of the International Astronomical Union appearing at 3-year intervals.

Quotations and short summaries of the entire astronomical literature up to the preceding year can be obtained from the annually published "Astronomy and Astrophysics Abstracts" (prepared by the Astronomisches Rechen-Institut in Heidelberg and published by Springer-Verlag).

We shall use the following abbreviations for important astronomical periodicals:

Astron. Astrophys. = Astronomy and Astrophysics (a European journal) (Springer, Berlin, Heidelberg, New York)

Astrophys. J. = The Astrophysical Journal (University of Chicago Press, Chicago, Ill.)

Astron. J. = The Astronomical Journal (American Institute of Physics, Inc., New York)

Astron. Nachr. = Astronomische Nachrichten (Akademie-Verlag, Berlin)

Mon. Not. R. Astron. Soc. = Monthly Notices of the Royal Astronomical Society (Blackwell Sci. Publ., London)

Proc. Astron. Soc. Aust. = Proceedings of the Astronomical Society of Australia (University Press, Sydney)

Publ. Astron. Soc. Jpn. = Publications of the Astronomical Society of Japan (Maruzen Co., Nihonbashi, Tokyo)

Publ. Astron. Soc. Pac. = Publications of the Astronomical Society of the Pacific (San Francisco, California)

Sov. Astron. = Soviet Astronomy, English edition of the Russian Astronomical Journal (Astronomicheskii Zhurnal) (American Institute of Physics, New York)

1. Introductory Survey

Hoskin, M. (1985): "The Milky Way from Antiquity to Modern Times", in *The Milky Way*, ed. by H. van Woerden, W.B. Burton, R.J. Allen, IAU Symposium, No. 106 (Reidel, Dordrecht) p. 11

Jaki, S.L. (1973): *The Milky Way* (Science History Publ., Neale Watson, New York) see especially Chap. 6

Struve, O., Zebergs, V. (1962): *Astronomy of the 20th Century* (MacMillan Co., New York, London)

2. Positions, Motions and Distances of the Stars – Concepts and Methods

2.1 Positions and Motions

General Accounts

Podobed, V.V. (1965): *Fundamental Astrometry* (University of Chicago Press, Chicago)

Smart, W.M. (1960): *Text-book on Spherical Astronomy* (Cambridge University Press, Cambridge)

Astronomical Yearbooks

The fundamental one is "The Astronomical Almanac", issued annually in London (The Government Bookshop, P.O. Box 569, London SE1 9NA) and in Washington (U.S. Government Printing Office, Washington, D.C. 20402).

Astronomical Constants (see list on p. 462)

Fricke, W. (1982): "The System of Astronomical Constants", in *Methods, Constants, Solar System*, ed. by K. Schaifers, H.H. Voigt, Landolt-Börnstein, Vol. 6/2a (Springer, Berlin, Heidelberg) p. 80

Lederle, T. (1980): Mitteilungen des Astronomischen Rechen-Instituts Heidelberg, Serie A, *Nr. 133*

Star Positions General

Eichhorn, H. (1974): *Astronomy of Star Positions* (Frederick Ungar Publ. Co., New York)

Positions, Proper Motions and Further Data on All Stars Brighter Than $V = 6^{\text{m}}5$ (Around 9100 Stars)

Hoffleit, D., Jaschek, C. (eds.) (1982): *Catalogue of Bright Stars* (Yale University Observatory, New Haven, Conn.)

Stars Brighter Than $8^{\text{m}}0$

Hirschfeld, A., Sinnot, R.W. (eds.) (1982): *Sky Catalogue 2000.0*, Vol. 1 (Sky Publishing Corporation, Cambridge, Mass.)

Positions and Proper Motions of Fainter Stars:
See list in
Schaifers, K., Voigt, H.H. (eds.) (1982): *Interstellar Matter. Galaxy. Universe*, Landolt-Börnstein, Vol. 6/2c (Springer, Berlin, Heidelberg) pp. 150, 157

Radial Velocities
Abt, H.A., Biggs, E.S. (1972): *Bibliography of Stellar Radial Velocities* (Kitt Peak Observatory, Tucson, Arizona)
Batton, A.H., Heard, J.F. (eds.) (1967): *Determination of Radial Velocities and Their Applications*, IAU Symposium, No. 30 (Reidel, Dordrecht)

Precession
Fricke, W. (1982): "The System of Astronomical Constants", in *Methods, Constants, Solar System*, ed. by K. Schaifers, H.H. Voigt, Landolt-Börnstein, Vol. 6/2a (Springer, Berlin, Heidelberg) p. 79
Gliese, W. (1982): "Precession Tables", in *Methods, Constants, Solar Systems*, ed. by K. Schaifers, H.H. Voigt, Landolt-Börnstein, Vol. 6/2a (Springer, Berlin, Heidelberg) p. 153

Fundamental Coordinate System and Radioastrometric Reference System
Fricke, W. (1972): Annu. Rev. Astron. Astrophys. **10**, 101
Counselman, Ch.C. (1976): Annu. Rev. Astron. Astrophys. **14**, 197

2.2 Distances

Trigonometric Parallaxes
Strand, K.A. (1963): "Trigonometric Stellar Parallaxes", in *Basic Astronomical Data*, ed. by G.P. Kuiper, B.M. Middlehurst, Stars Stellar Systems, Vol. III (The University of Chicago Press, Chicago) Chap. 6, p. 55
Vasilevskis, S. (1966): Annu. Rev. Astron. Astrophys. **4**, 57

Moving Cluster Parallaxes
Atanasijevic, I. (1971): *Selected Exercises in Galactic Astronomy*, Astrophysics and Space Science Library, Vol. 26 (Reidel, Dordrecht) Exercise III, p. 32
Bertiau, F.C. (1958): Astrophys. J. **128**, 533
Eggen, O.J. (1965): "Moving Groups of Stars", in *Galactic Structure*, ed. by A. Blaauw, M. Schmidt, Stars Stellar Systems, Vol. V (The University of Chicago Press, Chicago) Chap. 6, p. 111
Heckmann, O, Lübeck, K. (1956): Z. Astrophys. **40**, 1

Distance of the Hyades
Hanson, R.B. (1980): "The Hyades Cluster Distance", in *Star Clusters*, ed. by J.E. Hesser, IAU Symposium, No. 85 (Reidel, Dordrecht) p. 71

Photometric Distances
Becker, W. (1963): Z. Astrophys. **57**, 117
Blaauw, A. (1963): "The Calibration of Luminosity Criteria", in *Basic Astronomical Data*, ed. by G.P. Kuiper, B.M. Middlehurst, Stars Stellar Systems, Vol. III (The University of Chicago Press, Chicago) Chap. 20, p. 383

3. Structure and Kinematics of the Stellar System

General Accounts
Blaauw, A., Schmidt, M. (eds.) (1965): *Galactic Structure*, Stars Stellar Systems, Vol. V (The University of Chicago Press, Chicago)
Bok, B.J. (1970): "Basic Problems on the Structure and Dynamics of our Galaxy", in *Galactic Astronomy*, ed. by H.-Y. Chiu, A. Muriel, Vol. 1 (Gordon and Breach, New York)
Mihalas, D., Routly, P.M. (1968): *Galactic Astronomy* (Freeman, San Francisco)
Mihalas, D., Binney, J. (1981): *Galactic Astronomy, Structure and Kinematics* (Freeman, San Francisco)
Shuter, W.L.H. (ed.) (1983): Kinematics, Dynamics and Structure of the Milky Way (Reidel, Dordrecht)

Training Exercises
Atanasijevic, I. (1971): *Selected Exercises in Galactic Astronomy*, Astrophysics and Space Science Library, Vol. 26 (Reidel, Dordrecht)

3.1 Apparent Distribution of the Stars

Older Systematic Survey Catalogues Are Collected in
Voigt, H.H. (ed.) (1965): *Astronomy and Astrophysics*, Landolt-Börnstein, Vol. 6/1 (Springer, Berlin, Heidelberg) Sect. 8.2.1.3.1

Distribution of Individual Star Types, Integrated Starlight, Surface Brightness
Schaifers, K., Voigt, H.H. (eds.) (1982): *Interstellar Matter. Galaxy. Universe*, Landolt-Börnstein, Vol. 6/2c (Springer, Berlin, Heidelberg) Sect. 8.3.1

Wide-Angle Photographs of the Milky Way
Laustsen, S., Madsen, C., West, R.M. (1987): Exploring the Southern Sky - A Pictorial Atlas from the European Southern Observatory (Springer, Berlin, Heidelberg, New York)
Schlosser, W., Schmidt-Kaler, Th. (1977): Sky and Telescope **53**, 436
Schlosser, W., Schmidt-Kaler, Th. (1977): Vistas Astron. **21**, 447
Schlosser, W., Schmidt-Kaler, Th., Hünecke, W. (1979): *Atlas der Milchstraße (Kugelspiegelaufnahmen mit 140° Bildwinkel in vier Farben: U, B, V, R)* [Atlas of the Milky Way (spherical reflector photographs with 140° field of view in four colours U, B, V, R)] (Separate publication of the Astronomical Institute of the University of Bochum, Bochum, Fed. Rep. of Germany)

3.2 The Local Galactic Star Field

Methods of Density Analysis
Bahcall, J.N. (1986): Annu. Rev. Astron. Astrophys. **24**, 577
Ochsenbein, F. (1980): Astron. Astrophys. **86**, 321
Spaenhauer, A.M. (1978): Astron. Astrophys. **65**, 313

Regional Studies of the Star Density Distribution Are Collected in:
Schaifers, K., Voigt, H.H. (eds.) (1982): *Interstellar Matter. Galaxy. Universe*, Landolt-Börnstein, Vol. 6/2c (Springer, Berlin, Heidelberg) Sect. 8.3.2

Motions of the Stars in the Solar Neighbourhood
Delhaye, J. (1965): "Solar Motion and Velocity Distribution of Common Stars", in *Galactic Structure*, ed. by A. Blaauw, M. Schmidt, Stars Stellar Systems, Vol. V (The University of Chicago Press, Chicago)

Eggen, O.J. (1970): Vistas Astron. **12,** 367

Lindblad, P.O. (1974): "Gould's Belt", in *Stars and the Milky Way System,* ed. by L.N. Mavridis, Proceedings of the First European Astronomical Meeting, Vol. 2 (Springer, Berlin, Heidelberg) p. 65

Roman, N.G. (1965): "High-Velocity Stars", in *Galactic Structure,* ed. by A. Blaauw, M. Schmidt, Stars Stellar Systems, Vol. V (The University of Chicago Press, Chicago) Chap. 16, p. 345

Woolley, R. (1965): "Motions of Nearby Stars", in *Galactic Structure,* ed. by A. Blaauw, M. Schmidt, Stars Stellar Systems, Vol. V (The University of Chicago Press, Chicago)

Kinematics, Distribution and Chemical Composition

Basinska-Grzesik, E., Mayor, M. (eds.) (1977): *Chemical and Dynamical Evolution of Our Galaxy,* IAU Colloquium, No. 45 (Geneva Observatory, Sauverny, Switzerland)

3.3 Large Scale Distribution of the Stars

Recent Analyses of Star Counts

Bahcall, J.N. (1986): Annu. Rev. Astron. Astrophys. **24,** 577

Distribution of OB Stars, Young Open Clusters and Associations

Humphreys, R.M. (1979): "The Distribution of Young Stars, Clusters and Cepheides in the Milky Way and M 33 – a Comparison", in *The Large-Scale Characteristics of the Galaxy,* ed. by W.B. Burton, IAU Symposium, No. 84 (Reidel, Dordrecht) p. 93

Klare, G., Neckel, T. (1967). Z. Astrophys. **66,** 45

Catalogues of Star Clusters and Associations

Alter, C., Ruprecht, J., Vanysek, V. (1970): *Catalogue of Star Clusters and Associations* (in chart form), 2nd ed. (Czechoslovak Academy of Sciences, Prague) Contains 1039 open clusters, 5 motion groups, 11 "star groups", 70 OB associations and 125 globular clusters.

Review Articles on Star Clusters and Associations Are Contained in

Hesser, J.E. (ed.) (1980): *Star Clusters,* IAU Symposium, No. 85 (Reidel, Dordrecht)

Distribution of Globular Clusters

Hanes, D., Madore, M. (eds.) (1980): *Globular Clusters* (University Press, Cambridge)

Harris, W.A. (1976): Astron. J. **81,** 1095

Van den Bergh, S. (1979): Astron. J. **84,** 317

Other Galactic Windows Near $l = 0°$

Lynga, G. (1977): Astron. Astrophys. **54,** 71

Van den Bergh, S. (1974): Astrophys. J. **79,** 603

Stars in the Galactic Halo

Becker, W. (1980): Astron. Astrophys. **87,** 80

Fenkart, R.P. (1980): Astron. Astrophys. **91,** 352

Review of New Approaches to Obtaining a Model of the Galaxy on the Basis of Star Counts

Bahcall, J.N. (1986): Annu. Rev. Astron. Astrophys. **24,** 577

Spaenhauer, A. (1981): "Models of the Milky Way System", in *Galaktische Struktur und Entwicklung (Galactic Structure and Development),* ed. by R. Buser, Preprint No. 2 (Astronomical Institute of Basle University, Basle) p. 159

3.5 General Summary, Stellar Populations

Basinska-Grzesik, E., Mayor, M. (eds.) (1977): *Chemical and Dynamical Evolution of Our Galaxy*, IAU Colloquium, No. 45 (Geneva Observatory, Sauverny, Switzerland)

Blaauw, A. (1965): "The Concept of Stellar Populations", in: *Galactic Structure*, ed. by A. Blaauw, M. Schmidt, Stars Stellar Systems, Vol. V (The University of Chicago Press, Chicago), p. 435

Burton, W.B. (ed.) (1979): *The Large-Scale Characteristics of the Galaxy*, IAU Symposium, No. 84 (Reidel, Dordrecht)

O'Connell, D.J.K. (ed.) (1958): *Stellar Populations* (North-Holland, Amsterdam/Specola Vaticana Ric. Astron. 5)

Mould, J.R. (1982): Annu. Rev. Astron. Astrophys. **20**, 91

Sandage, A.R. (1986): Annu. Rev. Astron. Astrophys. **24**, 421

Tinsley, B.M., Larson, R.B. (1977): *The Evolution of Galaxies and Stellar Populations* (Yale University Observatory, New Haven)

Van den Bergh, S. (1975): Annu. Rev. Astron. Astrophys. **13**, 217

Van Woerden, H., Allen, R.J., Burton, W.B. (eds.) (1985): *The Milky Way*, IAU Symposium, No. 106 (Reidel, Dordrecht)

4. Interstellar Phenomena

General Accounts

Middlehurst, B.M., Aller, L.H. (eds.) (1968): *Nebulae and Interstellar Matter*, Stars Stellar Systems, Vol. VII (The University of Chicago Press, Chicago)

4.1 The Generally Distributed Medium

Interstellar Extinction, Polarisation and Scattering of Starlight

Aannestad, P.A., Purcell, E.M. (1973): Annu. Rev. Astron. Astrophys. **11**, 309

Johnson, H.L. (1968): "Interstellar Extinction", in *Nebulae and Interstellar Matter*, ed. by B.M. Middlehurst, L.H. Aller, Stars Stellar Systems, Vol. VII (The University of Chicago Press, Chicago) p. 167

McDonnell, J.A.M. (ed.) (1978): *Cosmic Dust* (Wiley, New York)

Savage, B.D., Mathis, J.S. (1979): Annu. Rev. Astron. Astrophys. **17**, 73

Interstellar Absorption Lines in Stellar Spectra

Münch, G. (1968): "Interstellar Absorption Lines", in *Nebulae and Interstellar Matter*, ed. by B.M. Middlehurst, L.H. Aller, Stars Stellar Systems, Vol. VII (The University of Chicago Press, Chicago) p. 365

Spitzer, L., Jenkins, E.B. (1975): Annu. Rev. Astron. Astrophys. **13**, 133

Interstellar Radiation Field

Greenberg, J.M. (1978): "Interstellar Dust", in *Cosmic Dust*, ed. by J.A.M. McDonnell (Wiley, New York) p. 208

Henry, R.C., Swandic, J.R., Shulman, S.D., Fritz, D. (1977): Astrophys. J. **212**, 707

Witt, A.N., Johnson, M.W. (1973): Astrophys. J. **181**, 363

Radio Observations, Magnetic Field

Heiles, C. (1976): Annu. Rev. Astron. Astrophys. **14**, 1

Meeks, M.L. (1976): *Astrophysics, Part C: Radio Observations* (Academic, New York, San Francisco, London)

Simard-Normandin, M., Kronberg, P.P. (1980): Astrophys. J. **242**, 74

Verschuur, G.L., Kellermann, K.I. (eds.) (1974): *Galactic and Extragalactic Radio Astronomy* (Springer, Berlin, Heidelberg)

**The Interstellar Scintillation of Cosmic Radiofrequency Radiation
Not Treated in This Book**
Rickett, B.J. (1977): Annu. Rev. Astron. Astrophys. **15**, 479

Cosmic X- and γ-Rays
Adams, D.J. (1980): *Cosmic X-Ray Astronomy* (Hilger, Bristol)
Baity, W.A., Peterson, L.E. (eds.) (1979): *X-Ray Astronomy*, COSPAR Symposium Series,
    Vol. 3 (Pergamon, Oxford)
Cowsik, R., Wills, R.D. (eds.) (1980): *Non-Solar Gamma-Rays* (Pergamon, Oxford)
Giacconi, R., Gursky, H. (eds.) (1974): *X-Ray Astronomy* (Reidel, Dordrecht)
Pinkau, K. (1979): Nature **277**, 17

3 K Background Radiation
Lequeux, J. (1980): "Cosmological Models Confronted to Observations", in *Physical Cosmol-
    ogy*, ed. by R. Balian, J. Audouze, D.N. Schramm (North-Holland, Amsterdam) p. 37
Weiss, R. (1980): Annu. Rev. Astron. Astrophys. **18**, 489

Cosmic Rays
Allkofer, O.C. (1975): *Introduction to Cosmic Radiation* (K. Thiemig, München)
Setti, G., Spada, G., Wolfendale, A.W. (eds.) (1981): *Origin of Cosmic Rays*, IAU Symposium,
    No. 94 (Reidel, Dordrecht)

see also Sect. 5.4.

4.2 Interstellar Clouds

Catalogues of Bright Nebulae Are Collected in:
Schaifers, K., Voigt, H.H. (eds.) (1982): *Interstellar Matter. Galaxy. Universe*, Landolt-
    Börnstein, Vol. 6/2c (Springer, Berlin, Heidelberg) Sect. 7.2.2

Radio and Infrared Observations of H II Regions
Habing, H.J., Israel, F.P. (1979): Annu. Astron. Astrophys. **17**, 345
Setti, G., Fazio, G.G. (eds.) (1978): *Infrared Astronomy* (Reidel, Dordrecht)
Shaver, P.A. (ed.) (1980): *Radio Recombination Lines* (Reidel, Dordrecht)
Wilson, T.L., Downes, D. (eds.) (1975): *H II Regions and Related Topics*, Lecture Notes in
    Physics, Vol. 42 (Springer, Berlin)

Catalogues of Dark Nebulae, See
Schaifers, K., Voigt, H.H. (eds.) (1982): *Interstellar Matter. Galaxy. Universe*, Landolt-
    Börnstein, Vol. 6/2c (Springer, Berlin, Heidelberg) Sect. 7.2.2

Molecular Clouds
Andrew, B.H. (ed.) (1980): *Interstellar Molecules*, IAU Symposium, No. 87 (Reidel, Dor-
    drecht)
Solomon, P.M., Edmunds, M. (eds.) (1980): *Giant Molecular Clouds in the Galaxy*, Third
    Gregynog Astrophysics Workshop (Pergamon, Oxford)
Peimbert, M., Jugaku, J. (eds.) (1987): *Star Forming Regions*, IAU Symposium, No. 115
    (Reidel, Dordrecht)
Serra, G. (ed.) (1985): *Nearby Molecular Clouds* (Springer, Berlin, Heidelberg)

5. Physics of the Interstellar Matter

General Accounts
Kaplan, S.A., Pikelner, S.B. (1970): *The Interstellar Medium* (Harvard University Press, Cambridge, Mass.)
Spitzer, L. (1978): Physical Processes in the Interstellar Medium (Wiley, New York)

5.1 Radiation in the Interstellar Gas

and

5.2 State of the Interstellar Gas

Radiation Transport, Excitation, Non-LTE, Ionisation, Radiation Processes
Balian, R., Encrenaz, P., Lequeux, J. (eds.) (1975): *Atomic and Molecular Physics and the Interstellar Matter* (North-Holland, Amsterdam)
Dalgarno, A., Masnou-Seeuws, F., McWhirter, R.W.P. (1975): *Atomic and Molecular Processes in Astrophysics* (Geneva Observatory, Sauverny, Switzerland)
Mihalas, D. (1978): *Stellar Atmospheres* (Freeman, San Francisco) Chaps. 1–6
Pacholczyk, A.G. (1970): *Radio Astrophysics* (Freeman, San Francisco)
Tucker, W.H. (1975): *Radiation Processes in Astrophysics* (MIT Press, Cambridge, Mass.)

References to the Parameters of the Individual Physical Processes:
energy states, ionisation energies, transition probabilities, collision rates, ionisation rates and recombination coefficients for interstellar atoms and ions together with dipole moments, dissociation-, recombination- and interchange-rates for interstellar molecules arranged as a compilation of numerical values and literature in:
Schaifers, K., Voigt, H.H. (eds.) (1982): *Interstellar Matter. Galaxy. Universe*, Landolt-Börnstein, Vol. 6/2c (Springer, Berlin, Heidelberg) Sects. 7.5.2 and 3

Synchrotron Radiation
Ginzburg, V.L., Syrovatskii, S.I. (1965): Annu. Rev. Astron. Astrophys. **3**, 297

Chemical and Thermal Equilibrium in the Clouds
Barsuhn, J., Walmsley, C.M.(1977): Astron. Astrophys. **54**, 345
Draine, B.T. (1978): Astrophys. J. Suppl. **36**, 595
Gerola, H., Glassgold, A.E. (1978): Astrophys. J. Suppl. **37**, 1
Goldsmith, P.F., Langer, W.D. (1978): Astrophys. J. **222**, 881
Iglesias, E. (1977): Astrophys. J. **218**, 697
Pickles, J.B., Williams, D.A. (1981): Mon. Not. R. Astron. Soc. **197**, 429

Heating and Cooling of the Interstellar Gas
Chevalier, R.A. (1977): Annu. Rev. Astron. Astrophys. **15**, 175
Dalgarno, A., McCray, R.A. (1972): Annu. Rev. Astron. Astrophys. **10**, 375
McCray, R., Snow, T.P. (1979): Annu. Rev. Astron. Astrophys. **17**, 213

State of the Molecular Clouds
Andrew, J. (ed.) (1980): *Interstellar Molecules*, IAU Symposium, No. 87 (Reidel, Dordrecht)
Kutner, M.L. (1984): Fundamentals of Cosmic Physics **9**, 233
Serra, G. (ed.) (1985): Nearby Molecular Clouds (Springer, Berlin, Heidelberg)

H II Regions
Felli, M. (1979): "Properties of H II-Regions", in *Stars and Star Systems*, ed. by B.E. Westerland (Reidel, Dordrecht) p. 195
Osterbrock, D. (1974): Astrophysics of Gaseous Nebulae (Freeman, San Francisco)
Wilson, T.L., Downes, D. (eds.) (1975): *H II-Regions and Related Topics*, Lectures Notes in Physics, Vol. 42 (Springer, Berlin, Heidelberg)

**Collections of the Characteristic Parameters of State of Numerous
H II Regions Are To Be Found in:**
Mezger, P.G., Pankonin, V., Schmidt-Burgk, J., Thum, C., Wink, J. (1979): Astron. Astrophys. Lett. **80**, L3
Smith, L.F., Biermann, P., Mezger, P.G. (1978): Astron. Astrophys. **66**, 65

Interstellar Masers
Cook, A.H. (1977): *Celestial Masers* (Cambridge University Press, Cambridge)
Genzel, R., et al. (1981): Astrophys. J. **244**, 844
Kegel, W.H. (1975): "Cosmic Masers", in *Problems in Stellar Atmospheres and Envelopes*, ed. by B. Baschek, W.H. Kegel, G. Traving (Springer, Berlin, Heidelberg) p. 257

Elemental Abundances in the Interstellar Gas
Aller, L.H. (1978): Proc. Astron. Soc. Austral. **3** (3), 213
Mezger, P.G., Pankonin, V., Schmidt-Burgk, J., Thum, C., Wink, J. (1979): Astron. Astropyhs. Lett. **80**, L3
Peimbert, M. (1979): "Chemical Evolution of the Galactic Interstellar Medium: Abundance Gradients", in *The Large-Scale Characteristics of the Galaxy*, ed. by W.B. Burton, IAU Symposium, No. 84 (Reidel, Dordrecht) p. 307
Snow, T.P., Meyers, K.A. (1979): Astrophys. J. **229**, 545
Thum, C. (1980): Vistas Astron. **24**, 355
Wannier, P.G. (1980): Annu. Rev. Astron. Astrophys. **18**, 399
Jenkins, E.B., Savage, B.D., Spitzer, L. (1986): Astrophys. J. **301**, 335

5.3 The Interstellar Dust Grains

Optics of Small Solid Particles
Greenberg, J.M. (1968): "Interstellar Grains", in *Nebulae and Interstellar Matter*, ed. by B.M. Middlehurst, L.H. Aller, Stars Stellar Systems, Vol. VII (The University of Chicago Press, Chicago) p. 221
Greenberg, J.M. (1978): "Interstellar Dust", in *Cosmic Dust*, ed. by J.A.M. McDonnell (Wiley, New York) p. 187
Van de Hulst, H.C. (1957): *Light Scattering by Small Particles* (Wiley, New York)

Interpretation of the Observations
Greenberg, J.M. (1978): "Physics and Astrophysics of Interstellar Dust", in *Infrared Astronomy*, ed. by G. Setti, G.G. Fazio (Reidel, Dordrecht) p. 51
Mathis, J.S. (1979): Astrophys. J. **232**, 747
Rogers, C., Martin, P.G. (1979): Astrophys. J. **228**, 450
Savage, B.D., Mathis, J.S. (1979): Annu. Rev. Astron. Astrophys. **17**, 73

**Temperature, Electric Charge, Formation and Decay, Orientation
in the Interstellar Magnetic Field**
Drain, B.T., Salpeter, E.E. (1979): Astrophys. J. **231**, 438
Greenberg, J.M. (1978): "Interstellar Dust", in *Cosmic Dust*, ed. by J.A.M. McDonnell (Wiley, New York) p. 187
Shull, J.M. (1978): Astrophys. J. **226**, 858

Thermal Emission of the Dust
Andriesse, C.D. (1977): Vistas Astron. **21**, 107
Israel, F.P. (ed.) (1986): Light on Dark Matter, Astrophysics and Space Science Library, Vol. 124 (Reidel, Dordrecht)

5.4 Distribution and Motion of the Interstellar Matter

Distribution of the Dust from the Interstellar Extinction
Knude, J. (1979): Astron. Astrophys. **71**, 344
Knude, J. (1981): Astron. Astrophys. **97**, 380
Knude, J. (1981): Astron. Astrophys. **98**, 74

Distribution of the Dust from Infrared Observations
Maihara, T., Oda, N., Shibai, H., Okuda, H. (1981): Astron. Astropyhs. **97**, 139
Israel, F.P. (ed.) (1986): Light on Dark Matter, Astrophysics and Space Science Library, Vol. 124 (Reidel, Dordrecht)

Analysis of the UV Observations of Hydrogen
Bohlin, R.C., Savage, B.D., Drake, J.F. (1978): Astrophys. J. **224**, 132

Analysis of the 21 cm Line, Cloud Structure, Inter-Cloud Medium
Kerr, F.J. (1968): "Radio-line Emission and Absorption by the Interstellar Gas", in *Nebulae and Interstellar Matter*, ed. by B.M. Middlehurst, L.H. Aller, Stars Stellar Systems, Vol. VII (The University of Chicago Press, Chicago) p. 575
Dickey, J.M., Salpeter, E.E., Terzian, Y. (1979): Astrophys. J. **228**, 465

Large Scale Distribution and Motion of the Neutral Atomic Hydrogen
Burton, W.B. (1974): "The Large Scale Distribution of Neutral Hydrogen in the Galaxy", in *Galactic and Extragalactic Radio Astronomy*, ed. by G.L. Verschuur, K.I. Kellermann (Springer, Berlin, Heidelberg, New York) p. 82
Burton, W.B. (1976): Annu. Rev. Astron. Astrophys. **14**, 275
Sinha, R.P. (1978): Astron. Astrophys. **69**, 227
Gunn, J.E., Knapp, G.R., Tremaine, S.D. (1979): Astron. J. **84**, 1181

CO and Molecular Hydrogen, Gamma Rays
Cohen, R.S., Cong, H., Dame, T.M., Thaddeus, P. (1980): Astrophys. J. Lett. **239**, L53
Gordon, M.A., Burton, W.B.: Astrophys. J. Lett. **239**, 271
Solomon, P.M., Sanders, D.B., Scoville, N.Z. (1979): "Giant Molecular Clouds in the Galaxy", in *The Large-Scale Characteristics of the Galaxy*, ed. by W.B. Burton, IAU Symposium, No. 84 (Reidel, Dordrecht) p. 35
Burton, W.B., Israel, F.P. (eds.) (1983): Surveys of the Southern Galaxy (Reidel, Dordrecht)

H II Regions
Georgelin, Y.M., Georgelin, Y.P. (1976): Astron. Astrophys. **49**, 57
Mezger, P.G. (1978): Astron. Astrophys. **70**, 565

Cosmic Radiation Particles
Contributions in: Setti, G., Spada, G., Wolfendale, W.L. (eds.) (1981): *Origin of Cosmic Rays*, IAU Symposium, No. 94 (Reidel, Dordrecht)
Schlickeiser, R., Thielheim, K.O. (1977): Astrophys. Space Sci. **47**, 415
Stecker, F.W. (1979): "Gamma-Ray Evidence for a Galactic Halo", in *The Large-Scale Characteristics of the Galaxy*, ed. by W.B. Burton, IAU Symposium, No. 84 (Reidel, Dordrecht) p. 475
Bogdan, T.J., Völk, H.J. (1983): Astron. Astrophys. **122**, 129

Galactic Central Region
Cohen, R.J., Davis, R.D. (1979): Mon. Not. R. Astron. Soc. **186**, 453
Güsten, R., Downes, D. (1980): Astron. Astrophys. **87**, 6

Mezger, P.G., Pauls, T. (1979): "H II Region and Star Formation in the Galactic Center", in *The Large-Scale Characteristics of the Galaxy*, IAU Symposium, No. 84 (Reidel, Dordrecht) p. 357

Oort, J.H. (1977): Annu. Rev. Astron. Astrophys. **15**, 295

Oort, J.H. (1985): "The Galactic Nucleus", in *The Milky Way Galaxy*, ed. by H. van Woerden, R.J. Allen, W.B. Burton, IAU Symposium, No. 106 (Reidel, Dordrecht) p. 349

Gaseous Galactic Halo

York, D.G. (1982): Annu. Rev. Astron. Astrophys. **20**, 221

6. Dynamics of the Galaxy

6.1 Stellar Dynamics

General Accounts
Chandrasekhar, S. (1942): *Principles of Stellar Dynamics* (The University of Chicago Press, Chicago; reprint: Dover Publ., New York, 1957)

Freeman. K.C. (1975): "Stellar Dynamics and the Structure of Galaxies", in *Galaxies and the Universe*, ed. by A. Sandage, M. Sandage, J. Kristian, Stars Stellar Systems, Vol. IX (The University of Chicago Press, Chicago) p. 409

Lindblad, B. (1959): "Galactic Dynamics", in *Astrophysik IV: Sternsysteme*, ed. by S. Flügge, Handbuch der Physik, Vol. LIII (Springer, Berlin, Heidelberg) p. 21

Mihalas, D. (1968): *Galactic Astronomy* (Freeman, San Francisco)

Ogorodnikov, K.F. (1965): *Dynamics of Stellar Systems* (Pergamon Press, Oxford)

Oort, J.H. (1965): "Stellar Dynamics", in *Galactic Structure*, ed. by A. Blaauw, M. Schmidt, Stars Stellar Systems, Vol. V (The University of Chicago Press, Chicago) p. 455

Omnibus Volumes with Individual Contributions on Recent Results
Contopoulos, G., Henon, M., Lynden-Bell, D. (eds.) (1973): *Dynamical Structure and Evolution of Stellar Systems* (Swiss Society of Astronomy and Astrophysics, Third Advanced Course, Saas-Fee)

Hayli, A. (ed.) (1975): *Dynamics of Stellar Systems*, IAU Symposium, No. 69

Lecar, M. (ed.) (1972): *The Gravitational N-Body Problem* (Reidel, Dordrecht)

Tapley, B.D., Szebehely, V. (1973): *Recent Advances in Dynamical Astronomy* (Reidel, Dordrecht)

Local Mass Density
House, F., Kilkenny, D. (1980): Astron. Astrophys. **81**, 251

Mass Models of the Galaxy
Caldwell, J.A.R., Ostriker, J.P. (1981): Astrophys. J. **251**, 61

Rohlfs, K., Kreitschmann, J. (1981): Astrophys. Space Sci. **79**, 289

Schmidt, M. (1965): "Rotation Parameters and Distribution of Mass in the Galaxy", in *Galactic Structure*, ed. by A. Blaauw, M. Schmidt, Stars Stellar Systems, Vol. V (The University of Chicago Press, Chicago) p. 513

Bahcall, J.N. (1986): Annu. Rev. Astron. Astrophys. **24**, 577

Stability of Disc Galaxies
Miller, R.H. (1978): Astrophys. J. **224**, 32

Ostriker, J.P., Peebles, P.J.L. (1973): Astrophys. J. **186**, 467

Sellwood, J.A. (1980): Astron. Astrophys. **89**, 296

6.2 Gravitational Theory of the Spiral Structure

Lin, C.C. (1970): "Theory of Spiral Structure", in *Galactic Astronomy*, Vol. 2, ed. by H.-Y. Chiu, A. Muriel (Gordon and Breach, New York) p. 1
Toomre, A. (1977): Annu. Rev. Astron. Astrophys. **15**, 437

6.3 Dynamics of the Interstellar Gas

General Accounts
Kaplan, S.A. (1966): *Interstellar Gas Dynamics* (Pergamon Press, Oxford)
Spitzer, L. (1978): *Physical Processes in the Interstellar Medium* (Wiley, New York)

Global Dynamic Equilibrium of the Gas
Cesarsky, C.J. (1980): Annu. Rev. Astron. Astrophys. **18**, 289
Fuchs, B., Thielheim, K.O. (1979): Astrophys. J. **227**, 801
Parker, E.N. (1979): *Cosmical Magnetic Fields* (Clarendon Press, Oxford)

Shock Waves in General
Landau, L.D., Lifshitz, E.M. (1976): *Fluid Mechanics* (Pergamon Press, New York) Chap. IX
Zeldovich, Ya.B., Raizer, Ya.P. (1968): *Elements of Gasdynamics and the Classical Theroy of Shock Waves* (Academic, New York, London)
Zeldovich, Ya.B., Raizer, Ya.P. (1967): *Physics of Shock Waves and High-Temperature Hydrodynamic Phenomena* (Academic, New York, San Francisco, London)

Shock Fronts in the Interstellar Gas
McKee, C.F., Hollenbach, D.J. (1980): Annu. Rev. Astron. Astrophys. **18**, 219

Gravitational Instability and Star Formation
Appenzeller, I., Lequeux, J., Silk, J. (1980): *Star Formation*, Tenth Advanced Course (Swiss Soc. of Astronomy and Astrophysics, Saas-Fee)
De Jong, T., Maeder, A. (eds.) (1977): *Star Formation*, IAU Symposium, No. 75 (Reidel, Dordrecht)
Woodward, P.R. (1978): Annu. Rev. Astron. Astrophys. **16**, 555

Evolution of the Galaxy
Field, G.B. (1975): "The Formation and Early Dynamical History of Galaxies", in *Galaxies and the Universe*, ed. by A. Sandage, M. Sandage, J. Kristian, Stars Stellar Systems, Vol. IX (The University of Chicago Press, Chicago) p. 359
Freeman, K.C., Larson, R.B., Tinsley, B. (1976): *Galaxies*, Sixth Advanced Course (Swiss Soc. of Astronomy and Astrophysics, Saas-Fee)
Mayor, M., Vigroux, L. (1981): Astron. Astrophys. **98**, 1
Vader, J.P., de Jong, T. (1981): Astron. Astrophys. **100**, 124

Classical Work on Element Synthesis in Stars
Burbidge, E.M., Burbidge, G., Fowler, W., Hoyle, F. (1957): Rev. Mod. Phys. **29**, 547
Wannier, P.G. (1980): Annu. Rev. Astron. Astrophys. **18**, 399

Chemical Evolution in the Cosmos Generally
Audouze, J., Vauclair, S. (1980): *An Introduction to Nuclear Astrophysics* (Reidel, Dordrecht)

Sources of Tables

Table 2.3   Becker, W., Fenkart, R. (1971): Astron. Astrophys. **4,** 241
Table 3.2   From Sharov, A.S., Lipaeva, N.A. (1973): Soviet Astronomy **17,** 69
Table 3.3   From Nort, H. (1950): Bull. Astron. Inst. Neth. **11,** 181
Table 3.4   From Table 4.1 in: Trumpler, R.J., Weaver, H.F. (1953): *Statistical Astronomy* (University of California Press, Berkeley, Los Angeles) p. 377
Table 3.5   Gliese, W. (1969): *Catalogue of Nearby Stars,* Veröff. Astron. Rechen-Inst., Heidelberg, No. 22
Table 3.6   Values in the second column from Jahreiss, H. (1974): "Die räumliche Verteilung, Kinematik und Alter der sonnennahen Sterne (The Spatial Distribution, Kinematics and Ages of the Stars near the Sun)"; Dissertation, Heidelberg. Values in the third and fifth columns from Allen, C.W. (1973): *Astrophysical Quantities,* 3rd ed. (Athlone, London) p. 248
Table 3.7   Upgren, A.R. (1962; 1963): Astron. J. **67,** 37; **68,** 194, 475
             Sturch, C.R., Helfer, H.L. (1972): Astron. J. **77,** 726
Table 3.8   Jahreiss, H. (1974): "Die räumliche Verteilung, Kinematik und Alter der sonnennahen Sterne (The Spatial Distribution, Kinematics and Ages of the Stars near the Sun)"; Dissertation, Heidelberg
Table 3.9   From Table 1 by Delhaye, J. (1965): in *Stars and Stellar Systems V: Galactic Structure,* ed. by A. Blaauw, M. Schmidt (University of Chicago Press, Chicago) pp. 64f.
Table 3.10   Jahreiss, H. (1974): "Die räumliche Verteilung, Kinematik und Alter der sonnennahen Sterne (The Spatial Distribution, Kinematics and Ages of the Stars near the Sun)", Dissertation, Heidelberg;
             Wielen, R. (1977): Astron. Astrophys. **60,** 263
Table 3.11   Normal high velocity stars: Fricke, W.(1949): Astron. Nachr. **277,** 241
             Sub-dwarfs: Eggen, O.J. (1970): Vistas Astron. **12,** 409
             Mira stars and RR-Lyrae stars: Smak, J.I. (1966): Annu. Rev. Astron. Astrophys. **4,** 26;
             Plaut, L. (1963): Bull. Astron. Inst. Neth. **17,** 81;
             van Herk, G. (1965): Bull. Astron. Inst. Neth. **18,** 71
             Globular clusters: Kinman, T.D. (1959): Mon. Not. R. Astron. Soc. **119,** 559;
             Woltjer, L. (1975): Astron. Astrophys. **42,** 109
Table 3.12   From Sandage, A. (1958): in *Stellar Populations,* ed. by J.K. O'Connell (Vatican Observatory Publication) p. 84
Table 3.14   From Table 11, Sect. 8.3.3.1 in Schaifers, K., Voigt, H.H. (eds.) (1982): *Interstellar Matter. Galaxy. Universe,* Landolt-Börnstein, Vol. VI/2c (Springer, Berlin, Heidelberg)
Table 4.1   Using data from: Schild, R.E. (1977): Astron. J. **82,** 337, Table II;
             Becklin, E.E., Neugebauer, G., Hackwell, J.A., Gehrz, R.D. (1974): Astrophys. J. **194,** 49, Table 4;
             Nandy, K. et al. (1975): Astron. Astrophys. **44,** 195
Table 4.2   Extract from Table 5, Sect. 7.1.4.1 in Schaifers, K., Voigt, H.H. (eds.) (1982): *Interstellar Matter. Galaxy. Universe,* Landolt-Börnstein, Vol. VI/2c (Springer, Berlin, Heidelberg)
Table 4.3   Extract from Table 2 in Lillie, C.F., Witt, A.N. (1976): Astrophys. J. **208,** 68
Table 4.5   Morton, D.C. (1975): Astrophys. J. **197,** 85

Table 4.6   Pottasch, S.R. (1974): in *The Interstellar Medium*, ed. by K. Pinkau (Reidel, Dordrecht) p. 133

Table 4.7   From Table 8.1 in Verschuur, G.L., Kellermann, K.I. (eds.) (1974): *Galatic and Extra-Galactic Radio Astronomy* (Springer, New York) p. 193

Table 4.8   From Witt, A.N., Johnson, M.W. (1973): Astrophys. J. **181**, 366, Table 4; Jura, M. (1974): Astrophys. J. **191**, 375; Zimmermann, H. (1965): Astron. Nachr. **288**, 95, 99; Henry, R.C. et al. (1977): Astrophys. J. **212**, 707

Table 4.10, 11,12   Extracts from Tables 4a, 5a, 5b of Sect. 7.3.4 in Schaifers, K., Voigt, H.H. (eds.) (1982): *Interstellar Matter. Galaxy. Universe*, Landolt-Börnstein, Vol. VI/2c (Springer, Berlin, Heidelberg)

Table 4.13   Extracts from Tables 2, 3 and 6 of Sect. 7.2 in Schaifers, K., Voigt, H.H. (eds.) (1982): *Interstellar Matter. Galaxy. Universe*, Landolt-Börnstein, Vol. VI/2c (Springer, Berlin, Heidelberg)

Table 4.14   After Winnewisser, G., Churchwell, E., Walmsley, C.M. (1979): "Astrophysics of Interstellar Molecules", in *Modern Aspects of Microwave Spectroscopy*, ed. by G.W. Chantry (Academic, London) p. 317

Table 4.15   Burke, B.F. (1975): in *H II Regions and Related Topics*, ed. by T.L. Wilson, D. Downes (Springer, Berlin, Heidelberg), p. 190

Table 5.1   From Table 4.2 by Spitzer, L. (1978): *Physical Processes in the Interstellar Medium* (Wiley, New York) p. 76; Last column: $\bar{\sigma}$ from Table 4.1 by Spitzer, L. (1968): *Diffuse Matter in Space* (Wiley, New York) p. 93

Table 5.2   From Burgess, A. (1958): Mon. Not. R. Astron. Soc. **118**, 477

Table 5.4   From Ginzburg, V.L., Syrovatskii, S.I. (1965): Annu. Rev. Astron. Astrophys. **3**, 321

Table 5.5   From Table 5.2 and Table 6.2 by Spitzer, L. (1978): *Physical Processes in the Interstellar Medium* (Wiley, New York) pp. 107, 136

Table 5.6   From Table 2.3 by Osterbrock, D.E. (1974): *Astrophysics of Gaseous Nebulae* (Freeman, San Francisco) p. 22

Table 5.7   From Table 5.4 by Spitzer, L. (1978): *Physical Processes in the Interstellar Medium* (Wiley, New York) p. 113

Table 5.8   Mitchell, G.F. et al. (1978): Astrophys. J. Suppl. **38**, 39

Table 5.9   Extracts from Tables 2, 3 and 8 of Sect. 7.2 in Schaifers, K., Voigt, H.H. (eds.) (1982): *Interstellar Matter. Galaxy. Universe*, Landolt-Börnstein, Vol. VI/2c (Springer, Berlin, Heidelberg)

Table 5.10   From Table 4b in Sect. 7.3.4 in Schaifers, K., Voigt, H.H. (eds.) (1982): *Interstellar Matter. Galaxy. Universe*, Landolt-Börnstein, Vol. VI/2c (Springer, Berlin, Heidelberg)

Table 5.11   After Table 1 by Churchwell, E. (1975): "Evolved H II Regions", in *H II Regions and Related Topics*, ed. by T.L. Wilson and D. Downes, Lecture Notes in Physics, Vol. 42 (Springer, Berlin, Heidelberg) p. 247

Table 5.13   From Table 6 by Hawley, S.A. (1978): Astrophys. J. **224**, 428

Table 5.15   From Table 2 by Mezger, P.G. (1978): Astron. Astrophys. **70**, 569

Table 6.3   From Table A by Lin, C.C. (1970), in *Galactic Astronomy*, Vol. 2, ed. by H.-Y. Chiu, A. Muriel (Gordon and Breach, New York) p. 60

Table 6.4   Data from Ostriker, J.P., Caldwell, J.A.R. (1979): "The Mass and Light Distribution of the Galaxy: A Three-Component Model", in *The Large-Scale Characteristics of the Galaxy*, ed. by W.B. Burton, IAU Symposium, No. 84 (Reidel, Dordrecht) p. 441; Einasto, J. (1979): "Galactic Mass Modelling", in *The Large-Scale Characteristics of the Galaxy*, ed. by W.B. Burton, IAU Symposium, No. 84 (Reidel, Dordrecht) p. 451

Sources of Figures

Fig. 1.2, 3   Max-Planck-Institut für Astronomie, Heidelberg, photographs Calar Alto Observatory

Fig. 1.4a, 10   Photographs by the Hale Observatories, Pasadena, California, USA

Fig. 1.4b   Kindly supplied by Dr. J. Solf, Max-Planck-Institut für Astronomie, Heidelberg

Fig.1.5   Kindly supplied by Prof. K. Mattila, Helsinki. Taken with the 3 inch Ross camera at Boyden Observatory in South Africa

Fig. 1.6   Sheet R1 from Schlosser, W., Schmidt-Kaler, Th., Hünecke, W. (1979): *Atlas of the Milky Way* (Astronomical Institute of the University of Bochum)

Fig. 1.7   Barnard, E.E. (1927): *A Photographic Atlas of Selected Regions of the Milky Way*, Part I, Plate 4 (Carnegie Institution of Washington)

Fig. 1.8   National Radio Astronomy Observatory, Green Bank, West Virginia, USA. Kindly supplied by Prof. P.G. Mezger and Dr. W.J. Altenhoff (Max-Planck-Institut für Radio Astronomie, Bonn)

Fig. 1.9   Fig. 3 by Hubble, E. (1934): Astrophys. J. **79**, 8

Fig. 1.10   Photograph by the Hale Observatories, Pasadena, California, USA

Fig. 1.11   Mathewson, D.S., Ford, V.L. (1970): Mem. R. Astron. Soc. **74**, 143

Fig. 1.12   Fig. 11 from Voigt, H.H. (ed.) (1965): *Astronomy and Astrophysics*, Landolt-Börnstein, Vol. VI/1 (Springer, Berlin, Heidelberg) p. 624. First published by Oort, J.H., Kerr, F.T., Westerhout, G. (1958): Mon. Not. R. Astron. Soc. **118**, 379

Fig. 1.13   From Altenhoff, W.J. et al. (1971): Astron. Astrophys. Suppl. **1**, 337

Fig. 2.13   From Fricke, W. et al. (1963): *Fourth Fundamental Catalogue*, Veröff. Astron. Rechen-Inst. Heidelberg, Vol. 10;
                   Boss, B. (1936; 1937): *General Catalogue*, Carnegie Inst. Washington Publ. No. 468, Vols. I–V;
                   Morgan, H.R. (1952): *Catalogue of 5268 Standard Stars*, Astron. Pap. Washington, Vol. 13/3

Fig. 2.15   From Fig. 4 by van de Kamp, P. (1956): Sproul Observatory Reprint, No. 95

Fig. 2.16   From Fig. 4 by van Bueren, H.G. (1952): Bull. Astron. Inst. Neth. XI 432, 390

Fig. 3.2   B0 to B5 stars from Charlier, C.V.L. (1926): Medd. Lund Ser. II, **34**;
                   Open star clusters from Collinder, P. (1931): Ann. Lund. **2**

Fig. 3.3   From Elsässer, H., Haug, U. (1960): Z. Astrophys. **50**, 121

Fig. 3.4   From Hayakawa, S. et al. (1979): *Near Infrared Surface Brightness of Southern Galactic Plane*, Preprint DPNU-4-79

Fig. 3.5   Fig. 1 by Becklin, E.E., Neugebauer, G. (1975): Astrophys. J. Lett. **200**, L72

Fig. 3.7   From Trumpler, R.J., Weaver, H.F. (1953): *Statistical Astronomy* (University of California Press, Berkeley, Los Angeles) Chap. 5.4

Figs. 3.10, 11, 18 From Jahreiss, H. (1974): "Die räumliche Verteilung, Kinematik und Alter der sonnennahen Sterne (The Spatial Distribution, Kinematics and Ages of the Stars near the Sun)"; Dissertation, Heidelberg

Fig. 3.14   From Figs. 7 and 9 by McCuskey, S.W. (1965): in *Stars and Stellar Systems V, Galactic Structure*, ed. by A. Blaauw, M. Schmidt (The University of Chicago Press, Chicago, pp. 15, 17

Figs. 3.15, 16   From Figs. 4, 8 and 9 in Gschwind, P. (1975): Astron. Astrophys. Suppl. **19**, 295, 301

Fig. 3.23   From Fig. 19 by Eggen, O.J. (1970): Vistas Astron. **12**, 410

Fig. 3.24   From Fig. 3 by Neckel, T., Klare, G. (1980): Astron. Astrophys. Suppl. **42**, 257

Fig. 3.25   Figure 4 by Klare, G., Neckel, T. (1967): Z. Astrophys. **66**, 54

Fig. 3.26   Above: Fig. 1 from Vogt, N., Moffat, A.F.J. (1975): Astron. Astrophys. **39**, 479;

Below: Fig. 2 from Humphreys, R.M. (1976): Publ. Astron. Soc. Pac. **88**, 651

Fig. 3.29   From Fig. 1 by Racine, R. (1973): Astron. J. **78**, 180

Fig. 3.30, 31   From Figs. 1, 2 and 6 by Harris, W.E. (1976): Astron. J. **81**, 1103, 1104, 1108

Fig. 3.33   Upper part from Fig. 2 by Oort, J.H. Plaut, L. (1975): Astron. Astrophys. **41**, 75

Fig. 3.34   From Fig. 3 by Tammann, G.A. (1970): "The Galactic Distribution of Young Cepheids", in *The Spiral Structure of our Galaxy*, ed. by W. Becker and G. Contopoulos, IAU Symposium, No. 38 (Reidel, Dordrecht) p. 244

Fig. 3.35   From Fig. 1c by Fenkart, R.P. (1968): Z. Astrophys. **68**, 91

Fig. 3.36   From Fenkart, R.P. (1967): Z. Astrophys. **66**, 390

Fig. 3.38   From Fig. 5 by Elsässer, H., Haug, U. (1960): Z. Astrophys. **50**, 131 and Fig. 2 by Hayakawa, S. et al. (1979): *Near Infrared Surface Brightness of Southern Galactic Plane*, Preprint DPNU-4-79

Fig. 3.40   From Fig. 3 by Neckel, T. (1968): Z. Astrophys. **69**, 119

Fig. 3.45   (a) From Fig. 5 by Joy, A.H. (1939): Astrophys. J. **89**, 356;

(b) From Fig. 4 by Feast, M.W., Thackeray, A.D. (1958): Mon. Not. R. Astron. Soc. **118**, 134

Fig. 4.2   From Plate 13 by Barnard, E.E. (1927): *A Photographic Atlas of Selected Regions of the Milky Way*, Part I (Carnegie Institution of Washington)

Fig. 4.5   From Fig. 1 by Danks, A.C., Lambert, D.L. (1975): Astron. Astrophys. **41**, 456

Fig. 4.7   From Serkowski, K., Mathewson, D.S., Ford, V.L. (1975): Astrophys. J. **196**, 261

Fig. 4.8   From Fig. 1 by Dorschner, J., Gürtler, J. (1965): Astron. Nachr. **289**, 58

Fig. 4.11   After Smith, A.M. (1972): Astrophys. J. **172**, 129

Fig. 4.12   From Fig. 1 by Savage, B.D., Bohlin, R.C., Drake, J.F., Budich, W. (1977): Astrophys. J. **216**, 295

Fig. 4.13   Kindly supplied by Prof. G. Münch, Max-Planck-Institut für Astronomie, Heidelberg

Fig. 4.14   Figure 2 by Münch, G. (1965), in *Stars and Stellar Systems V, Galactic Structure*, ed. by A. Blaauw, M. Schmidt (The University of Chicago Press, Chicago) p. 208

Fig. 4.17   Figure 4.1 by Burton, W.B. (1974): in *Galactic and Extra-Galactic Radio Astronomy*, ed. by G.L. Verschuur, K.I. Kellermann (Springer, New York) p. 83

Fig. 4.19   From Fig. 1 by Hughes, M.P., Thompson, A.R., Colvin, R.S. (1971): Astrophys. J. Suppl. **23**, 323

Fig. 4.20   From Fig. 1 by Burton, W.B., Gordon, M.A. (1978): Astron. Astrophys. **63**, 7

Fig. 4.21   Landecker, T.L., Wielebinski, R. (1970): Aust. J. Phys., Astrophys. Suppl. No. 16, 1

Fig. 4.24   From Fig. 3 by Berkhuijsen, E.M., Brouw, W.N. (1963): Bull. Astron. Inst. Neth. **17**, 187

Fig. 4.25   From Fig. 5–4 by Kraus, J.D. (1966): *Radio Astronomy* (McGraw-Hill, New York) p. 146

Fig. 4.26   From Fig. 3 by Gursky, H. (1973): "Observations of X-Ray Sources", in *Black Holes*, ed. by C. DeWitt, B.S. DeWitt (Gordon and Breach, New York) p. 302

Fig. 4.27   From Fig. 10.2 by Schwartz, D., Gursky, H. (1974): "The Cosmic X-Ray Background", in X-Ray Astronomy, ed. by R. Giacconi, H. Gursky (Reidel, Dordrecht) p. 364

Fig. 4.28   From Mayer-Hasselwander, H.A. et al. (1982): Astron. Astrophys. **105**, 164

Fig. 4.29   From Fig. 2 by Paul, J.A. et al. (1978): Astron. Astrophys. **63**, L32

Fig. 4.30   Adapted from Figs. 1.2 and 1.3 by Mezger P.G. (1972): "Interstellar Matter – An Observer's View", in *Interstellar Matter*, ed. by N.C. Wickrama-

	singhe, F.D. Kahn, P.G. Mezger (Published by Geneva, Observatory, Sauverny, Switzerland) pp. 12, 13
Fig. 4.31	Photograph by H.-E. Schuster with the Schmidt telescope of the European Southern Observatory (ESO). Kindly supplied by Mr. Claus Madsen, European Southern Observatory, Photographic Service, Garching near Munich
Fig. 4.32	Hale Observatories, Pasadena, California, USA
Fig. 4.33	Max-Planck-Institut für Astronomie, Heidelberg, Photograph Calar Alto Observatory
Fig. 4.34	Figure 1 by Elliot, K.H., Meaburn, J. (1974): Astrophys. Space Sci. **28**, 352
Fig. 4.35	Kindly supplied by Dr. H. Hippelein, Max-Planck-Institut für Astronomie, Heidelberg
Fig. 4.36	Georgelin, Y.P., Georgelin, Y.M. (1970): Astron. Astrophys. **6**, 349
Fig. 4.37	Terzian, Y., Parrish, A. (1970): Astrophys. Lett. **5**, 261
Fig. 4.38	Schraml, J., Mezger, P.G. (1969): Astrophys. J. **156**, 269
Fig. 4.39	Figure 1 by Sullivan, W.T., Downes, D. (1973): Astron. Astrophys. **29**, 370. Kindly supplied by Dr. D. Downes, Max-Planck-Institut für Radioastronomie, Bonn
Fig. 4.40 a	From Fig. 1 by Lo, K.Y., Claussen, M.J. (1983): Nature **306**, 648
Fig. 4.40 b	From Fig. 1 by Mezger, P.G., Pauls, T. (1979): "H II Regions and Star Formation in the Galactic Center", in *The Large-Scale Characteristics of the Galaxy*, ed. by W.B. Burton, IAU Symposium, No. 84 (Reidel, Dordrecht) p. 357
Fig. 4.41	Lemke, D., Low, F. (1972): Astrophys. J. Lett. **177**, L53
Fig. 4.42	From Fig. 6 by Wynn-Williams, C.G., Becklin, E.E. (1974): Publ. Astron. Soc. Pac. **86**, 15
Fig. 4.43	From Fig. 3 by Mezger, P.G., Wink, J.E. (1975): "The Giant H II Regions W3", in *H II Regions and Related Topics*, ed. by T.L. Wilson, D. Downes, Lecture Notes in Physics, Vol. 42 (Springer, Berlin, Heidelberg) p. 415
Fig. 4.45	From the National Geographic Society (1960): Palomar Observatory Sky Survey (California Institute of Technology, Pasadena, USA)
Fig. 4.47	Neckel, T., Klare, G. (1980): Astron. Astrophys. Suppl. **42**, 251
Fig. 4.49	From Fig. 1 by Milman, A.S. et al. (1975): Astron. J. **80**, 107
Fig. 4.50	From Fig. 4 by Myers, P.C. et al. (1978): Astrophys. J. **220**, 869
Fig. 4.51	Left: Fig. 1 by Liszt, H.S. et al. (1974): Astrophys. J. **190**, 558;
Fig. 4.51	Right: Fig. 2 by Thaddeus, P. et al. (1971): Astrophys. J. Lett. **168**, L61
Fig. 4.52	From Scoville, N.Z., Solomon, P.M., Penzias, A.A. (1975): Astrophys. J. **201**, 352
Fig. 4.53	From Fig. 6 by Genzel, R., et al. (1978): Astron. Astrophys. **66**, 19
Fig. 4.54	Adapted from Fig. 1 by Kutner, M.L., Tucker, K.D., Chiu, G., Thaddeus, P. (1977): Astrophys. J. **215**, 522
Fig. 4.55	From Zuckerman, B. (1973): Astrophys. J. **183**, 863
Fig. 4.56	Blair, G.N. et al. (1978): Astrophys. J. **219**, 896
Fig. 4.57	Max-Planck-Institut für Astronomie, Heidelberg. Photograph Calar Alto Observatory
Fig. 4.58	Molecule observations by Little, L.T. et al. (1979): Mon. Not. R. Astron. Soc. **188**, 429
Fig. 5.3	Sejnowski, T.J., Hjellming, R.M. (1969): Astrophys. J. **156**, 915
Fig. 5.4	Adapted from Fig. 5.3 by Osterbrock, D.E. (1974): *Astrophysics of Gaseous Nebulae* (Freeman, San Francisco) p. 112
Fig. 5.7	From Fig. 5 by Barsuhn, J., Walmsley, C.M. (1977): Astron. Astrophys. **54**, 350
Fig. 5.8	From Fig. 4.1 by Spitzer, L. (1968): *Diffuse Matter in Space* (Wiley, New York) p. 134
Fig. 5.11	From Fig. 3 by Linke, R.A., Goldsmith, P.F. (1980): Astrophys. J. **235**, 445
Fig. 5.14	From Fig. 5 by Morton, D.C. (1975): Astrophys. J. **197**, 101
Fig. 5.15	Figure 4 by Münch, G. (1968): "Interstellar Absorption Lines", in *Stars and Stellar Systems VII: Nebulae and Interstellar Matter*, ed. by B.M. Middlehurst, L.H. Aller (The University of Chicago Press, Chicago) p. 389
Fig. 5.16	From Fig. 5 by Spitzer, L., Jenkins, E.B. (1975): Annu. Rev. Astron. Astrophys. **13**, 147

Fig. 5.18	Figure 3 in Wickramasinghe, N.C., Kahn, F.D., Mezger, P.G. (1972): *Interstellar Matter* (Published by Geneva Observatory, Sauverny, Switzerland) p. 221
Fig. 5.19	van de Hulst, H.C. (1957): *Light Scattering by Small Particles* (Wiley, New York)
Fig. 5.20	(a) From Fig. 42; (b) from Fig. 51 by Greenberg, J.M. (1968): "Interstellar Grains", in *Stars and Stellar Systems VII: Nebulae and Interstellar Matter*, ed. by B.M. Middlehurst, L.H. Aller (The University of Chicago Press, Chicago) pp. 282, 290
Fig. 5.22	From Fig. 2 by Drapatz, S. (1979): Astron. Astrophys. **75**, 27
Fig. 5.23	From Fig. 21 in Wickramasinghe, N.C., Kahn, F.D., Mezger, P.G. (1972): *Interstellar Matter* (Published by Geneva Observatory, Sauverny, Switzerland) p. 288
Fig. 5.26	Figure 9a by Neckel, T., Klare, G. (1980): Astron. Astrophys. Suppl. **42**, 281
Fig. 5.30	Sinha, R.P. (1978): Astron. Astrophys. **69**, 227
Figs. 5.31, 32	From Figs. 4.4 and 4.10 by Burton, W.B. (1974): "The Large-Scale Distribution of Neutral Hydrogen in the Galaxy", in *Galactic and Extra-Galactic Radio Astronomy*, ed. by G.L. Verschuur, K.I. Kellermann (Springer, New York) pp. 87, 98
Fig. 5.34	Figure 5 by Burton, W.B., Liszt, H.S. (1978): Astrophys. J. **225**, 826
Fig. 5.35	Adapted from Fig. 1 by Burton, W.B., Liszt, H.S. (1978): Astrophys. J. **225**, 817
Fig. 5.36	From Fig. 6 by Georgelin, Y.M., Georgelin, Y.P., Sivan, J.-P. (1979): "Optical H II Regions", in *The Large-Scale Characteristics of the Galaxy*, ed. by W.B. Burton, IAU Symposium, No. 84 (Reidel, Dordrecht) p. 69
Fig. 5.37	(a) Burton, W.B., Gordon, M.A. (1978): Astron. Astrophys. **63**, 7; (b) Lockman, F.J. (1975): Astrophys. J. **209**, 429; (c) Solomon, P.M., Sanders, D.B., Scoville, N.Z. (1979): "Giant Molecular Clouds in the Galaxy; Distribution, Mass, Size and Age", in *The Large-Scale Characteristics of the Galaxy*, ed. by W.B. Burton, IAU Symposium, No. 84 (Reidel, Dordrecht) p. 35; (d) Caraveo, P.A., Paul, J.A. (1979): Astron. Astrophys. **75**, 340
Fig. 6.3	From Fig. 10 by Wielen, R. (1974): "The Gravitational N-Body Problem for Star Clusters", in *Stars and the Milky Way System*, ed. by L.N. Mavridis, Proceedings of the First European Astronomical Meeting, Vol. 2 (Springer, Berlin, Heidelberg) p. 349
Fig. 6.4	From Fig. 1 by Hohl, F. (1975): "N-Body Simulations of Disks", in *Dynamics of Stellar Systems*, ed. by A. Hayli, IAU Symposium, No. 69 (Reidel, Dordrecht) p. 350
Fig. 6.5	From Figs. 6 and 7 by Sellwood, J.A. (1980): Astron. Astrophys. **89**, 302, 303
Fig. 6.9	From Fig. 1 by Schmidt, M. (1956): Bull Astron. Inst. Neth. XII, No. 468, 30
Figs. 6.11, 12	Lin, C.C., Yuan, C., Shu, F.H. (1969): Astrophys. J. **155**, 721
Fig. 6.17	From Figs. 1 and 2 by McKee, C.F., Ostriker, J.P. (1977): Astrophys. J. **218**, 159
Fig. 6.18	From Fig. 1 by Tenorio-Tagle, G. (1979): Astron. Astrophys. **71**, 60
Fig. 6.19	From Fig. 7 by Woodward, P.R. (1975): Astrophys. J. **195**, 68
Fig. 6.21	Figure 8 by Woodward, P.R. (1976): Astrophys. J. **207**, 493
Fig. 6.22	From Fig. 1 by Gerola, H., Seiden, P.E. (1978): Astrophys. J. **223**, 131
Fig. 6.24	From Fig. 28 by Freeman, K.C. (1975): "Stellar Dynamics and the Structure of Galaxies", in *Stars and Stellar Systems IX: Galaxies and the Universe*, ed. by A. Sandage, M. Sandage, J. Kristian (The University of Chicago Press, Chicago) p. 469
Fig. 6.25	From Fig. 9 by Larson, R.B. (1976): Mon. Not. R. Astron. Soc. **176**, 43
Fig. B.1	From Fig. 1 by Field, G.B., Sommerville, W.B., Dressler, K. (1966): Annu. Rev. Astron. Astrophys. **4**, 210

Subject Index

Aberration   34f.
Absorption coefficient   241, 244f.
Ages of stars   93
−−, kinematic   102
Albedo   318, 322f.
Alfvén velocity   420
Alignment, degree of   326, 337
Antenna temperature   176
Apex   40, 95f.
−, standard   97
Aries, First Point of   see Vernal Equinox
Associations   109f.
Astronomical unit   33, 462
Asymmetric drift   97, 102, 105, 142, 389
Asymmetry parameter   318, 319, 322f.
Azimuth   25

Background radiation, 3K   197, 201, 255
−−, galactic   186, 265f.
−−, −, polarisation   187f., 268
Basic solar motion   99
Becklin-Neugebauer object   155, 220
Bipolar nebula   239f.
Boltzmann formula   244, 245, 256, 460
− equation   368
Bonner Durchmusterung   61
Brightness temperature   176, 243
Bulge component   398f.

C II regions   218
Central region, galactic   68f.
−−, −, gas mass   357, 362
−−, −, H I distribution   356f.
−−, −, infrared emission   67f., 130f., 222f.
−−, −, molecular clouds   361f.
−−, −, radio emission   180f., 215
−−, −, ratio gas : star mass   362
−−, −, star concentration   130, 145
Centre, galactic   see Galactic centre
Centroid   75, 94
Cepheids, classical   113
−, −, absolute magnitudes   113
−, −, ages   123
−, −, distribution   122f.
−, −, velocity dispersion   101
Champagne effect   428
Chemical evolution of the galaxy   446

Circular orbit velocity   139
CO molecules, interstellar   183, 232, 237, 360f.
Coalsack   10, 11, 166
Cocoon stars   222
Collision
− efficiency   250
− rates   247f.
Collision-free stellar systems   366, 373, 377f.
Colour excess   56, 58f., 152f.
Column density   309
Compact H II regions   212f.
Convergence point   50
Cooling function   293
− rate   291, 293f.
− time   286, 292f., 297
Cordoba Durchmusterung   61
Corona gas   see Hot component
Corona of the galaxy   398f.
Cosmic rays   19, 199, 265f.
−−, ionisation by   270, 278
− particle radiation   see Cosmic rays
Crossing time   368, 373
Culmination   25
Curve of growth   409

Dark clouds   8, 151, 226
−−, CO emission   232f.
−−, infrared sources in   226
−−, optical data   223f.
−−, scattered light from   166
Declination   26
−, measurement of absolute   42
Density wave theory   529f.
Dielectronic recombination   280
Diffuse galactic light   166
Diffuse interstellar bands   156
−−−, interpretation   327
−− clouds   173, 349, 362f., 420, 424f.
− nebulae   8, 201f.
−−, radial velocities   209f.
−−, spectrum   9, 206f.
Dilatation of the stellar system   135
Dilution factor   193, 251
Dispersion equation   410
−, interstellar   188f.
− measure   190

Rayleigh-Taylor instability 423, 428, 431, 439
Reddening, interstellar 152f.
–, –, correlation with H I 345
Reflection nebulae 8, 163f.
Relaxation time 372, 373
Right ascension 26
– –, measurement of absolute 43
Rotation curve 350f., 397, 399
– measure 189, 191
– parallax 54
RR Lyrae stars 104, 105, 111
– –, absolute magnitudes 113f.
– –, distribution 120f.
– –, kinematics 141

Sagittarius Arm 92, 111, 210, 353
Saha Equation 244, 256, 460
SAO Catalogue 46
Schmidt mass model 396
Scorpio-Centaurus group 49, 51
Selected Areas 62
Shock fronts 416f., 426, 427, 439
– –, hydromagnetic 430
– –, spiral-form 432
Sidereal time 27f.
Solar motion 40, 94f., 98
– time 28
Sonic velocity 416, 420
Source function 242, 244
Spatial velocity 39f.
Spectral index 186
Spectrophotometric distance 55
Spiral arm tracers 401
– arms 4, 5, 91, 108, 128, 173, 186, 210, 355, 358
– structure of galaxies 4, 5, 401
– – –, density wave theory 401f.
– – –, formation 377f., 403, 441
– – in the interstellar gas 173, 186, 210, 353f., 412, 431f.
– –, models 410
– –, stability 409f., 412
Stability conditions 400
– number 409
Star clusters, globular see Globular clusters
– –, open 109f.
– –, –, distribution 110
– –, –, dynamics 374f.
– –, –, evaporation time 375
Star counts 61
Star density 70
– –, distribution of 89ff.
– –, galactic gradients 142, 390
– –, local 82
Star formation 22, 237f., 432f., 437f.
– –, chain reaction of 440
– –, in central region 446
– –, halo stars 443
– – rate, local 87

Statistical equations 247
Stellar associations 109
– –, distribution 110
Stellar dynamics 365
– –, analytical theory 376, 379
– –, fundamental equation of 368, 381
– –, numerical experiments 374, 377f.
Stellar populations 7, 143
– –, classification of 144
– –, formation 145f., 447
Stellar statistics, fundamental equation of 71
– –, history of 1f., 72
Stokes's parameter 159f.
Strömgren radius 275, 427
Structure function 341, 342
Sub-dwarfs 104, 105
Sub-systems 6, 141f.
Supernova explosion waves 429f.
– remnants 185f., 202, 431
– shells 425, 429
Surface brightness 63
– – of the Milky Way 66f., 127
Synchrotron radiation 187, 265f.
– self-absorption 268

Tangential velocity 39
Thermal instability 295f., 438
– balance 285
– radio continuum 263
– – sources 211
Third integral 380
Three component model 425, 433
Toomre's stability number 409
Trapezium Nebula 219, 221
– stars 206, 208, 209, 220
Two-colour diagram 59
– – for globular clusters 115
– –, fractionated 74, 124
Two-phase model 295
Two-photon emission 265

Ultraviolet excess 102, 104, 123

Velocity ellipsoid 76, 99, 381, 385
– distribution of the stars 70, 74f., 95, 103
– – –, asymmetry 103f., 389
– – –, description 74
– – –, results 95, 98, 103
– – –, theory 381ff.
– dispersion of the stars 98, 100f., 105, 142
– – –, age dependence 102, 387
– – –, significance for stability 399, 408f.
– – –, theory 381f.
Vernal equinox 26
Vertex of stellar motions 50
– deviation 94, 98, 100, 387
Vertical, prime 25